Cavity Quantum Electrodynamics

WILEY SERIES IN LASERS AND APPLICATIONS

D. R. VIJ, Editor
Kurukshetra University

OPTICS OF NANOSTRUCTURED MATERIALS • Vadim Markel

LASER REMOTE SENSING OF THE OCEAN: METHODS AND
APPLICATIONS • Alexey B. Bunkin and Konstantin Voliak

COHERENCE AND STATISTICS OF PHOTONS AND ATOMS • Jan Peřina,
Editor

METHODS FOR COMPUTER DESIGN OF DIFFRACTIVE OPTICAL
ELEMENTS • Victor A. Soifer

THREE-DIMENSIONAL HOLOGRAPHIC IMAGING • Chung J. Kuo and
Meng Hua Tsai

ANALYSIS AND DESIGN OF VERTICAL CAVITY SURFACE EMITTING
LASERS • Siu Fung and F. Yu

CAVITY QUANTUM ELECTRODYNAMICS: THE STRANGE THEORY OF
LIGHT IN A BOX • Sergio M. Dutra

Cavity Quantum Electrodynamics
The Strange Theory of Light in a Box

Sergio M. Dutra

A JOHN WILEY & SONS, INC., PUBLICATION

Library of Congress Cataloging-in-Publication Data:

Dutra, S. M. (Sergio M.)
  Cavity quantum electrodynamics : the strange theory of light in a box / S. M. Dutra.
    p. cm — (Wiley series in lasers and applications)
  Includes bibliographical references and index.
  ISBN 0-471-44338-7 (cloth : acid-free paper)
    1. Quantum optics.   2. Quantum electrodynamics.   I. Title.   II. Series.

  QC446.2.D87 2005
  535'.15—dc22   2004050922

Printed in the United States of America

10 9 8 7 6 5 4 3 2 1

To
Salvador P. Dutra,
from whom I inherited a fondness for books

My parents, who nurtured it

Kati, Dani, Anna, and Nani,
who have so patiently endured the writing of this book

Contents

Preface   *xi*

Acknowledgments   *xv*

1   Introduction   *1*
   1.1   What is light?   *3*
     1.1.1 Geometrical optics   *3*
     1.1.2 Wave optics   *4*
     1.1.3 Classical electrodynamics and relativity   *5*
     1.1.4 Quantum mechanics and quantum electrodynamics   *7*
   1.2   A brief history of cavity QED   *9*
   1.3   A map of the book   *11*
   1.4   How to read this book   *14*

2   Fiat Lux!   *17*
   2.1   How to quantize a theory   *18*
   2.2   Why the radiation field is special   *24*
   2.3   What is a cavity?   *27*
     2.3.1 What is resonance?   *29*
     2.3.2 Confinement is the key   *34*
   2.4   Canonical quantization of the radiation field   *39*

2.4.1 Quantization in a cavity   39
2.4.2 Quantization in free space   43
2.5 The Casimir force   47
2.5.1 Zero-point potential energy   48
2.5.2 Maxwell stress tensor   54
2.5.3 The vacuum catastrophe   62
Recommended reading   63
Problems   65

3 The photon's wavefunction   69
3.1 Position in relativistic quantum mechanics   71
3.2 Extreme quantum theory of light with a twist   74
3.3 The configuration space problem   89
3.4 Back to vector notation   90
3.5 The limit of vanishing rest mass   94
3.6 Second quantization   99
Recommended reading   106
Problems   109

4 A box of photons   113
4.1 The classical limit   113
4.1.1 Coherent states   114
4.1.2 The density matrix   118
4.1.3 The diagonal coherent-state representation   121
4.2 Squeezed states   124
4.2.1 The squeezing operator   125
4.2.2 Generating squeezed states   130
4.2.3 Geometrical picture   133
4.2.4 Homodyne detection   139
Recommended reading   142
Problems   143

5 Let matter be!   145
5.1 A single point dipole   146
5.2 An arbitrary charge distribution   154
5.3 Matter–radiation coupling and gauge invariance   161
Recommended reading   163

6 Spontaneous emission   165

6.1 Emission in free space   166
6.2 Emission in a cavity   176
      Recommended reading   182

7 Macroscopic QED   183
7.1 The dielectric JCM   185
7.2 Polariton–photon dressed excitations   191
7.3 Quantum noise of matter and macroscopic averages   193
7.4 How a macroscopic description is possible   196
7.5 The Kramers–Kronig dispersion relation   198
7.6 Including absorption in the dielectric JCM   198
7.7 Dielectric permittivity   202
7.8 Huttner–Barnett theory   205
      7.8.1 The matter Hamiltonian   206
      7.8.2 Diagonalization of the total Hamiltonian   208
      Recommended reading   210
      Problems   211

8 The maser, the laser, and their cavity QED cousins   213
8.1 The ASER idea   214
8.2 How to add noise   220
      8.2.1 Einstein's approach to Brownian motion   221
      8.2.2 Langevin's approach to Brownian motion   222
      8.2.3 The modern form of Langevin's equation   223
      8.2.4 Ito's and Stratonovich's stochastic calculus   225
8.3 Rate equations with noise   231
8.4 Ideal laser light   235
8.5 The single-atom maser   238
8.6 The thresholdless laser   245
8.7 The one-and-the-same atom laser   250
      Recommended reading   251
      Problems   253

9 Open cavities   255
9.1 The Gardiner–Collett Hamiltonian   258
9.2 The radiation condition   269
9.3 Natural modes   270
9.4 Completeness in general   271
      9.4.1 Whittaker's scalar potentials   272

9.4.2 General formulation of the problem   274
Recommended reading   276
Problems   277

Appendix A Perfect cavity modes   283

Appendix B Perfect cavity boundary conditions   291

Appendix C Quaternions and special relativity   293
C.1 What are quaternions?   295
C.2 Quaternion calculus   300
C.3 Biquaternions and Lorentz transformations   304

Appendix D The Baker–Hausdorff formula   311

Appendix E Vectors and vector identities   313
E.1 Relation between vector products and determinants   314
E.2 Vector products and the Levy–Civita tensor   315
E.3 The product of two Levy–Civita tensors as a determinant   316
E.4 The vector product of three vectors   316
E.5 Vectorial expressions involving del   317
E.6 Some useful integral theorems   318

Appendix F The Good, the Bad, and the Ugly   321
F.1 Connections   322
F.1.1 The rectangular barrier   323
F.1.2 The sinc function   324
F.1.3 The Lorentzian representation and the principal part   324
F.1.4 The Gaussian   326
F.1.5 The Laplacian of $1/r$ 327
F.1.6 The comb function   327
F.1.7 A general rule to find representations   328
F.2 Product of two principal parts   331
F.3 Discontinuous functions   331

References   335

Index   381

Preface

First, then, I will lay down some general rules, most of which, I believe, you have considered already; but if any of them be new to you, they may excuse the rest; if none at all, yet is my punishment more in writing than yours' in reading.

—Sir Isaac Newton's letter to a certain Francis Aston (May 18, 1669)

Cavity quantum electrodynamics is about what happens to light and, in particular to its interaction with matter when it is trapped inside a box. By *light* I mean not only visible light (i.e., the optical region of the spectrum) but also the infrared and microwave regions. This book has two main aims. The first is to serve as an introductory text for Ph.D. students, postdocs, and more experienced physicists wishing to enter this field. I have worked hard to write the book that I dreamed of reading when I started my Ph.D. The second aim is to serve as a supplementary text for an advanced undergraduate course in quantum mechanics.

Why would cavity quantum electrodynamics, which sounds like a highly specialized topic, be of any use as a supplement to an undergraduate lecture course in quantum mechanics? Because it provides a bridge between the bread-and-butter quantum mechanics that is usually taught to undergraduates and most of the exciting new physics and technology that is being done today. Quantum electrodynamics is the basis of all other quantum field theories, which together with general relativity, form the fundamental core of the whole of contemporary physics. Considering quantum electrodynamics in a cavity makes it more accessible both from a pedagogical point of view and from that of an experimentalist who wishes to observe the various quantum effects predicted by the theory. Many conceptual issues of elementary quantum mechanics, such as the EPR and Schrödinger cat paradoxes, have been investigated in cavity QED experiments that I wish to bring within reach of undergraduate students.

We also address many of the exciting questions that students often ask themselves when they are starting to learn modern physics, such as: What would happen if the photon had a rest mass? Does the photon have a wavefunction? Could it be that the electromagnetic field itself is the wavefunction of the photon? If in the microscopic world ruled by quantum mechanics both matter and light are made up of particles that sometimes behave as waves, why is it that in the macroscopic world ruled by classical mechanics, matter behaves only as particles and light only as waves? The main theme of this book is a question that even laypeople sometimes ask: What happens to light if we trap it inside a box with mirrored walls? Connected with this question, this book also deals with a more subtle question that is normally not addressed in quantum optics textbooks: What happens when the walls move?

Some popular science books touch on a few of these questions but only superficially. Serious introductory textbooks tend to avoid them, leaving the students rather frustrated. The persistent student can find these questions discussed in more advanced books and papers. But these discussions are all scattered in the literature and are often difficult for the uninitiated to follow. So one of the subsidiary aims has been to gather all these interesting issues together in a form that can be understood by someone who is just starting to think about them with fresh and genuine excitement. This is a serious physics textbook written in the spirit of those popular science books that often excite the imagination of would-be scientists.

The core structure of this book came out of a lecture course on cavity QED I gave at the University of Campinas (SP, Brazil) and at the Instituto Nacional de Astrofísica, Óptica y Electrónica (Puebla, Mexico). But as the book began to mature, a number of extra topics were added, not to mention some key advances of this field that happened while the book was being written and had to be included (such as the realization of the first single-atom laser in the strong-coupling regime).

There are several appendixes. These are intended to help our occasional "top-down" and "just-in-time" approach to some of the subjects. Instead of the traditional approach, where students first have to go through a lot of apparently pointless learning of apparently unconnected mathematical techniques before they can hopefully "put everything back together again" at the end, whenever appropriate, this book aims at going straight to the point, introducing the math along the way as we need it. Most of the details are in the appendixes for later study, after the student is already motivated by the main text. I believe that this style of presentation stands a better chance of keeping a student's enthusiasm alive than the traditional approach does.

Another feature of this book is the attention that has been paid to historical context. It is sad to see students learn a particular technique, such as canonical quantization, without knowing who proposed it first, why, and in which context. So I tried to write a little about the historical origins of some of the ideas and techniques discussed in this book, and to give references not only to the more recent literature but also to some of the historical breakthroughs. Nowadays, some references are freely available on the World Wide Web. So to make life easier for the reader, whenever I knew of such availability, I mention not only the printed source but also the URL where the free online version can be found. I am well aware that URLs change quite often, and even though I have checked them all prior to publication, there is a good chance that many will be out of date by the time you read this [149]. Still, I think it is worth providing them, as they can help you find the updated URLs more quickly.

Last but not least, I would like to mention specifically some of the topics this book covers that are not usually found in quantum optics textbooks. These topics include, for instance:

1. A deduction of a first-quantized theory of the photon from relativistic invariance alone, with Maxwell's equations being deduced at the end rather than being used as the starting point.

2. A discussion about photon mass.

3. A brief introduction to the vacuum catastrophe, a very important open problem where quantum field theory diverges from experiment by more than 100 orders of magnitude. This problem might turn out to hide the key to the long-sought unification of gravity and quantum mechanics. Here it is seen under the new insight brought by the discovery in 1998 that the expansion of the universe is accelerating rather than slowing down.

4. A discussion about cavities with movable walls (the dynamic Casimir effect).

5. Quantization of the radiation field in material media from first principles.

6. Quantization of radiation with matter without using electromagnetic potentials in the dipole approximation (this is all that you need for most of quantum optics).

7. A pedagogical account of some mathematical subtleties and techniques, including, multiplication of distributions (e.g., the product of a delta function and a step function), and quaternions.

8. A first-principles discussion about cavity QED in leaky cavities using Fox–Li modes.

It is also worth pointing out that even though this book deals primarily with fundamental ideas in modern physics, many of the mathematical techniques presented here can be quite useful to someone who is not going to follow an academic career. For instance, quaternions, which are seldom taught to undergraduates, but to which I devote an entire appendix, are very important in the computer graphics and animation industries. Langevin equations, or stochastic differential equations as they are often referred to by mathematicians, are very important in quantitative finance and insurance.

SERGIO M. DUTRA

Acknowledgments

The scientific profession is a bit of a craft and it is still best learned from a master craftsman as in the Middle Ages. The ultimate origin of this book is perhaps my M.Sc. and Ph.D. theses, which were supervised by Luiz Davidovich and Peter L. Knight, respectively. I was lucky to have been able to learn many of our "trade secrets" from these two highly skilled masters. Without them this book would not have been possible and I am forever grateful to them. I am also thankful to Han Woerdman, Moysés Nussenzveig, Amir Caldeira, Gerard Nienhuis, and Nicim Zagury, from whom I have learned many things.

Many of the ideas presented in this book have evolved through scientific collaborations. I am grateful to Peter L. Knight, Hector M. Moya-Cessa, Luiz Davidovich, Nicim Zagury, Amir O. Caldeira, Gerard Nienhuis, Antonio Vidiella-Barranco, Kyoko Furuya, José Roversi, Eric Eliel, Martin van Exter, Krista Joosten, Alexander van der Lee, Klaasjan van Druten, Dagoberto S. Freitas, Jorrit Visser, Marcelo Fraça, Luiz Guilherme Lutterbach, and Marcelo Roseneau. I would also like to thank all the students who attended my cavity QED lectures, but in particular, three who gave me a lot of feedback: Marcelo Terra-Cunha (here are the expanded lecture notes you have constantly asked me to finish), Alessandro Moura, and Daniel Jonathan. I have also profited from discussions with Paulo Nussenzveig, Brian Dalton, Gregory Surdotovich, Rodney Loudon, and Ottavia Jedrkiewicz. I am very grateful to Gabriel Barton, Iwo Bialynicki-Birula, Han Woerdman, Gerard Nienhuis, Sumant Oemrawsingh, Andrea Aiello, Hayk Haroutyunyan, Fábio Reis, Marco Moriconi, Antonio Vidiella-Barranco, and Henry Baker for helping me obtain some key papers. Gerard Nienhuis, José Arruda de Oliveira Freire, Marcelo Terra Cunha, and Kati read various drafts of the manuscript and helped me to improve it by spotting mistakes and making many useful suggestions.

I am very grateful to Benedict Mihaly, Czirják Attila, and Földi Peter for their hospitality and help during my visit to Szeged University. Besides kindly granting me access to their computing and library facilities, they also helped me to find some key historical references in Hungary.

I thank Charles H. Townes, Serge Haroche, Herbert Walther, Bernard Richardson, Gary S. Chottiner, and Joe Wolfe, who have kindly given me permission to use images derived from their work. I am also grateful to Gilles Tran[1] for granting me permission to use his (late nineteenth century-style) Parisian lamppost macro in the POV-Ray[TM2] script I wrote to produce the picture on which the cover of the book is based.[3]

The Wiley team was also very helpful and professional. Angioline Loredo, Rachel Witmer, Brendan Codey, Donna Namorato, George Telecki, and the editor of this series, Dr. Vij, patiently put up with my constantly missing the postponed deadlines for handing in the finished manuscript of the book. I am very grateful to them.

Last but not least, my warmest thanks to Kati for her patience and for many suggestions to improve the text. I am also very grateful to Erzsi néni for helping out with the kids and to Sogi for helping to set up my home LAN.

I am sure that many more people contributed in one way or another to make this book a reality. It is impossible to keep track of everything. So there might be many unintentional omissions in this acknowledgment list. If you find such an omission, please accept my apologies.

S. M. D.

---

[1]Gilles has an eye-catching homepage: *http://www.oyonale.com.*
[2]"POV-Ray" is a trademark of Persistence of Vision Raytracer Pty. Ltd. All other trademarks are acknowledged as the property of their respective owners.
[3]The cover picture shows a lamppost in a box with mirrored walls, seen from inside through a fisheye lens of $100°$.

1

Introduction

An author ought to consider himself, not as a gentleman who gives a private or eleemosynary treat, but rather as one who keeps a public ordinary, at which all persons are welcome for their money. In the former case, it is well known that the entertainer provides what fare he pleases; and though this should be very indifferent, and utterly disagreeable to the taste of his company, they must not find any fault; ... good breeding forces them outwardly to approve and to commend whatever is set before them. Now the contrary of this happens to the master of an ordinary. Men who pay for what they eat will insist on gratifying their palates, however nice and whimsical these may prove; and if everything is not agreeable to their taste, will challenge a right to censure, to abuse, and to d—n their dinner without control.

To prevent, therefore, giving offence to their customers by any such disappointment, it hath been usual with the honest and well-meaning host to provide a bill of fare which all persons may peruse at their first entrance into the house ... we ... shall prefix not only a general bill of fare to our whole entertainment, but shall likewise give the reader particular bills to every course which is to be served up in this and the ensuing volumes.

—Henry Fielding [208]

In this introduction we give you the "general bill of fare" for the entire book. The best way to start is to say a few words about the main aim of the book, its raison d'être. This is twofold. First, it is to provide an introduction to the field of cavity quantum electrodynamics for postgraduates and any researcher in general who wishes to enter this field or make use of some of its techniques. If you fit this description, you will find that this book deals with some topics often avoided by quantum optics textbooks.[1] There is, however, a less obvious raison d'être for this book.

In his autobiographical notes [184], Einstein wrote that one of the reasons he chose to work in physics rather than mathematics was that in physics he could clearly differentiate what was fundamentally important and basic from "the rest of the more or less dispensable erudition." Mathematics, he wrote, "was split up into numerous specialities, each of which could easily absorb the short lifetime granted to us. . . . True enough, physics also was divided into separate fields, each of which was capable of devouring a short lifetime of work without having satisfied the hunger for deeper knowledge." However, in physics, he soon learned to scent out the problems that were capable of leading to fundamentals and to avoid "the multitude of things which clutter up the mind and divert it from the essential." Well, today, physics is much more fragmented into a multitude of specialities than it was in Einstein's time.[2] This makes it even more difficult today than it was then to pass to students an idea of the fundamental issues in physics.[3]

Despite being only one of many specialities within physics, cavity quantum electrodynamics can serve as a window to the fundamental issues of physics from which undergraduates can greatly profit. Light was a key ingredient of the second great unification in physics,[4] Maxwell's unification of electricity, magnetism, and optics with his theory of classical electrodynamics. Light also held the secret to the two great fundamental discoveries in physics in the twentieth century: relativity and quantum mechanics. As we see in Chapter 3, the photon is an intrinsically relativistic particle that, as far as we know, always travels at the speed of light and has no rest mass; it is never in a nonrelativistic regime. Quantum electrodynamics is the basis of all other field theories, which together with general relativity form the fundamental core of contemporary physics. Cavity quantum electrodynamics is a "tamed" form of quantum electrodynamics that undergraduates can easily tackle after seeing electromagnetism and basic quantum mechanics.

So this book is also aimed at undergraduates. If you are an undergraduate, I think you will find this book quite different from the usual physics textbooks that you have come across. I hope you find it inspiring. There are several pointers to the literature in each chapter, mainly in the sections on recommended reading. Some of the primary historical papers are also mentioned there. You will find it very instructive, and hopefully entertaining, to read at least some of these remarkable papers. And please, remember that even though my main aim has been to show you how fascinating and

---

[1]For a brief list of some of these topics, see the preface.

[2]In Einstein's time, it was still possible for a single person to make contributions to the whole of physics, whereas nowadays this is very rare.

[3]Already in his day, Einstein complained [184] about having to cram into one's mind for examinations all this "multitude of things which clutter up the mind and divert it from the essential."

[4]The first being Newton's unification of celestial and earth mechanics with his theory of classical mechanics and gravity.

interesting fundamental issues in physics really are, I have also included material that will be useful for those who do not wish to work in scientific research.[5]

Next, we will take a "panoramic view" of the entire book, beginning by asking a simple, yet quite fundamental question: What is light? There is no straight or easy answer, but on pursuing this question we introduce you to most of what we now know about light. We review briefly how our understanding of what light is has evolved from simple everyday concepts to the more subtle ones brought about by quantum mechanics. We consider some deep questions concerning the nature of light. Why is the speed of light always the same even when the source of light is moving? Why in the classical mechanics regime is light a wave, whereas electrons are particles? We briefly review the history of cavity QED. Last but not least, I explain what the various chapters are about and suggest some ways in which the book can be read.

1.1 WHAT IS LIGHT? THE DEVELOPMENT OF OUR MENTAL PICTURE OF LIGHT

Light is a funny thing. You cannot catch it with your hands. It appears to be very light (please forgive the pun), even weightless. Light looks completely different from the substances (matter) that make up the world around us, such as air, sand, water, and so on. It is almost as if it is not material at all— as if it is something celestial, even divine. For centuries light has caught the attention and imagination of children, poets, artists, philosophers, and scientists.

Apart from being such a curious phenomenon, there is another, stronger reason why studying light is both interesting and important. Light holds the key to some of nature's deepest secrets. In this section we give an overview of how we got to know, to a better and better approximation,[6] what light is. On the way, we will see how this better understanding of the nature of light helped to unlock some of the strangest hidden features of our world, now described by the two great theories of modern physics: relativity and quantum mechanics.

1.1.1 The crudest picture: Geometrical optics

Our limited everyday experience of light is explained almost entirely by the theory of geometrical optics.[7] Its origins lie in antiquity, but most of it was developed in the seventeenth century by a number of people, including René Descartes, Willebrord

---

[5]The preface mentions some of this material.

[6]Our commonsense picture of light is a very crude caricature. Nevertheless, modern scientific knowledge of light is nothing but an extension of this caricature. Instead of our limited everyday experience, it accesses a much broader one made possible by instruments. Instead of our imprecise language and reasoning, it uses the precision and rigor of mathematics. The end result, however, is just a model of reality; a picture much more accurate than what common sense can ever aspire to be, but still only a picture.

[7]The image of a lamppost on the cover of this book, for example, was generated by a computer using the principles of geometrical optics.

Snell, Christian Huygens, and Isaac Newton. The modern mathematical theory of geometrical optics was developed mainly in the nineteenth century by William R. Hamilton. For a thorough historical account, see Mach's book [422].

As readers might recall from their basic physics lectures, geometrical optics considers light as being composed of rays whose path is determined by Fermat's principle of least time. So in vacuum, these rays travel along straight lines. But when they meet a medium with a different refractive index, such as a lens, they bend (refraction). Extended objects can be decomposed into an infinite number of points, each emitting rays in all directions. Geometrical optics can explain many everyday phenomena, including the reflection of light off mirrors, the workings of lenses, refraction, and the formation of images. Still, however, it tells us nothing about the nature of rays or why they always take the path of least time.

1.1.2 An improvement: Wave optics

In the seventeenth century there were two rival theories about the nature of light. Newton proposed that light rays were made out of particles that followed his laws of mechanics. Huygens, on the other hand, proposed that they were made out of waves. Geometrical optics does not allow us to decide between these two theories. As Newton carried much more authority than Huygens, the particle view prevailed until the nineteenth century. It is curious, however, that at the time these two theories were proposed, there was a discovery that could have settled the dispute in favor of Huygens. Hooke and Grimaldi had already observed diffraction. Diffraction signals the breakdown of the geometrical optics approximation and indicates that, more accurately, light must be regarded as a wave. The breakdown happens when the relevant dimensions (size of an aperture, for example) are so small that they become comparable to the wavelength of the light wave. But Huygens apparently did not know about this discovery and never attempted to describe it with his theory. It was only in the early part of the nineteenth century, with the discovery of interference by Young and with the work of Fresnel, Arago, and others, that the need for a wave theory of light was finally accepted. The transition was not made without a struggle, though.

There is an interesting anecdote that clearly describes the reluctance of even some of the most eminent scientists of the time to give up their old beliefs and adopt the wave theory of light. When Fresnel finished a talk at the French Academy of Sciences about his wave theory of diffraction, Poisson stood up and said that he could prove by reduction ad absurdum that Fresnel's theory was wrong. Poisson claimed that Fresnel's theory implied that the shadow cast by a circular object illuminated by a point source should have a bright spot in the center. This was, according to Poisson, absurd: Obviously, no shadow can have a bright spot in its center. Arago, however, was not so sure and did the experiment to find out. It turned out that there was indeed a bright spot in the center, which is now known as the *spot of Arago*.

1.1.3 Great guns and fresh insight into the nature of space and time: Classical electrodynamics and relativity

So if we look closer at light rays, we find that they are really waves. But what is the nature of these waves? In mechanics, waves are oscillations that propagate in a medium, such as water waves or sound in air. But light also propagates through a vacuum. Nineteenth-century physicists tried to solve this puzzle by postulating the existence of an elastic medium, the ether, that permeated everything, even the vacuum.

A big problem with the ether was that ordinary elastic media support both longitudinal and transverse waves. Since the early part of the nineteenth century, however, physicists were convinced that light waves were purely transverse, like the waves on a stretched string. This problem was solved by James MacCullagh. In 1839 he showed that if instead of a solid that resists compression and distortion, the ether is assumed to have a potential energy depending only on the rotation of its volume elements, it would support only transverse waves.

But the major breakthrough came from the apparently totally unrelated field of electricity and magnetism. In the second half of the nineteenth century, James Clerk Maxwell developed a theory of electromagnetism predicting that light is an electromagnetic wave. Classical electrodynamics was born and the nature of the waves in the wave theory of light was clarified. In a letter dated January 5, 1865, to his cousin Charles Cay, Maxwell wrote [87]: "I have also a paper afloat, containing an electromagnetic theory of light, which, till I am convinced of the contrary, I hold to be great guns." And great guns it was, indeed! It took 20 years from Maxwell's original paper to Hertz's experimental confirmation; this is one of the greatest predictions of theoretical physics.

Maxwell's theory not only unified electricity, magnetism, and optics, but also introduced the key idea of *field* in physics. In Newtonian mechanics there is the mysterious instantaneous action at a distance. A field, on the other hand, is a local property of space that can propagate from one point to another only at a finite speed. The idea of field was probably originated by Faraday's lines of force, but it was Maxwell who put it in a sound mathematical form. With the development of the theory of relativity, we now know that there is no ether, just the field: It is the field, permeating even the vacuum, that oscillates, not a material medium.

A very important clue to relativity and the overthrow of the ether came from Maxwell's puzzling prediction that electromagnetic waves propagate at the same speed, the speed of light, regardless of the speed of the source of the waves or of the observer. The fact that this speed is determined by the electric permittivity and magnetic permeability of the material was at first thought to be a signal that Maxwell's theory was valid only in a special frame where the ether was at rest. But years later, Michaelson and Morley [445] showed that the speed of light is really the same in any reference frame. Relativity is built on this invariance of the speed of light and explains it by changing the Newtonian notions of absolute time and space. The invariance of the speed of light showed us that the true nature of space and time is completely different from what is naively apparent to us from our everyday experience: Time for someone who is moving closer to the speed of light in relation to you passes much more slowly than it does for you, and his or her space (along the direction of motion) contracts in relation to your space. Electrodynamics is such a great theory that it had relativity already built into it many years before relativity itself was discovered by Einstein!

It is curious that even though in the long run, Maxwell's theory helped to revolutionize the Newtonian view of the world, Maxwell himself used theoretical mechanical models to build his theory. One of his very distinctive characteristics was his habit of working by analogy [87], where he recognized mathematical similarities between quite distinct physical problems and tried to apply the successes of a well-tested theory to the mathematical analogous but physically very different situation. He imagined, for instance, that the magnetic field was made up of vortex tubes filling space. Between these vortices, to prevent friction, there were ball bearings, which Maxwell identified with electric particles (see Figure 1.1).

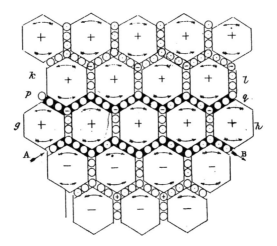

Figure 1.1 Maxwell's mechanical model for the production of lines of force by an electric current. He imagine that the ether was filled with molecular vortices (represented by the hexagons). Between these vortices there were charged particles (the little balls). AB represents an electric current going from A to B. As this current starts flowing, the row of vortices gh above AB will rotate counterclockwise. Assuming that the row of vortices kl is still at rest, the layer of particles between these rows will be acted on by the row gh below, making them rotate clockwise and move opposite from the current (i.e., from right to left). These charges moving in the opposite direction to the initial current would then constitute an *induced current*. (Reproduced from [435].)

1.1.4 Quantum mechanics and quantum electrodynamics: The best picture so far

Classical electrodynamics is a great theory, but there are some problems with it, and these problems point the way to a new and more encompassing theory: *quantum electrodynamics* (QED). As the name says, this is the theory that joins electrodynamics with quantum mechanics.

To most physicists, quantum mechanics introduced an even greater departure from the old Newtonian ideas than relativity did. There were many puzzling phenomena that, left unexplained by classical physics, pointed the way to the brave new world of the quantum. These are reviewed in most elementary quantum mechanics textbooks and we do not wish to go into details here. Let us just say that a lot of them came from spectroscopy and can be well accounted for by a semiclassical theory where only matter is quantized, not light. Others, such as the Compton effect, however, called for a quantum description of light, too.

In quantum mechanics everything behaves sometimes as a particle, sometimes as a wave. So according to quantum electrodynamics, light is made up of particles called *photons* that sometimes behave as waves! As this entire book is about this best approximation that we have to what light really is, we will not say much more about QED in this introduction. We will say something about this funny wave–particle duality, though.

Strange as it is, the wave–particle duality revealed by quantum mechanics must be accepted as a fact of life. It just turns out that nature is that strange deep inside. The reason that we find it so weird is because we are used to our everyday macroscopic experience, where particles always behave as particles and waves always as waves. So instead of asking the pointless question why the photon sometimes behaves as a wave, what we should really ask is why in our macroscopic world photons only behave as waves, whereas electrons only behave as particles. This question was answered in 1933 by Pauli [477] (see also "Part VI, Final Remarks: The New Particles" by Peierls in Sec. 1.3 of [485], and [486], pp. 471–473). He showed that what determines the particle-only or wave-only behavior of a quantum particle in our macroscopic world is its rest mass and quantum statistics.

A classical particle always has a definite position and momentum. This introduces the well-known problem of *complementarity,* our inability to measure the position and momentum of a quantum mechanical particle simultaneously. In Chapter 4 we see how complementarity effectively disappears in our macroscopic world. But even more basic than this is the problem that to talk about position and momentum, both must be observables. As we see in Chapter 3, although the momentum of a quantum particle can be measured to any accuracy, its position cannot be determined more accurately than to locate it inside a volume the size of the cube of its Compton wavelength. Any attempt to determine the position more accurately than this involves an interaction energy high enough to generate particle–antiparticle pairs out of the vacuum. This then necessarily takes us from a single-particle description to a many-particle one, where the idea of position of an indistinguishable particle among many others of its kind makes no sense. So position is an observable only in the nonrelativistic regime where the Compton wavelength is negligible in comparison with the de Broglie wavelength. Unfortunately, quantum particles with a vanishing rest mass do not have a nonrelativistic regime. That is why the photon does not behave as a classical particle in our macroscopic world.[8]

What about the classical wave behavior? A wave is characterized by both a phase and an amplitude. Ideally, a phase operator $\hat{\phi}$ for the electromagnetic radiation field would be the canonical conjugate of the photon number operator $\hat{N}$, just as position is the canonical conjugate of momentum; that is, the commutator of $\hat{N}$ and $\hat{\phi}$ would be given by

$$[\hat{N}, \hat{\phi}] = -i. \tag{1.1}$$

Unlike position in nonrelativistic quantum mechanics, however, there are serious difficulties with the introduction of a phase operator in quantum mechanics (see [89], [484], Sec. 1.4 of [485], and [490]). One such difficulty arises because unlike the momentum operator, $\hat{N}$ has a discrete spectrum of eigenvalues. This can be seen by sandwiching (1.1) between a bra–ket pair of eigenstates of $\hat{N}$, $\langle n|$ and $|n'\rangle$:

$$(n - n')\langle n|\hat{\phi}|n'\rangle = -i\delta_{nn'}. \tag{1.2}$$

Now take $n' = n$ in (1.2). The left-hand side vanishes while the right-hand side becomes $-i$. Attempts to avoid this problem usually encounter another difficulty, this time associated with the lack of negative eigenvalues in the spectrum of $\hat{N}$ [i.e., it starts at $N = 0$ rather than at $N = -\infty$ (see Chapter 4)]. A further difficulty is connected with the fact that a phase of ϕ and another of $\phi + 2\pi$ should be equivalent. Fortunately, for our purposes here, we can use the well-defined operator $\exp(i\hat{\phi})$ rather than $\hat{\phi}$ itself.[9] The commutator between $\hat{N}$ and $\exp(i\hat{\phi})$ is given by [89, 485]

$$[\hat{N}, e^{i\hat{\phi}}] = e^{i\hat{\phi}}. \tag{1.3}$$

Repetition of the steps that yielded (1.2) now yields

$$(n - n')\langle n|e^{i\hat{\phi}}|n'\rangle = \langle n|e^{i\hat{\phi}}|n'\rangle. \tag{1.4}$$

Unlike (1.2), equation (1.4) does not lead to any contradiction. Its consequence is merely that $\langle n| \exp(i\hat{\phi})|n'\rangle$ can only be nonvanishing for $n' = n - 1$ [i.e., $\exp(i\hat{\phi})$ is similar to a raising or creation operator].

From (1.3) we can obtain the following uncertainty relation for the accuracy ΔN in determining $\hat{N}$ and the accuracy $|\Delta \exp(i\hat{\phi})|$ in $\exp(i\hat{\phi})$ [89]:

$$\Delta N \left|\Delta e^{i\hat{\phi}}\right| \geq \frac{1}{2}\left|\left\langle e^{i\hat{\phi}}\right\rangle\right|. \tag{1.5}$$

---

[8]In the geometrical optics regime, however, we can regard photons as classical particles (Newton's view of the nature of rays). This is because the position of such a "classical" photon is never determined more accurately than a ray (i.e., in geometrical optics, a photon is at most localized to a volume on the order of the cube of its wavelength).

[9]In terms of creation and annihilation operators for a single field mode or an harmonic oscillator, $\exp(i\hat{\phi})$ is given by $\hat{a}^\dagger(\hat{a}^\dagger\hat{a} + 1)^{-1/2}$ (see Carruthers and Nieto [89]).

If we call $\Delta\phi$ a measure of the phase uncertainty defined by[10]

$$\Delta\phi \equiv \frac{\left|\Delta e^{i\hat{\phi}}\right|}{\left|\left\langle e^{i\hat{\phi}}\right\rangle\right|} = \sqrt{\frac{(\Delta C)^2 + (\Delta S)^2}{\langle\hat{C}\rangle^2 + \langle\hat{S}\rangle^2}}, \qquad (1.6)$$

where $\hat{C}$ and $\hat{S}$ are the sine and cosine Hermitian operators[11]

$$\hat{C} \equiv \frac{e^{i\hat{\phi}} + e^{-i\hat{\phi}}}{2} \quad \text{and} \quad \hat{S} \equiv \frac{e^{i\hat{\phi}} - e^{-i\hat{\phi}}}{i2}, \qquad (1.7)$$

we have the following simple uncertainty relation for ΔN and $\Delta\phi$

$$\Delta N \Delta\phi \geq \frac{1}{2}. \qquad (1.8)$$

Now it is clear from (1.8) that to measure the phase of a field with accuracy, we must change the number of particles. For light, this is not a big problem, as photons can be created easily because they have no rest mass. For a massive particle, however, this will require a large amount of energy. That is why massive particles do not behave as waves in our macroscopic world.

In a classical wave, however, it is not only the phase that must have negligible uncertainty, but also the amplitude. In other words, the uncertainty in the phase should be small compared to unity, but the uncertainty in N should also be small compared to N itself. From (1.8), we see that this requires

$$N \gg 1. \qquad (1.9)$$

Photons can easily fulfill this requirement, as they are bosons. In fact, for a 1-kW transmitter at 1 MHz, it can be estimated that only at large distances exceeding 10^{10} km from the transmitter does the number of photons drop to about 1 [485]. Another example is a typical 10-mW HeNe laser, whose number of photons in the lasing mode can easily be estimated to be on the order[12] of 10^{10}. For fermions, however, the Pauli exclusion principle limits N to 1. Thus, fermions cannot behave as classical waves in our macroscopic world.

1.2 A BRIEF HISTORY OF CAVITY QED

The first known reference to a cavity QED effect dates back to the 1940s, when the *Physical Review* published the abstract of a paper presented by Edward M. Purcell

---

[10]Heuristically, if $\hat{\phi} = \bar{\phi} + \Delta\phi$, for $\Delta\phi \ll 1$ we can write $\exp(i[\bar{\phi} + \Delta\phi]) \approx (1 + i\Delta\phi)\exp(i\bar{\phi})$, so that $\left|\Delta e^{i\hat{\phi}}\right| / \left|\left\langle e^{i\hat{\phi}}\right\rangle\right| \approx \Delta\phi$.

[11]Notice that $\langle\hat{C}\rangle^2 + \langle\hat{S}\rangle^2$ can be different from 1. That is why $\left|\left\langle e^{i\hat{\phi}}\right\rangle\right|$ is in the denominator of the second term on the left-hand side of (1.6).

[12]An even more striking example is a typical CO_2 laser operating with an output power of 10 kW. The number of photons in the cavity mode is then on the order of 10^{16}.

at the 1946 Spring Meeting of the American Physical Society [502]. His paper was about the changes in the spontaneous emission rate for nuclear magnetic moment transitions at radio frequency, when a spin system is coupled to a resonant electrical circuit. At room temperature (300 K), for a frequency of 10 MHz and a $\mu = 1$ nuclear magneton, he found that the spontaneous emission relaxation time was about 5×10^{21} seconds. If, however, small metallic particles with a diameter of 10^{-3} cm were mixed with the nuclear magnetic medium, he found that the spontaneous emission relaxation time dropped to only a few minutes.

The 1950s saw the realization of the first maser [242], which stimulated further research in the interaction between matter and the radiation field in cavities. During this decade, modification of spontaneous emission rates of electron spin transitions was predicted [115] and confirmed experimentally [191]. The mechanism of spontaneous emission rate enhancement deduced by Purcell was predicted to hold as well for collective spontaneous emission in magnetic resonance experiments [61]. Another significant development came in 1958, when Schawlow and Townes [536], Prokhorov [500], and Dicke [152] independently proposed use of the Fabry–Perot interferometer as a resonator for the realization of a maser that operated in the optical regime (i.e., a laser). This generated a lot of interest in the theory of open cavities.

The 1960s began with the realization of the first laser [424]. A number of theoretical papers were published on open resonators and their modes, with important contributions by Fox and Li [220, 221] and Vaĭnshteĭn [605–607].[13] The 1960s also saw Jaynes and Cummings develop a fundamental model, which now bears their names, for the interaction between an atom/molecule and a single mode of the quantized radiation field [323, 559]. Last but not least were Drexhage's beautiful experiments where the florescence of dye molecules placed at precisely controlled distances from a mirror was measured [100, 164].

In the 1970s, change in the spontaneous emission rate of an atom when it is inside a Fabry–Perot interferometer was calculated using full-blown quantum electrodynamics [39, 577]. These two papers marked the beginning of a great theoretical interest in cavity QED that has resulted in a large number of papers being published on this subject ever since.

After those primordial times, the field of cavity QED entered a period that one of its leading researchers has named the *weak-coupling-regime age* [503]. It was then that spontaneous emission enhancement [245] and inhibition [303] for an *atom* in an *empty* cavity was experimentally demonstrated.[14]

Then there came what that researcher calls *the strong-coupling-regime age*. It was then that one-photon [440] and two-photon [81] single-atom masers (micromasers) were realized, and quantum Rabi oscillations [82] and vacuum Rabi splitting[15] [592] were observed. Strong coupling can also be used to manipulate entanglement, and this has given rise to a number of applications in quantum information.

---

[13]For a detailed review of the developments in resonator theory during this time, see [563].

[14]Drexhage had worked with *molecules* in material *media*.

[15]This is normal-mode splitting due to the oscillatory exchange of energy quanta between an atom and a quantized cavity mode.

At the time of writing, in parallel with the ongoing basic research, we are experiencing an "industrial age," where cavity QED ideas are being applied to optoelectronic devices. This is being done mainly through *photonic crystals* [326, 655], new materials whose dielectric permittivity varies in a regular way analogous to the spatial arrangement of atoms in a crystal. As in a crystal, these structures give rise to bands (allowed values of the wave vector) and gaps (forbidden values of the wave vector) that change the mode structure of free space. A single mode-cavity, for example, can be constructed in a photonic crystal by adding a defect in the regular variation of the dielectric permittivity. This defect acts as an impurity that when properly engineered will create a discrete level in the middle of a gap: a single mode where a radiation field can exist.

1.3   A MAP OF THE BOOK: AN OVERVIEW OF WHAT IS TO COME

This book is about theory, mostly theory connected with fundamental ideas in physics.[16] There is almost no description of experiments. You can find these in experimental papers, review articles, and other books, some of which are listed in the References. Whenever possible, this theory has been presented using very simple models that emphasize only the key points we wish to focus our attention on, neglecting everything else. This model-building approach has been used by Maxwell and many other British physicists, often to the horror of their counterparts in continental Europe. Model building is not only a useful aid to highlight, develop, and illustrate theoretical ideas, but also a skill in high demand in the nonacademic job market. One of the aims of this book is to help train students, especially undergraduates, in this often neglected art.

The topics covered in this book were chosen using four basic criteria. First, there were those topics that were picked because they are in some way connected to fundamental ideas in physics. Second, there were topics which the author regards as important and which often rouse the curiosity of students but that somehow have not been addressed in other quantum optics textbooks (see the preface for a list of some of these topics). Third, there were topics which can also be found in other textbooks, but which this author addresses from a completely different point of view, with an original approach that cannot be found anywhere else. Fourth, everything had to fit in a coherent story, with one topic related to the others and following naturally in the sequence of our discussions (i.e., no nonsequiters). These criteria, of course, do not determine a list of topics uniquely. Ultimately, a large component of personal taste was involved in the choice.

Chapter 2 starts our incursion into cavity QED country with an introduction to field quantization. It goes through quantization in cavities as well as in free space. But before dealing with quantization, we discuss three issues that are essential for

---

[16]This is also reflected on the choice of units. The CGS units adopted throughout the book are better suited for presenting theoretical ideas than SI units are.

most of what is to come. The first is canonical quantization (here we also review Dirac's representation-free quantum mechanics). The second is why only the radiation field is quantized. The third is what constitutes a resonator. Here, we discuss the physics behind a cavity's resonator features rather than just presenting a calculation of cavity modes.[17] This is also where we introduce the model-building approach that is used many times in the book. The chapter closes with a look at one of the few consequences of field quantization that does not need interaction with matter in its theoretical description: the Casimir force. The last subsection deals with a very important problem related not only to quantum electrodynamics but also to the whole of field theory and general relativity. This is the problem of the vacuum catastrophe, which may well point the way to a new theory where gravitation and quantum mechanics will be unified.

In Chapter 3 we present an alternative approach to quantum electrodynamics, where rather than using canonical quantization, we follow the route usually taken in elementary nonrelativistic quantum mechanics: going from a single-particle wavefunction description to a many-particle second-quantized theory. We address various fundamental issues in quantum electrodynamics, including why the idea of a wavefunction of the photon in configuration space is plagued with problems, and the connection between nonlocalizability and vanishing rest mass. Instead of adopting the traditional abstract group-theoretical approach, these issues are discussed using concrete examples. The chapter also offers insight into the way that first-order-in-time relativistic wave equations such as the Dirac equation can be derived from the relativistic energy–momentum relation.

In Chapter 4 we end our discussion on light without matter by looking at some of the things that we can do with light by itself in a cavity. First we will see, in more detail than was presented in this introduction, how our ordinary "classical" experience of the radiation field as a wave can arise from quantum electrodynamics. This will take us to coherent states and statistical mixtures. Statistical mixtures will naturally lead us to introduce the density matrix. Then we will show how quantum mechanics can be exploited to effectively reduce the noise of light below its fundamental quantum limit experienced in coherent states. Rather than just postulating the form of the squeezing operator, we deduce it from its intended properties. We discuss in great detail the usual graphical representation of squeezed states, where they are depicted as phasors with an "ellipse of noise" attached to the tip of the phasor. We review how squeezed states can be detected using homodyne detection and how they can be generated (in an unusual way) by suddenly changing the size of a cavity.

The two shortest chapters in the book are Chapters 5 and 6. The former starts on our examination of the interaction between quantized field and matter by introducing the coupling between the two. This is done in three different ways. First, we see how the simple free-space quantization without electromagnetic potentials and associated gauge complications can easily be extended to include interaction with matter, as long as this interaction is only within the dipole approximation. Second, we will

---

[17]Such a calculation is left for Appendix A.

go through the complete theory of the interaction with matter beyond the dipole approximation but using electromagnetic potentials in the Coulomb gauge. Finally, we will see how the minimal coupling term describing the interaction between matter and radiation can also be understood as a consequence of gauge invariance. Having managed to include matter in our big picture, then in chapter 6 we look at an important application: spontaneous emission. We see how a cavity can change the spontaneous emission rate and even make spontaneous emission become a reversible process, turning the exponential decay experienced in free space into Rabi oscillations. The famous Jaynes–Cummings model is also discussed.

Ordinary QED is about individual atoms and molecules sitting in the vacuum interacting with the quantized electromagnetic field. In many important situations, however, the number of individual atoms/molecules is so large that this approach is just not feasible. This is the case, for example, for cavity QED applications to solid-state devices such as semiconductor lasers, where the entire system is built in a material medium (e.g., gallium arsenide). Chapter 7 shows how we can construct a macroscopic version of quantum electrodynamics, where a medium can be described by dielectric permittivity as it is done in the classical Maxwell equations. Starting from ordinary QED for atoms/molecules in the vacuum interacting with the field, we see how a macroscopic version of QED with dielectric permittivity can emerge as an approximation of the "exact" theory, as Lorentz did in classical theory. We review the Huttner–Barnett modification of Hopfield's model to account for absorption and calculate the dielectric permittivity explicitly. We also introduce the useful technique of Fano diagonalization.

The laser, the maser, and their cavity QED counterparts are discussed in Chapter 8. The laser was the first source of coherent electromagnetic radiation of optical frequency: almost as pure a sinusoidal wave as that generated by a radio oscillator. Here we see how an attempt to realize the principles of a sustained radio oscillator in the optical domain leads us naturally to the laser. Then we go on to explore phenomena that are negligible in radio transmitters but not when the frequency is in the optical range. First, we look at spontaneous emission, and to account for that in a simple "handwaving" way, we introduce a powerful mathematical technique with applications even in analysis of the stock market: Langevin equations or stochastic differential equations. We then go from this "back of the envelope" description of a laser to a full quantum theory. After showing that the radiation generated by an "ideal" laser would be in a coherent state, we address the microscopic regime achieved in cavity QED devices. We show how the Jaynes–Cummings model can act as the basic building block of a microscopic laserlike device operating in a highly quantum regime: the micromaser. The connection between the first-principles cavity QED theory used to describe micromasers and ordinary lasers is sketched with a simple example, where it is shown how incoherent processes lead to the usual features we associate with macroscopic lasers. As a bonus, we see how the laser threshold can disappear in the cavity QED regime. The chapter closes with a quick look at a recent breakthrough in cavity QED: realization of the one-and-the-same atom laser.

Chapter 9 deals with the problem of open cavities. Although the definition of cavity mode is quite straightforward for perfect closed cavities, it is subtle when the cavity is leaky or open. Then it is a question of how to define what constitutes the cavity and what is the outside. Such open cavities gained prominence with the advent of the laser, which had to adopt an open cavity: the Fabry–Perot. In this chapter we show that the proper generalization of the idea of cavity mode to open cavities is that of natural modes of oscillation. These are the modes in which the cavity resonates naturally when its driving is withdrawn. They go back at least as far as to 1884, when Joseph J. Thomson (the discoverer of the electron) studied the natural electrical vibrations of a spherical shell. Mathematically, they are defined as solutions to the Helmholtz equation which satisfy the boundary conditions on the cavity walls and Sommerfeld's radiation condition at infinity. Then the problem of constructing a quantum theory using these natural modes reduces to that of how to complement these solutions of Helmholtz equation to get a complete basis (i.e., how to describe vacuum fluctuations that do not have to satisfy Sommerfeld's radiation condition). The chapter starts with a first-principles derivation of the famous Gardiner–Collett model used to describe losses in high-Q cavities quantum mechanically. Then a general formulation of the problem of an open cavity is given together with a review of Sommerfeld's radiation condition.[18]

The appendixes provide supplementary mathematical material. There are appendixes on electromagnetic boundary conditions, how to calculate the modes of a perfect cavity, the Baker–Hausdorff formula, quaternions, vector-calculus tools, and on principal parts, step, and delta functions. All this material is presented at a level accessible to advanced undergraduates.

1.4   HOW TO READ THIS BOOK: SUGGESTIONS ON HOW TO FIND YOUR OWN OPTIMAL PATH THROUGH THE CHAPTERS

This book is self-contained and can even be used for independent study. It is impossible, however, to cover everything in the field of cavity quantum electrodynamics in a book of this size. So, depending on your aims, you might wish to supplement this book with some other books and review papers. In particular, if you are a graduate student or research scientist, you should look at the current literature for descriptions of relevant experiments. The theory presented here should leave you in a good position to understand most of the experiments in this field. A book that is focused on experiments in quantum optics and that can complement this one is [26].

A course for advanced undergraduates could cover Chapter 2 without Section 2.5.2 but including Appendixes A, E, and F. Then the whole of Chapter 3 with Appendix C is suggested. The material on squeezed states in Chapter 4 can be omitted completely. Only the first section of Chapter 5 needs to be covered to understand the next chapters. Then the course can cover all of Chapter 6 and the first four sections of Chapter 7. Given the technological importance of lasers, I would recommend covering all of

---

[18] As at the time of writing, this is still an open problem, I only point out what the problem is and what its solution would involve.

Chapter 8. If there is time, the first section of Chapter 9 can also be covered. As a course project, students could give special seminars about key experiments in this field.

A course for postgraduate students starting research in this field can cover the entire book except for the appendixes, which should be left for students to read on their own according to their needs, and parts of Chapter 2, depending on the student's background.

A last word for undergraduates and people wishing to use the book for self-study. I have included intermediate steps in the calculations. But if you still get stuck somewhere, do not worry. Just carry on reading and there is a good chance that you will understand everything at a later stage. One of the philosophies behind this book is Silvanus P. Thompson's motto [593]: "What one fool can do, another can."[19]

---

[19] An ancient Chinese proverb.

2

Fiat Lux! A free tasting of field quantization

God said

$$\nabla \cdot \hat{\mathbf{E}} = 0,$$

$$\nabla \cdot \hat{\mathbf{B}} = 0,$$

$$\nabla \wedge \hat{\mathbf{E}} = -\frac{1}{c}\frac{\partial \hat{\mathbf{B}}}{\partial t},$$

$$\nabla \wedge \hat{\mathbf{B}} = \frac{1}{c}\frac{\partial \hat{\mathbf{E}}}{\partial t}, \quad \text{with } [\hat{E}_i(\mathbf{r}), \hat{B}_j(\mathbf{r}')] = -i2hc \sum_{l=1}^{3} \varepsilon_{ijl}\frac{\partial}{\partial r_l}\delta(\mathbf{r} - \mathbf{r}').$$

and then there was light![1]

—Popular physics joke[2]

This chapter is an introduction to the quantization of the electromagnetic radiation field. Here we deal with the simple case of radiation alone, without any sources or sinks that its interaction with matter generates. One of the reasons that the quantization of the free radiation field is simpler than that of the field with sources is that there is no need to use electromagnetic potentials and to choose a gauge.[3]

---

[1] As Einstein *would* [64] have said: God does not use SI units; She or He prefers CGS.

[2] The vanishing sources, operator hats, and commutator are the author's own additions.

[3] If you are not new to this field and cannot believe what you have just read about no potentials or gauge choice, go to Section 2.4 and check it out now. The gauge-independent part (i.e., transverse component) of the vector potential eventually appears but in a natural way, without any need to introduce electromagnetic potentials in the beginning. A similar approach was adopted by Knight and Allen [359] and by Gray and Kobe [249].

In the next section we review the canonical quantization method. This is one of the finest "wines" in physics. To help even newcomers appreciate it fully, we establish the connection with the more familiar elementary quantum mechanics formalism by using canonical quantization to derive the Schrödinger equation description of a nonrelativistic particle from classical mechanics. In Section 2.2 we explain why it is only the radiation field, rather than the total field, that is quantized. In Section 2.3 we review cavities. Rather than just presenting a formal calculation of cavity modes (such a calculation is done in Appendix A), we discuss the idea of a cavity or resonator and the physics it involves. We start with an elementary introduction to the phenomenon of resonance. Then we go on to show how the features we associate with resonators emerge when an oscillating field is confined spatially. The actual quantization is done in Section 2.4 for a perfect cavity (2.4.1) and for free space (2.4.2). In Section 2.5 we discuss an amazing effect of the free field that can be regarded as a direct consequence of the quantum nature of light and of its vacuum having zero-point energy: the Casimir force.

In this free tasting all the wines are quite traditional. If, as I do, you like to sample some of the alternative wines, make sure not to miss Chapter 3. If nothing else, that free tasting of alternative wines will at least change forever the way you think about the Dirac equation for the electron.

2.1   A BRIEF REVIEW OF QUANTUM MECHANICS: HOW TO QUANTIZE A THEORY

Dirac remarked once [158] that there was a time during the development of quantum mechanics when even the average physicist could do outstanding work. According to him, this happened because of his discovery of a rule for going from classical to quantum mechanics. Then people could play a little game, where they could take the various classical models of dynamical systems and transform them into the new quantum mechanics. This game, called *canonical quantization,* is what we review here.

One of the most striking features of quantum mechanics is that it requires an algebra that differs from the one we are used to from classical physics: an algebra where dynamical variables in general do not commute. So the main hurdle someone faces when trying to construct a quantum version of a classical theory is how to generalize the classical theory to include noncommutability. Dirac realized that Hamilton's formulation of classical mechanics is in a form that makes adoption of noncommutability extremely easy. He noticed that the Poisson brackets,[4] in terms of which Hamilton's formulation becomes simpler, have the same algebra as that of the commutators of quantum mechanics. So he saw that he could obtain the quantum equations for vir-

---

[4]If you are not familiar with Poisson brackets, have a look at the recommended reading at the end of the chapter. Even if you are already familiar with Poissonian brackets, the historical accounts might be worth reading.

tually any physical system for which he could write down Hamilton's equations just by replacing the Poisson brackets by commutators multiplied by a certain constant.

Here is how the analogy between Poisson brackets and commutators comes about.[5] The *Poisson bracket* of any two dynamical variables, u and v, in classical mechanics is defined by

$$(u, v) = \sum_n \left\{ \frac{du}{dq_n} \frac{dv}{dp_n} - \frac{du}{dp_n} \frac{dv}{dq_n} \right\}, \tag{2.1}$$

where q_n and p_n are any set of canonical coordinates and momenta that describe the system. In terms of Poisson brackets, the classical equation of motion of any dynamical variable η is given simply by

$$\dot{\eta} = (\eta, H), \tag{2.2}$$

where H is the Hamiltonian. From (2.1) it follows that Poisson brackets satisfy the identity

$$(u, (v, w)) + (v, (w, u)) + (w, (u, v)) = 0 \tag{2.3}$$

and have the following properties:

$$(u, v) = -(v, u), \tag{2.4}$$

$$(u, C) = 0, \tag{2.5}$$

$$(u_1 + u_2, v) = (u_1, v) + (u_2, v), \tag{2.6}$$

$$(u, v_1 + v_2) = (u, v_1) + (u, v_2), \tag{2.7}$$

$$(u_1 u_2, v) = (u_1, v)u_2 + u_1(u_2, v), \tag{2.8}$$

$$(u, v_1 v_2) = (u, v_1)v_2 + v_1(u, v_2), \tag{2.9}$$

where C in equation (2.5) is a c-number (i.e., it does not depend on any of the canonical coordinates).

Let us suppose that there is an analog of the Poisson bracket in quantum mechanics from which equations of motion can be obtained as in (2.2). Such an analog would have all the properties of a classical Poisson bracket described in equations (2.3) to (2.9) except that the dynamical variables would not commute. We have taken care when writing equations (2.3) to (2.9) to keep the same ordering of dynamical variables on each side. To find what this quantum analog of the Poisson bracket is, let us consider $(u_1 u_2, v_1 v_2)$. What can we deduce by forgetting (2.1) and using only equations (2.3) to (2.9)? If we first use equation (2.8) and then (2.9), we obtain

$$\begin{aligned}(u_1 u_2, v_1 v_2) = &(u_1, v_1)v_2 u_2 + v_1(u_1, v_2)u_2 \\ &+ u_1(u_2, v_1)v_2 + u_1 v_1(u_2, v_2),\end{aligned} \tag{2.10}$$

---

[5]This derivation follows closely the treatment in [159].

and if we use first equation (2.9) and then (2.8), we obtain

$$(u_1 u_2, v_1 v_2) = (u_1, v_1) u_2 v_2 + v_1 (u_1, v_2) u_2$$
$$+ u_1 (u_2, v_1) v_2 + v_1 u_1 (u_2, v_2). \tag{2.11}$$

Therefore,

$$(u_1, v_1)[u_2, v_2] = [u_1, v_1](u_2, v_2), \tag{2.12}$$

where $[A, B] \equiv AB - BA$ is the commutator of A and B. Now as condition (2.12) must hold for u_1 and v_1 regardless of what u_2 and v_2 might be, and vice versa, it follows that

$$[u_1, v_1] = r(u_1, v_1) \tag{2.13}$$

and

$$[u_2, v_2] = r(u_2, v_2), \tag{2.14}$$

where r is a numerical constant. In order that the Poisson bracket of a pair of Hermitian operators also be Hermitian, r must be imaginary. In order to agree with experiment, this imaginary constant must be given by

$$r = i\hbar, \tag{2.15}$$

where $\hbar$ is Planck's constant divided by 2π.

So the problem of determining the commutator reduces to that of finding its quantum Poisson bracket. Now if we assume that the quantum Poisson bracket has the same value as its classical counterpart, we can obtain the commutator for any quantum system with a classical analog making the correspondence

$$\underbrace{(u, v)}_{\text{classical}} \rightarrow \frac{1}{i\hbar} \underbrace{[u, v]}_{\text{quantum}}. \tag{2.16}$$

As Hamilton's theory is based on the idea of canonical variables, this procedure is known as *canonical quantization*. In the next subsection we use canonical quantization to obtain a quantum theory of electromagnetic radiation.

Before going to the next subsection, there are two things we should mention. First, we have to stress that once the Hamiltonian and the commutation rules are obtained, the quantum theory is complete in the sense that the dynamics can be worked out at least in principle. But most people are more familiar with a version of quantum mechanics where one just writes down Schrödinger's equation and solves it to find the wavefunction. What is the connection between that and this more general and abstract formulation of quantum mechanics? The connection is that the ordinary wavefunction language is a rewriting of everything in terms of spatial coordinates. It is very much like the difference between dealing with vectors in an abstract way and taking a specific coordinate system, say Cartesian, and writing down the components of each vector in this coordinate system. The coordinate-free approach is more compact, less clumsy, and helps to make important physical ideas more transparent, but the coordinate approach is often better for practical calculations.

To clarify this point, let us consider a specific example. Let us use the canonical quantization procedure to derive the wavefunction quantum mechanics for a particle of mass m moving in a potential $V(x)$ in one dimension. The quantum Hamiltonian is then the same as the classical one, but with the classical momentum p and position x replaced by the operators[6] $\hat{p}$ and $\hat{x}$,

$$\hat{H} = \frac{\hat{p}^2}{2m} + V(\hat{x}). \qquad (2.17)$$

From equation (2.1) we find that the Poisson bracket of x and p is $(x, p) = 1$. So the commutator of $\hat{x}$ and $\hat{p}$ must be

$$[\hat{x}, \hat{p}] = i\hbar. \qquad (2.18)$$

The most natural way to proceed from here is to adopt *Heisenberg's convention*[7] that all the time dependence is contained in the operators themselves, with the state of the system remaining constant in time. Then the dynamics of the system are obtained as in Hamilton's formalism [240] but with the Poisson brackets replaced by the commutators divided by $i\hbar$. In other words, the equation of motion for the operator $\hat{O}_H$ in this Heisenberg convention is

$$\frac{\partial}{\partial t}\hat{O}_H = \frac{1}{i\hbar}[\hat{O}_H, \hat{H}]. \qquad (2.19)$$

Schrödinger, however, adopted a different convention, where the time dependence is all contained in the state vectors. The key to finding the relation between these conventions is to notice that observable quantities should come out the same regardless of the convention. From (2.19) we find that (assuming $\hat{H}$ is independent of time) $\hat{O}_H$ is given by

$$\hat{O}_H(t) = \exp\left(\frac{i}{\hbar}\hat{H}t\right)\hat{O}_H(0)\exp\left(-\frac{i}{\hbar}\hat{H}t\right). \qquad (2.20)$$

Calling $|\psi_H\rangle$ the quantum state in Heisenberg's convention, we can write the expectation value $\langle\hat{O}\rangle$ of $\hat{O}$ as

$$\begin{aligned}
\langle\hat{O}\rangle &= \langle\psi_H|\hat{O}_H(t)|\psi_H\rangle \\
&= \underbrace{\langle\psi_H|\exp\left(\frac{i}{\hbar}\hat{H}t\right)}_{\langle\psi_S(t)|}\underbrace{\hat{O}_H(0)}_{\hat{O}_S}\underbrace{\exp\left(-\frac{i}{\hbar}\hat{H}t\right)|\psi_H\rangle}_{|\psi_S(t)\rangle} \\
&= \langle\psi_S(t)|\hat{O}_S|\psi_S(t)\rangle, \qquad (2.21)
\end{aligned}$$

---

[6]About the notation: In this and Section 2.4 we use hats to denote operators. In some other parts of this book, such as Section 2.5, hats are reserved to denote unit vectors.

[7]Also known as the *Heisenberg picture*.

where the subscript S refers to Schrödinger's convention. So the Schrödinger equation for the arbitrary quantum state $|\psi_S(t)\rangle$ can easily be derived like this:

$$i\hbar\frac{\partial}{\partial t}|\psi_S(t)\rangle = i\hbar\frac{\partial}{\partial t}\left\{\exp\left(-\frac{i}{\hbar}\hat{H}t\right)|\psi_H\rangle\right\}$$

$$= \hat{H}\exp\left(-\frac{i}{\hbar}\hat{H}t\right)|\psi_H\rangle$$

$$= \hat{H}|\psi_S(t)\rangle. \tag{2.22}$$

Now what is usually called the wavefunction $\psi_S(x)$ are the components of $|\psi_S\rangle$ on the basis of eigenstates of position[8] $|x\rangle$:

$$\psi_S(x) = \langle x|\psi_S\rangle. \tag{2.23}$$

The eigenstates of position are defined by

$$\hat{x}|x\rangle = |x\rangle x. \tag{2.24}$$

This equation also implies that they are orthogonal. To see that, we take the Hermitian conjugate of (2.24) and replace x by x':

$$\langle x'|\hat{x} = x'\langle x'|. \tag{2.25}$$

Then we multiply (2.24) by $\langle x'|$ from the left, (2.25) by $|x\rangle$ from the right, and subtract the two results, finding

$$(x - x')\langle x'|x\rangle = 0. \tag{2.26}$$

This equation means that whenever $x' \neq x$, $\langle x'|x\rangle$ must vanish. As $\hat{x}$ has continuous rather than discrete eigenvalues (i.e., the particle can assume any intermediate position between, say, x_1 and x_2), we find that $\langle x'|x\rangle = C\delta(x - x')$, where C is a constant. For simplicity, the eigenstates of $\hat{x}$ are chosen to be normalized in such a way that $C = 1$, and then

$$\langle x'|x\rangle = \delta(x - x'). \tag{2.27}$$

Physically, we expect to be able to write any arbitrary state of the particle as a linear combination of these position eigenstates. In other words, the position eigenstates form a complete basis:

$$\int dx\, |x\rangle\langle x| = 1. \tag{2.28}$$

So, if we then multiply (2.22) by $\langle x|$ from the left, with H given by (2.17), and using completeness and orthonormality, we obtain

$$i\hbar\frac{\partial}{\partial t}\psi_S(x) = \frac{1}{2m}\int dx'\int dx''\,\langle x|\hat{p}|x'\rangle\langle x'|\hat{p}|x''\rangle\psi_S(x'') + V(x)\psi_s(x). \tag{2.29}$$

---

[8]I am referring to nonrelativistic quantum mechanics here. As we will see in Chapter 3, in the relativistic regime there are problems with the idea of position eigenstates.

This equation does not look much like the well-known Schrödinger equation for the wavefunction from elementary quantum mechanics. However, we still have to work out what the momentum operator is in the x representation (i.e., $\langle x|\hat{p}|x'\rangle$).

Multiplying (2.18) by $\langle x|$ from the left and by $|x'\rangle$ from the right, we find that

$$(x - x')\langle x|\hat{p}|x'\rangle = i\hbar\delta(x - x'). \tag{2.30}$$

From this equation we can obtain $\langle x|\hat{p}|x'\rangle$ in three simple steps by rewriting $\delta(x - x')$ in terms of its Fourier expansion and integrating by parts:

$$\begin{aligned} \langle x|\hat{p}|x'\rangle &= i\hbar \int \frac{dk}{2\pi} \frac{e^{i(x-x')k}}{x - x'} \\ &= -\frac{i\hbar}{2\pi} \int dk \, e^{i(x-x')k} ik \\ &= \frac{\hbar}{i} \frac{\partial}{\partial x}\delta(x - x'). \end{aligned} \tag{2.31}$$

The delta function in (2.31) is omitted in some elementary textbooks, where $\hat{p}$ in the x representation is often taken to be just $(i/\hbar)\partial/\partial x$. But the delta function in (2.31) is very important because it tells us that $\hat{p}$ is diagonal in the x representation. This diagonality leads to the elimination of the integrals in (2.29) when we use (2.31), yielding the familiar Schrödinger equation for a particle under the action of a potential V:

$$i\hbar\frac{\partial}{\partial t}\psi_S(x) = -\frac{\hbar^2}{2m}\frac{\partial^2}{\partial x^2}\psi_S(x) + V(x)\psi_s(x). \tag{2.32}$$

Some operators in quantum mechanics might not be diagonal in the x representation. This is the case of a nonlocal potential considered by Gottfried (see Prob. 4 on p. 75 of [243]), for example.

The second thing we should mention is that there is another way in which a classical theory can be quantized. This other way uses the older Lagrange formulation of classical mechanics rather than Hamilton's. It was developed by Feynman based on a hint he found in one of Dirac's papers. For some systems, where no Hamiltonian exists, this kind of quantization is the only alternative. Feynman's formulation of quantum mechanics is also very interesting because it shows us what lies beyond the minimal action principle, and in doing so sort of explains why the minimum action principle comes about. The *minimal action principle* is the fascinating idea behind the Lagrangian formulation that the actual motion of a particle, say, is always the one that minimizes a certain "magical" quantity called the *action*. Feynman showed that in quantum mechanics, a particle actually follows all possible paths and that it is only in the classical limit that most of them interfere destructively, leaving only that single path where the action is minimum! Here we always use Hamiltonians, so this is all that we are going to say about Feynman's formulation of quantum mechanics. The curious reader might wish to have a look at the recommended reading at the end of the chapter.

2.2   WHY THE RADIATION FIELD IS SPECIAL

In quantum electrodynamics it is the radiation field that is quantized[9] rather than the Coulomb field of the charged particles. The radiation field is special because it has a "life" of its own. In more technical terms: It has its own degrees of freedom, whereas the Coulomb field has none. The Coulomb field is determined completely by the degrees of freedom of the charged particles.

To illustrate this point, we derive an important result in electrodynamics called the *Helmholtz theorem.* We start by noticing that according to Section F.1.5 of Appendix F, given a square-integrable vector field $\mathbf{F}(\mathbf{r})$, we can write

$$\mathbf{F}(\mathbf{r}) = -\frac{1}{4\pi} \int d^3r' \; \mathbf{F}(\mathbf{r}') \nabla^2 \frac{1}{|\mathbf{r} - \mathbf{r}'|}. \tag{2.33}$$

As ∇^2 acts only on $\mathbf{r}$, we can take it outside the integral sign. Then using (E.21) with $\mathbf{a} = \mathbf{b} = \nabla$ and $\mathbf{c}$ being the integral, we get

$$\mathbf{F}(\mathbf{r}) = \frac{1}{4\pi} \left\{ \nabla \wedge \int d^3r' \; \nabla \wedge \left[\frac{\mathbf{F}(\mathbf{r}')}{|\mathbf{r} - \mathbf{r}'|} \right] - \nabla \int d^3r' \; \nabla \cdot \left[\frac{\mathbf{F}(\mathbf{r}')}{|\mathbf{r} - \mathbf{r}'|} \right] \right\}$$

$$= -\frac{1}{4\pi} \nabla \wedge \int d^3r' \; \mathbf{F}(\mathbf{r}') \wedge \nabla \frac{1}{|\mathbf{r} - \mathbf{r}'|}$$

$$- \frac{1}{4\pi} \nabla \int d^3r' \; \mathbf{F}(\mathbf{r}') \cdot \nabla \frac{1}{|\mathbf{r} - \mathbf{r}'|}. \tag{2.34}$$

Using the fact that

$$\mathbf{F}(\mathbf{r}') \wedge \nabla \frac{1}{|\mathbf{r} - \mathbf{r}'|} = -\mathbf{F}(\mathbf{r}') \wedge \nabla' \frac{1}{|\mathbf{r} - \mathbf{r}'|}$$

$$= \frac{\nabla' \wedge \mathbf{F}(\mathbf{r}')}{|\mathbf{r} - \mathbf{r}'|} - \nabla' \wedge \left[\frac{\mathbf{F}(\mathbf{r}')}{|\mathbf{r} - \mathbf{r}'|} \right], \tag{2.35}$$

$$\mathbf{F}(\mathbf{r}') \cdot \nabla \frac{1}{|\mathbf{r} - \mathbf{r}'|} = -\mathbf{F}(\mathbf{r}') \cdot \nabla' \frac{1}{|\mathbf{r} - \mathbf{r}'|}$$

$$= \nabla' \cdot \left[\frac{\mathbf{F}(\mathbf{r}')}{|\mathbf{r} - \mathbf{r}'|} \right] - \frac{\nabla' \cdot \mathbf{F}(\mathbf{r}')}{|\mathbf{r} - \mathbf{r}'|}, \tag{2.36}$$

and that (from the integral theorems in Appendix E and the fact that $\mathbf{F}$ is square integrable)

$$\int d^3r' \; \nabla' \cdot \left[\frac{\mathbf{F}(\mathbf{r}')}{|\mathbf{r} - \mathbf{r}'|} \right] = \oint_{\text{infinity}} d\mathbf{S} \cdot \frac{\mathbf{F}(\mathbf{r}')}{|\mathbf{r} - \mathbf{r}'|}$$

$$= 0, \tag{2.37}$$

---

[9]To obtain a relativistically covariant theory, extra degrees of freedom are introduced, leading to the quantization of the longitudinal static field. Physically meaningful results, however, can be extracted only after these extra degrees of freedom are eliminated.

$$\int d^3r' \; \boldsymbol{\nabla}' \wedge \left[\frac{\mathbf{F}(\mathbf{r}')}{|\mathbf{r} - \mathbf{r}'|} \right] = \oint_{\text{infinity}} d\mathbf{S} \wedge \frac{\mathbf{F}(\mathbf{r}')}{|\mathbf{r} - \mathbf{r}'|}$$
$$= \mathbf{0}, \qquad (2.38)$$

we obtain the Helmholtz theorem:

$$\mathbf{F}(\mathbf{r}) = \frac{1}{4\pi} \left\{ \boldsymbol{\nabla} \wedge \int d^3r' \; \frac{\boldsymbol{\nabla}' \wedge \mathbf{F}(\mathbf{r}')}{|\mathbf{r} - \mathbf{r}'|} - \boldsymbol{\nabla} \int d^3r' \; \frac{\boldsymbol{\nabla}' \cdot \mathbf{F}(\mathbf{r}')}{|\mathbf{r} - \mathbf{r}'|} \right\}. \qquad (2.39)$$

The first term on the right-hand side of (2.39) is the curl of a vector, so its divergence vanishes. Vector fields with vanishing divergence are called *transverse fields* for reasons that will become obvious soon. The second term on the right hand side of (2.39) is the gradient of a scalar, so its curl vanishes. For reasons that will become apparent soon, vector fields with vanishing curls are called *longitudinal fields*. The Helmholtz theorem says that any square-integrable vector field can be expressed as the sum of a transverse and a longitudinal vector field. The transverse component can be obtained if we know the value of the curl of $\mathbf{F}$ in the whole of space and the longitudinal component of $\mathbf{F}$ can be obtained if we know the value of the divergence of $\mathbf{F}$ in the whole of space. In particular, if both the curl and the divergence of $\mathbf{F}$ vanish everywhere, $\mathbf{F}$ will vanish everywhere, too.

Now suppose that all the charged particles in the universe were kept at rest[10] and that there were no radiation in the entire universe. Then Maxwell's equations would reduce to

$$\boldsymbol{\nabla} \cdot \mathbf{E} = 4\pi\rho,$$
$$\boldsymbol{\nabla} \cdot \mathbf{B} = 0,$$
$$\boldsymbol{\nabla} \wedge \mathbf{E} = \mathbf{0}, \qquad (2.40)$$
$$\boldsymbol{\nabla} \wedge \mathbf{B} = \mathbf{0},$$

where ρ is the charge density associated with the charged particles. Using Helmholtz theorem (2.39), we find that $\mathbf{B}$ vanishes everywhere and that

$$\mathbf{E}(\mathbf{r}) = \int d^3r' \; \frac{\rho(\mathbf{r}')}{|\mathbf{r} - \mathbf{r}'|^3} (\mathbf{r} - \mathbf{r}'). \qquad (2.41)$$

This is the familiar electrostatic Coulomb field. It has no life of its own. If for some reason the charges disappear, so does the Coulomb field. It is not a big surprise that there is no static field when there are no charges to produce it. What might really come as a surprise, however, is that with the dynamic radiation field, things are completely different. Maxwell's equations for time-dependent fields are

$$\boldsymbol{\nabla} \cdot \mathbf{E} = 4\pi\rho, \qquad (2.42)$$

---

[10]According to Earnshaw's theorem, to keep these charged particles at rest, other forces that are not electrostatic must be applied. We will just assume that such forces can be supplied.

$$\nabla \cdot \mathbf{B} = 0, \tag{2.43}$$

$$\nabla \wedge \mathbf{E} = -\frac{1}{c}\frac{\partial}{\partial t}\mathbf{B}, \tag{2.44}$$

$$\nabla \wedge \mathbf{B} = \frac{1}{c}\frac{\partial}{\partial t}\mathbf{E} + 4\pi\mathbf{J}, \tag{2.45}$$

where $\mathbf{J}$ is the current density. So for time-dependent fields, ρ and $\mathbf{J}$ can actually vanish everywhere without the Helmholtz theorem automatically implying that the fields also have to vanish everywhere. As long as there are both an electric and a magnetic field with nonvanishing first time derivatives, their curls will not vanish. Suppose that a radiating electron suddenly hits a positron and the two annihilate each other. To vanish, their fields must vary in time from whatever value they had to zero. But a magnetic field varying in time will generate an electric field varying in time which in turn will generate another magnetic field varying in time. This will keep the radiation field going even after its source (the electron and the positron) disappears.[11] Moreover, as only their curls will be nonvanishing, these self-perpetuating fields must be transverse. In other words, the reason why only transverse fields can be self-perpetuating fields (i.e., radiation fields, fields that keep existing even if their source disappears) is that in the four Maxwell equations, the electric and magnetic fields couple through the two curl equations rather than through the two divergence equations.

Another way to see what we mentioned above is to consider (2.42)–(2.45) with vanishing ρ and $\mathbf{J}$ and to show explicitly that they admit a solution with nonvanishing $\mathbf{E}$ and $\mathbf{B}$. These equations become much simpler in Fourier space. Let $\mathcal{E}$ be the Fourier transform of $\mathbf{E}$, defined by

$$\mathcal{E}(\mathbf{k}) = \frac{1}{(2\pi)^3}\int d^3r\,\mathbf{E}(\mathbf{r})e^{-i\mathbf{k}\cdot\mathbf{r}}, \tag{2.46}$$

and analogously for $\mathbf{B}$. In Fourier space, (2.42)–(2.45) relate only the fields at the *same* point rather than at all points in the neighborhood of a point as in ordinary coordinate space, where these equations involve spatial derivatives of the fields[12]:

$$i\mathbf{k}\cdot\mathcal{E}(\mathbf{k}) = 0, \tag{2.47}$$

$$i\mathbf{k}\cdot\mathcal{B}(\mathbf{k}) = 0, \tag{2.48}$$

$$i\mathbf{k}\wedge\mathcal{E}(\mathbf{k}) = -\frac{1}{c}\dot{\mathcal{B}}(\mathbf{k}), \tag{2.49}$$

$$i\mathbf{k}\wedge\mathcal{B}(\mathbf{k}) = \frac{1}{c}\dot{\mathcal{E}}(\mathbf{k}). \tag{2.50}$$

---

[11]It also explains where the energy goes: After both electron and positron cease to exist, their rest energies and their kinetic energies are carried away in the radiation that is created.

[12]The significance of the terms *longitudinal* and *transverse* becomes very clear in reciprocal space. A vector field with vanishing divergence is orthogonal (transverse) to $\mathbf{k}$ in reciprocal space, whereas one with vanishing curl is parallel (longitudinal) to $\mathbf{k}$.

Taking the vector product of (2.49) with **k** from the left and using (2.47) and (2.50), we find that

$$\ddot{\mathcal{E}}(\mathbf{k}) + c^2 k^2 \mathcal{E}(\mathbf{k}) = 0. \tag{2.51}$$

Proceeding in an analogous way but starting with (2.50) rather than (2.49), we find an identical equation for $\mathcal{B}$:

$$\ddot{\mathcal{B}}(\mathbf{k}) + c^2 k^2 \mathcal{B}(\mathbf{k}) = 0. \tag{2.52}$$

Equations (2.51) and (2.52) are harmonic oscillator equations, and they certainly have nonvanishing solutions as long as the electric and magnetic fields do not both vanish initially. This amazing result is also the reason why we can study light on its own in this chapter.

2.3 WHAT IS A CAVITY AND HOW DO WE FIND ITS MODES?

In this chapter we are concerned only with closed cavities, such as the one shown in Figure 2.1. There are also open cavities. These are discussed in detail in Chapter 9. Here we only introduce the main idea of cavity. For a detailed calculation of the modes of a perfect cavity, see Appendix A.

Figure 2.1 This is a closed superconducting microwave cavity of the type used in the first (two-photon) micromaser experiments.[13] There is actually a tiny hole across this cavity through which the atoms go, but as it is smaller than the wavelength of the mode, its effect is negligible.[14] This cavity belongs to Serge Haroche's cavity QED group of the Laboratoire Kastler Brossel at the École Normale Supérieure in Paris, France, where the first two-photon micromaser was realized.

---

[13]For the curious, the hand belongs to Dr. Paulo A. Nussenzveig.
[14]This is true for superconducting microwave cavities [51] but curiously enough not in the visible region of the electromagnetic spectrum [14, 432].

In quantum optics, the word *cavity* means a resonator for electromagnetic radiation. Acoustic resonators have been known since antiquity. They are the basis of most musical instruments. In ancient China, for example, vessels filled with water up to certain levels were used as resonators to give standard tones [527]. But what exactly is a resonator? Broadly speaking, a *resonator* is a device that is resonant only to certain frequencies, whose vibrating fields at these frequencies form spatial patterns called *resonator modes*. A stretched string with both ends fixed is a closed one-dimensional resonator. The body of a violin is an open three-dimensional acoustic resonator. The vibrating field in the string is the vertical displacement of the string at each point along it. String modes can be seen with the naked eye. Figure 2.2 shows the mode corresponding to the second lowest resonance frequency. In the violin, the

Figure 2.2 The output of a function generator fed through an audio amplifier controls an electromechanical device that drives one end of a string with a precise frequency. The other end of the string is kept fixed. In this figure, the frequency was set in such a way that the wavelength of the standing-wave pattern approximately coincides with the length of the string. This is a good way of measuring v, the speed of wave propagation along the string, as the frequency here should be equal to $2\pi v/l$, with l being the length of the string, which is also known. This picture was taken by Dr. Gary Chottiner from Case Western Reserve University, where this experiment is part of their introductory laboratory courses for undergraduates.

vibrating field is the small variation in air pressure at each point inside the violin. A two-dimensional picture of a mode can be obtained by spreading a fine powder at the back of the violin and then sounding one of its resonances. The powder will settle at the nodes of the mode on the two-dimensional surface of the back of the violin because these are the places where the surface will not vibrate. Figure 2.3 shows the resulting picture, and Figure 2.4 shows the actual standing-wave patterns in a classic guitar.

In this section we use simple models to explain the essence of what a resonator is, its modes, and how they arise. But to understand resonators we must first understand the phenomenon of resonance. This is the topic of the next subsection. Readers familiar with that might wish to skip it and move to the second and last subsection of this section, where we address resonators directly.

Figure 2.3 Here you can see the nodes of a vibration mode on the back plate of a violin. These are called *Chladni patterns* [529]. This picture was taken by Emmanuel Bossy, Renaud Carpentier, and Joe Wolfe at the University of New South Wales, Sydney, Australia. They attached a magnet to the back plate of the violin and using a coil attached to an audio amplifier and a function generator, produced an oscillating magnetic field that drove the back plate with the exact frequency of one of its modes. The Chladni patterns were made visible by spreading black sand on the back plate before driving it. Once it starts vibrating, the sand only settles at the nodes of the mode, which remain at rest.

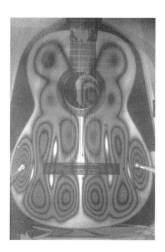

Figure 2.4 Here we see one of the modes of vibration of a classic guitar. This picture was taken by Prof. Bernard Richardson of Cardiff University using the technique of laser holographic interferometry. This technique offers a better way of visualizing the modes than the sand method, but unfortunately, it does not yield images of such high quality for the violin because its surface vibrations are smaller and its body is less reflective than the guitar's. The technique works this way: First, we excite one of the modes of the guitar. Then we make a hologram of the instrument. It will show the instrument overlaid with optical interference fringes (the dark and bright lines in this figure) produced by the vibration pattern of the mode on the instrument's body. Adjacent fringes represent a height change of about 0.1 μm.

2.3.1 What is resonance? A simple physical system with a single resonance

A look at a dictionary yields a definition of resonance along the following lines:

> **Resonance** (physics): The state of a system in which an abnormally large vibration is produced in response to an external stimulus, occurring when the frequency of the stimulus is the same, or nearly the same, as the natural vibration frequency of the system.[15]

So first, the system must have natural vibration frequencies of its own. This happens when there is some sort of feedback mechanism whereby energy is converted back and forth between two different forms, say potential and kinetic energy. Let us consider

---

[15]From the online dictionary at *http://www.infoplease.com.*

a simple example, a system that has a single resonance: the typical swing found in children's playgrounds. The swing is essentially a pendulum. A mass m, say the child and the seat, is constrained to move along an arch by a rope (the chains on both sides of the seat) of length l that is attached at one end to the mass and at the other to the swing hangers. The net force on the mass is the result of the force of gravity acting on it and the tension of the rope. The net force is always tangent to the arch and in such a way as to push the mass toward its point of minimum height (the middle of the arch). So when we give the mass some gravitational potential energy, the net force on it will make it move toward its point of lowest height. Then it will lose potential energy and gain kinetic energy. At its point of lowest height, its potential energy will be minimum but its kinetic energy will be maximum. Its momentum will then impel it to gain height again, transferring its kinetic energy into potential energy until it reaches a point where its kinetic energy is zero and the potential energy is maximum. At this point it will again start moving toward the point of minimum height, transforming its potential energy into kinetic energy once more, and so on. This feedback mechanism establishes a rhythm, a frequency that is characteristic of the system: its natural frequency of vibration. If we try to push the swing with a rhythm different from its natural one, we will find ourselves pushing it when it is moving toward us. As a result, we will take energy from the swing, damping its movement instead of giving energy to it to increase its movement. But if we push it in its own rhythm, the swing can reach considerable heights very fast if we are not careful. Anyone who has ever tried to push a child on a swing knows that. This is the phenomenon of resonance. Now that we have seen it in words, let us have a look at the mathematics of resonance.

The magnitude of the net force is $mg \sin \phi$, where g is the local acceleration constant due to gravity and ϕ is the angle the rope forms with the vertical (see Figure 2.5). If the amplitude of the swings is kept small so that ϕ is small, we can approximate

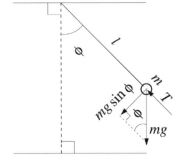

Figure 2.5 A swing is essentially a pendulum. The mass of the seat and of the child on it is represented by a single-point particle of mass m placed at the center of mass. The mass of the chains is neglected, and they are described by a nonstretchable string of length l whose tension T exactly balances the projection of the weight of the particle on the direction of the string. The net force on the particle is then $mg \sin \phi$, where ϕ is the angle that the rope forms with the vertical. This net force is always tangent to the arch described by the motion of the particle.

$\sin\phi$ by ϕ. This is a very important special case, because then the force is proportional to the displacement and always directed against it: The basic ingredient for a harmonic oscillator. This is a case that is often found in many different branches of physics and also, fortunately, a case that leads to a simple analytical solution of the equations of motion. The equation of motion of such a frictionless swing is simply

$$\ddot{\phi} + \frac{g}{l}\phi = 0. \tag{2.53}$$

Now let us suppose that the swing was somehow set in oscillatory motion with frequency ω in the remote past and that it has been swinging with this frequency ever since; for example,

$$\phi(t) = \phi_0 \sin\omega t, \tag{2.54}$$

where ϕ_0 is the amplitude of the oscillations. Then, if we substitute (2.54) in (2.53), we find that

$$\frac{g}{l} - \omega^2 = 0, \tag{2.55}$$

whose only solution[16] is $\omega = \sqrt{g/l}$. In other words, the frictionless swing can oscillate forever only at the single frequency $\sqrt{g/l}$. For this reason, $\sqrt{g/l}$ is said to be its *natural frequency of oscillation*.

Any real swing, however, also has friction that will eventually stop its motion. Let us suppose that this is a sort of viscous friction proportional to the velocity, say $-m\gamma l\dot{\phi}$, where γ is a constant. The effect of the person pushing the swing will be taken into account by adding a driving force: say, $mlF(t)$, for convenience. Then the equation of motion is

$$\ddot{\phi} + \gamma\dot{\phi} + \frac{g}{l}\phi = F. \tag{2.56}$$

Consider now what happens if we push the swing with an arbitrary frequency ω [e.g., we take $F(t) = A\sin\omega t$, where A is a real number describing the strength of our pushes]. To simplify the math, we can define a new complex function $\varphi(t)$ such that[17] $\phi(t) = \Im\{\varphi(t)\}$; then, as (2.56) is linear, $\varphi(t)$ obeys the simpler equation

$$\ddot{\varphi} + \gamma\dot{\varphi} + \frac{g}{l}\varphi = Ae^{i\omega t}. \tag{2.57}$$

Equation (2.57) can be solved very quickly using an intuitive method invented by Oliver Heaviside called *operational calculus* [282, 283, 610]. First, we rewrite (2.57) as

$$\left\{\frac{d^2}{dt^2} + \gamma\frac{d}{dt} + \frac{g}{l}\right\}\varphi(t) = Ae^{i\omega t}. \tag{2.58}$$

The expression inside the braces on the left-hand side of (2.58) is an operator that acts on $\varphi(t)$. If we multiply both sides of (2.58) from the left by the inverse of this

---

[16]Assuming that ω is always a positive quantity.
[17]$\Im$ denotes the operation of taking the imaginary part.

operator, we find the (particular) solution of (2.58):

$$\varphi(t) = \left\{ \frac{d^2}{dt^2} + \gamma \frac{d}{dt} + \frac{g}{l} \right\}^{-1} A e^{i\omega t}. \tag{2.59}$$

But what is the right-hand side of (2.59)? Fractional expressions involving differential operators are defined in terms of their power series expansions. For example, $(1 - d/dt)^{-1}$ is just $\sum_{n=0}^{\infty} (d/dt)^n$, where $(d/dt)^n$ just means applying d/dt to something n times (i.e., d^n/dt^n). So going back to (2.59), as each d/dt applied to $\exp(i\omega t)$ is the same as multiplication by $i\omega$, the solution is given by

$$\varphi(t) = -\frac{A e^{i\omega t}}{\omega^2 - i\gamma\omega - g/l}. \tag{2.60}$$

An interesting case is when the friction is weak (i.e., $\sqrt{g/l} \gg \gamma/2$). Then (2.60) leads to

$$\phi(t) \approx \frac{A\left[(\omega^2 - g/l)\sin\omega t + \omega\gamma\cos\omega t\right]}{\left[\left(\omega + \sqrt{g/l}\right)^2 + (\gamma/2)^2\right]\left[\left(\omega - \sqrt{g/l}\right)^2 + (\gamma/2)^2\right]}. \tag{2.61}$$

If we push the swing at its natural frequency,[18] $\sqrt{g/l}$, the swing oscillates like this:

$$\phi(t) \approx A\frac{1}{\gamma}\sqrt{\frac{l}{g}}\cos\left(\sqrt{\frac{g}{l}}t\right). \tag{2.62}$$

In other words, the swing oscillates 90 degrees out of phase with our pushing. Our push is stronger when the swing is at its point of minimum height and weaker when it is at its point of maximum height. This is exactly what you experience when pushing a child on a swing. As the swing approaches you, you must decrease the intensity of the force so that it vanishes when the swing reaches its point of maximum height.

Far away from resonance (i.e., $|\omega - \sqrt{g/l}| \gg \sqrt{A}$), $\phi(t) \approx 0$ and the swing does not move. Figure 2.6 plots the average of the square of (2.61) over one oscillation period against ω for a value of γ 100 times smaller than the natural frequency. Notice that it decreases quickly as the frequency becomes different from the natural frequency. The sharp peak in Figure 2.6 is typical of resonant phenomena.

To solve (2.56) for any driving force, it is convenient to solve it first for an impulse force $\delta(t - t')$. This solution is called the *Green's function*, $G(t|t')$, after George Green, a nineteenth-century mathematician from Nottingham, England. The equation for the Green's function is

$$\ddot{G} + \gamma\dot{G} + \frac{g}{l}G = \delta(t - t'). \tag{2.63}$$

---

[18]For weak damping the natural frequency is approximately the same as the natural frequency with no damping.

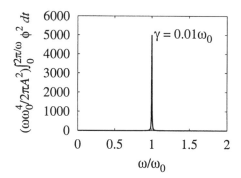

Figure 2.6 This is a plot of the average over one period of oscillation of the square of (2.61) as a function of the frequency w of the driving force $F(t) = A \sin wt$. All quantities involved have been made dimensionless by expressing them in units of key parameters. The driving frequency w is in units of the natural frequency of oscillation of the swing $w_0 = \sqrt{g/l}$. The angle ϕ was scaled by the dimensionless factor w_0^2/A and γ was taken to be 100 times smaller than w_0. The sharp peak centered at the natural frequency is a well-known signature of resonance.

Once we solve (2.63) and find $G(t|t')$, the solution $\phi(t)$ of (2.56) for an arbitrary driving force $F(t)$ can be obtained by doing the integral

$$\phi(t) = \int_{-\infty}^{\infty} dt' \, G(t|t')F(t') \tag{2.64}$$

because

$$\ddot{\phi} + \gamma\dot{\phi} + \frac{g}{l}\phi = \int_{-\infty}^{\infty} dt' \, \delta(t - t')F(t')$$
$$= F(t). \tag{2.65}$$

Let us use a Fourier transform to solve (2.63):

$$G(t|t') = \int_{-\infty}^{\infty} dw \, e^{iwt}\tilde{G}(w, t'). \tag{2.66}$$

Substituting (2.66) in (2.63), multiplying by $(1/2\pi)\exp(-iwt)$, integrating over t from $-\infty$ to ∞, and using (F.5), we find that

$$\tilde{G}(w, t') = -\frac{1}{2\pi}\frac{e^{-iwt'}}{w^2 - i\gamma w - g/l}. \tag{2.67}$$

The denominator of (2.67) can also be written as

$$w^2 - i\gamma w - \frac{g}{l} = \left\{w - \left[i\frac{\gamma}{2} - \sqrt{\frac{g}{l} - \left(\frac{\gamma}{2}\right)^2}\right]\right\}\left\{w - \left[i\frac{\gamma}{2} + \sqrt{\frac{g}{l} - \left(\frac{\gamma}{2}\right)^2}\right]\right\}. \tag{2.68}$$

Let us suppose that the friction is small (i.e., $\gamma \ll \sqrt{g/l}$). Then (2.68) becomes

$$\omega^2 - i\gamma\omega - \frac{g}{l} \approx \left\{\omega - \left[i\frac{\gamma}{2} - \sqrt{\frac{g}{l}}\right]\right\}\left\{\omega - \left[i\frac{\gamma}{2} + \sqrt{\frac{g}{l}}\right]\right\}. \qquad (2.69)$$

Substituting (2.69) in (2.67) and then back in (2.66), we find that (see Problem 2.1)

$$G(t|t') \approx \Theta(t - t') \sqrt{\frac{g}{l}} e^{-(t-t')\gamma/2} \sin\left([t - t']\sqrt{\frac{g}{l}}\right), \qquad (2.70)$$

where $\Theta(u)$ is the Heaviside step function, which vanishes for negative u and is unity for positive u.

Now suppose that we are driving our swing with a force of frequency ω from $t = 0$ until $t = T$; for example,

$$F(t) = \Theta(t)\Theta(T - t) \sin\omega t. \qquad (2.71)$$

I leave it for the reader to work out integral (2.64) in this case (Problem 2.2). As we have this unrealistic sudden switching on and off of the force, the general solution is not so informative. But the following particular cases are. Assuming that $\omega \gg \sqrt{g/l}$, the solution is given by

$$\phi(t) \approx \begin{cases} \sqrt{l/(\omega^2 g)} \sin\omega t, & \text{for } 0 < t < T \\ 0, & \text{for } 0 < T < t. \end{cases} \qquad (2.72)$$

Assuming that $\omega = \sqrt{g/l}$, the solution is given by

$$\phi(t) \approx \begin{cases} [l/(4g)]\left(1 + e^{-\gamma t/2}\right)\sin\left(\sqrt{g/l}t\right) \\ \quad + \sqrt{l/(\gamma^2 g)}\left(1 - e^{-\gamma t/2}\right)\cos\left(\sqrt{g/l}t\right) & \text{for } 0 < t < T \\ e^{-\gamma t/2}\left([l/(4g)]\left[e^{\gamma T/2}\sin\left([2T - t]\sqrt{g/l}\right)\right.\right. \\ \quad \left. + \sin\left(\sqrt{g/l}t\right)\right] + \sqrt{l/(\gamma^2 g)} \\ \quad \times \left[e^{\gamma T/2}\cos\left(\sqrt{g/l}t\right) - \cos\left(\sqrt{g/l}t\right)\right]\right) & \text{for } 0 < T < t. \end{cases} \qquad (2.73)$$

2.3.2 Confinement is the key: How resonator features arise whenever we attempt to trap dynamic fields

In Section 2.3.1 we introduced the phenomenon of resonance. But as we mentioned earlier to have resonances is only one of the essential features of a resonator. The other essential feature is that the fields acquire particular shapes in space when they vibrate at a resonator's resonance frequency (i.e., the mode patterns). If the reader wishes, he or she can even regard the simple physical example given in Section 2.3.1 as a *zero-dimensional resonator:* a pathological case that has a resonance, yet lacks the mode pattern feature. To have a proper resonator, we must at least consider a one-dimensional system. This is what is done in this subsection, where we consider what is perhaps the simplest kind of resonator, the one-dimensional resonator obtained by holding fixed both ends of a stretched string. We will show that the resonator is created by the finiteness of the length of the string.

In this one-dimensional world, the equivalent of having just the open air (i.e., no violin body as an acoustic resonator) is having an infinite string stretched from $x = -\infty$ to $x = +\infty$. The string is elastic, as the air is compressible and supports waves. At any position x, its vertical displacement y obeys the one-dimensional wave equation (see Sec. 8.1 of [587])

$$\frac{\partial^2 y}{\partial t^2} - v^2 \frac{\partial^2 y}{\partial x^2} = 0. \tag{2.74}$$

This infinite string is at rest. Now suppose that there was a force $F(t)$ driving it. For simplicity, let us also suppose that this force oscillates at a definite frequency. This is a bit of a problem, because we also would like it to have a beginning: An oscillation at a definite frequency can have no beginning. It must be something like a perfect sine wave or cosine wave which was always on. We make a compromise and choose our force to be $F_\epsilon(t) = A \exp(\epsilon t) \sin \omega t$, where A is the amplitude of the force and ϵ is a positive real number, small enough so that the force may still be regarded approximately as oscillating with a definite frequency ω even though it starts adiabatically at the remote past. At the end of our calculation, we take the limit[19] of $\epsilon \to 0$. We apply this force to a specific point along the string, say $x = x_0$. Then the equation of motion of the string becomes

$$\frac{\partial^2 y_\epsilon}{\partial t^2} - v^2 \frac{\partial^2 y_\epsilon}{\partial x^2} = F_\epsilon(t)\delta(x - x_0). \tag{2.75}$$

I will show now that for an infinite string, we can vary the frequency ω of the driving force at will without finding even a single resonance. To do that, we must solve (2.75). This equation of motion is a partial differential equation for x and t. But we can simplify it, turning it into an ordinary differential equation for t alone, by applying a Fourier transform in x. Let

$$y_\epsilon(x, t) = \int_{-\infty}^{\infty} dk' \, e^{ik'x} \mathcal{Y}_\epsilon(k', t). \tag{2.76}$$

Substituting (2.76) into (2.75), then multiplying the resulting equation by $(1/2\pi) \exp(-ikx)$ and integrating over x along the entire infinite string, we find that

$$\ddot{\mathcal{Y}}_\epsilon + v^2 k^2 \mathcal{Y}_\epsilon = \frac{e^{-ikx_0}}{2\pi} F_\epsilon(t), \tag{2.77}$$

---

[19]This "trick," known as *regularization,* is used again in this chapter, in the last section, when we compute the Casimir force between two parallel plates.

where the overdots stand for the second derivative with respect to time. To obtain
(2.77) we have also used (F.5), and the basic property of the delta function: that given
a function $g(u)$, continuous at $u = u_0$,

$$\int_{-\infty}^{\infty} du \, g(u)\delta(u - u_0) = g(u_0). \tag{2.78}$$

We leave the detailed solution of this problem as an exercise (Problem 2.3). The
solution, $y(x, t) = \lim_{\epsilon \to 0} y_\epsilon(x, t)$ is given by[20]

$$y(x, t) = \begin{cases} -[A/(2\omega v)] \cos([x - x_0 + vt]\omega/v) & \text{for } x < x_0 \\ -[A/(2\omega v)] \cos([x - x_0 - vt]\omega/v) & \text{for } x > x_0. \end{cases} \tag{2.79}$$

This is a wave of frequency ω traveling to the left from the left side of the point x_0
and to the right from the right side of the point x_0. It is the one-dimensional analog
of what happens when you throw a pebble into a pond. But even more important for
our discussion now, there are no resonances in the entire spectrum of frequencies.[21]

Now consider what happens if we reduce our infinite string world to a finite length
L with both ends (i.e., $x = -L/2$ and $x = L/2$) fixed, so that they are always at rest.
This means that instead of a Fourier transform as before, we should use a Fourier
series. Moreover, this Fourier series will be a sine series because of the fixed ends
[i.e., $y_\epsilon(-L/2, t) = y_\epsilon(L/2, t) = 0$] [433]. So instead of (2.76), we now write

$$y_\epsilon(x, t) = \sum_{m=1}^{\infty} y_{m,\epsilon}(t) \sin\left(\left[x + \frac{L}{2}\right]\frac{\pi}{L}m\right). \tag{2.80}$$

Substituting (2.80) into (2.75), then multiplying by $(2/L)\sin(n\pi x/L)$ and integrating
over x along the entire string (i.e., from $x = 0$ to $x = L$ now), we find that

$$\ddot{y}_{n,\epsilon} + \left(\frac{\pi}{L}n\right)^2 v^2 y_{n,\epsilon} = \frac{2}{L}\sin\left(\left[x_0 + \frac{L}{2}\right]\frac{\pi}{L}n\right)F_\epsilon(t). \tag{2.81}$$

To obtain (2.81), we also used the basic property of the delta function (2.78) and the
Fourier series result:

$$\int_{-L/2}^{L/2} dx \, \sin\left(\left[x + \frac{L}{2}\right]\frac{\pi}{L}n\right)\sin\left(\left[x + \frac{L}{2}\right]\frac{\pi}{L}m\right) = \frac{L}{2}\delta_{nm}. \tag{2.82}$$

I also leave it as an exercise to the reader working out the solution to this equation
(Problem 2.5). In closed form, that is, after summing up the Fourier series (Problem

---

[20]Even though the force density is a delta and becomes infinite at $x = x_0$, the displacement of the string
is finite at $x = x_0$ (see Problem 2.4).
[21]Solution (2.79) holds only for $\omega \neq 0$. The pathological case of vanishing frequency, $\omega = 0$, yields
$y(x, t) = 0$ as $F_\epsilon(t) = A\exp(\epsilon t)\sin\omega t$ vanishes for $\omega = 0$.

2.6), and taking the limit $\epsilon \to 0$ (problem 2.7), the displacement of the string is given by[22]

$$y(x,t) = \frac{A}{\omega v} \frac{\sin \omega t}{\sin(L\omega/v)}$$

$$\times \begin{cases} \cos\left([x+x_0]\omega/v\right) - \cos\left([x-x_0+L]\omega/v\right) & \text{for } -L/2 < x < x_0 < L/2 \\ \cos\left([x+x_0]\omega/v\right) - \cos\left([x-x_0-L]\omega/v\right) & \text{for } -L/2 < x_0 < x < L/2. \end{cases}$$

$$(2.83)$$

Looking at (2.83), we can see that unlike the infinite string, there are no traveling waves, only a standing wave.[23] This happens because the ends of the string are fixed and reflect back the waves. Moreover, there is now a $\sin(\omega L/v)$ in the denominator of the second term in (2.83). When this sine vanishes, the displacement actually becomes infinite. This happens for $\omega L/v = n\pi$, where n is an integer. These frequencies $\omega_n = (\pi/L)nv$ are the resonance frequencies of our one-dimensional resonator.

We have seen that the confinement makes the resonances appear. Being resonant to specific discrete frequencies is a very important characteristic of resonators. It is so important that it gave them their name. The second important feature is the specific field patterns that form inside the resonator when it vibrates at one of its resonance frequencies (i.e., the modes). Unfortunately, we cannot see these patterns using (2.83), as this solution diverges when the driving force vibrates at one of the resonator resonances. In practice, the displacement can become very large but never really diverges because of damping. Real resonators are never perfect and always let the fields leak a bit from their interior (or even absorb them a little), which leads to the damping of the vibrations. We could introduce damping, as is often done in textbook treatments of the harmonic oscillator, for example. However, then the field patterns that we would see would not correspond to a mode of a perfect resonator, which clearly show the ideas we are discussing, but to a more complicated sort of mode. We discuss damping in resonators later (you will get a flavor of it in Chapter 6 and a more detailed discussion will be given in Chapter 9). To see the patterns that develop in our one-dimensional string resonator, we will simply remove the driving force and assume that we have somehow set the string in oscillation so that it oscillates as a perfect sine at a point $x = x_0$: for example,

$$y(x_0, t) = A \sin \omega t. \tag{2.84}$$

As there is no driving force, the displacement satisfies (2.74). Then the general solution[24] must be of the form

$$y(x,t) = \sum_{n=1}^{\infty} \left\{ a_n \sin\left(n\frac{\pi}{L}vt\right) + b_n \cos\left(n\frac{\pi}{L}vt\right) \right\} \sin\left(n\frac{\pi}{L}\left[x + \frac{L}{2}\right]\right). \tag{2.85}$$

---

[22]The previous solution can be recovered by taking the limit $L \to \infty$ and then $\epsilon \to 0$ (Problem 2.8).

[23]In fact, this standing wave can be written as the sum of two traveling waves: one that travels to the right and another that travels to the left.

[24]The solution obtained by imposing only the boundary conditions $y(-L/2, t) = y(L/2, t) = 0$.

At $x = x_0$ (2.85) must coincide with (2.84):

$$\sum_{n=1}^{\infty}\left\{a_n\sin\left(n\frac{\pi}{L}vt\right) + b_n\cos\left(n\frac{\pi}{L}vt\right)\right\}\sin\left(n\frac{\pi}{L}\left[x_0 + \frac{L}{2}\right]\right) = A\sin\omega t.$$

(2.86)

Now, as this is a solution for all times, we can multiply (2.86) by $(1/2\pi)\exp(-i\omega't)$ and integrate over t from $-\infty$ to ∞. We find that

$$\sum_{n=1}^{\infty}a_n\sin\left(n\frac{\pi}{L}\left[x_0 + \frac{L}{2}\right]\right)\left\{\delta\left(n\frac{\pi}{L}v - \omega'\right) - \delta\left(n\frac{\pi}{L}v + \omega'\right)\right\}$$

$$+i\sum_{n=1}^{\infty}b_n\sin\left(n\frac{\pi}{L}\left[x_0 + \frac{L}{2}\right]\right)\left\{\delta\left(n\frac{\pi}{L}v - \omega'\right) + \delta\left(n\frac{\pi}{L}v + \omega'\right)\right\}$$

$$= A\left\{\delta\left(\omega - \omega'\right) - \delta\left(\omega + \omega'\right)\right\}.$$

(2.87)

As a_n, b_n, and A are all real, it follows that

$$b_n = 0$$

(2.88)

and

$$\sum_{n=1}^{\infty}a_n\sin\left(n\frac{\pi}{L}\left[x_0 + \frac{L}{2}\right]\right)\left\{\delta\left(n\frac{\pi}{L}v - \omega'\right) - \delta\left(n\frac{\pi}{L}v + \omega'\right)\right\}$$

$$= A\left\{\delta\left(\omega - \omega'\right) - \delta\left(\omega + \omega'\right)\right\}.$$

(2.89)

Now if $\omega \neq n(\pi/L)v$ for all $n = 1, 2, \ldots$ there is an ε such that the interval $(\omega - \varepsilon, \omega + \varepsilon)$ does not contain any of the points $n(\pi/L)v$. Integrating over ω' on this interval, we find that A and all the a_n must vanish for (2.86) to hold. In other words, our simple one-dimensional resonator can only support free persistent oscillations of frequencies $\omega_n = n(\pi/L)v$. Having both $\omega \neq \omega_n$ and $A \neq 0$ in (2.86) is inconsistent. When $\omega = \omega_{\bar{n}}$ for a given $\bar{n}$, the integration over ω' mentioned above yields

$$a_{\bar{n}}\sin\left(\bar{n}\frac{\pi}{L}\left[x_0 + \frac{L}{2}\right]\right) = A$$

(2.90)

and $a_n = 0$ for all other integers $n \neq \bar{n}$. Thus,

$$y(x,t) = \frac{A}{\sin\left([x_0 + L/2]\,\bar{n}\pi/L\right)}\sin\left(\bar{n}\frac{\pi}{L}vt\right)\sin\left(\bar{n}\frac{\pi}{L}\left[x + \frac{L}{2}\right]\right).$$

(2.91)

You should note that as $-L/2 < x_0 < L/2$, the sine in the denominator of (2.91) never vanishes. Also notice that even though the general solution (2.85) is not necessarily separable, (2.91) is [i.e., (2.91) separates into a spatially dependent factor $\sin([\pi/L]\bar{n}[x + L/2])$ and a time-dependent factor $\sin([\pi/L]\bar{n}vt)$]. The spatially dependent factor is what we call the *mode*. So our one-dimensional resonator produces the spatial field pattern (i.e., mode) $\sin([\pi/L]\bar{n}[x + L/2])$ when its $\bar{n}$th resonance $\omega_{\bar{n}} = (\pi/L)\bar{n}v$ is excited.

The important message for us here is this. In a perfect resonator, not only is the frequency spectrum of the radiation reduced to the discrete cavity resonances, but the fields themselves must be linear combinations of the corresponding mode patterns. So the cavity modes form a *complete* basis in terms of which *any* intracavity radiation field can be expanded.

2.4   CANONICAL QUANTIZATION OF THE RADIATION FIELD

This is the staple wine of our free tasting. We will start with field quantization in a perfect cavity because it is easier. Even though the important results of the mode calculation are summarized briefly at the beginning of Section 2.4.1, the reader might find it useful to have a look at Appendix A. After that, we are ready to tackle the free-space case in Section 2.4.2. This is just a bit more sophisticated because it involves continuum modes with lots of delta functions instead of Kronecker deltas and complex canonical variables, but the basic idea is the same of Section 2.4.1.

Finally, a disclaimer. We do not talk a lot here about the photon as a particle. In this section it appears simply as the quantum of excitation of the modes of the electromagnetic radiation field. A more vivid description of the photon as a quantum particle is given in Chapter 3.

2.4.1   Canonical quantization of the free field in a cavity

The key results of our mode calculation in Appendix A that we will need here are:

1. The orthonormality of the electric field modes

$$\int d^3r \, \chi_{n',s'} \cdot \chi_{n,s} = \delta_{n'n}\delta_{s's},$$
(2.92)

and of the magnetic field modes

$$\int d^3r \, \frac{1}{k_{n'}} \left(\nabla \wedge \chi_{n',s'} \right) \cdot \frac{1}{k_n} \left(\nabla \wedge \chi_{n,s} \right) = \delta_{n'n}\delta_{s's}.$$
(2.93)

2. The expansion of the fields in terms of field modes

$$\mathbf{E} = \sum_{n,s} \chi_{n,s} e_{n,s} \quad \text{and} \quad \mathbf{B} = \sum_{n,s} \frac{1}{k_n} \left(\nabla \wedge \chi_{n,s} \right) b_{n,s},$$
(2.94)

where

$$e_{n,s} = \int d^3r \, \chi_{n,s} \cdot \mathbf{E} \quad \text{and} \quad b_{n,s} = \int d^3r \, \frac{1}{k_n} \left(\nabla \wedge \chi_{n,s} \right) \cdot \mathbf{B}.$$
(2.95)

3. The equations of motion obeyed by the expansion coefficients

$$\dot{b}_{n,s} = -ck_n e_{n,s} \quad \text{and} \quad \dot{e}_{n,s} = ck_n b_{n,s}.$$
(2.96)

Now inspecting (2.96), we see that if we define the new variables

$$q_{n,s} = \frac{\Upsilon}{ck_n} b_{n,s} \tag{2.97}$$

$$p_{n,s} = -\Upsilon e_{n,s}, \tag{2.98}$$

where Υ is any constant, they acquire the form of Hamilton's equations for a set of *uncoupled* harmonic oscillators

$$\dot{q}_{n,s} = p_{n,s} \tag{2.99}$$

$$\dot{p}_{n,s} = -c^2 k_n^2 q_{n,s}, \tag{2.100}$$

whose Hamiltonian is

$$H = \sum_{n,s} H_{n,s}, \tag{2.101}$$

with

$$H_{n,s} = \frac{1}{2} p_{n,s}^2 + \frac{c^2 k_n^2}{2} q_{n,s}^2. \tag{2.102}$$

So $q_{n,s}$ is the position variable of oscillator n,s, and $p_{n,s}$ is its canonically conjugate momentum. We can specify Υ by requiring that the Hamiltonian (2.20) be equal to the energy of the radiation field. Substituting (2.94) in the expression for the field energy

$$\mathcal{E} = \frac{1}{8\pi} \int d^3r \; (\mathbf{E}^2 + \mathbf{B}^2), \tag{2.103}$$

and using (2.97) and (2.98), we find that

$$\mathcal{E} = \frac{1}{8\pi} \sum_{n,s} \left\{ \frac{p_{n,s}^2}{\Upsilon^2} + \frac{c^2 k_n^2}{\Upsilon^2} q_{n,s}^2 \right\}. \tag{2.104}$$

Then choosing $\Upsilon = 1/\sqrt{4\pi}$ guarantees that the Hamiltonian is also the energy.

This analogy between the electromagnetic radiation field and a set of uncoupled harmonic oscillators is very much in the spirit of Maxwell's original method of working in theoretical physics [87]. It can simplify enormously the treatment of several problems in classical electrodynamics [90, 233]. But it became really popular as a simple way of quantizing the radiation field because of a review paper by Fermi[25] [195]. In this approach, the field can be quantized simply by following the canonical quantization procedure and replacing the Poisson brackets of $q_{n,s}$ and $p_{n,s}$ by quantum commutators as in (2.16). The quantized oscillators positions $\hat{q}_{n,s}$ and momenta $\hat{p}_{n,s}$ then have the following commutators:

$$[\hat{q}_{n,s}, \hat{p}_{n',s'}] = i\hbar \delta_{n,n'} \delta_{s,s'}. \tag{2.105}$$

---

[25]Born and Jordan [68] originally exploited this analogy to quantize the radiation field in the 1940s, but the oscillator approach only became really popular as a simple way to get to QED when it appeared in Fermi's review paper.

As is usually done for the quantum harmonic oscillator, we can now define pairs of ladder operators

$$\hat{a}_{n,s} = (ck_n \hat{q}_{n,s} + i\hat{p}_{n,s}) \frac{1}{\sqrt{2\hbar ck_n}}, \tag{2.106}$$

$$\hat{a}_{n,s}^{\dagger} = (ck_n \hat{q}_{n,s} - i\hat{p}_{n,s}) \frac{1}{\sqrt{2\hbar ck_n}}, \tag{2.107}$$

whose commutators are

$$[\hat{a}_{n,s}, \hat{a}_{n',s'}^{\dagger}] = \delta_{n,n'} \delta_{s,s'}, \tag{2.108}$$

$$[\hat{a}_{n,s}, \hat{a}_{n',s'}] = 0. \tag{2.109}$$

The quantized fields are given in terms of these ladder operators by

$$\hat{\mathbf{E}} = i \sum_{n,s} \sqrt{2\pi \hbar ck_n} \left(\hat{a}_{n,s} - \hat{a}_{n,s}^{\dagger} \right) \boldsymbol{\chi}_{n,s}, \tag{2.110}$$

$$\hat{\mathbf{B}} = \sum_{n,s} \sqrt{\frac{2\pi \hbar c}{k_n}} \left(\hat{a}_{n,s} + \hat{a}_{n,s}^{\dagger} \right) \boldsymbol{\nabla} \wedge \boldsymbol{\chi}_{n,s}. \tag{2.111}$$

The Hamiltonian is given by

$$\hat{H} = \sum_{n,s} \frac{\hbar ck_n}{2} \left(\hat{a}_{n,s}^{\dagger} \hat{a}_{n,s} + \hat{a}_{n,s} \hat{a}_{n,s}^{\dagger} \right). \tag{2.112}$$

For those who are not familiar with the algebraic treatment of the quantum harmonic oscillator using ladder operators, here is what it is all about.[26] As the field oscillators are all uncoupled, we can consider each oscillator independently. Let us then take any such oscillator and drop the indices n, s for simplicity. The Hamiltonian of the oscillator of frequency $\omega = ck$ is given by

$$\hat{H}_{\text{osc}} = \frac{\hbar \omega}{2} \left(\hat{a}^{\dagger} \hat{a} + \hat{a} \hat{a}^{\dagger} \right)$$

$$= \hbar \omega \left(\hat{a}^{\dagger} \hat{a} + \frac{1}{2} \right), \tag{2.113}$$

where we have used $[\hat{a}, \hat{a}^{\dagger}] = 1$ to eliminate[27] $\hat{a}\hat{a}^{\dagger}$.

Now suppose that $|n\rangle$ is an eigenstate of $\hat{a}^{\dagger}\hat{a}$:

$$\hat{a}^{\dagger}\hat{a}|n\rangle = |n\rangle n. \tag{2.114}$$

---

[26]See also [337].

[27]Notice that the same can, of course, be done in (2.112). Then we will end up with a Hamiltonian given by the sum of the energies of each photon in the cavity plus an infinite term: $\sum_{n,s} \hbar ck_n/2$. This infinite term is the zero-point energy of the vacuum and has an important physical consequence that is discussed in the next section.

Then $|n\rangle$ is also an energy eigenstate of (2.113) with eigenenergy $(n + 1/2)\hbar\omega$. As the norm of a state vector cannot be negative, we must have $|\hat{a}|n\rangle| \geq 0$. Thus, taking $|n\rangle$ as normalized, we find that

$$\langle n|\hat{a}^\dagger\hat{a}|n\rangle = n \geq 0. \tag{2.115}$$

Now as

$$[\hat{a}^\dagger\hat{a}, \hat{a}] = -\hat{a}, \tag{2.116}$$

it follows that

$$\hat{a}^\dagger\hat{a}\,(\hat{a}|n\rangle) = (\hat{a}|n\rangle)\,(n - 1) \tag{2.117}$$

(i. e., $\hat{a}|n\rangle$ is an eigenstate of $\hat{a}^\dagger\hat{a}$ corresponding to eigenvalue $n - 1$). So if we keep applying $\hat{a}$ to $|n\rangle$ successively, we will eventually hit a negative eigenvalue. As according to (2.115), negative eigenvalues are not allowed, there must be an integer m for which $\hat{a}^m|n\rangle$ does not vanish but $\hat{a}^{m+1}|n\rangle$ does. Then

$$\hat{a}^\dagger\hat{a}\,(\hat{a}^m|n\rangle) = (n - m)\,(\hat{a}^m|n\rangle), \tag{2.118}$$

so that $n = m$. In other words, the eigenvalues of $\hat{a}^\dagger\hat{a}$ are all positive integers. The lowest eigenvalue is $n = 0$, with the corresponding eigenstate defined by

$$\hat{a}|0\rangle = 0. \tag{2.119}$$

From this eigenstate we can generate any other eigenstate by applying $\hat{a}^\dagger$ a sufficient number of times. To see this, notice that from

$$[\hat{a}^\dagger\hat{a}, \hat{a}^\dagger] = \hat{a}^\dagger, \tag{2.120}$$

it follows that

$$\hat{a}^\dagger\hat{a}\,(\hat{a}^\dagger|n\rangle) = (\hat{a}^\dagger|n\rangle)\,(n + 1). \tag{2.121}$$

Now from (2.121) we find that

$$\frac{1}{\sqrt{n!}}\left(\hat{a}^\dagger\right)^n|0\rangle = |n\rangle. \tag{2.122}$$

For the quantized radiation field, these ladder operators aquire an interesting new meaning. The energy eigenstates $|n\rangle$ are states with a well-defined number of electromagnetic energy quanta in the given field mode. These quanta are called *photons*.[28] They can carry not only energy but also linear and angular momentum. Destruction and creation of a photon means subtracting from or adding to the field the full amount of each of these quantities carried by the photon. It is in this sense that photons

---

[28]The name *photon* was coined by the chemist Gilbert N. Lewis in a letter to *Nature* [407]. He wrote: "I therefore take the liberty of proposing for this hypothetical new atom, which is not light but plays an essential part in every process of radiation, the name photon." Curiously enough, he was not writing about the electromagnetic energy quantum but rather speculating about a new kind of subatomic particle that would act as a carrier of radiant energy.

behave like particles; they are like grains of these physical quantities.[29] As the ladder operator $\hat{a}$ annihilates photons, it is called the *annihilation operator.* Analogously, $\hat{a}^\dagger$ is called the *creation operator.*

2.4.2 Canonical quantization of the free field in free space

In free space it is natural to use plane waves in the mode expansion. Then the mode expansion is just an ordinary inverse Fourier transform that reconstructs the radiation fields in configuration space from their reciprocal-space counterparts. We have already written the reciprocal-space Maxwell equations without charges in Section 2.2. These are equations (2.47)–(2.50), which have the nice property of involving only quantities at a given point $\mathbf{k}$ in reciprocal space without any reciprocal-space derivatives. The reciprocal-space fields are defined by a Fourier transform such as (2.46). Unlike the cavity case, where the mode functions were all real, plane waves are complex. As a consequence, the fields in reciprocal space are also complex. But as the configuration space fields have to be real, their counterparts in reciprocal space are not all independent. Their complex conjugates are related to their values at the negative of their reciprocal-space position by

$$\mathcal{E}^*(\mathbf{k}) = \mathcal{E}(-\mathbf{k}) \quad \text{and} \quad \mathcal{B}^*(\mathbf{k}) = \mathcal{B}(-\mathbf{k}). \tag{2.123}$$

So, to deal only with truly independent degrees of freedom, we use (2.123) to write E and B only in terms of half of the reciprocal space. We indicate that by writing a prime over the integral signs to indicate that they cover only half of reciprocal space:

$$\mathbf{E}(\mathbf{r}) = \int' d^3k \, \mathcal{E}(\mathbf{k}) e^{i\mathbf{k}\cdot\mathbf{r}} + \text{c.c.} \quad \text{and} \quad \mathbf{B}(\mathbf{r}) = \int' d^3k \, \mathcal{B}(\mathbf{k}) e^{i\mathbf{k}\cdot\mathbf{r}} + \text{c.c.} \tag{2.124}$$

where c.c. stands for *complex conjugate.*

From (2.47) and (2.48) we see that both electric and magnetic reciprocal-space fields are transverse, with $\mathcal{E}$ and $\mathcal{B}$ having just two polarizations, both orthogonal to k. Let $\epsilon_s(\mathbf{k})$ with $s = 1, 2$ be the two linear and real polarization vectors of the electric field $\mathcal{E}(\mathbf{k})$:

$$\mathcal{E}(\mathbf{k}) = \sum_{s=1,2} \epsilon_s(\mathbf{k}) \mathcal{E}_s(\mathbf{k}). \tag{2.125}$$

The magnetic field can also be written as a linear combination of two unit polarization vectors multiplied by scalar amplitudes:

$$\mathcal{B}(\mathbf{k}) = \sum_{s=1,2} \beta_s(\mathbf{k}) \mathcal{B}_s(\mathbf{k}). \tag{2.126}$$

---

[29] An excellent account of this can be found in Chapter 8 of Gottfried's book [243]. In the same chapter there is a beautiful discussion of the physical significance of polarization and its connection to the spin of the photon. We discuss this connection in Chapter 3.

Using the reciprocal-space counterpart (2.49) of Faraday's law, we define the magnetic field polarization vectors $\beta_s(\mathbf{k})$ as pure *imaginary* unit vectors given in terms of $\epsilon_s(\mathbf{k})$ by

$$\beta_s(\mathbf{k}) = i\frac{\mathbf{k}}{k} \wedge \epsilon_s(\mathbf{k}). \tag{2.127}$$

By defining the two magnetic field polarization components $B_s(\mathbf{k})$ in this way with an imaginary unit in their polarization vectors, Faraday's law (2.49) and Maxwell's modification of Ampère's law (2.50) yield the following dynamic equations for the field components:

$$\dot{B}_s = -ck\mathcal{E}_s, \tag{2.128}$$
$$\dot{\mathcal{E}}_s = ckB_s. \tag{2.129}$$

In other words, in this way the real (imaginary) component of $\mathcal{E}_s(\mathbf{k})$ will couple to the real (imaginary) component of $B_s(\mathbf{k})$ rather than having the real (imaginary) component of $\mathcal{E}_s(\mathbf{k})$ couple to the imaginary (real) component of $\mathcal{E}_s(\mathbf{k})$.

Let us define the variables

$$q_{\alpha,s} = \frac{\aleph}{ck}\, B_{\alpha,s} \quad \text{and} \quad p_{\alpha,s} = -\aleph\, \mathcal{E}_{\alpha,s}, \tag{2.130}$$

where $\alpha = R, I$ with R standing for real component and I for imaginary component. With these new variables, for each $\mathbf{k}$ and polarization s, we have four equations of motion

$$\dot{q}_{\alpha,s} = p_{\alpha,s} \quad \text{and} \quad \dot{p}_{\alpha,s} = -c^2k^2 q_{\alpha,s}. \tag{2.131}$$

These are Hamilton's dynamical equations of motion for two ($\alpha = R, I$) harmonic oscillators of frequency ck and unit mass. The Hamiltonian from which these equations of motion can be derived is given by

$$H = \int' d^3k \sum_{\alpha,s} \frac{1}{2}\left\{p_{\alpha,s}^2 + c^2k^2 q_{\alpha,s}^2\right\}. \tag{2.132}$$

For the Hamiltonian also to be the energy,

$$\mathcal{E} = \frac{1}{8\pi}\int d^3r\,\left\{\mathbf{E}^2 + \mathbf{B}^2\right\}$$
$$= \pi^2 \int' d^3k\,\left\{|\mathcal{E}|^2 + |\mathcal{B}|^2\right\}$$
$$= \pi^2 \int' d^3k\,\frac{1}{\aleph^2}\sum_{\alpha,s}\left\{p_{\alpha,s}^2 + c^2k^2 q_{\alpha,s}^2\right\}, \tag{2.133}$$

we must choose $\aleph$ to be

$$\aleph = \pi\sqrt{2}. \tag{2.134}$$

The canonical variables $q_{R,s}$ and $p_{R,s}$ are independent of $q_{I,s}$ and $p_{I,s}$. Their Poisson brackets are given by[30]

$$[q_{\alpha,s}(\mathbf{k}), p_{\alpha',s'}(\mathbf{k}')]_{PB} = \delta_{\alpha,\alpha'}\delta_{s,s'}\delta(\mathbf{k}' - \mathbf{k}) \qquad (2.135)$$

and

$$[q_{\alpha,s}(\mathbf{k}), q_{\alpha',s'}(\mathbf{k}')]_{PB} = [p_{\alpha,s}(\mathbf{k}), p_{\alpha',s'}(\mathbf{k}')]_{PB} = 0. \qquad (2.136)$$

Now, notice that if we define the complex canonical position and momentum variables by [108]

$$Q_s = \frac{1}{\sqrt{2}}(q_{R,s} + iq_{I,s}) \quad \text{and} \quad P_s = \frac{1}{\sqrt{2}}(p_{R,s} + ip_{I,s}), \qquad (2.137)$$

these complex canonical conjugate variables will have the following simple Poisson brackets:

$$[Q_s(\mathbf{k}), P_{s'}^*(\mathbf{k}')]_{PB} = [Q_s^*(\mathbf{k}), P_{s'}(\mathbf{k}')]_{PB} = \delta_{s,s'}\delta(\mathbf{k}' - \mathbf{k}) \qquad (2.138)$$

and

$$[Q_s(\mathbf{k}), P_{s'}(\mathbf{k}')]_{PB} = [Q_s^*(\mathbf{k}), P_{s'}^*(\mathbf{k}')]_{PB} = 0. \qquad (2.139)$$

As we go to the quantum regime, the commutators of the $q_{\alpha,s}$ and $p_{\alpha,s}$ will be given by the value of their Poisson brackets multiplied by $i\hbar$ (canonical quantization). Then the complex canonical position and momentum operators will have the following commutators:

$$[\hat{Q}_s(\mathbf{k}), \hat{P}_{s'}^*(\mathbf{k}')] = [\hat{Q}_s^*(\mathbf{k}), \hat{P}_{s'}(\mathbf{k}')] = i\hbar\delta_{s,s'}\delta(\mathbf{k}' - \mathbf{k}) \qquad (2.140)$$

and

$$[\hat{Q}_s(\mathbf{k}), \hat{P}_{s'}(\mathbf{k}')] = [\hat{Q}_s^*(\mathbf{k}), \hat{P}_{s'}^*(\mathbf{k}')] = 0. \qquad (2.141)$$

We can define the creation and annihilation operators just as in Section 2.4.1 but now using non-Hermitian canonical position and momentum operators rather than Hermitian ones. The reader can easily verify using (2.140) and (2.141) that defining the annihilation operator by

$$\hat{a}_s(\mathbf{k}) \equiv \frac{1}{\sqrt{2\hbar ck}}\left\{ck\hat{Q}_s(\mathbf{k}) + i\,\hat{P}_s(\mathbf{k})\right\}, \qquad (2.142)$$

we have the expected commutation relations between creation and annihilation operators:

$$[\hat{a}_s(\mathbf{k}), \hat{a}_{s'}^\dagger(\mathbf{k}')] = \delta_{s,s'}\delta(\mathbf{k}' - \mathbf{k}) \qquad (2.143)$$

and

$$[\hat{a}_s(\mathbf{k}), \hat{a}_{s'}(\mathbf{k}')] = [\hat{a}_s(\mathbf{k}), \hat{a}_{s'}(-\mathbf{k}')] = [\hat{a}_s^\dagger(\mathbf{k}), \hat{a}_{s'}(-\mathbf{k}')] = 0. \qquad (2.144)$$

---

[30]Otherwise, the equations of motion (2.131) will not be derivable from the Hamiltonian (2.132).

From (2.130), (2.137), (2.134), and (2.142), we find that in terms of the quantized reciprocal-space field components, $\hat{a}_s(\mathbf{k})$ is given by

$$\hat{a}_s(\mathbf{k}) = \frac{\pi}{\sqrt{2\hbar ck}} \left\{ \hat{\mathcal{B}}_s(\mathbf{k}) - i\,\hat{\mathcal{E}}_s(\mathbf{k}) \right\}. \tag{2.145}$$

Inverting (2.145), using (2.125)–(2.127), and then substituting in (2.124), we find the following expressions for the quantized free-space radiation fields[31]:

$$\mathbf{E}(\mathbf{r}) = i \int' d^3k \, \frac{1}{\pi} \sqrt{\frac{\hbar ck}{2}} \sum_s \{\hat{a}_s(\mathbf{k}) - \hat{a}_s^\dagger(-\mathbf{k})\} \; \boldsymbol{\epsilon}_s(\mathbf{k}) \, e^{i\mathbf{k}\cdot\mathbf{r}} + \text{H.c.}$$

$$\tag{2.146}$$

$$\mathbf{B}(\mathbf{r}) = i \int' d^3k \, \frac{1}{\pi} \sqrt{\frac{\hbar ck}{2}} \sum_s \{\hat{a}_s(\mathbf{k}) + \hat{a}_s^\dagger(-\mathbf{k})\} \; \frac{\mathbf{k}}{k} \wedge \boldsymbol{\epsilon}_s(\mathbf{k}) \, e^{i\mathbf{k}\cdot\mathbf{r}} + \text{H.c.}$$

$$\tag{2.147}$$

where H.c. stands for *Hermitian conjugate*. Substituting (2.146) and (2.147) in (2.133), we find the following expression for the Hamiltonian:

$$
\begin{aligned}
\hat{H} &= \frac{1}{8\pi} \int d^3r \, \left\{ \hat{\mathbf{E}}^2 + \hat{\mathbf{B}}^2 \right\} \\
&= \int' d^3k \, \frac{\hbar ck}{2} \sum_s \{[\hat{a}_s(\mathbf{k}) - \hat{a}_s^\dagger(-\mathbf{k})][\hat{a}_s^\dagger(\mathbf{k}) - \hat{a}_s(-\mathbf{k})] \\
&\qquad\qquad + [\hat{a}_s(\mathbf{k}) + \hat{a}_s^\dagger(-\mathbf{k})][\hat{a}_s^\dagger(\mathbf{k}) + \hat{a}_s(-\mathbf{k})]\} \\
&= \int' d^3k \, \hbar ck \sum_s \{\hat{a}_s^\dagger(-\mathbf{k})\hat{a}_s(-\mathbf{k}) + \hat{a}_s(\mathbf{k})\hat{a}_s^\dagger(\mathbf{k})\} \\
&= \frac{1}{2} \int d^3k \, \hbar ck \sum_s \{\hat{a}_s^\dagger(\mathbf{k})\hat{a}_s(\mathbf{k}) + \hat{a}_s(\mathbf{k})\hat{a}_s^\dagger(\mathbf{k})\}. \tag{2.148}
\end{aligned}
$$

As you can see, everything is very much similar to what we have done before for a cavity except that now $\mathbf{k}$ and the frequency can assume any value in a continuum. In particular, if we use the commutator relation (2.143) to get rid of the $\hat{a}_s(\mathbf{k})\hat{a}_s^\dagger(\mathbf{k})$ term in the Hamiltonian (2.148), just as in the cavity case, we will get a Hamiltonian that equals the sum of the energies of each photon plus an infinite term. But now this infinite term is made up of an integral over a continuum rather than an infinite discrete sum. As we will see in the next section, this can have an interesting physical consequence.

---

[31]If you are surprised by (2.146) and (2.147) involving the creation operator of a photon with $-\mathbf{k}$ when the plane wave has wave vector $\mathbf{k}$, do not miss Chapter 3. There, in Section 3.6, you will see that this can be understood as arising from the contribution of the antiphoton to the fields.

2.5 A PHYSICAL EFFECT DUE TO ZERO-POINT FLUCTUATIONS: THE CASIMIR FORCE[32]

There are many electromagnetic phenomena that cannot be accounted for by classical electrodynamics, only by a full quantum theory of light. To mention a few, we have: Compton scattering, revivals of the atomic population inversion in the Jaynes–Cummings model, sub-Poissonian light, the Lamb shift, and so on. Funny enough, the photoelectric effect, whose explanation by introducing the idea of quantum of light gave Einstein a Nobel prize, is not one of those. It can be explained by a semiclassical theory where only matter is quantized[33] [376]. But most of these phenomena require us to take the interaction with matter into account in order to develop their theories. So they would be unsuitable to include in a chapter where this interaction is not being discussed. One of the few QED phenomena that does not require interaction with matter in its theory is the *Casimir effect*.[34] In the original situation considered by Casimir [91], this is an attractive force between two parallel, perfectly conducting plates standing close to each other. The plates are not charged and there are no real thermal photons. So the force is interpreted as resulting from the radiation pressure on the plates by the same virtual photons that give rise to the zero-point energy: The virtual photons between the plates give a weaker net kick than that of those outside, resulting in an attractive force.

Thus we have chosen to talk about this striking phenomenon here. However, beware that this is more an illustration of a cavity effect that does not require the coupling with matter than of a truly quantum effect. Even though the Casimir force is derived in QED as a direct physical consequence of the vacuum zero-point fluctuations, a noisy but classical background field of the type considered in stochastic electrodynamics can produce the same effect. In fact, it was found that there is a completely classical analog of the Casimir effect, known by sailors at least since the early nineteenth century. In his book *Album of the Mariner*, Causseé describes a curious situation where there is no wind but still a big swell running [93]. He then warns of a certain attractive force that will pull two ships at close quarters toward each other. In 1996, Boersma suggested that this mysterious force can be understood as a maritime analog of the Casimir force, where the kicks are due to a classical random field (i.e., the sea waves) rather than zero-point fluctuations [63].

Casimir was doing applied research at Philips in the stability of colloidal suspensions used to deposit films in the manufacture of lamps and cathode ray tubes when he discovered the effect that is now named after him. J. T. G. Overbeek and E. J. W.

---

[32]A reminder about our notation: Operators will have no hats in this section. Hats are reserved to denote unit vectors.

[33]The same can be said of the blackbody radiation problem, which led to the introduction of Planck's constant. In his original paper, Planck quantizes the material oscillators only, not the radiation field.

[34]That is not to say that there is no interaction with matter there. The parallel metal plates that the Casimir force attracts toward each other are, after all, matter. But then again, as we shall soon see, for perfectly reflecting plates, the Casimir force turns out to involve only two fundamental constants: $\hbar$ and c. It does not involve the charge of the electron, suggesting that this is an effect that does not involve coupling between matter and radiation.

Verwey at Philips Research had developed a theory of colloidal stability to explain the properties of suspensions of quartz powder. But their experiments indicated that the theory could not be entirely correct. Realizing that a better agreement between theory and experiment could be obtained if the interparticle interaction somehow fell off more rapidly at large distances than had been supposed, Overbeek hinted that this might be related to the finiteness of the speed of light. So Casimir and Polder started to investigate what would happen to the theory of the van der Waals interaction when retardation was included. They found that retardation makes the interaction at large intermolecular separations ΔR go as $1/\Delta R^7$ rather than just $1/\Delta R^6$ as was usually assumed. Overbeek was right. However, Casimir was still intrigued by the simplicity of the result. Then, as he told Milonni and Shih [455]:

> Summer or autumn 1947 (but I am not absolutely certain that it [was] not somewhat earlier or later) I mentioned my results to Niels Bohr, during a walk. That is nice, he said, that is something new. I told him that I was puzzled by the extremely simple form of the expressions for the interaction at very large distance and he mumbled something about zero-point energy. That was all, but it put me on a new track.
>
> I found that calculating changes of zero-point energy really leads to the same results as the calculations of Polder and myself....
>
> On May 29, 1948 I presented my paper "On the attraction between two perfectly conducting plates" at a meeting of the Royal Netherlands Academy of Arts and Sciences. It was published in the course of the year. ...

In his autobiography [92], he hardly mentions it except to make a low-key remark so typical of the Dutch: It is "of some theoretical significance." Yet the Casimir effect has turned out to be one of the most interesting effects in physics. It is important not only from a theoretical point of view but also in technology. Proper consideration of the Casimir force is crucial in the design of nanomachinery [83, 309, 352, 353]. There is so much interest in the Casimir effect that the amount of literature on it is just overwhelming, and we could not possibly cover it all in one section. This section will be restricted to the bare essentials. In Section 2.5.1 we derive the Casimir force between two parallel plates from a zero-point "potential energy," similarly to the way Casimir himself originally derived it [91]. Then we present a somewhat more complex, yet more physical derivation in terms of the Maxwell stress tensor. Section 2.5.2 deals with the implications of the vacuum zero-point energy to gravity, leading to the largest disagreement between theory and experiment ever known in the history of physics: the "vacuum catastrophe."

2.5.1 Zero-point potential energy[35]

The two parallel plates of infinite area considered by Casimir [91] can be obtained from the closed cavity geometry shown in Figure 2.7 by taking the limit of $L \to \infty$. This allows us to use here the closed cavity formalism developed in Appendix A. You can easily convince yourself (see Problem 2.9) that the allowed mode frequencies

---

[35]This subsection closely follows Casimir's approach in [91].

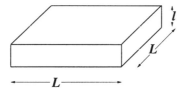

Figure 2.7 This cavity is a parallelepiped box with perfectly conducting walls. The length of each side is specified in the figure. The height l of the box corresponds to the plate separation. The bottom and top squares, whose sides have length L, correspond to the plates themselves, which will have an infinite area when we take the limit $L \to \infty$.

are given by

$$\omega_{qmn} = \pi c \sqrt{\left(\frac{q}{L}\right)^2 + \left(\frac{m}{L}\right)^2 + \left(\frac{n}{l}\right)^2}, \tag{2.149}$$

where q, m, and n are positive integers. Thus, the zero-point energy in the box is given by

$$\mathcal{E}(l) = \sum_{q,m,n} \mathcal{N}_{qmn} \frac{\hbar\omega_{q,m,n}}{2}, \tag{2.150}$$

where $\mathcal{N}_{qmn}$ is the number of independent polarization modes with frequency ω_{qmn}. This is always 2 when none of the integers q, m, and n vanishes. When any one of them vanishes, there is just one polarization (see Problem 2.9) and $\mathcal{N}_{qmn} = 1$.

As we are interested in the case where $L \to \infty$, we can replace the sums over q and m by integrals using

$$\lim_{L \to \infty} \sum_m \Delta k_m = \int_0^\infty dk, \tag{2.151}$$

where $\Delta k_m = k_{m+1} - k_m = \pi(m+1)/L - \pi m/L = \pi/L$. Then

$$\mathcal{E}(l) = \frac{\hbar c L^2}{\pi^2} \sum_n \int_0^\infty dk_x \int_0^\infty dk_y \, \mathcal{N}_n \sqrt{k_x^2 + k_y^2 + \left(\frac{\pi n}{l}\right)^2}, \tag{2.152}$$

where, again, $\mathcal{N}_n$ stands for the number of independent polarizations with frequency $c\sqrt{k_x^2 + k_y^2 + (\pi n/l)^2}$. Notice that it depends only on n because $k_x = 0$ and also $k_y = 0$ are now points of measure zero in a continuum. Going to cylindrical coordinates, we find that

$$\begin{aligned}
\mathcal{E}(l) &= \frac{\hbar c L^2}{\pi^2} \sum_n \int_0^\infty d\kappa \int_0^{\pi/2} \kappa \, d\theta \, \mathcal{N}_n \sqrt{\kappa^2 + \left(\frac{\pi n}{l}\right)^2} \\
&= \frac{\hbar c L^2}{4\pi} \sum_n \int_0^\infty du \, \mathcal{N}_n \sqrt{u + \left(\frac{\pi n}{l}\right)^2} \\
&= \frac{\hbar c L^2 \pi^2}{4l^3} \sum_n \int_0^\infty dv \, \mathcal{N}_n \sqrt{v + n^2}.
\end{aligned} \tag{2.153}$$

Now consider the situation where $l \to \infty$, too. Using (2.151) with L/π replaced by 1, we find that[36]

$$\mathcal{E}(\infty) = \frac{\hbar c L^2 \pi^2}{4 l^3} \int_0^\infty dv \int_0^\infty dw \sqrt{v + w^2}. \tag{2.154}$$

This zero-point energy is *different* from the zero-point energy (2.153) when the plates are only a finite distance l apart. The difference is due to the change in the mode spectral distribution when l was varied. This situation is analogous to that in electrostatics, where there is also a difference between the total electrostatic energy of a system of charges when these charges are at infinity and when they are arranged in space a finite distance from each other. As this energy difference depends only on the spatial configuration and not on how the charges are brought from infinity, it defines a potential energy that gives rise to a force. Here, also, the energy difference depends only on the spatial configuration[37] (i.e., on l). So we can define a potential zero-point energy as

$$U(l) = \mathcal{E}(l) - \mathcal{E}(\infty)$$
$$= \frac{\hbar c L^2 \pi^2}{4 l^3} \left\{ \sum_n \int_0^\infty dv \, \mathcal{N}_n \sqrt{v + n^2} - \int_0^\infty dv \int_0^\infty dw \sqrt{v + w^2} \right\}. \tag{2.155}$$

Before we try to get a closed-form expression from (2.155), let us pause to think about an import physical point that has been overlooked so far. No material known to us can reflect all frequencies equally well. Moreover, when the frequency becomes large, all materials become transparent. In particular, our plates cannot remain perfect reflectors all the way to infinite frequency. So we must include a cutoff in (2.155). However, we do not want to know the details of the fall-off of reflectivity with frequency. This is possible if the relevant physics happens for frequencies much lower than the ones where the reflectivity starts to fall off. We will see later that that is exactly what happens here, but for now let us just assume that it is so. In that case, we can use a mathematical trick called *regularization* to obtain a closed expression for (2.155) without having to specify precisely the general form of the cutoff function.

Let us define a cutoff function $f_a(\omega/c)$ that falls off sufficiently rapidly when the frequency becomes large and that depends on a parameter a in such a way that when $a \to 0$ the cutoff function becomes 1 for any frequency:

$$\lim_{k \to \infty} f_a(k) \to 0 \quad \text{and} \quad \lim_{a \to 0} f_a(k) \to 1. \tag{2.156}$$

---

[36] Notice that $\mathcal{N}_n$ disappears as we go to the continuum because then $w = 0$ is a point of measure zero.

[37] As long as the plates are brought from infinity in a adiabatic way. In Chapter 4 we will see what happens when cavity walls are moved in a nonadiabatic way, an effect that is often called the *dynamic Casimir effect*.

Including this cutoff function, (2.155) becomes

$$U(l) = \frac{\hbar c L^2 \pi^2}{4l^3} \left\{ \sum_n \int_0^\infty dv \, \mathcal{N}_n \, \sqrt{v + n^2} f_a(\sqrt{v + n^2}) \right.$$
$$\left. - \int_0^\infty dv \int_0^\infty dw \, \sqrt{v + w^2} f_a(\sqrt{v + w^2}) \right\}. \tag{2.157}$$

Let us examine the integrals over v in (2.157). They have the general form

$$\phi_a(\xi) \equiv \int_0^\infty dv \, \sqrt{v + \xi^2} f_a(\sqrt{v + \xi^2}). \tag{2.158}$$

Changing the variable of integration from v to $v' \equiv (2/3)(v + \xi^2)^{3/2}$, we obtain the simpler expression,

$$\phi_a(\xi) = \int_{2\xi^3/3}^\infty dv' \, f_a\left(\sqrt[3]{\frac{3}{2} v'} \right). \tag{2.159}$$

If we take the limit $a \to 0$ after doing the integral in (2.159), the various derivatives of $\phi_a(\xi)$ are given by

$$\lim_{a \to 0} \frac{d\phi_a}{d\xi} = -2\xi^2, \quad \lim_{a \to 0} \frac{d^2\phi_a}{d\xi^2} = -4\xi, \quad \lim_{a \to 0} \frac{d^3\phi_a}{d\xi^3} = -4, \quad \lim_{a \to 0} \frac{d^4\phi_a}{d\xi^4} = 0, \ldots \tag{2.160}$$

Now if we expand (2.157) into an Euler–Maclaurin series [433], we find that

$$U(l) = \frac{\hbar c L^2 \pi^2}{4l^3} \left\{ -\frac{1}{2}\phi_a(0) - \frac{1}{12}\frac{d\phi_a}{d\xi}\bigg|_{\xi=0} + \frac{1}{720}\frac{d^3\phi_a}{d\xi^3}\bigg|_{\xi=0} + \cdots \right\}. \tag{2.161}$$

If the relevant physics is independent of the manner in which the reflectivity falls off with frequency, we are allowed to take the limit $a \to 0$ in (2.161). Then all derivatives of ϕ_a vanish at $\xi = 0$, except the third derivative, and we find that

$$U(l) = -\frac{\hbar c L^2 \pi^2}{720 l^3}. \tag{2.162}$$

This potential implies an attractive force per unit area, or pressure on the plates, given by

$$-\frac{1}{L^2}\frac{dU}{dl} = -\frac{\hbar c \pi^2}{240 l^4}. \tag{2.163}$$

This is the Casimir force, or more precisely, pressure. For a plate separation of 1 cm, this pressure is only 1.3×10^{-18} dyn/cm². This is so weak that it is even less than half of the electrostatic pressure on the plates if they were charged with a single electron [i.e., $4\pi \bar{e}^2/L^4 \approx 2.9 \times 10^{-18}$ dyn/cm² (where $\bar{e}$ is the charge of the electron)]. But on the scale of nanomachinery (millimeter to nanometer range) the Casimir force is dominant [83, 309, 352, 353].

The first experiment on the Casimir force was done by Sparnaay in 1957, nine years after Casimir's paper was published. Unfortunately, the experimental error was far too large for this to be a decisive confirmation of Casimir's prediction. Since then, the experimental accuracy has improved immensely. But there are still very few experiments in comparison with the number of theoretical papers on this subject. For a list of the experimental papers on the Casimir force, see the recommended reading at the end of the chapter.

Last but not least, we promised to show you the physical reason why the regularization trick works. To do so, we must calculate first the mode spectral density for both free space and for the parallel-plate case. The mode spectral density is the function $\rho(\omega)$ of the frequency ω that gives the number of modes having a frequency between ω and $\omega + d\omega$. So, in terms of the mode spectral density, the zero-point energy (without including the cutoff) is given by

$$\mathcal{E} = \int_0^\infty d\omega \, \rho(\omega) \frac{1}{2} \hbar \omega. \tag{2.164}$$

For the free-space case (i.e., when both $L \to \infty$ and $l \to \infty$), (2.164) yields

$$\int_0^\infty d\omega \, \rho_{FS}(\omega) \frac{1}{2} \hbar \omega = \frac{V}{\pi^3} \int_0^\infty dk_x \int_0^\infty dk_y \int_0^\infty dk_z \, 2\frac{1}{2} \hbar \omega. \tag{2.165}$$

Changing the integration variables on the right-hand side of (2.165) from Cartesian to spherical, we find that

$$\int_0^\infty dk_x \int_0^\infty dk_y \int_0^\infty dk_z \, 2\frac{1}{2} \hbar \omega = \int_0^\infty k^2 \, dk \, \hbar \omega \int_0^{\pi/2} d(-\cos\theta) \int_0^{2\pi} d\varphi$$

$$= \frac{2\pi}{c^3} \int_0^\infty d\omega \, \omega^2 \hbar \omega. \tag{2.166}$$

Substituting (2.166) in (2.165), we see that we can take $\rho_{FS}(\omega)$ as given by[38]

$$\rho_{FS}(\omega) = \frac{4V}{\pi^2 c^3} \omega^2. \tag{2.167}$$

For the parallel-plate case, (2.164) yields

$$\int_0^\infty d\omega \, \rho_{PP}(\omega) \frac{\hbar \omega}{2} = \frac{Vc}{\pi^2 l} \int_0^\infty dk_x \int_0^\infty dk_y \, \hbar \left\{ \frac{\kappa_\perp}{2} + \sum_{n=1}^\infty \sqrt{\kappa_\perp^2 + \left(\frac{n\pi}{l}\right)^2} \right\}, \tag{2.168}$$

---

[38]Sometimes in the literature (e.g., in [274] and [305]), $\rho_{FS}(\omega)$ is quoted without the factor 4 seen in (2.166). The reason for the missing factor 4 in those references is that unlike here, they use periodic boundary conditions for k_x and k_y so that these components of the wave vector are discretized in units of $L/(2\pi)$ instead of L/π, as we here.

where $\kappa_\perp \equiv \sqrt{k_x^2 + k_y^2}$. Using delta functions to transform the discrete sum into an integral over k_z and then changing from Cartesian to spherical coordinates, we get

$$\int_0^\infty dk_x \int_0^\infty dk_y \, \hbar \left\{ \frac{\kappa_\perp}{2} + \sum_{n=1}^\infty \sqrt{\kappa_\perp^2 + \left(\frac{n\pi}{l}\right)^2} \right\}$$

$$= \int_0^\infty k^2 \, dk \int_0^1 du \int_0^{2\pi} d\varphi \left\{ \frac{1}{2}\delta(uk) + \sum_{n=1}^\infty \delta\left(uk - \frac{n\pi}{l}\right) \right\} \hbar\omega$$

$$= 2\pi \int_0^\infty k \, dk \int_0^1 du \, \frac{\partial}{\partial u} \left\{ \frac{1}{2}\Theta(uk) + \sum_{n=1}^\infty \Theta\left(uk - \frac{n\pi}{l}\right) \right\} \hbar\omega$$

$$= 2\pi \int_0^\infty k \, dk \left\{ \frac{1}{2} + \sum_{n=1}^\infty \Theta\left(k - \frac{n\pi}{l}\right) \right\} \hbar\omega, \tag{2.169}$$

where we have used that ω is independent of u and that the delta function $\delta(x)$ is the derivative of the step function $\Theta(x)$. Substituting (2.169) in (2.168), we find that we can take in this case the following expression for the spectral density

$$\rho_{PP}(\omega) = \rho_{FS}(\omega) \frac{\omega_0}{\omega} \left\{ \frac{1}{2} + \sum_{n=1}^\infty \Theta(\omega - n\omega_0) \right\}, \tag{2.170}$$

where $\omega_0 \equiv \pi c/l$. Notice that when $l \to \infty$, we can replace $\sum_n$ by $(l/\pi)\int_0^\infty dk_z$ as in (2.151), and then

$$\lim_{l\to\infty} \rho_{PP}(\omega) = \rho_{FS}(\omega)\frac{\omega_0}{\omega}\frac{l}{\pi}\int_0^\infty dk_z \, \Theta(\omega - ck_z)$$

$$= \rho_{FS}(\omega)\frac{\omega_0}{\omega}\frac{l}{\pi}\frac{\omega}{c}\int_0^1 dv$$

$$= \rho_{FS}(\omega), \tag{2.171}$$

as it should.

Now if we compare the cavity spectral mode density $\rho_{PP}(\omega)$ with the free-space density $\rho_{FS}(\omega)$ (see Figure 2.8), we see that they become very similar for $\omega \gg \omega_0$. In other words, they differ from each other significantly, only in the region around ω_0 [274]. This means that the rearrangement of modes in different frequencies that gives rise to the Casimir pressure is only significant around ω_0. Thus, as long as the fall-off of the reflectivity with frequency happens only at frequencies much larger than ω_0, so that for frequencies around ω_0 the plates are nearly perfect reflectors, the effect of this fall-off (which the cutoff models) is just to make the integrals finite without affecting the result. This is why our regularization trick works: As the pressure arises from the rearrangement of modes around ω_0 only, we can do the integral with the cutoff to make it converge and later neglect any further effects of the cutoff by taking the limit $a \to 0$ at the end of the calculation.

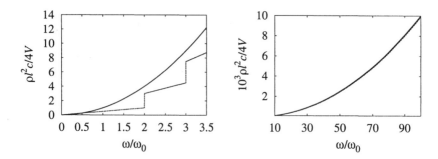

Figure 2.8 On the left, we plot the free-space spectral mode density (thick line) and the parallel plates spectral mode density (thiner dotted line) around ω_0. The same plot is repeated on the right but then for frequencies much larger than ω_0. The difference between the free-space (thick line) and parallel-plate (thin dotted line) spectral mode densities is so small in comparison with their magnitudes that it is barely visible in the plot.

2.5.2 Maxwell stress tensor

Whenever you derive a surprising result, it is often useful, both as a reassurance check and as a possible source of further insight, to rederive the result with a different approach.[39] Feynman was a master at rederiving things using different approaches.

There are two features of the previous derivation of the Casimir pressure that we would like to improve on in this new derivation. First, we used a closed cavity when the actual cavity (i.e., the two parallel plates) has no sidewalls. It is true that we took the limit of $L \to \infty$ in the end, but in the actual situation there is no reflecting surface at infinity. In this case, having a different boundary condition at infinity does not affect the main result. But we will see in a later chapter that the boundary conditions at infinity are usually very important. So we would like to rederive the Casimir pressure from a calculation where the actual parallel-plate geometry is adopted from the start, without having to convert discrete mode sums into integrals as before. Second, we mentioned the nice physical interpretation of the Casimir pressure as a result of the radiation pressure of the vacuum zero-point photon fluctuations, but this radiation pressure did not appear explicit in the previous calculation. What we would like to do is to derive the Casimir pressure directly from a calculation of the vacuum radiation pressure.

To make such a calculation, we must write down the quantized electromagnetic field using the modes for the parallel-plate geometry. This gives us the opportunity to show another way of determining the modes: the scattering method [169].

---

[39] You might be thinking now that the canonical quantization of the radiation field is also a surprising result that deserves to be derived in a different way. Please be patient: We present a different derivation of that in Chapter 3. For yet another derivation, you might wish to have a look at [199].

Let the z axis be perpendicular to the plates, with one of the plates located at $z = 0$ and the other at $z = l$. We will assume that the plate at $z = 0$ is a perfect reflector but only for the moment, we will think of the other plate as having a nonvanishing transmissivity. In the end of our mode calculation, we will take the limit of a vanishing transmissivity (i.e., the perfect reflector limit). This way of determining the cavity modes is a concrete example of a suggestion by Michael Berry[40] [50]. He suggested that every confined mode would correspond to the continuation to the interior of the cavity of an external superposition of plane waves for which the cavity is effectively transparent (i.e., for which there is no reflected wave). This way of thinking about cavity modes will turn out to be very useful when the cavity is no longer perfect, and we will come back to it in a later chapter. For now, just notice the key idea of the scattering method. In the absence of the plates, we have seen that the free-space field can be written in the form of a plane-wave expansion. Then if we determine how each plane-wave component is modified as it is scattered by the plates, we just have to sum the modified components to obtain an expression for the total quantized field.

Due to the nature of the boundary conditions (see Appendix B), it is convenient to decompose the electric field of each plane wave in two parts: a part perpendicular to the plane of incidence, $\alpha = \perp$, and another lying on the plane of incidence, $\alpha = \|$. So we write the electric field of such a plane-wave component as

$$\mathbf{E}_{I,\alpha} = \hat{\mathbf{v}}_{I,\alpha} E^0_{I,\alpha} e^{i\mathbf{k}_{I,\alpha} \cdot \mathbf{r}}, \tag{2.172}$$

where $\hat{\mathbf{v}}_{I,\alpha}$ gives the direction of this electric field component and $E^0_{I,\alpha}$ its amplitude. Just as in the usual theory of the Fabry–Perot interferometer [69], we can obtain the field inside the cavity due to the plane-wave component $\mathbf{E}_{I,\alpha}$ simply by adding the multiple reflections depicted in Figure 2.9. Plane waves with their electric fields perpendicular to the plane of incidence are multiplied by $m_\perp = -1$ upon reflection by the plate at $z = 0$ (see Appendix B). Plane waves whose electric field is on the plane of incidence get multiplied by $m_\| = 1$ upon reflection by the plate at $z = 0$. The reason that a plane wave with an electric field on the plane of incidence does not get multiplied by -1 on reflection is best understood by looking at Figure 2.10.

As we have just mentioned, we do not assume yet that the plate at $z = l$ is a perfect reflector. So for the moment, we take its amplitude reflectivities and transmissivities to be the following generic functions of the incidence angle θ: $r_\perp(\theta)$, $r_\|(\theta)$, $t_\perp(\theta)$, and $t_\|(\theta)$, where r stands for reflectivity, t for transmissivity, $\perp$ for being perpendicular to the plane of incidence, and $\|$ for being on the plane of incidence.

Using this notation, between the plates, the multiple scattering makes the electric field of the incident plane wave (2.172) change into

$$\mathbf{E}_{\text{cav},\mathbf{k},\alpha} = \mathbf{E}_{C1,\alpha} + \mathbf{E}_{C2,\alpha} + \cdots$$

$$= E^0_{I,\alpha} t_\alpha(\theta) \left(\hat{\mathbf{v}}_{I,\alpha} \left\{ \sum_{n=0}^{\infty} \left[e^{2i\delta} m_\alpha r_\alpha(\theta) \right]^n \right\} e^{i\mathbf{k}_{I,\alpha} \cdot \mathbf{r}} \right)$$

---

[40]See also [154, 162, 163].

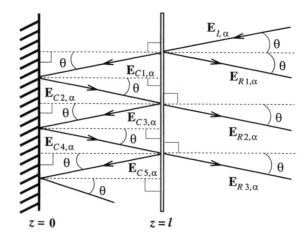

Figure 2.9 Multiple reflections in the planar Fabry–Perot cavity with the variables used in the calculations indicated.

$$+ \hat{\mathbf{v}}_{R,\alpha} \left\{ m_\alpha \sum_{n=0}^{\infty} \left[e^{2i\delta} m_\alpha r_\alpha(\theta) \right]^n \right\} e^{i\mathbf{k}_{R,\alpha} \cdot \mathbf{r}} \Bigg)$$

$$= E^0_{I,\alpha} t_\alpha(\theta) \left[\frac{\hat{\mathbf{v}}_{I,\alpha} e^{i\mathbf{k}_{I,\alpha} \cdot \mathbf{r}}}{1 - m_\alpha r_\alpha(\theta) e^{2i\delta}} + \frac{m_\alpha \hat{\mathbf{v}}_{R,\alpha} e^{i\mathbf{k}_{R,\alpha} \cdot \mathbf{r}}}{1 - m_\alpha r_\alpha(\theta) e^{2i\delta}} \right],$$

(2.173)

where $\hat{\mathbf{v}}_{R,\alpha}$ gives the direction of the electric field of the reflected wave and

$$\delta = \frac{\omega}{c} l \cos\theta$$

(2.174)

is the extra phase gained on each trip inside the cavity. In free space (i.e., above the plate at $z = l$), the field is the sum of $\mathbf{E}_{I,\alpha}$, its reflection on the plate at $z = l$, and the partial transmissions to the outside of each multiple scattered wave between the plates:

$$\mathbf{E}_{\text{out},k,\alpha} = \mathbf{E}_{I,\alpha} + \mathbf{E}_{R1,\alpha} + \mathbf{E}_{R2,\alpha} + \cdots$$

$$= E^0_{I,\alpha} \left\{ \hat{\mathbf{v}}_{I,\alpha} e^{i\mathbf{k}_{I,\alpha} \cdot \mathbf{r}} \right.$$

$$\left. + \hat{\mathbf{v}}_{R,\alpha} \left[r_\alpha(\theta) + \frac{t_\alpha(\theta)^2 m_\alpha e^{2i\delta}}{1 - r_\alpha(\theta) m_\alpha e^{2i\delta}} \right] e^{-i2\delta} e^{i\mathbf{k}_{R,\alpha} \cdot \mathbf{r}} \right\}$$

$$= E^0_{I,\alpha} \left\{ \hat{\mathbf{v}}_{I,\alpha} e^{i\mathbf{k}_{I,\alpha} \cdot \mathbf{r}} \right.$$

$$\left. + \hat{\mathbf{v}}_{R,\alpha} \frac{r_\alpha(\theta) + \left[t_\alpha(\theta)^2 - r_\alpha(\theta)^2 \right] m_\alpha e^{i2\delta}}{1 - r_\alpha(\theta) m_\alpha e^{i2\delta}} e^{-i2\delta} e^{i\mathbf{k}_{R,\alpha} \cdot \mathbf{r}} \right\}.$$

(2.175)

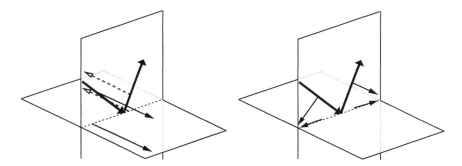

Figure 2.10 The diagram on the left represents the reflection on the perfectly reflecting plate of a plane wave whose electric field is perpendicular to the plane of incidence. If you follow this wave along, you see from the diagram that to satisfy the boundary conditions (Appendix B), the electric field must invert its direction relative to the wave on reflection: Before reflection, the electric field is pointing to the right side of the plane of incidence (for someone facing the direction of propagation); after reflection it points to the left side of the incidence plane. The diagram on the right shows that when the electric field is on the plane of incidence, its projection on the surface of the plate changes direction on reflection (as demanded by the boundary conditions) without the electric field changing direction relative to the wave.

We take the amplitude of the incident positive-frequency plane-wave component to be [169]

$$E^0_{I,\alpha} = \mathcal{E}_{\text{vac}}(\omega)a_\alpha(\mathbf{k}), \tag{2.176}$$

where $a_\alpha(\mathbf{k})$ is the annihilation operator of the mode with wave vector $\mathbf{k}$ and polarization α, and

$$\mathcal{E}_{\text{vac}}(\omega) = \frac{\sqrt{\hbar\omega}}{\pi} \tag{2.177}$$

is the modified vacuum field strength.[41] We obtain the positive-frequency part of the electric field, summing up all the modified plane waves, by varying the direction and polarization of the incident plane wave,

$$\mathbf{E}^+_\beta(\mathbf{r}) = \int' d^3k \left\{ \boldsymbol{\mathcal{E}}^\beta_\perp(\mathbf{r},\mathbf{k})a_\perp(\mathbf{k}) + \left[\boldsymbol{\mathcal{E}}^\beta_{\|,1}(\mathbf{r},\mathbf{k}) + \boldsymbol{\mathcal{E}}^\beta_{\|,2}(\mathbf{r},\mathbf{k})\right] a_\|(\mathbf{k})\right\}, \tag{2.178}$$

where the integration is restricted to half space (i.e., positive k_z), because of the perfect plate at $z = 0$, and the annihilation and creation operators obey the usual continuum commutation relations

$$[a_\alpha(\mathbf{k}), a_{\alpha'}(\mathbf{k}')] = 0 \quad \text{and} \quad [a_\alpha(\mathbf{k}), a^\dagger_{\alpha'}(\mathbf{k}')] = \delta_{\alpha'\alpha}\delta(\mathbf{k}' - \mathbf{k}). \tag{2.179}$$

---

[41]There is an extra factor 2 in comparison to the free-space case because of the perfect plate at $z = 0$ that eliminates half the space [35, 169, 448, 495].

The label β stands either for *cav*, the field between the plates, or for *out*, the field outside. The mode functions obtained by our scattering method, also known as the *modes of the universe,* are

$$\mathcal{E}_\perp^{cav}(\mathbf{r}, \mathbf{k}) = i\hat{z} \wedge \hat{\kappa}\mathcal{E}_{vac}e^{i\boldsymbol{\kappa}\cdot\mathbf{r}}\mathcal{L}_\perp \sin k_z z, \tag{2.180}$$

$$\mathcal{E}_{\|,1}^{cav}(\mathbf{r}, \mathbf{k}) = i\hat{\kappa}\mathcal{E}_{vac}e^{i\boldsymbol{\kappa}\cdot\mathbf{r}}\mathcal{L}_\| \cos\theta \sin k_z z, \tag{2.181}$$

$$\mathcal{E}_{\|,2}^{cav}(\mathbf{r}, \mathbf{k}) = -\hat{z}\mathcal{E}_{vac}e^{i\boldsymbol{\kappa}\cdot\mathbf{r}}\mathcal{L}_\| \sin\theta \cos k_z z, \tag{2.182}$$

$$\mathcal{E}_\perp^{out}(\mathbf{r}, \mathbf{k}) = i\hat{z} \wedge \hat{\kappa}\mathcal{E}_{vac}e^{i\boldsymbol{\kappa}\cdot\mathbf{r}}\mathcal{L}_\perp \left[\sin k_z z + i2\frac{r_\perp(\theta)}{t_\perp(\theta)} \sin\delta \sin(k_z z - \delta)\right], \tag{2.183}$$

$$\mathcal{E}_{\|,1}^{out}(\mathbf{r}, \mathbf{k}) = i\hat{\kappa}\mathcal{E}_{vac}e^{i\boldsymbol{\kappa}\cdot\mathbf{r}}\mathcal{L}_\| \left[\sin k_z z - i2\frac{r_\|(\theta)}{t_\|(\theta)} \sin\delta \sin(k_z z - \delta)\right] \cos\theta, \tag{2.184}$$

$$\mathcal{E}_{\|,2}^{out}(\mathbf{r}, \mathbf{k}) = -\hat{z}\mathcal{E}_{vac}e^{i\boldsymbol{\kappa}\cdot\mathbf{r}}\mathcal{L}_\| \left[\cos k_z z - i2\frac{r_\|(\theta)}{t_\|(\theta)} \sin\delta \cos(k_z z - \delta)\right] \sin\theta, \tag{2.185}$$

where the functions

$$\mathcal{L}_\alpha(\mathbf{k}) \equiv \frac{t_\alpha(\theta)}{1 - m_\alpha \, r_\alpha(\theta) \, \exp(i2\delta)} \tag{2.186}$$

describe the resonances of the parallel-plate cavity.

Now let us take the limit of perfect reflectivity. In this limit $r_\alpha \to m_\alpha$ and $t_\alpha \to 0$. Then we can easily see that the field outside the parallel-plate cavity is given by

$$\lim_{\substack{t_\perp \to 0 \\ t_\| \to 0}} \mathbf{E}_{out}^+(\mathbf{r}) = \int d^3k \, \mathcal{E}_{vac} \, e^{i\boldsymbol{\kappa}\cdot\mathbf{r}}e^{-i\delta}\bigg\{ i\hat{z} \wedge \hat{\kappa}a_\perp(\mathbf{k}) \sin(k_z z - \delta)$$

$$+ \Big[i\hat{\kappa}\cos\theta \sin(k_z z - \delta) - \hat{z}\sin\theta \cos(k_z z - \delta)\Big]a_\|(\mathbf{k}) \bigg\}. \tag{2.187}$$

This limit is more subtle for the field between the plates. If you are not careful, you might think that the field will vanish because the transmissivity in the nominator of $\mathcal{L}_\alpha$ vanishes. But $\mathcal{L}_\alpha$ is not what it seems at first sight. Remember that we mentioned above that it describes the cavity resonances. Notice that

$$|\mathcal{L}_\alpha|^2 = \frac{[1 - |t_\alpha|^2/2 + \sqrt{1 - |t_\alpha|^2} \, m_\alpha \cos(2\delta + \delta_\alpha)]|t_\alpha|^2/2}{(1 - |t_\alpha|^2)\sin^2(2\delta + \delta_\alpha) + |t_\alpha|^4/4}, \tag{2.188}$$

where we have used that $|r_\alpha|^2 + |t_\alpha|^2 = 1$ (i.e., the plate does not absorb any radiation) and called δ_α the phase of r_α [i.e., $r_\alpha = |r_\alpha| \exp(\delta_\alpha)$]. In the limit of perfect reflectivity,[42]

$$\lim_{t_\alpha \to 0} |\mathcal{L}_\alpha|^2 = \lim_{t_\alpha \to 0} \frac{[1 + m_\alpha \cos(2\delta + \delta_\alpha)]\,|t_\alpha|^2/2}{\sin^2(2\delta + \delta_\alpha) + |t_\alpha|^4/4}$$

$$= \pi\Big\{1 + m_\alpha \cos(2\delta + \Delta_\alpha)\Big\}\delta\big(\sin(2\delta - \Delta_\alpha)\big), \qquad (2.189)$$

where Δ_α is the perfect reflectivity limit of δ_α, which is π for $\alpha = \perp$ and 0 for $\alpha = \parallel$. From (2.189) it follows that (see Problem 2.10)

$$\lim_{t_\alpha \to 0} |\mathcal{L}_\alpha|^2 = \frac{\pi}{l} \sum_{n=-\infty}^{\infty} \delta\left(k_z - \frac{\pi}{l}n\right). \qquad (2.190)$$

To make $|\mathcal{L}_\alpha|^2$ appear explicitly in the expression of the cavity field, we define a new operator $A_\alpha(\mathbf{k})$ such that

$$a_\alpha(\mathbf{k}) = \mathcal{L}_\alpha^* A_\alpha(\mathbf{k}). \qquad (2.191)$$

We can determined the commutation relations obeyed by $A_\alpha(\mathbf{k})$ substituting (2.191) in (2.179). The first commutation relation (2.179) tells us that $A_\alpha(\mathbf{k})$ commutes with $A_{\alpha'}(\mathbf{k}')$. The second commutation relation (2.179) yields

$$\Big[A_\alpha(\mathbf{k}), A_{\alpha'}^\dagger(\mathbf{k}')\Big]\,\mathcal{L}_\alpha^*(\mathbf{k})\mathcal{L}_{\alpha'}(\mathbf{k}') = \delta_{\alpha'\alpha}\delta(\mathbf{k}' - \mathbf{k}). \qquad (2.192)$$

Multiplying both sides by $\mathcal{L}_\alpha(\mathbf{k})\mathcal{L}_{\alpha'}^*(\mathbf{k}')$ and taking the limit $t_\alpha \to 0$, we find that

$$\frac{\pi}{l}\sum_{n,n'} \delta\,(k_z - k_{z,n})\,\delta\,(k_z' - k_{z,n'}) \Big[A_\alpha(\mathbf{k}), A_{\alpha'}^\dagger(\mathbf{k}')\Big]$$

$$= \delta_{\alpha'\alpha}\sum_{m} \delta\,(k_z - k_{z,m})\,\delta\,(\mathbf{k}' - \mathbf{k})\,, \qquad (2.193)$$

where $k_{z,n} \equiv n\pi/l$. If we now integrate equation (2.193) over k_z from $k_{z,n} - \epsilon$ to $k_{z,n} + \epsilon$ and over k_z' from $k_{z,n'} - \epsilon$ to $k_{z,n'} + \epsilon$, with ϵ being small enough to include only $k_{z,n}$ and $k_{z,n'}$, we obtain

$$\Big[A_\alpha\,(k_{z,n}\hat{\mathbf{z}} + \boldsymbol{\kappa})\,, A_{\alpha'}^\dagger\,(k_{z,n'}\hat{\mathbf{z}} + \boldsymbol{\kappa})\Big]$$

$$= \frac{l}{\pi}\delta_{\alpha'\alpha}\delta(\boldsymbol{\kappa}' - \boldsymbol{\kappa}) \int_{k_{z,n'}-\epsilon}^{k_{z,n'}+\epsilon} dk_z'\,\delta\,(k_z' - k_{z,n})$$

$$= \frac{l}{\pi}\delta_{\alpha'\alpha}\delta(\boldsymbol{\kappa}' - \boldsymbol{\kappa})\,\delta_{n',n}. \qquad (2.194)$$

---

[42]See equation (F.7) in Appendix F.

So, apart from a normalization factor of $\sqrt{\pi/l}$, the operator $A_\alpha\left(k_{z,n}\hat{z}+\kappa\right)$ behaves as a continuum annihilation operator in κ and as a discrete annihilation operator in k_z. Then let us we define

$$a_{\alpha,n}(\kappa) \equiv \sqrt{\frac{\pi}{l}} A_\alpha\left(k_{z,n}\hat{z}+\kappa\right), \tag{2.195}$$

where

$$\left[a_{\alpha,n}(\kappa), a^\dagger_{\alpha',n'}(\kappa')\right] = \delta_{\alpha'\alpha}\delta_{n',n}\,\delta(\kappa'-\kappa) \quad \text{and} \quad [a_{\alpha,n}(\kappa), a_{\alpha',n'}(\kappa')] = 0. \tag{2.196}$$

Thus, in the limit of perfect reflectivity, the positive-frequency part of the electric field in the cavity is given by

$$\lim_{\substack{t_\perp \to 0 \\ t_\parallel \to 0}} \mathbf{E}^+_{\text{cav}}(\mathbf{r}) = i\sqrt{\frac{\pi}{l}} \sum_{n=0}^{\infty} \int d^2\kappa\, \mathcal{E}_{\text{vac}} e^{i\boldsymbol{\kappa}\cdot\mathbf{r}} \left[\hat{z}\wedge\hat{\kappa}\, a_{\perp,n}(\kappa)\, \sin k_{z,n} z \right.$$
$$\left. + \left(\hat{\kappa}\frac{k_{z,n}}{k}\sin k_{z,n}z + i\hat{z}\frac{\kappa}{k}\cos k_{z,n}z\right)a_{\parallel,n}(\kappa)\right]. \tag{2.197}$$

The corresponding magnetic fields are given by (Problem 2.11)

$$\lim_{\substack{t_\perp \to 0 \\ t_\parallel \to 0}} \mathbf{B}^+_{\text{cav}}(\mathbf{r}) = \sqrt{\frac{\pi}{l}} \sum_{n=0}^{\infty} \int d^2\kappa\, \mathcal{E}_{\text{vac}} e^{i\boldsymbol{\kappa}\cdot\mathbf{r}} \left[\hat{z}\wedge\hat{\kappa}\, \cos(k_{z,n}z)\, a_{\parallel,n}(\kappa) \right.$$
$$\left. + i\left(\hat{z}\frac{\kappa}{k}\sin k_{z,n}z + i\hat{\kappa}\frac{k_{z,n}}{k}\cos k_{z,n}z\right)a_{\perp,n}(\kappa)\right] \tag{2.198}$$

and

$$\lim_{\substack{t_\perp \to 0 \\ t_\parallel \to 0}} \mathbf{B}^+_{\text{out}}(\mathbf{r}) = \int' d^3k\, \mathcal{E}_{\text{vac}} e^{i\boldsymbol{\kappa}\cdot\mathbf{r}} e^{-i\delta} \left\{\hat{z}\wedge\hat{\kappa}\cos(k_z z - \delta)\, a_\parallel(\mathbf{k}) \right.$$
$$\left. + i\left[\hat{z}\sin\theta\sin(k_z z - \delta) + i\hat{\kappa}\cos\theta\cos(k_z z - \delta)\right]a_\perp(\mathbf{k})\right\}. \tag{2.199}$$

Expressions equivalent to these were obtained by a number of authors, including [293].

Now we are ready to calculate the Casimir pressure. From here onward it is understood that all fields are taken in the limit of perfect reflectivity so there is no need to use the cumbersome $\lim_{t_\perp, t_\parallel \to 0}$ notation any more.

The force on the plate at l due to the radiation pressure of the vacuum fluctuations is given by the expectation value in the vacuum state of

$$\mathbf{F} = \sum_{n=1}^{3} \hat{\mathbf{x}}_n \int_{-\infty}^{\infty} dx \int_{-\infty}^{\infty} dy\, \{\mathcal{T}_{n3}(x,y,l_+) - \mathcal{T}_{n3}(x,y,l_-)\}, \tag{2.200}$$

where l_+ indicates that T_{n3} is evaluated on the side of the plate facing the outside and l_- that it is evaluated on the side that faces the other plate, $\hat{\mathbf{x}}_1 \equiv \hat{\mathbf{x}}$, $\hat{\mathbf{x}}_2 \equiv \hat{\mathbf{y}}$, $\hat{\mathbf{x}}_3 \equiv \hat{\mathbf{z}}$, and

$$T_{ij} = \frac{1}{4\pi} \left\{ E_i E_j + B_i B_j - \frac{\delta_{ij}}{2}(\mathbf{E}^2 + \mathbf{B}^2) \right\} \tag{2.201}$$

is the Maxwell stress tensor. As the pressure is the force per unit area, it is given at a point (x, y) on the plate by

$$\mathbf{P}_{\text{cas}}(x, y) = \sum_{n=1}^{3} \hat{\mathbf{x}}_n \langle T_{n3}(x, y, l_-) - T_{n3}(x, y, l_+) \rangle. \tag{2.202}$$

To calculate the vacuum expectation value, note that as number states of the continuum are orthogonal to different number states, only terms containing products of annihilation operators on the left and creation operators on the right for the *same mode* and in *this precise order* will survive. All other terms will vanish. The calculation of (2.202) is quite straightforward, then, and I leave it as an exercise to the reader (Problem 2.12), quoting only the result here:

$$\mathbf{P}_{\text{cas}}(x, y) = \frac{\hat{\mathbf{z}}}{4\pi} \left\{ \frac{\pi}{l} \sum_{n=0}^{\infty} \int d^2\kappa \left(\frac{k_{z,n}}{k} \right)^2 \mathcal{E}_{\text{vac}}^2 - \int d^3k\, \mathcal{E}_{\text{vac}}^2 \cos^2\theta \right\}. \tag{2.203}$$

Using cylindrical coordinates and changing the radial variable from $\kappa = \sqrt{k_x^2 + k_y^2}$ to $\kappa l/\pi$, we can rewrite the first term inside the braces on the right-hand side of (2.203) as

$$\frac{\pi}{l} \sum_{n=0}^{\infty} \int d^2\kappa \left(\frac{k_{z,n}}{k} \right)^2 \mathcal{E}_{\text{vac}}^2 = \frac{\pi^3 \hbar c}{l^4} \sum_{n=0}^{\infty} n \int_0^\infty \frac{dv}{\sqrt{v + n^2}}, \tag{2.204}$$

where $v \equiv (\kappa l/\pi)^2$. Using cylindrical coordinates and changing the variable $\kappa = \sqrt{k_x^2 + k_y^2}$ to $\kappa l/\pi$ and k_z to $w \equiv k_z l/\pi$, we can rewrite the second term inside the braces on the right-hand side of (2.203) as

$$\int d^3k\, \mathcal{E}_{\text{vac}}^2 \cos^2\theta = \frac{\pi^3 \hbar c}{l^4} \int_0^\infty dw\, w^2 \int_0^\infty \frac{dv}{\sqrt{v + w^2}}, \tag{2.205}$$

where again $v \equiv (\kappa l/\pi)^2$. Thus, introducing the cutoff, we can rewrite (2.203) as

$$\mathbf{P}_{\text{cas}}(x, y) = \hat{\mathbf{z}} \frac{\pi^2 \hbar c}{4l^4} \lim_{a \to 0} \left\{ \sum_{n=0}^{\infty} n \int_0^\infty \frac{dv}{\sqrt{v + n^2}} - \int_0^\infty dw\, w^2 \int_0^\infty \frac{dv}{\sqrt{v + w^2}} \right\}. \tag{2.206}$$

Using the method of Section 2.5.1, we can easily show (Problem 2.13) that

$$\mathbf{P}_{\text{cas}}(x, y) = -\hat{\mathbf{z}} \frac{\pi^2 \hbar c}{240 l^4}, \tag{2.207}$$

which confirms the result of Section 2.5.1, showing that the pressure on the plate at $z = l$ is directed toward the other plate (i.e., it is attractive) and has a magnitude of $\pi^2 \hbar c / 240 l^4$.

2.5.3 The vacuum catastrophe: Vacuum zero-point energy and gravity

As in most of physics only energy differences are important, one frequently eliminates the zero-point energy by shifting the origin of energy [i.e., by adding a constant (but infinite) term to the Hamiltonian of the electromagnetic field]. In general relativity, however, one cannot shift the origin of energy, as according to Einstein, energy is equivalent to mass and must generate a gravitational field. So not only the electromagnetic zero-point energy, but the zero-point energies of all the fundamental boson fields of nature (including that of the graviton if it really exists) should give rise to a gravitational field. Interestingly, fermion fields would tend to cancel this effect, as their zero-point energy is negative. Some versions of supersymmetry that predict a fermion partner for every boson would then lead to no gravitational effect [5]. But despite intense experimental effort, no supersymmetry partners of the known elementary particles have been observed yet. If they do exist at large energy scales than are now accessible, the effect would be the same as that of a cutoff to the zero-point spectrum at these energies.

In empty space, the electromagnetic vacuum zero-point energy is uniform. The same can be assumed for the zero-point energies of the other fields in nature. What kind of observable gravitational effect would such uniform energy distribution produce? As a simple example, we can consider the effect of this uniform density on the motion of the outer planets of our solar system [5]. Unlike the inner planets, which are significantly affected by the other planets (and in the case of our Earth, by the Moon), the motion of the outer planets is dominated by the Sun.

Now a uniform density outside a sphere does not produce any gravitational force inside. So for a planet undergoing a *circular* orbit of radius r centered around the Sun, only the zero-point energy inside a sphere of radius r will contribute to the total gravitational force pulling it toward the Sun. Moreover, the resulting gravitational force will be the same as if the mass equivalent to this energy would be all located at the Sun. Assuming that astronomical observations agree with Newton's theory of gravity, any gravitational force due to the vacuum zero-point energy must be very small in comparison with that due to the Sun. An estimate using the average orbital radius of Pluto yields [5] an upper bound for the total zero-point energy density of the vacuum of about 10^{18} GeV/m^3.

Although the simple estimation above give us a pedagogical example of an observable effect of a uniform mass density filling the whole of space, there are several problems with it. First, the orbit of Pluto is actually very far from being circular. At its closest approach, Pluto is less than 30 astronomical units[43] from the Sun, whereas half an orbit later, Pluto is almost 50 astronomical units from the Sun. Moreover, it is well known in the literature that there are unexplained irregularities in the orbit of Pluto. In fact, we can get a *lower*[44] bound to the cosmological constant this way (i.e.,

---

[43]The *astronomical unit* is the average distance between the Sun and Earth (i.e., about 149,597,870 km).
[44]But this lower bound is higher than the upper bound that can be estimated from other observations.

calculating how large the cosmological constant must be in order to explain these irregularities) [523]. A more rigorous upper bound than that found in the preceding paragraph was obtained by Anderson et al. [21] using data from the *Voyager* spacecraft to calculate the fractional change in period of Uranus and Neptune. Their analysis shows that the energy density of the vacuum must be smaller than 10^{14} GeV/m^3.

To estimate the gravitational effect of the electromagnetic zero-point energy predicted by theory, we can adopt the Planck energy as a cutoff. This is the energy at which the gravitational interaction becomes as strong as the other three fundamental forces of nature (i.e., the scale at which we expect the current theory to break down). This energy is about 10^{19} GeV. This yields a zero-point energy density of about 10^{121} GeV/m^3; that is, the disagreement between this and the experimental upperbound derived in the preceding paragraph is of 107 orders of magnitude: the largest disagreement between theory and experiment ever seen in the history of physics! In analogy with the ultraviolet catastrophe of blackbody radiation that lead Planck to introduce his constant in physics, this has been called the *vacuum catastrophe.*

In 1967, Zeldovich [663] pointed out that the vacuum zero-point energy could account for the cosmological constant [483, 629]. The cosmological constant is a term that Einstein introduced to his field equations in 1917 when he was trying to apply them to the entire universe. At that time it was generally believed that the universe would be static. Without this constant term to counterbalance the gravitational attraction of matter, the universe would collapse. The connection between the vacuum zero-point energy and the cosmological constant is particularly relevant today, because in 1998 new astronomical observations revealed that the expansion of the universe is actually accelerating[45] [482, 519, 538]. These observations point to a vacuum zero-point energy of about 70% of the total energy density of the universe. Unfortunately, this is still too low to be accounted for by the current fundamental field theory. Some researchers consider the vacuum catastrophe a sign of the breakdown of our current physical theories, something that could only be resolved within the context of a bigger theory, where gravity and quantum mechanics would be unified [27].

RECOMMENDED READING

- To review the role of Poisson brackets in classical mechanics, see Goldstein's book [240]. You might also like to read Lanczos's interesting historical account [385] of the origins of the Poisson bracket and its introduction in quantum mechanics, as well as have a look at [62] and [143].

- Our introduction to canonical quantization followed closely Dirac's own approach in [159]. For an account of how Dirac hit upon the idea of canonical quantization, there is nothing better than getting it straight "from the horse's mouth" in [158].

---

[45]See [357] for a recent review.

- If you are interested in knowing how canonical quantization can be applied to systems with constraints, check out [160].

- For a discussion on the shortcomings of Dirac's formalism in quantum mechanics, see [231].

- Even though position was represented by an operator in Section 2.1, time was only a parameter. There have been attempts to promote time to an operator, too [10, 85]. But the modern approach in quantum field theory is to make position also a parameter, as in Section 2.4. The lack of a time operator implies that the time–energy uncertainty relation does not have the same status as the position–momentum uncertainty relation in nonrelativistic quantum mechanics, for it cannot be derived from fundamental commutation relations [649]. We will get back to this issue in Chapter 3.

- Some of the key papers on the old quantum theory are collected in [591]. Key papers on QED can be found in [544]. For the early history of QED and a nice description of Born and Jordan's seminal work [68], see [447]. Key papers from the early days of quantum optics can be found in [359]. The latter reference also has a very pedagogical introduction to field quantization.

- Two very good introductions to Feynman's version of quantum mechanics are the marvelous book by Feynman and Hibbs [204] and the very interesting paper [430]. For a popular version of Feynman's approach to QED, see [201] (another interesting popular book related to this chapter is Podolny's book about the vacuum [493]). For a more detailed account of Feynman's approach to QED, see [203].

- If you would like to know more about Green's functions, reference [642] gives a brief elementary account of them. For a more complete account, see Barton's excellent book [40].

- Chapter 22 of [206] shows how cavities can be seen as the high-frequency counterparts of ordinary electronic lumped-circuit oscillators (e.g., LC oscillators). For a detailed calculation of perfect cavity modes, see Appendix A and also Chapter 2 of Haus's book [276].

- The formal analogy between Poisson brackets and commutators can also be used to do calculations in classical electrodynamics very similar to the way they are done in quantum electrodynamics (i.e., with mode expansions and using the harmonic oscillator Poisson brackets satisfied by the expansion coefficients). This *field oscillator* approach is described and applied in [90, 233].

- If you have trouble following the various manipulations of vector expressions involving ∇ in Section 2.2 and in Appendix A, have a look at Appendix E.

- For more on the delta function, see Appendix F and the references mentioned there.

- Milonni's book [451] offers a quite comprehensive pedagogical account of the Casimir effect, including a discussion of the curious model that Casimir proposed for the electron. Most of the important literature on the Casimir force is listed in [377], with key advances discussed in detail in the recent review [66]. For a recent measurement of the Casimir force, see [97]. For a calculation of the fluctuations around the expected value of the Casimir force (i.e., variances), see the chapter by Barton in [49].

- For more on the vacuum catastrophe, see [5, 316, 483, 511, 629].

Problems

2.1   Do the Fourier integral (2.66) using (2.67) with the approximation (2.69). This integral is wellsuited for contour integration. Note that for $t - t' < 0$, the contour must be closed on the lower half of the complex plane not to get any contribution from the semicircle arch when its radius is made infinitely large. The converse holds for $t - t' > 0$.

2.2   Calculate how the pendulum oscillates with the driving force given by (2.71), which is switched on at $t = 0$ and off at $t = T$. Use the Green's function method (2.64), with the approximate Green's function (2.70).

2.3   Show that the solution of (2.75) with the driving force $F_\epsilon(t) = A \exp(\epsilon t) \sin \omega t$ is given by (2.79). Start from equation (2.77) and note that its solution is simply

$$\mathcal{Y}_\epsilon(k, t) = \frac{A}{i4\pi} e^{-ikx_0} \left\{ \frac{e^{i(\omega - i\epsilon)t}}{v^2 k^2 - (\omega - i\epsilon)^2} - \frac{e^{-i(\omega + i\epsilon)t}}{v^2 k^2 - (\omega + i\epsilon)^2} \right\}.$$

Then substitute in (2.76) and do the integral by contour integration.

2.4   Redo Problem 2.3 with the delta function in (2.75) replaced by the Lorentzian function

$$\Delta(x - x_0) = \frac{1}{\pi} \frac{\Lambda}{(x - x_0)^2 + \Lambda^2}.$$

Show that $y(x)$ remains finite at $x = x_0$ even in the limit of $\Lambda \to 0^+$, in which the Lorentzian $\Delta(x - x_0)$ turns into a delta function (F.7) and we recover (2.79).

2.5   Solve (2.81) and substitute the solution back in (2.80) to obtain $y_\epsilon(x, t)$ for the finite string.

2.6   Work out the Fourier series for $\cos ku$ with $-L < x < L$. Use this result to sum into a closed-form expression the series solution obtained in Problem 2.5. Show that the closed-form solution is given by

$$y_\epsilon(x, t) = \begin{cases} (A/2v)\Im\left[f(x + x_0) - f(x - x_0 + L)\right] & \text{for } -L < x - x_0 < 0, \\ (A/2v)\Im\left[f(x + x_0) - f(x - x_0 - L)\right] & \text{for } 0 < x - x_0 < L, \end{cases}$$

where

$$f(\xi) \equiv \frac{e^{i(\omega - i\epsilon)t} \cos\left([\omega - i\epsilon]\xi/v\right)}{(\omega - i\epsilon) \sin\left([\omega - i\epsilon]L/v\right)}$$

and $\Im(z)$ stands for the imaginary part of the complex number z [e.g., if $z = a + ib$, $\Im(z) = b$].

2.7   Take the limit $\epsilon \to 0$ of the result of Problem 2.6 and show that $y(x, t)$ is given by (2.83).

2.8   Go back to the result of Problem 2.6 and take the limit $L \to \infty$, and only afterward, the limit $\epsilon \to 0$. Show that this recovers (2.79).

2.9   For the cavity shown in Figure 2.7, the boundary conditions (see Appendix B) are given by

$$\hat{z} \wedge \mathbf{E}(x, y, l, t) = \mathbf{0},$$
$$-\hat{z} \wedge \mathbf{E}(x, y, 0, t) = \mathbf{0},$$
$$\hat{y} \wedge \mathbf{E}(x, L, z, t) = \mathbf{0},$$
$$-\hat{y} \wedge \mathbf{E}(x, 0, z, t) = \mathbf{0}$$
$$\hat{x} \wedge \mathbf{E}(L, y, z, t) = \mathbf{0},$$
$$-\hat{x} \wedge \mathbf{E}(0, y, z, t) = \mathbf{0}.$$

Using the ansatz $\chi = \hat{x} f_x(x, y, z) + \hat{y} f_y(x, y, z) + \hat{z} f_z(x, y, z)$, where

$$f_x(x, y, z) = \mathcal{N}_x \cos k_x x \sin k_y y \sin k_z z,$$
$$f_y(x, y, z) = \mathcal{N}_y \sin k_x x \cos k_y y \sin k_z z,$$
$$f_z(x, y, z) = \mathcal{N}_z \sin k_x x \sin k_y y \cos k_z z$$

show that these boundary conditions and (A.10) imply that $k_x = n_x \pi / L$, $k_y = n_y \pi / L$, and $k_z = n_z \pi / l$, where n_x, n_y, and n_z are integers. Now use the divergence condition $\nabla \cdot \chi = 0$ to show that if any one of k_x, k_y, and k_z vanishes, there is only one polarization, but otherwise there are two polarizations. See Sec. 10.2 of [537] for a nice pedagogical account of the modes in a box.

2.10   Use the well-known delta function identity

$$\int dx\, g(x) \delta\big(f(x)\big) = \frac{g(x)}{|df/dx|}\bigg|_{f(x)=0},$$

$\delta = k_z l$, $\Delta_\perp = \pi$, and $\Delta_\| = 0$ to obtain (2.190) from (2.189).

2.11   Derive the expressions (2.198) and (2.199) for the positive-frequency part of the quantized magnetic fields from the corresponding expressions (2.187) and (2.197) for the electric fields using Faraday's law. The calculation becomes quite simple, once you notice that the positive-frequency components (2.187) and (2.197) of the electric field have the general form

$$\mathbf{E}_\beta^+(\mathbf{r}) = \int d^2\kappa \, \mathcal{E}_{\text{vac}} e^{i\boldsymbol{\kappa}\cdot\mathbf{r}} \mathbf{U}_\beta^+(\boldsymbol{\kappa}, z),$$

so that

$$\mathbf{B}_\beta^+(\mathbf{r}) = \int d^2\kappa \, \mathcal{E}_{\text{vac}} e^{i\boldsymbol{\kappa}\cdot\mathbf{r}} \left(-\frac{i}{k}\right) \left\{ i\boldsymbol{\kappa} + \hat{\mathbf{z}}\frac{\partial}{\partial z} \right\} \wedge \mathbf{U}_\beta^+(\boldsymbol{\kappa}, z).$$

2.12   Using (2.187), (2.197), (2.198), and (2.199), show that

$$\langle (\hat{\mathbf{x}}_n \cdot \mathbf{E}_{\text{cav}})(\hat{\mathbf{x}}_3 \cdot \mathbf{E}_{\text{cav}}) \rangle = \delta_{n3} \frac{\pi}{l} \sum_{m=0}^{\infty} \int d^2\kappa \, \mathcal{E}_{\text{vac}}^2 \left(\frac{\kappa}{k}\right)^2 \cos^2 k_{z,m} z,$$

$$\langle (\hat{\mathbf{x}}_n \cdot \mathbf{B}_{\text{cav}})(\hat{\mathbf{x}}_3 \cdot \mathbf{B}_{\text{cav}}) \rangle = \delta_{n3} \frac{\pi}{l} \sum_{m=0}^{\infty} \int d^2\kappa \, \mathcal{E}_{\text{vac}}^2 \left(\frac{\kappa}{k}\right)^2 \sin^2 k_{z,m} z,$$

$$\langle \mathbf{E}_{\text{cav}}^2 \rangle = \frac{\pi}{l} \sum_{m=0}^{\infty} \int d^2\kappa \, \mathcal{E}_{\text{vac}}^2 \left\{ \left[1 + \frac{k_{z,m}^2}{k^2}\right] \sin^2 k_{z,m} z + \left(\frac{\kappa}{k}\right)^2 \cos^2 k_{z,m} z \right\},$$

$$\langle \mathbf{B}_{\text{cav}}^2 \rangle = \frac{\pi}{l} \sum_{m=0}^{\infty} \int d^2\kappa \, \mathcal{E}_{\text{vac}}^2 \left\{ \left[1 + \frac{k_{z,m}^2}{k^2}\right] \cos^2 k_{z,m} z + \left(\frac{\kappa}{k}\right)^2 \sin^2 k_{z,m} z \right\},$$

$$\langle (\hat{\mathbf{x}}_n \cdot \mathbf{E}_{\text{out}})(\hat{\mathbf{x}}_3 \cdot \mathbf{E}_{\text{out}}) \rangle = \delta_{n3} \int' d^3k \, \mathcal{E}_{\text{vac}}^2 \sin^2\theta \cos^2(k_z z - \delta),$$

$$\langle (\hat{\mathbf{x}}_n \cdot \mathbf{B}_{\text{out}})(\hat{\mathbf{x}}_3 \cdot \mathbf{B}_{\text{out}}) \rangle = \delta_{n3} \int' d^3k \, \mathcal{E}_{\text{vac}}^2 \sin^2\theta \sin^2(k_z z - \delta),$$

$$\langle \mathbf{E}_{\text{out}} \rangle = \int' d^3k \, \mathcal{E}_{\text{vac}}^2 \left\{ [1 + \cos^2\theta] \sin^2(k_z z - \delta) + \sin^2\theta \cos^2(k_z z - \delta) \right\},$$

$$\langle \mathbf{B}_{\text{out}} \rangle = \int' d^3k \, \mathcal{E}_{\text{vac}}^2 \left\{ [1 + \cos^2\theta] \cos^2(k_z z - \delta) + \sin^2\theta \sin^2(k_z z - \delta) \right\},$$

where $\mathbf{E}_\beta = \mathbf{E}_\beta^+ + \mathbf{E}_\beta^-$ and $\mathbf{B}_\beta = \mathbf{B}_\beta^+ + \mathbf{B}_\beta^-$. Then substitute these expressions in (2.202) using (2.201) and derive (2.203).

2.13   Use the Euler–Maclaurin formula, as was done in Section 2.5.1, to show that (2.206) yields (2.207).

2.14   Using (2.175), show that the multiple reflections between the plates transmitted to the outside exactly cancel the reflection of the incident wave by the plate at $z = l$, when $k_z = n\pi/l$ and $t_\alpha \to 0$. In other words, at the cavity resonances there is no reflected wave, just as Michael Berry suggested [50].

3

The alternative free tasting:
First quantization of light
and the photon's
wavefunction

We shall begin by developing an extreme light quantum theory, and forget for the time all connection with Maxwell equations. . . . Our present problem is to find the wave equation for the de Broglie waves of the quantum; the theory of the field we shall then obtain by a suitable (second) quantization of the de Broglie amplitudes.

—Julius Robert Oppenheimer [472]

Welcome to our free tasting of alternative wines. So far we have presented the traditional route to QED. From the classical Maxwell equations, canonical quantization takes us straight to many-body (i.e., second-quantized) quantum electrodynamics. This is quite a shortcut on the usual route taken in elementary quantum mechanics of nonrelativistic point particles, where one first has to construct a single-particle first-quantized theory (e.g., Schrödinger equation) from which to build the many-body second-quantized theory. Why not follow the ordinary longer route from first to second quantization for electromagnetic radiation as well? The main problem is that first quantization is really good only in nonrelativistic quantum mechanics. Pair creation out of the vacuum implies that we always have a many-body problem in the relativistic case. Now, unlike matter, there is no nonrelativistic limit for radiation. The photon has no rest mass and it always travels at the speed of light. There is no inertial frame where a photon is at rest. As a consequence, any interaction with matter, even at very small energies, can lead to the creation of photons and hence the need for a many-body theory. To create an electron–positron pair, for example, one needs at least twice the rest energy of the electron (i.e., $2mc^2$, where m is the rest mass of the electron). Even though m is very small (about 9.1×10^{-31} kg) as c is large, $2mc^2$ is still a respectable amount of energy: about 1 MeV. Compare this to 1 eV, the energy of a photon of visible light produced in a typical spontaneous-emission event when an excited atom decays to its ground state.

But in this chapter, as in Chapter 2, we are not looking at the interaction with matter yet. What about a first-quantized theory of light without sources, then? There is still a problem. Another consequence of being a truly relativistic particle with zero rest mass is that the photon cannot be localized with infinite precision: There is no position operator for the photon![1] This means that it impossible to define a proper probability density for finding a photon at a given point in space, making the idea of a photon wavefunction in configuration space meaningless.

So the first quantization of light is plagued with problems. Still, this is an idea to which people return over and over again because of its appeal and the subtleness of its problems. It is very tempting, for instance, to imagine that the wavefunction of the photon would be the classical electromagnetic field itself. In this chapter we look into all these issues in more detail. In the next section we discuss the problem of measuring the position and momentum of a particle in the relativistic regime. We will see that the momentum, but *not* the position, can be determined with arbitrary precision just as in nonrelativistic quantum mechanics. There is a general group theory treatment by Newton and Wigner [464] about the localizability of elementary systems which predicts that there is no position operator for the photon. We do not present this treatment here because even though it is very general, it is also very abstract and mathematical. The curious reader can have a look at the references recommended at the end of the chapter. We chose instead to base our discussion on a concrete and specific case: What would happen if the photon had a rest mass? In the spirit of Oppenheimer's "extreme quantum theory of light" [472], we pretend that we know nothing about the classical limit of the theory and trace the reverse route to that of the usual quantization approaches: We start from a quantum wave equation and go from there to its classical limit. The key difference between our extreme quantum theory of light and Oppenheimer's is that we do not assume that the photon's rest mass vanishes. Our derivation of the photon's wave equation uses Lanczos's generalization of Dirac's method[2] to get from the Klein–Gordon equation to an equation that is first order in time. This leads us to Proca equations. The problems with a configuration space wavefunction become evident. Then we show how, when the rest mass does not vanish, the problems disappear in the nonrelativistic limit. But as the underlying logic of the present-day physical theory as well as experimental evidence seems to suggest that the photon has no rest mass, we take the limit of vanishing rest mass and write down a first-quantized theory whose classical limit is Maxwell equations. Finally, in the last section, we construct a second-quantized theory, starting from the first-quantized one. We show then that the second-quantized theory coincides with the usual quantum electrodynamics derived in Chapter 2 through canonical quantization.

---

[1]I mean, in the sense of the Newton–Wigner–Wightman group theory approach [464, 643].
[2]That is the method Dirac developed to derive the Dirac equation for the electron.

3.1 THE PROBLEM OF THE POSITION OPERATOR IN
RELATIVISTIC QUANTUM MECHANICS[3]

One of the consequences of having a maximum speed allowed in nature is that we can no longer assume that certain physical observables can be determined exactly. In particular, it turns out that although the momentum of a single particle can be determined with arbitrary accuracy, there is a limit to the accuracy that can be achieved in a measurement of position. This is deeply connected with the problem of pair creation out of the vacuum mentioned above. Let us now look more closely into these issues.

The natural way to measure the momentum of a particle is by doing a scattering experiment (i.e., we let it collide with a control object). Ideally, we would like to have a control object that behaves like a perfect mirror, bouncing the particle off itself. As this is a control object, we assume that we know its energy ϵ and momentum $\mathbf{p}$ before and after (denoted by primes) the collision. Momentum conservation applies rigorously, but energy will only be conserved within the limits prescribed by the time–energy uncertainty relation. This uncertainty relation has a status different from that of the position–momentum uncertainty relation as, unlike that relation, it is not derived from a commutator relation (see the recommended reading at the end of the chapter), and it has been often misinterpreted. It does *not* mean that the energy cannot be known exactly at a given time, nor that the energy cannot be measured with arbitrary accuracy within a fixed time interval. What it does mean is that the measurement makes the state after the measurement differ from that before the measurement. This difference causes an energy uncertainty in *both* the states before and after the measurement on the order of at least $\hbar/\Delta t$, where Δt is the time the measurement lasts [388].

So if we denote by $\mathbf{P}$ the momentum of the particle before the collision, by E its energy before the collision, and use the same symbols primed for these quantities after the collision, we can write the following conservation laws

$$\mathbf{P} + \mathbf{p} = \mathbf{P}' + \mathbf{p}', \tag{3.1}$$

$$|\Delta\epsilon + \Delta E - \Delta\epsilon' - \Delta E'| \geq \frac{\hbar}{\Delta t}. \tag{3.2}$$

Assuming that there is no uncertainty in $\mathbf{p}, \mathbf{p}'$, ϵ, and ϵ' (our macroscopic measurement apparatus), we can rewrite (3.1) and (3.2) as

$$\Delta\mathbf{P} = \Delta\mathbf{P}', \tag{3.3}$$

$$|\Delta E - \Delta E'| \geq \frac{\hbar}{\Delta t}. \tag{3.4}$$

Now if the rest mass of the particle is M_0 (it can vanish),

$$E = \sqrt{M_0^2 c^4 + P^2 c^2}, \tag{3.5}$$

---

[3]This section follows closely the approach of [388] and that of Sec. 2 of [480].

so that

$$\frac{\partial E}{\partial P_n} = \frac{P_n c^2}{E}. \qquad (3.6)$$

As $E = Mc^2$ and $P_n = Mv_n$, where M is the mass and v_n is the nth component of the particle's velocity, we find that

$$\frac{\partial E}{\partial P_n} = v_n. \qquad (3.7)$$

The same relation obviously holds after the collision, too. So if v and v' are the particle's speed before and after the collision, it follows from (3.3), (3.4), and (3.7) that

$$|v - v'|\Delta P \geq \frac{\hbar}{\Delta t}. \qquad (3.8)$$

According to relativity, $|v - v'|$ can at most be on the order of c, the speed of light. Then the accuracy with which the momentum of the particle can be determined must obey the uncertainty relation

$$\Delta P \geq \frac{\hbar}{c\,\Delta t}. \qquad (3.9)$$

Thus for arbitrarily long measuring times Δt, the momentum can be determined with arbitrary accuracy.[4]

For a position measurement, we can again use a scattering experiment. But this time we probe the particle with γ-rays instead of using a mirror. The γ-rays then enter a Heisenberg microscope[5] and the position of the particle can be determined by simple geometric optics. It is well known from optics that the inaccuracy Δx with which a position can be discriminated in this way cannot be smaller than $\lambda'/\sin(\alpha/2)$, where λ' is the wavelength the γ-rays after the collision[6] and α is the angle of the aperture of the objective. If we take the largest aperture possible (180 degrees), this inaccuracy cannot be smaller than λ'. So apparently the higher the energy of the γ-rays, the lower their wavelength and therefore the smaller this inaccuracy in the position measurement. Unfortunately, as we show next, even when the wavelength λ of γ-rays before the collision is made arbitrarily small, their wavelength λ' *after* the collision does not become arbitrarily small.

Let us assume that we know the single particle's energy before the measurement. Then, as we can in principle determine with arbitrary accuracy the energy of our probe before and after the measurement, there can be no uncertainty in the energy

---

[4]The uncertainty relation (3.9) also holds for the state after the measurement. In that case it can be derived very easily [388] from the position–momentum uncertainty relation $\Delta P \Delta X \geq \hbar$. Assuming that the particle was in a position eigenstate before the measurement, the uncertainty in its position after a time Δt will be $v' \Delta t$, where v' is its speed. As v' can be at most c, we find that the position–momentum uncertainty relation yields (3.9).

[5]This microscope will have special lenses for γ-rays that we do not know yet how to make but that we will assume have been produced somehow for the purposes of this gedanken experiment.

[6]Due to energy and momentum conservation (as in the Compton effect), the wavelengths before and after the collision are usually different.

of the single particle after the collision (asymptotic state), for this would violate the conservation of energy. So in this case we have the energy E and momentum $\mathbf{P}$ of the single particle, and the energy ϵ and momentum $\mathbf{p}$ of the γ-ray photon that collided with it are related to their primed counterparts after the collision by[7]

$$E + \epsilon = E' + \epsilon', \tag{3.10}$$

$$\mathbf{P} + \mathbf{p} = \mathbf{P}' + \mathbf{p}'. \tag{3.11}$$

Calling the rest mass of the single particle M_0, we have

$$\epsilon' = cp', \tag{3.12}$$

$$E' = c\sqrt{M_0^2 c^2 + P'^2}, \tag{3.13}$$

$$\epsilon = cp, \tag{3.14}$$

$$E = c\sqrt{M_0^2 c^2 + P^2}. \tag{3.15}$$

If the beam is very intense, $\mathbf{P}' = \mathbf{p} - \mathbf{p}'$. Now it is clear from simple geometry that the position can best be determined when the scattering angle is 90 degrees (i.e., when $\mathbf{p} \cdot \mathbf{p}' = 0$). Thus, let us assume that this is so. In this case, $E' = \sqrt{\epsilon^2 + \epsilon'^2 + M_0^2 c^4}$. Substituting this in (3.10), rearranging, and taking the square of the resulting equation, we find that

$$\epsilon' = \frac{E^2 + 2E\epsilon - M_0^2 c^4}{2(E + \epsilon)}. \tag{3.16}$$

As $\epsilon \to \infty$, we see from (3.16) that $\epsilon' \to E$. In other words, even as the energy of the probe particle is made arbitrarily large, after the collision the probe particle's energy is just that of the probed particle.

Using the relation $\lambda' = hc/\epsilon'$, we find that the inaccuracy in the position measurement will satisfy the inequality

$$\Delta x \geq \frac{hc}{E}. \tag{3.17}$$

The right-hand side of (3.17) is the Compton wavelength of the observed particle. So a position measurement cannot determine the particle's position with an uncertainty lower than its Compton wavelength. For nonrelativistic particles, the Compton wavelength is much smaller than the de Broglie wavelength $\lambda_B \equiv h/|P|$, so there is effectively no limit to the accuracy of a position measurement in that case. For the photon, however, which is always a relativistic particle, as the relation $E = cP$ indicates, the Compton and the de Broglie wavelength coincide, and the limit to

---

[7]Most other treatments [243, 480] borrow from Compton's original analysis [117] at this point. I do not wish to do so, for Compton's analysis assumes that the electron is initially at rest. We obviously would not be able to do the same for a photon. So the final choice was to consider a collision between γ-rays (the probe) and a particle of arbitrary rest mass, which is not initially at rest in the reference frame of the laboratory. This allows us to discuss both the case of a massive particle and that of a particle with no rest mass. Note that the scattering of light by light is allowed by QED.

the accuracy of a position measurement can never be ignored. A photon cannot be localized to a volume smaller than the cube of its wavelength.

Our assumption of asymptotic states means that we have taken E' and ϵ' as the energies a very long time after the collision, to ensure the exact enforcement of conservation of energy. But clearly we do not have to wait an arbitrarily long time. The time needed to determine the energies with enough accuracy is [480]

$$\Delta t \sim \frac{h}{E}. \tag{3.18}$$

This is the time it takes to measure the particle's position. As this is also the period of the scattered γ-rays, we see that a position measurement cannot be faster than the time taken for the probing radiation to perform one oscillation, which is physically reasonable.

The time duration of the measurement is important because it determines whether the results of a measurement now can be used to make predictions about the outcome of future repetitions of the same measurement. Predictability is central to quantum mechanics and is what allows the existence of eigenstates [388].

3.2 EXTREME QUANTUM THEORY OF LIGHT WITH A TWIST: PROCA EQUATIONS FOR A MASSIVE PARTICLE OF SPIN 1

Instead of going from Maxwell equations to a quantum description of light as we have done so far, let us take the reverse route. Let us forget about classical electrodynamics for the moment and try to obtain a wave equation for a free[8] quantum of light. This is what Oppenheimer called an *extreme quantum theory of light* [472]. This section is very much in the spirit of [472], with the crucial exception that here we will allow for the possibility that the light quantum might have a nonvanishing rest mass. As discussed in Section 3.1, this will allow us to define a position operator that will have localized eigenstates in the nonrelativistic limit. So an ordinary first-quantization wavefunction description will be possible. On the other hand, we will see that instead of Maxwell equations, this approach leads us to a different description where there is no gauge freedom for the potentials: Proca equations. Maxwell equations are recovered only if we take the limit of vanishing rest mass.

Our derivation of the Proca equations will be quite different from the usual textbook derivation, where a so-called *mass term* is added to the ordinary Lagrangian of the electromagnetic field. Even though this popular textbook derivation is how Proca eventually derived his equations in [498], we think that his earlier failed approach to the problem in [497] is more enlightening. In [497], Proca starts to look for a wave equation for the photon the same way as Dirac did for the electron. But then he follows an idea by Lanczos [383, 384] that both generalizes and gives some insight into Dirac's method to get an equation of first order in time from an equation that is

---

[8]That is, without sources or sinks.

second order in time (i.e., the Klein–Gordon equation). So we present a derivation of Proca equations along the lines of [497] but not quite.[9]

If the photon has a nonvanishing rest mass m, however small, its energy–momentum relation is given by

$$\left(\frac{E}{c}\right)^2 - p_1^2 - p_2^2 - p_3^2 = m^2 c^2, \tag{3.19}$$

where E is the energy, c is the speed of light, and p_1, p_2, and p_3 are the Cartesian components of the momentum. Now, as we explained earlier, the spatial dependence of the wavefunction, like its time dependence, will be treated as a parameter not to be confused with the position of the particle.

Recalling that the three-momentum **p** is the generator of spatial translations, we can write for the wavefunction in configuration space $\psi(\mathbf{r}, t)$ the equation

$$\exp\left(\frac{i}{\hbar}\mathbf{a} \cdot \hat{\mathbf{p}}\right)\psi(\mathbf{r}, t) = \psi(\mathbf{r} + \mathbf{a}, t), \tag{3.20}$$

where **a** is any vector in ordinary three-dimensional space. Comparing (3.20) with the expression for the Taylor series expansion of $\psi(\mathbf{r} + \mathbf{a}, t)$ around $\psi(\mathbf{r}, t)$, we find that the momentum operator in configuration space must be given by[10]

$$\hat{\mathbf{p}} \rightarrow \frac{\hbar}{i}\nabla. \tag{3.21}$$

Similarly, the energy cp_0 is the generator of time translations, and

$$\exp\left(-\frac{i}{\hbar}c\hat{p}_0 \, \Delta t\right)\psi(\mathbf{r}, t) = \psi(\mathbf{r}, t + \Delta t). \tag{3.22}$$

Comparing (3.22) with the Taylor expansion of $\psi(\mathbf{r}, t + \Delta t)$ around $\psi(\mathbf{r}, t)$, we find that the time component of the four-momentum in configuration space must be given by

$$\hat{p}_0 \rightarrow i\frac{\hbar}{c}\frac{\partial}{\partial t}. \tag{3.23}$$

So if we use (3.19) as it is, we obtain the following quantum-mechanical wave equation:

$$-\frac{\hbar^2}{c^2}\frac{\partial^2}{\partial t^2}\psi + \hbar^2\nabla^2\psi = m^2 c^2 \psi. \tag{3.24}$$

This is the Klein–Gordon equation.[11] Unfortunately, it is second order in the time,[12] which leads to a probability density that is not positive definite. There are very

---

[9]One of the problems in [497], for instance, is that the point of departure is the energy–momentum relation for a particle of vanishing rest mass, but the conclusion is that the rest mass must be nonvanishing. Also, the wave equations obtained in [497] are not those that are now known as Proca equations.

[10]We notice that unlike Section 2.1, there is no delta function here, because we are no longer assuming that $\psi(\mathbf{r}, t) = \langle\mathbf{r}|\psi\rangle$, with $|\mathbf{r}\rangle$ being the eigenstates of the particle's position operator. Here **r** is a parameter like t.

[11]It apparently was discovered by Schrödinger.

[12]Nevertheless, there are ways to avoid these problems, and the Klein–Gordon equation is still used to describe particles of spin zero.

general arguments leading to the conclusion that a suitable quantum-mechanical wave equation should be first order in time (e.g., see Sec. 27 in [159]).

Dirac factorized (3.19) to express it as the square of a relation linear in E. But instead of simply writing

$$\sqrt{\left(\frac{E}{c}\right)^2 - p_1^2 - p_2^2 - p_3^2}\, \psi = \pm mc\, \psi, \tag{3.25}$$

which introduces the double sign[13] and nonlinear terms in E/c (the Taylor expansion of the square root has all even powers of E/c), Dirac wrote

$$\frac{E}{c}\psi = \{\alpha_1 p_1 + \alpha_2 p_2 + \alpha_3 p_3 + \beta mc\}\,\psi, \tag{3.26}$$

with the constraint that the square of (3.26) must yield the Klein–Gordon equation (3.24)

$$\left(\frac{E}{c}\right)^2 \psi = (\alpha_1 p_1 + \alpha_2 p_2 + \alpha_3 p_3 + \beta mc)^2 \,\psi$$
$$= \{p_1^2 + p_2^2 + p_3^2 + m^2 c^2\}\,\psi. \tag{3.27}$$

As a consequence of (3.27), α_1, α_2, α_3, and β cannot be simple numbers, as they must satisfy the relations

$$\alpha_r^2 = 1, \qquad\qquad \alpha_r \alpha_s + \alpha_s \alpha_r = 0,$$
$$\beta^2 = m^2 c^2, \qquad\qquad \alpha_r \beta + \beta \alpha_r = 0,$$

where r and s stand for 1, 2, or 3. Dirac's solution was to pick them as matrices and the wavefunction as a spinor. As is well known [159], the lowest possible dimension for these spinors is four. Moreover, Dirac's choice of spinors and matrices only allows even dimensions. This is all right for electrons[14] but is not what we need for light. The selection rules observed for absorption and emission of light by matter tell us that the light quantum must have an intrinsic angular momentum (i.e., spin) of one unit of Planck's constant [472].

Lanczos [383, 384] developed a generalization of Dirac's approach to factorize (3.19). Lanczos's idea was to use biquaternions instead of matrices and spinors.[15] Biquaternions are complex quaternions, and quaternions are, in a nutshell, a four-dimensional generalization of complex numbers introduced by Hamilton in 1844 [263, 264]. Quaternions can be thought as a scalar plus an axial vector [567]. The

---

[13]There is no escape from the double sign in the end.

[14]They turn out to have spin 1/2, which can be described by a four-component spinor and those 4×4 matrices. The extra two dimensions are needed to account for the positrons, which appear in the theory because of the negative energy solutions that can be traced back to the minus sign in (3.25) coming back to haunt us.

[15]For a very readable introduction to this way of writing field equations, see [427].

reader who is not familiar with quaternions might wish to have a look at Appendix C, where we give a gentle introduction to quaternions. But here we just use the following properties of biquaternions:

1. A biquaternion $\mathcal{Z}$ (quaternions and biquaternions are denoted by script capital letters) is the generalized four-dimensional complex number (called a *hyper-complex number*) $\mathcal{Z} = z_0 + z_1 i_1 + z_2 i_2 + z_3 i_3$, where z_0, z_1, z_2, and z_3 are ordinary complex numbers and i_1, i_2, and i_3 are new generalized imaginary units[16] whose multiplication, unlike the ordinary imaginary unit i, is noncommutative. The multiplication rules are given by

   (a) $i_1^2 = i_2^2 = i_3^2 = -1$

   (b) $i_1 i_2 = i_3, i_2 i_3 = i_1, i_3 i_1 = i_2$

   (c) $i_2 i_1 = -i_3, i_3 i_2 = -i_1, i_1 i_3 = -i_2$

   which, as noted in Appendix C, can be summarized by the relation

   $$i_n i_m = \sum_{l=1}^{3} \epsilon_{nml} i_l - \delta_{n,m}, \tag{3.28}$$

   where ϵ_{nml} is the Levy–Civita symbol and $\delta_{n,m}$ is the Kronecker delta.

2. Two biquaternions $\mathcal{Z} = z_0 + z_1 i_1 + z_2 i_2 + z_3 i_3$ and $\mathcal{Z}' = z_0' + z_1' i_1 + z_2' i_2 + z_3' i_3$ are equal only if their components are equal (i.e., $z_0 = z_0'$, $z_1 = z_1'$, $z_2 = z_2'$, and $z_3 = z_3'$).

3. The quaternion conjugate of $\mathcal{Z} = z_0 + z_1 i_1 + z_2 i_2 + z_3 i_3$ is given by $\bar{\mathcal{Z}} = z_0 - z_1 i_1 - z_2 i_2 - z_3 i_3$.

4. The quaternion modulus of $\mathcal{Z}$ is given by[17]
   $|\mathcal{Z}| \equiv \sqrt{\bar{\mathcal{Z}}\mathcal{Z}} = \sqrt{z_0^2 + z_1^2 + z_2^2 + z_3^2}$.

5. The complex conjugate of $\mathcal{Z} = z_0 + z_1 i_1 + z_2 i_2 + z_3 i_3$ is, of course, given by $\mathcal{Z}^* = z_0^* + z_1^* i_1 + z_2^* i_2 + z_3^* i_3$.

6. The Hermitian conjugate, or as Hamilton called it, the *biconjugate,* of $\mathcal{Z} = z_0 + z_1 i_1 + z_2 i_2 + z_3 i_3$, is given by $\mathcal{Z}^\dagger = \bar{\mathcal{Z}}^* = z_0^* - z_1^* i_1 - z_2^* i_2 - z_3^* i_3$.

7. The scalar part of a biquaternion (or quaternion) $\mathcal{Z}$ is given by $S(\mathcal{Z}) = (\mathcal{Z} + \bar{\mathcal{Z}})/2$, and for a product $\mathcal{Z}_1 \mathcal{Z}_2$, $S(\mathcal{Z}_1 \mathcal{Z}_2) = S(\mathcal{Z}_2 \mathcal{Z}_1)$.

   In Lanczos's method, the four-momentum becomes a biquaternion and so does the wavefunction. Let us try to factorize (3.24) as Dirac did but using biquaternions

---

[16]Hamilton called them i, j, and k, but we will use i_1, i_2, and i_3 instead to avoid confusing the first generalized imaginary unit with the old commutative i.

[17]If the components of $\mathcal{Z}$ are not real, the quaternion modulus can vanish, be negative, or even, be complex.

instead of matrices and spinors. We use a four-sided diamond to represent the four-dimensional biquaternion gradient[18]

$$\lozenge \equiv \frac{i}{c} \frac{\partial}{\partial t} + \sum_{n=1}^{3} i_n \frac{\partial}{\partial x_n}, \tag{3.29}$$

and as we know from experiment that light has spin 1, we let the wavefunction be a pure biquaternion (i.e., a vector wavefunction)

$$\mathcal{F} \equiv \sum_{n=1}^{3} i_n i f_n, \tag{3.30}$$

where i is the ordinary commutative imaginary number, not to be confused with i_1, i_2, and i_3, the generalized noncommutative imaginary quantities of quaternion theory.

Notice that

$$\bar{\lozenge} \lozenge = \lozenge \bar{\lozenge} = \nabla^2 - \frac{1}{c^2} \frac{\partial^2}{\partial t^2} = \Box, \tag{3.31}$$

where $\Box$ is the D'Alembertian.[19] So if we write (3.24) in the form

$$\hbar^2 \bar{\lozenge} \lozenge \mathcal{F} = mc\,mc\,\mathcal{F}, \tag{3.32}$$

it is tempting to try the factorized equation

$$\lozenge \mathcal{F} = \mu \mathcal{F}, \tag{3.33}$$

where

$$\mu \equiv \frac{mc}{\hbar}. \tag{3.34}$$

But (3.33) is not covariant.

Under a Lorentz transformation, $\lozenge$ transforms as[20] a four-vector: that is, $\lozenge' = U \lozenge U^\dagger$, where U is a biquaternion such that

$$\bar{U}U = 1. \tag{3.35}$$

Now then, to turn $\lozenge \mathcal{F}$ into $\lozenge' \mathcal{F}'$, we must multiply it on the left by U, on the right by T, and $\mathcal{F}'$ must be obtained from $\mathcal{F}$ by $\mathcal{F}' = U^* \mathcal{F}T$, where T is still to be determined:

$$U \lozenge \mathcal{F}T = U \lozenge \underbrace{U^\dagger U^*}_{1} \mathcal{F}T$$

$$= \lozenge' \mathcal{F}'. \tag{3.36}$$

---

[18]In Appendix C, we only talk about the quaternion gradient (i.e., a gradient when the coefficients of the quaternion $\mathcal{R} = x_0 + \sum_n i_n x_n$ describing a point in four-dimensional space are all real). For Minkowsky's space-time, we use biquaternions (last section of Appendix C) (i.e., use complex coefficients, with $x_0 = ct$ and $x_1 = ix$, $x_2 = iy$, $x_3 = iz$ in the previous expression for $\mathcal{R}$). Then it is convenient to define the biquaternion gradient as i multiplied by the gradient defined in Appendix C.

[19]Sometimes the D'Alembertian is written as $\Box^2$, perhaps because it is composed of square derivatives (i.e., second derivatives). But we feel that the superscript 2 is an unnecessary emphasis, for a square is already a square.

[20]See Appendix C or [506] or Chapter 9 of Lanczos's book on mechanics [386].

Then we see that regardless of what T is, the right-hand side of (3.33) does not transform properly unless U is a quaternion with real coefficients rather than a biquaternion for U^* to be equal to U. But if $U^* = U$, the transformation can only be an ordinary spatial rotation (i.e., the time component will be left unchanged).[21]

If our wave equation is to transform accordingly with respect to a general Lorentz transformation that also transforms the time, we must have another biquaternion multiplying $\mathcal{F}$ on the right-hand side of (3.33)

$$\Diamond \mathcal{F} = \mu X \mathcal{F}, \tag{3.37}$$

where

$$X' = U X U^\dagger, \tag{3.38}$$

(i.e., X must transform like $\Diamond$). Moreover, if we multiply (3.37) by $\bar{\Diamond}$ from the left, we must recover (3.24); that is,

$$\begin{aligned} \bar{\Diamond}\Diamond\mathcal{F} &= \mu\bar{\Diamond}X\mathcal{F} \\ &= \mu^2 \mathcal{F}. \end{aligned} \tag{3.39}$$

Then

$$\bar{\Diamond}X = \mu \tag{3.40}$$

and X must be given by

$$X = \mu\bar{\Diamond}^{-1}. \tag{3.41}$$

Let us define $\mathcal{A}$ such that $\mathcal{F} = \bar{\Diamond}\mathcal{A}$; then the wave equation we are looking for is

$$\Diamond \mathcal{F} = \mu^2 \mathcal{A}, \tag{3.42}$$
$$\bar{\Diamond}\mathcal{A} = \mathcal{F}. \tag{3.43}$$

These are Lanczos's equations.[22] Instead of a single biquaternion wavefunction candidate $\mathcal{F}$, we were forced by relativity to have two: $\mathcal{F}$ and $\mathcal{A}$. To find out how they transform, we must now determine T.

We want $\mathcal{F}$ to remain a pure biquaternion in all Lorentz frames; otherwise, this formulation will not be covariant. The simplest way for this to hold is if our $\mathcal{F}$ is what is called a *six-vector*. For instance, T cannot be $\bar{U}$; otherwise, $\mathcal{F}$ will transform as the conjugate of a four-vector and will develop a scalar component in some Lorentz frames. What we want is to find a T such that for a given biquaternion $\mathcal{Z} = z_0 + \sum_{n=1}^{3} i_n z_n$, the same biquaternion $\mathcal{Z}' = U^* \mathcal{Z} T$ in a different Lorentz frame will have its scalar component unchanged (i.e., $z_0' = z_0$ must be a true scalar). The simplest solution (i.e., where $\mathcal{F}$ is a six-vector) turns out[23] to be $T = U^\dagger$.

---

[21] See Appendix C.

[22] Actually, Lanczos's equations are more general in that the wavefunction $\mathcal{F}$ does not need to be a pure biquaternion and can transform in a different way. Lanczos's equations can describe all scalar, all vector, and all pseudovector particles.

[23] This can be seen by noticing that (3.35) also implies that $U\bar{U} = 1$. Taking the complex conjugate of the former equation and of (3.35), we find that $U^\dagger U^* = 1$ and $U^* U^\dagger = 1$. Then z_0 is not changed by

We know now that $\mathcal{F}$ is a six-vector and that $T = U^\dagger$. What about $\mathcal{A}$? As $\mathcal{A}$ must transform in the same way as $\Diamond$, it is a four-vector. This leaves us, in principle, with eight real quantities to determine in $\mathcal{A}$, as each of its four components is complex. Let us now make a bold and simplifying assumption. Let us assume that $\mathcal{A}$ is anti-Hermitian (i.e., $\mathcal{A}^\dagger = -\mathcal{A}$). This reduces the number of real quantities in $\mathcal{A}$ by half: from eight, we have only four real functions to determine now. This bold assumption is consistent with relativity, as the property of a four-vector of being Hermitian or anti-Hermitian does not change from one Lorentz frame to another. But the ultimate test will come in the end, when we will see that such assumption does not lead to any inconsistency.

To sum up our discussion so far: We have a double biquaternion wavefunction candidate, consisting of the six-vector $\mathcal{F}$ and the four-vector $\mathcal{A}$. These biquaternions obey Lanczos equations (3.42) and (3.43) and have the following additional properties:

$$\bar{\mathcal{F}} = -\mathcal{F} \quad \text{(i.e., } \mathcal{F} \text{ is a pure biquaternion or vector),} \tag{3.44}$$

$$\mathcal{A}^\dagger = -\mathcal{A} \quad \text{(i.e., } \mathcal{A} \text{ is anti-Hermitian).} \tag{3.45}$$

Another interesting feature of (3.42) and (3.43) is that like the Klein–Gordon equation and the Dirac equation, they connect $\hbar$ to the rest mass through the single parameter μ. In the nonrelativistic Schrödinger equation, this is not so, as $\hbar$ appears linearly in the time derivative term and quadratically in the spatial derivative terms. Relativity is revealing a deep secret of nature here: If the rest mass vanishes, the quantum-mechanical wave equation will not contain $\hbar$. This means that for vanishing rest mass, when we take the limit of $\hbar \to 0$, the same wave equation remains valid in the realm of classical physics. We will see later that when we take $\mu \to 0$, (3.42) and (3.43) coincide with the Maxwell equations, which are *both* the quantum and the classical equations for a photon (if it is, indeed, massless).

Now before we can call this a proper quantum wave equation, we must show that it leads to a probability density and that the probability is conserved in time. To do so, we must find a continuity equation relating the probability density to a probability current density. It is this continuity equation that will allow us to identify what is the probability density. Moreover, the probability density must be positive definite and the probability interpretation must not depend on the choice of Lorentz frames. For the latter to hold, the probability density and the probability current density must be, respectively, the time and spatial components of a four-vector. This can be seen by noticing that if the Hermitian quaternion $\mathcal{Z} = D + \sum_{n=1}^{3} i_n i C_n$,

---

the transformation because $U^* z_0 U^\dagger = z_0 U^* U^\dagger = z_0$. Moreover, the pure biquaternion part of $\mathcal{Z}$ (i.e., $z_1 i_1 + z_2 i_2 + z_3 i_3$) transforms them into another pure biquaternion. An easy way to see this is to take the square of any transformed quaternion-imaginary unit. The square of the transformed quaternion-imaginary unit i_1, for example, is given by

$$\left(U^* i_1 U^\dagger\right)^2 = U^* i_1 \underbrace{U^\dagger U^*}_{1} i_1 U^\dagger = U^* i_1 i_1{}^2 U^\dagger = -\underbrace{U^* U^\dagger}_{1} = -1.$$

So, as its square is -1, $U^* i_1 U^\dagger$ cannot be 1 but rather, must be one of the quaternion-imaginary units i_1, i_2, or i_3. The same holds for the transformation of the other two quaternion-imaginary units, i_2 and i_3.

where D, C_1, C_2, and C_3 are real, is a four-vector, then $\bar{\Diamond} \mathcal{Z}$ is a six-vector (i.e., the scalar component of $\bar{\Diamond} \mathcal{Z}$ is the same in all Lorentz frames). This scalar component is just $i \{ (1/c) \partial D / \partial t + \nabla \cdot \mathbf{C} \}$, where $\mathbf{C}$ is the vector formed by C_1, C_2, and C_3. So if $S \left(\bar{\Diamond} \mathcal{Z} \right) = 0$, the time and spatial components of the Hermitian four-vector $\mathcal{Z}$ will satisfy the continuity equation $(1/c) \partial D / \partial t + \nabla \cdot \mathbf{C} = 0$ in every Lorentz frame.

If we are looking for something bilinear in the wavefunction that forms a four-vector, as $\mathcal{F}$ is a six-vector and $\mathcal{A}$ is a four-vector, the only combinations that would obviously lead to a four-vector are $\mathcal{A}^\dagger \mathcal{F}$, $\mathcal{A} \mathcal{F}$, and their Hermitian conjugates. Unfortunately, none of them leads to a positive-definite time component.[24] So we must look for something else.

Consider the biquaternion

$$\mho \equiv \mathcal{F}^\dagger \Diamond \mathcal{F} + \mu^2 \mathcal{A}^\dagger \bar{\Diamond} \mathcal{A}. \tag{3.46}$$

As $\mathcal{A}$ and $\Diamond$ are four-vectors and $\mathcal{F}$ is a six-vector, $\mho$ is a four-vector. Moreover, as $\hbar \Diamond$ is the quaternionic four-momentum operator, $\mho$ has dimensions of momentum in units of $\hbar$. From (3.42) and (3.43), we find that

$$\mathcal{F}^\dagger \Diamond \mathcal{F} = \left(\mathcal{F}^\dagger \Diamond \right) \mathcal{F} + \mathcal{F}^\dagger \left(\Diamond \mathcal{F} \right)$$
$$= -\mu^2 \mathcal{A}^\dagger \mathcal{F} + \mu^2 \mathcal{F}^\dagger \mathcal{A} \tag{3.47}$$

and

$$\mathcal{A}^\dagger \bar{\Diamond} \mathcal{A} = \left(\mathcal{A}^\dagger \bar{\Diamond} \right) \mathcal{A} + \mathcal{A}^\dagger \left(\bar{\Diamond} \mathcal{A} \right)$$
$$= -\mathcal{F}^\dagger \mathcal{A} + \mathcal{A}^\dagger \mathcal{F}, \tag{3.48}$$

where the parentheses denote the limits beyond which the differential operator no longer acts [e.g., in $\left(\mathcal{A}^\dagger \bar{\Diamond} \right) \mathcal{A}$ $\bar{\Diamond}$ operates only on $\mathcal{A}^\dagger$, not on $\mathcal{A}$]. Using (3.47) and (3.48) in (3.46), we find that $\mho$ always vanishes:

$$\mathcal{F}^\dagger \Diamond \mathcal{F} + \mu^2 \mathcal{A}^\dagger \bar{\Diamond} \mathcal{A} = 0. \tag{3.49}$$

Now we will use the quaternion version of the Gauss integral theorem (see [146] or Appendix C)

$$\int_{\delta \sigma} dQ \, \mathcal{Z} = \int_\sigma \left(\Diamond \mathcal{Z} \right) d^4 \mathcal{R} \quad \text{and} \quad \int_{\delta \sigma} \mathcal{Z} \, dQ = \int_\sigma \left(\mathcal{Z} \Diamond \right) d^4 \mathcal{R}, \tag{3.50}$$

where σ stands for the four-volume closed by the boundary hypersurface $\delta \sigma$ and $d^4 \mathcal{R}$ for the infinitesimal four-volume element. Let the four-volume consist of the entire three-dimensional space and the time interval from t_0 to t_1. This is a hyperbox, and the bounding hypersurface is spacelike (the entire three-dimensional space at times t_0 and t_1). Using (3.50), their quaternion conjugates, and (3.49), we find that

$$\int \left(\mathcal{F}^\dagger \mathcal{F} + \mu^2 \mathcal{A}^\dagger \mathcal{A} \right)_{t=t_0} d^3 r = \int \left(\mathcal{F}^\dagger \mathcal{F} + \mu^2 \mathcal{A}^\dagger \mathcal{A} \right)_{t=t_1} d^3 r. \tag{3.51}$$

---

[24]Despite that, some authors [256, 257] suggest defining the probability four-current as $\mathcal{A}^\dagger \mathcal{F} + \mathcal{F}^\dagger \mathcal{A}$.

So $\int \left(\mathcal{F}^\dagger \mathcal{F} + \mu^2 \mathcal{A}^\dagger \mathcal{A} \right) d^3r$ is constant in time and is a four-vector because $\mho$ is a four-vector. It also has the same dimensions as $\mho$ (i.e., momentum). So this conserved quantity must be the particle's four-momentum, which is conserved for a free particle [44, 238]. Notice that $\mathcal{F}^\dagger \mathcal{F} + \mu^2 \mathcal{A}^\dagger \mathcal{A}$ *is not* a four-vector[25] because the three-dimensional volume element in the integral is not Lorentz invariant.[26] So $\mathcal{F}^\dagger \mathcal{F} + \mu^2 \mathcal{A}^\dagger \mathcal{A}$ cannot be regarded as a probability four-current and $\mathcal{F}$ and $\mathcal{A}$ cannot be wavefunctions. But what are they then? As the expression (3.51) for the conserved four-momentum is typical of fields [42], we will regard $\mathcal{F}$ and $\mathcal{A}$ as fields.

To get a four-vector from $\mathcal{F}^\dagger \mathcal{F} + \mu^2 \mathcal{A}^\dagger \mathcal{A}$, we must multiply it by something that transforms under a Lorentz transformation the same way as a three-dimensional volume element. We also want this multiplication factor to yield a properly normalized probability four-current. Now the particle's energy (i.e., the timelike component of the four-momentum) transforms as $1/d^3r$. So one way to get a four-vector from $\mathcal{F}^\dagger \mathcal{F} + \mu^2 \mathcal{A}^\dagger \mathcal{A}$ is to divide it by the energy. When $\mu \to 0$, this approach yields the *Riemann–Silberstein wavefunction* [53–55, 346, 364, 446, 514, 564–566, 625]. Here we will not follow this route, for two reasons. The first reason is because it leads to an unusual probabilistic interpretation. According to Oppenheimer [472], the usual probabilistic interpretation "is essential to a truly corpuscular theory."[27] The second reason is that as mentioned in Section 3.1, we can determine the particle's momentum with arbitrary precision but not its position. This suggests that momentum space is the right place to construct a wavefunction rather than configuration space. There,[28] an alternative approach that preserves the usual probabilistic interpretation can be formulated in a simple way. When $\mu \to 0$, the latter approach yields in configuration space the *Landau–Peierls wavefunction*[29] [12, 119, 120, 387].

---

[25]It is a tensor, the energy–momentum tensor of the Proca field [256].

[26]Moreover, the Lorentz covariance of the integral (i.e., the conclusion that the three-volume integral of $\mathcal{F}^\dagger \mathcal{F} + \mu^2 \mathcal{A}^\dagger \mathcal{A}$ is a four-vector) depends on the vanishing of $\mho$. Exactly the same thing happens to the electromagnetic four-momentum derived from Maxwell equations. It is only a four-vector in the absence of charges, for then the four-divergence of the relativistic energy–momentum tensor vanishes. In fact, as we will soon see, (3.42) and (3.43) reduce to Maxwell equations when $\mu \to 0$. For a proof that the electromagnetic energy and three-momentum are the time and space components, respectively, of a four-vector in the absence of charges, see Secs. 29 and 32 of [389] or Sec. 4-9 of [526]. Both references use ordinary tensor calculus rather than quaternions.

[27]Oppenheimer pointed out that this is a problem common to every system in which the wavefunction is a four-vector or tensor. He traced the problem to the Lagrangian, from which such wave equations are derived, not being a scalar density despite the wave equations themselves being covariant. See pages 734 and 735 in [472].

[28]To be more precise, we should say reciprocal space-time. A point in reciprocal space-time corresponds to a wave four-vector, which is just the momentum in units of $\hbar$.

[29]As we will see later, the Landau–Peierls wavefunction is nonlocal in the fields. This shows very clearly the problems in trying to define a configuration-space wavefunction for the photon. Bialynicki-Birula [53–55], Good and co-workers [241, 268, 463], and others manage to use a wavefunction that is local in the fields, the Riemann–Silberstein wavefunction, at the expense of having to adopt a nonlocal inner product. This only hides the problem in the inner product. In fact, if we try to regard as a probability density the integrand in their expression for the inner product of a wavefunction with itself, we are frustrated by the fact that this integrand is not positive definite (see [387]). There is no escape from the problems inherent in the idea of a photon's wavefunction in configuration space. With that in mind, we see that the Landau–Peierls wavefunction has two main advantages: first, it makes very clear the problem with the idea of a photon's

We can define a covariant Fourier transform as

$$\mathcal{G}(\mathcal{R}) = \int d^4\mathcal{K} \, e^{iS(\bar{\mathcal{K}}\mathcal{R})}\underline{\mathcal{G}}(\mathcal{K}),$$ (3.52)

where $\mathcal{R}$ is the Hermitian four-position,

$$\mathcal{R} = ct + \sum_{n=1}^{3} i_n \, ix_n,$$ (3.53)

$\mathcal{K}$ is the Hermitian wave four-vector or the four-position in reciprocal space-time,

$$\mathcal{K} = \frac{\omega}{c} + \sum_{n=1}^{3} i_n \, ik_n,$$ (3.54)

and $d^4\mathcal{K}$ is the Lorentz invariant four-volume in reciprocal space-time $d(\omega/c) \, dk_1 \, dk_2 \, dk_3$. The covariant Fourier transform of a field will be denoted by underlining their original configuration space-time symbol. The inverse transform is given by

$$\underline{\mathcal{G}}(\mathcal{K}) = \int \frac{d^4\mathcal{R}}{(2\pi)^4} e^{-iS(\bar{\mathcal{K}}\mathcal{R})}\mathcal{G}(\mathcal{R}),$$ (3.55)

where we have used a straightforward four-dimensional generalization of the well-known three-dimensional delta function identity (F.4)

$$\int \frac{d^4\mathcal{K}}{(2\pi)^4} e^{iS(\bar{\mathcal{K}}\mathcal{R})} = \delta^4(\mathcal{R}).$$ (3.56)

Now let us see how (3.42) and (3.43) look in reciprocal space-time. The quaternion four-gradient $\Diamond$ becomes $\mathcal{K}$ in reciprocal space-time. Then (3.42) and (3.43) are given in reciprocal space-time by

$$\mathcal{K}\underline{\mathcal{F}} = \mu^2\underline{\mathcal{A}},$$ (3.57)
$$\bar{\mathcal{K}}\underline{\mathcal{A}} = \underline{\mathcal{F}}.$$ (3.58)

Multiplying (3.57) by $\bar{\mathcal{K}}$ on the left and using (3.58), we find that

$$|\mathcal{K}|^2\underline{\mathcal{F}} = \mu^2\underline{\mathcal{F}}.$$ (3.59)

Analogously,

$$|\mathcal{K}|^2\underline{\mathcal{A}} = \mu^2\underline{\mathcal{A}}.$$ (3.60)

Equations (3.59) and (3.60) are the reciprocal space-time counterparts of the Klein–Gordon equations for $\mathcal{F}$ and $\mathcal{A}$, respectively. Rearranging (3.59), we can write

$$\left(|\mathcal{K}|^2 - \mu^2\right)\underline{\mathcal{F}} = 0.$$ (3.61)

wavefunction. Second, as we will see in Section 3.6, it makes the connection with the usual canonical quantization formalism become very simple.

An analogous equation holds for $\underline{A}$. From these two equations we see that whenever $|\mathcal{K}|^2$ differs from μ^2, $\underline{\mathcal{F}}$ and $\underline{A}$ will vanish. So $\underline{\mathcal{F}}$ and $\underline{A}$ must be given by

$$\underline{\mathcal{F}}(\mathcal{K}) = \delta\left(|\mathcal{K}|^2 - \mu^2\right) \mathcal{U}(\mathcal{K}), \tag{3.62}$$

$$\underline{A}(\mathcal{K}) = \delta\left(|\mathcal{K}|^2 - \mu^2\right) \mathcal{V}(\mathcal{K}), \tag{3.63}$$

where $\mathcal{U}(\mathcal{K})$ and $\mathcal{V}(\mathcal{K})$ are two quaternion functions of $\mathcal{K}$ yet to be determined. The delta function forces $\mathcal{K}$ always to have a quaternion modulus square of μ^2. This gives rise to the dispersion relation [238]

$$\left(\frac{\omega}{c}\right)^2 = \mu^2 + k^2. \tag{3.64}$$

So for each **k**, we have two solutions: one with positive frequency,

$$\omega = c\sqrt{k^2 + \mu^2}, \tag{3.65}$$

and another with negative frequency,

$$\omega = -c\sqrt{k^2 + \mu^2}. \tag{3.66}$$

As the frequency is the energy in units of $\hbar$, the solution with positive frequency is a solution with positive energy and that with negative frequency is a solution with negative energy. The negative-energy solutions correspond to antiparticles and the positive-energy solutions to particles.

We must distinguish the particle solutions (positive frequency) from their antiparticle counterparts (negative frequency). We will do so by using the subscript $+$ for the positive-frequency solutions and $-$ for the negative-frequency solutions. From (3.52) and (3.62), we find that

$$\begin{aligned}
\mathcal{F}_+(\mathcal{R}) &= \int d^3k \, \frac{e^{i(\mathbf{k}\cdot\mathbf{r}-\omega t)}}{2\sqrt{k^2 + \mu^2}} \, \mathcal{U}_+(\mathcal{K}) \\
&= \int d^3k \, e^{i\mathbf{k}\cdot\mathbf{r}} \underline{F}_+(\mathbf{k}, t),
\end{aligned} \tag{3.67}$$

where

$$\underline{F}_+(\mathbf{k}, t) \equiv \frac{\mathcal{U}_+(\mathcal{K})e^{-i\omega t}}{2\sqrt{k^2 + \mu^2}} \tag{3.68}$$

is the ordinary three-dimensional Fourier transform of $\mathcal{F}_+$. We have written $\underline{F}_+$ as a function of **k** and t alone because ω is determined completely by **k** according to (3.65). Analogously,

$$\mathcal{A}_+(\mathcal{R}) = \int d^3k e^{i\mathbf{k}\cdot\mathbf{r}} \, \underline{A}_+(\mathbf{k}, t), \tag{3.69}$$

where

$$\underline{A}_+(\mathbf{k}, t) \equiv \frac{\mathcal{V}_+(\mathcal{K})e^{-i\omega t}}{2\sqrt{k^2 + \mu^2}}. \tag{3.70}$$

Now let us calculate the particle's four-momentum in terms of $\mathcal{U}_+$ and $\mathcal{V}_+$. Substituting (3.67) and (3.69) in (3.51) and using the well-known three-dimensional delta function identity (F.4), we find that

$$\int d^3r \left(\mathcal{F}_+^\dagger \mathcal{F}_+ + \mu^2 \mathcal{A}_+^\dagger \mathcal{A}_+ \right) = 2\pi^3 \int \frac{d^3k}{\sqrt{k^2 + \mu^2}} \frac{\mathcal{U}_+^\dagger \mathcal{U}_+ + \mu^2 \mathcal{V}_+^\dagger \mathcal{V}_+}{\sqrt{k^2 + \mu^2}}. \quad (3.71)$$

As $d^3k/\sqrt{k^2 + \mu^2}$ is an invariant quantity,[30] and $\int d^3r\, (\mathcal{F}_+^\dagger \mathcal{F}_+ + \mu^2 \mathcal{A}_+^\dagger \mathcal{A}_+)$ is a four-vector, it follows that $(\mathcal{U}_+^\dagger \mathcal{U}_+ + \mu^2 \mathcal{V}_+^\dagger \mathcal{V}_+)/\sqrt{k^2 + \mu^2}$ is a four-vector. This shows a way of defining *densities* of probability and of probability current that form a four-vector[31]: Instead of insisting on a wavefunction composed of $\mathcal{F}$ and $\mathcal{A}$, we should take as our wavefunction (3.67) and (3.69) with $\underline{F}_+$ and $\underline{A}_+$ replaced by $\underline{F}_+/\sqrt[4]{k^2 + \mu^2}$ and $\underline{A}_+/\sqrt[4]{k^2 + \mu^2}$, respectively. So we define the wavefunction components as

$$\Psi_\alpha(\mathcal{R}) = \int d^3k\, e^{ik\cdot r} \frac{F_\alpha(\mathbf{k}, t)}{\sqrt[4]{k^2 + \mu^2}}, \quad (3.72)$$

$$\Phi_\alpha(\mathcal{K}) = \int d^3k\, e^{ik\cdot r} \frac{A_\alpha(\mathbf{k}, t)}{\sqrt[4]{k^2 + \mu^2}}, \quad (3.73)$$

where α stands for the sign of the frequency.

We now have a four-vector candidate to be our probability four-current density: $\Psi^\dagger \Psi + \mu^2 \Phi^\dagger \Phi$. Moreover, from (3.57) and (3.58), we see that Ψ and Φ satisfy the same equations as $\mathcal{F}$ and $\mathcal{A}$; that is,

$$\Diamond \Psi = \mu^2 \Phi, \quad (3.74)$$

$$\bar{\Diamond} \Phi = \Psi. \quad (3.75)$$

They also share the properties (3.44) and (3.45) of $\mathcal{F}$ and $\mathcal{A}$, as you can easily check for yourself (see Problem 3.2):

$$\bar\Psi = -\Psi \quad \text{and} \quad \Phi^\dagger = -\Phi. \quad (3.76)$$

But Ψ and Φ transform in a different way from $\mathcal{F}$ and $\mathcal{A}$. If they transformed in the same way as the fields, $\Psi^\dagger \Psi + \mu^2 \Phi^\dagger \Phi$ would not be a four-vector, just as $\mathcal{F}^\dagger \mathcal{F} + \mu^2 \mathcal{A}^\dagger \mathcal{A}$ is not. Cook [120], Hammer and Good [268], and Nelson and Good [463] worked out how they transform from one Lorentz frame to another when $\mu \to 0$. The general case is discussed by Weaver et al. [624].

The next condition that our four-vector candidate to probability four-current must satisfy is the continuity equation. As we have mentioned before, the continuity

---

[30]See Sec. 10 of [389].

[31]As we discussed earlier, it is the *densities* rather than their three-dimensional volume integrals, the probability and current, that must form a four-vector. Otherwise, the probability interpretation will not hold in all Lorentz frames, as required by relativity.

equation is equivalent to $S(\bar{\Diamond}[\Psi^\dagger\Psi + \mu^2\Phi^\dagger\Phi]) = 0$. We will now show that this is satisfied. Using $S(\mathcal{Z}_1\mathcal{Z}_2) = S(\mathcal{Z}_2\mathcal{Z}_1)$, (3.74), (3.75), (3.76), and Feynman's trick of splitting an operator that acts on a product of two terms into the sum of one that acts only on the first term and another that acts only on the second (see Appendix E), we find that

$$\begin{aligned} S\left\{\bar{\Diamond}\left(\Psi^\dagger\Psi\right)\right\} &= S\left\{\left[\bar{\Diamond}_{\Psi^\dagger} + \bar{\Diamond}_\Psi\right]\Psi^\dagger\Psi\right\} \\ &= S\left\{\bar{\Diamond}_{\Psi^\dagger}\Psi^\dagger\mathcal{F} + \Psi^\dagger\Psi\bar{\Diamond}_\Psi\right\} \\ &= S\left\{\left(\bar{\Diamond}\Psi^\dagger\right)\Psi + \Psi^\dagger\left(\Psi\bar{\Diamond}\right)\right\} \\ &= S\left\{\mu^2\Phi^*\Psi - \mu^2\Psi^\dagger\bar{\Phi}\right\}, \end{aligned} \tag{3.77}$$

where $\bar{\Diamond}_{\Psi^\dagger}$ only acts on $\Psi^\dagger$, $\bar{\Diamond}_\Psi$ only acts on Ψ, and the parentheses denote the limit within which $\bar{\Diamond}$ operates (i.e., it does not operate on anything outside the parentheses).

Analogously, we find that

$$\begin{aligned} S\left\{\bar{\Diamond}\left(\Phi^\dagger\Phi\right)\right\} &= S\left\{\left[\bar{\Diamond}_{\Phi^\dagger} + \bar{\Diamond}_\Phi\right]\Phi^\dagger\Phi\right\} \\ &= S\left\{\left(\bar{\Diamond}\Phi^\dagger\right)\Phi + \Phi^\dagger\left(\Phi\bar{\Diamond}\right)\right\} \\ &= S\Big\{\underbrace{-\Psi\Phi}_{\Psi\Phi^\dagger} + \underbrace{\Phi^\dagger\Psi^\dagger}_{-\Phi\Psi^\dagger}\Big\} \\ &= S\left\{\Phi^\dagger\Psi - \Psi^\dagger\Phi\right\}. \end{aligned} \tag{3.78}$$

Multiplying (3.78) by μ^2 and adding to (3.77), we obtain

$$\begin{aligned} S\left\{\bar{\Diamond}\left(\Psi^\dagger\Psi + \mu^2\Phi^\dagger\Phi\right)\right\} &= \mu^2 S\Big\{\underbrace{\left(\Phi^* + \bar{\Phi}^*\right)}_{2S(\Phi^*)}\Psi - \Psi^\dagger\underbrace{\left(\Phi + \bar{\Phi}\right)}_{2S(\Phi)}\Big\} \\ &= 0, \end{aligned} \tag{3.79}$$

as both Ψ and $\Psi^\dagger$ do not have a scalar part (i.e., no time component).

We have a wave equation for a candidate wavefunction of our massive photon, and a four-vector, formed from it, candidate to probability four-current that satisfies a continuity equation. But we have said nothing yet about how observable quantities are to be calculated in this theory. This is a very important point. After all, this is the link between the theoretical framework and experiment. If the timelike component of $\Psi^\dagger\Psi + \mu^2\Phi^\dagger\Phi$ is to be a probability density, the expectation values of physical observables should be given by the averages of the corresponding operators weighed by this probability density. We will show now that this is so.

Using (3.72) and (3.73), we can easily check that for positive-energy solutions,[32]

$$\int d^3r\, S\left\{\Psi^\dagger\left(\frac{i}{c}\frac{\partial}{\partial t}\Psi\right) + \mu^2\Phi^\dagger\left(\frac{i}{c}\frac{\partial}{\partial t}\Phi\right)\right\} = \int d^3r\, S\left\{\mathcal{F}^\dagger\mathcal{F} + \mu^2 A^\dagger A\right\}. \tag{3.80}$$

---

[32]Negative-energy solutions need to be rewritten as positive-energy antiparticles before their energy can be compared to the field energy. When this is done, we find that the expectation value of their energy coincides with the field energy, as it should.

As $\hbar \int d^3r \{ \mathcal{F}^\dagger \mathcal{F} + \mu^2 \mathcal{A}^\dagger \mathcal{A} \}$ is the four-momentum of the field, it is clear that the expectation value of the energy $\mathcal{E}$ must be given by

$$\mathcal{E} = \int d^3r \ S \left\{ \Psi^\dagger \left(i\hbar \frac{\partial}{\partial t} \Psi \right) + \mu^2 \Phi^\dagger \left(i\hbar \frac{\partial}{\partial t} \Phi \right) \right\}, \tag{3.81}$$

which has the proper form that we promised above.

What about the momentum? The kth component of the field's three-momentum $\mathcal{P}_k$ is given by

$$\mathcal{P}_k = \frac{\hbar}{i} \int d^3r \ S \left\{ (-i_k) \left[\mathcal{F}^\dagger \mathcal{F} + \mu^2 \mathcal{A}^\dagger \mathcal{A} \right] \right\}. \tag{3.82}$$

Now using

$$(-i_k)\mathcal{F}^\dagger = \mathcal{F}^\dagger i_k + i2f_k^*, \tag{3.83}$$

and that as $\mathcal{F}$ is a pure biquaternion, $S\{f_k^* \ \mathcal{F}\} = 0$, we find that

$$S \left\{ (-i_k)\mathcal{F}^\dagger \mathcal{F} \right\} = S \left\{ \mathcal{F}^\dagger i_k \mathcal{F} \right\}. \tag{3.84}$$

Moreover, using $S\{\mathcal{Z}_1 \mathcal{Z}_2\} = S\{\mathcal{Z}_2 \mathcal{Z}_1\}$ and $\mathcal{A}^\dagger = -\mathcal{A}$, we find that

$$S \left\{ (-i_k)\mathcal{A}^\dagger \mathcal{A} \right\} = S \left\{ \mathcal{A}^\dagger (-i_k)\mathcal{A} \right\}. \tag{3.85}$$

Thus, substituting (3.84) and (3.85) in (3.82), we see that we can rewrite the kth component of the field's three-momentum as

$$\mathcal{P}_k = \frac{\hbar}{i} \int d^3r \ S \left\{ \mathcal{F}^\dagger \left[i_k \right] \mathcal{F} + \mu^2 \mathcal{A}^\dagger \overline{[i_k]} \mathcal{A} \right\}. \tag{3.86}$$

From (3.72) and (3.73) we find that for positive frequencies, and in terms of Ψ and Φ, (3.86) becomes

$$\mathcal{P}_k = \frac{\hbar}{i} \int d^3r \ S \left\{ \Psi^\dagger \left(i_k \frac{i}{c} \frac{\partial}{\partial t} \Psi \right) + \mu^2 \Phi^\dagger \left(-i_k \frac{i}{c} \frac{\partial}{\partial t} \Phi \right) \right\}. \tag{3.87}$$

To show that (3.87) is equivalent to the expectation value of the kth component of the momentum operator $(\hbar/i)(\partial/\partial x_k)$, note that

$$\frac{i}{c} \frac{\partial}{\partial t} = \frac{1}{2} \{ \lozenge + \bar{\lozenge} \} \tag{3.88}$$

and that

$$i_k \{ \lozenge + \bar{\lozenge} \} = i_k \lozenge + \lozenge i_k + 2 \frac{\partial}{\partial x_k} = i_k \bar{\lozenge} + \bar{\lozenge} i_k - 2 \frac{\partial}{\partial x_k}. \tag{3.89}$$

Using (3.88) and (3.89) in (3.87), we obtain

$$\mathcal{P}_k = \int d^3r \ S \left\{ \Psi^\dagger \left(\frac{\hbar}{i} \frac{\partial}{\partial x_k} \Psi \right) + \mu^2 \Phi^\dagger \left(\frac{\hbar}{i} \frac{\partial}{\partial x_k} \Phi \right) \right\}$$
$$+ \frac{\hbar}{i2} \int d^3r \ S \left\{ \Psi^\dagger \left([i_k \lozenge + \lozenge i_k] \Psi \right) + \mu^2 \Phi^\dagger \left([-i_k \bar{\lozenge} - \bar{\lozenge} i_k] \Phi \right) \right\}, \tag{3.90}$$

where we have used the first expression for $i_k \{\Diamond + \bar{\Diamond}\}$, which contains no $\bar{\Diamond}$ in (3.89), to substitute for $(i/c)(\partial/\partial t)$ in the term involving $\Psi^\dagger$ and Ψ in (3.87), but we have used the *second* expression for $i_k \{\Diamond + \bar{\Diamond}\}$, which contains no $\Diamond$ in (3.89), to substitute for $(i/c)(\partial/\partial t)$ in the term involving $\Phi^\dagger$ and Φ in (3.87). Now, we will show that the second line in (3.90) vanishes identically.

As $\Psi^\dagger \Diamond \Psi + \mu^2 \Phi^\dagger \bar{\Diamond} \Phi = 0$ and $\Psi^\dagger \bar{\Diamond} \Psi + \mu^2 \Phi^\dagger \Diamond \Phi = 0$, it follows that

$$\Psi^\dagger \left(\frac{i}{c} \frac{\partial}{\partial t} \Psi \right) + \mu^2 \Phi^\dagger \left(\frac{i}{c} \frac{\partial}{\partial t} \Phi \right) = - \left(\frac{i}{c} \frac{\partial}{\partial t} \Psi^\dagger \right) \Psi - \mu^2 \left(\frac{i}{c} \frac{\partial}{\partial t} \Phi^\dagger \right) \Phi. \quad (3.91)$$

Now, using the same arguments that we used to derive (3.86) from (3.82), we find

$$S \left\{ \Psi^\dagger i_k \left(\frac{i}{c} \frac{\partial}{\partial t} \Psi \right) + \mu^2 \Phi^\dagger (-i_k) \left(\frac{i}{c} \frac{\partial}{\partial t} \Phi \right) \right\}$$

$$= -S \left\{ \left(\frac{i}{c} \frac{\partial}{\partial t} \Psi^\dagger \right) i_k \Psi + \mu^2 \left(\frac{i}{c} \frac{\partial}{\partial t} \Phi^\dagger \right) (-i_k) \Phi \right\}. \quad (3.92)$$

Using (3.92) and integrating by parts the spatial components of $\Diamond i_k$ and $\bar{\Diamond} i_k$, we find that

$$\int d^3r \, S \left\{ \Psi^\dagger \left([i_k \Diamond + \Diamond i_k] \Psi \right) + \mu^2 \Phi^\dagger \left([-i_k \bar{\Diamond} - \bar{\Diamond} i_k] \Phi \right) \right\}$$

$$= \int d^3r \, S \left\{ \Psi^\dagger i_k \left(\Diamond \Psi \right) - \left(\Psi^\dagger \Diamond \right) i_k \Psi - \mu^2 \Phi^\dagger i_k \left(\bar{\Diamond} \Phi \right) + \mu^2 \left(\Phi^\dagger \bar{\Diamond} \right) i_k \Phi \right\}$$

$$= \int d^3r \, S \left\{ \mu^2 \Psi^\dagger i_k \Phi + \mu^2 \Phi^\dagger i_k \Psi - \mu^2 \Phi^\dagger i_k \Psi - \mu^2 \Psi^\dagger i_k \Phi \right\}$$

$$= 0, \quad (3.93)$$

where we have used (3.74) and (3.75) to get the third line of (3.93) from the second.

So we have shown that the expectation value of the nth component of three-momentum is given by

$$\mathcal{P}_n = \int d^3r \, S \left\{ \Psi^\dagger \left(\frac{\hbar}{i} \frac{\partial}{\partial x_n} \Psi \right) + \mu^2 \Phi^\dagger \left(\frac{\hbar}{i} \frac{\partial}{\partial x_n} \Phi \right) \right\}. \quad (3.94)$$

A similar calculation shows that the value of the nth component of angular momentum $J_n = (\hbar/i) \int d^3r S\{\mathcal{F}^\dagger[\sum_{mq} \epsilon_{nmq} x_m i_q]\mathcal{F} + \mu^2 \mathcal{A}^\dagger[\sum_{mq} \epsilon_{nmq} x_m i_q]\mathcal{A}\}$ is given, as an expectation value in terms of the wavefunction, by (see Problem 3.4)

$$J_n = \int d^3r \, S \left\{ \Psi^\dagger \sum_{mq} \epsilon_{nmq} x_m \left(\frac{\hbar}{i} \frac{\partial}{\partial x_q} \Psi \right) + \mu^2 \Phi^\dagger \sum_{mq} \epsilon_{nmq} x_m \left(\frac{\hbar}{i} \frac{\partial}{\partial x_q} \Phi \right) \right\}$$

$$+ \int d^3r \, S \left\{ \Psi^\dagger i\hbar \, i_n \Psi + \mu^2 \Phi^\dagger i\hbar \, i_n \Phi \right\}. \quad (3.95)$$

The first line on the right-hand side of (3.95) is the expectation value of the nth component of the orbital angular momentum. The second line is the expectation value of the nth component of the intrinsic angular momentum or spin of the photon. Here, as the expression that appears sandwiched between $\Psi^\dagger$ and Ψ and between $\Phi^\dagger$ and Φ turned out to be one of the quaternion units i_n, we must be very careful. In particular, as we are taking the scalar part of the whole integrand, we cannot say that $i\hbar i_n$ is the nth component of the spin operator. We go back to this point in Section 3.5.

3.3 THE PROBLEM WITH THE WAVEFUNCTION IN CONFIGURATION SPACE

We have thus succeeded in finding a wave equation [i.e., (3.74) and (3.75)] and a four-vector (i.e., $\Psi^\dagger\Psi + \mu^2\Phi^\dagger\Phi$) that satisfies a continuity equation. Moreover, the expectation values of physical observables are given as averages of the corresponding operator, weighed by the timelike component of this four-vector. In other words, the expectation values are given in exactly the way that they should be given if the timelike component of this four-vector is the probability density. Can we then interpret this four-vector as the probability four-current and Ψ and Φ as our particle's wavefunctions? If we want to do so in *configuration space*, there is a serious problem.

To show what this problem is, let us write down the relation between the ordinary three-dimensional Fourier transform of the wavefunctions (Ψ and Φ) and that of the fields ($\mathcal{F}$ and $\mathcal{A}$). Using (3.72), we find the following expression for the ordinary three-dimensional Fourier transform of Ψ_α:

$$
\begin{aligned}
\underline{\Psi}_\alpha(\mathbf{k}, t) &= \int \frac{d^3r}{(2\pi)^3}\, e^{-i\mathbf{k}\cdot\mathbf{r}} \Psi_\alpha(\mathcal{R}) \\
&= \int \frac{d^3r}{(2\pi)^3}\, e^{-i\mathbf{k}\cdot\mathbf{r}} \int d^3k'\, e^{i\mathbf{k}'\cdot\mathbf{r}} \frac{\underline{F}_\alpha(\mathbf{k}', t)}{\sqrt[4]{k'^2 + \mu^2}} \\
&= \int d^3k\, \delta^3(\mathbf{k}' - \mathbf{k}) \frac{\underline{F}_\alpha(\mathbf{k}', t)}{\sqrt[4]{k'^2 + \mu^2}} \\
&= \frac{\underline{F}_\alpha(\mathbf{k}, t)}{\sqrt[4]{k^2 + \mu^2}}.
\end{aligned}
\tag{3.96}
$$

Analogously, the ordinary three-dimensional Fourier transform of Φ_α is given by

$$
\underline{\Phi}_\alpha(\mathbf{k}, t) = \frac{\underline{A}_\alpha(\mathbf{k}, t)}{\sqrt[4]{k^2 + \mu^2}}.
\tag{3.97}
$$

Now, the four-momentum is a three-dimensional volume integral of the quaternion $\mathcal{F}^\dagger\mathcal{F} + \mu^2\mathcal{A}^\dagger\mathcal{A}$. So, in a given Lorentz frame, the scalar component of this quaternion is the energy density of our particle. As we can see, the energy density at a point depends on the values of the fields at that point and nowhere else. But from (3.96) and (3.97), we see that the fields $\mathcal{F}$ and $\mathcal{A}$ depend on Ψ and Φ in a nonlocal way:

$$
\mathcal{F}_\alpha(\mathcal{R}) = \int d^3k\, e^{i\mathbf{k}\cdot\mathbf{r}} \underline{F}_\alpha(\mathbf{k}, t)
$$

$$= \int d^3k \; e^{i\mathbf{k}\cdot\mathbf{r}} \sqrt[4]{k^2 + \mu^2} \; \underline{\Psi}_\alpha(\mathbf{k}, t)$$

$$= \int d^3k \; e^{i\mathbf{k}\cdot\mathbf{r}} \sqrt[4]{k^2 + \mu^2} \int \frac{d^3r'}{(2\pi)^3} e^{-i\mathbf{k}\cdot\mathbf{r}'} \Psi_\alpha(\mathcal{R}')$$

$$= \int d^3r' \; G(\mathbf{r}|\mathbf{r}')\Psi_\alpha(\mathcal{R}'), \tag{3.98}$$

where[33]

$$G(\mathbf{r}|\mathbf{r}') = \int \frac{d^3k}{(2\pi)^3} \; \sqrt[4]{k^2 + \mu^2} \; e^{i(\mathbf{r}-\mathbf{r}')\cdot\mathbf{k}}. \tag{3.99}$$

Analogously,

$$\mathcal{A}_\alpha(\mathcal{R}) = \int d^3r' \; G(\mathbf{r}|\mathbf{r}')\Phi_\alpha(\mathcal{R}'). \tag{3.100}$$

If it were not for $\sqrt[4]{k^2 + \mu^2}$ in (3.99), $G(\mathbf{r}|\mathbf{r}')$ would be a three-dimensional delta function and $\mathcal{F}_\alpha(\mathcal{R})$ and $\mathcal{A}_\alpha(\mathcal{R})$ would depend on Ψ_α and Φ_α, respectively, at the space-time point $\mathcal{R}$ only. But as it stands, the fields depend on the wavefunctions in a nonlocal way. This nonlocal dependency implies that the wavefunction can even vanish at a point in space where the energy density does not. So we cannot interpret the scalar component of the quaternion $\Psi^\dagger\Psi + \mu^2\Phi^\dagger\Phi$ at a point in space as the probability of finding the photon (the energy quantum) at that point.[34] Moreover, as we will see in Chapter 5, the interaction with an electron, say, at a given point in space depends on the value of the *fields* at that point. But due to the nonlocal relation between the fields and the wavefunction, this interaction will be nonlocal in the wavefunction. So the electron will interact with the photon even if it sits at a place where the wavefunction vanishes (see Sec. 25(d) in [480]).

This difficulty is not present in momentum space, where the fields depend on the wavefunction in a local way. It is perfectly legitimate to talk about the wavefunction of the photon in momentum space [12]. In fact, the photon wavefunction in momentum space is often used in high-energy physics. We can also see immediately that this problem does not exist in the nonrelativistic limit either. For then, the particle's momentum $\hbar\mathbf{k}$ is always much smaller than μ, so that $G(\mathbf{r}|\mathbf{r}')$ reduces to $\sqrt{\mu}\delta^3(\mathbf{r}-\mathbf{r}')$, making the fields become local in Ψ and Φ.

3.4 BACK TO VECTOR NOTATION

Quaternions are very powerful and we have been able to go quite a long way with them. For some purposes, however, as Cayley once put it [461], "a quaternion formula

---

[33] Another way to write (3.98) is to define the operator $\sqrt[4]{\mu^2 - \nabla^2}$. Then, as $\sqrt[4]{\mu^2 - \nabla^2}e^{i\mathbf{k}\cdot\mathbf{r}} = \sqrt[4]{\mu^2 + k^2}e^{i\mathbf{k}\cdot\mathbf{r}}$, we can rewrite (3.98) as $\mathcal{F}_\alpha(\mathcal{R}) = \sqrt[4]{\mu^2 - \nabla^2}\Psi_\alpha(\mathcal{R})$. See [387].

[34] We will see later that this interpretation holds in a course-grained way [119, 311, 428]. It also holds in the paraxial approximation [151].

is like a pocket-map that for use must be unfolded." Before going back to the more familiar vector notation, it is useful to find out which of the vectors in $\mathcal{F}$ and $\mathcal{A}$ are polar and which are axial. A polar vector remains the same when we change from a right-handed coordinate system to a left-handed one. An axial vector reverses its direction (i.e., goes into minus itself) under such a change of coordinates.

To go from our right-handed coordinate system to a left-handed one, we can change x_1 into $-x_1$. Let us denote the new coordinates and all quantities in the new coordinate system in general by adding a prime to their old names. The chainrule tells us that the partial derivatives in the new coordinates relate to those in the old ones by

$$\frac{\partial}{\partial t'} = \frac{\partial}{\partial t}, \quad \frac{\partial}{\partial x'_1} = -\frac{\partial}{\partial x_1}, \quad \frac{\partial}{\partial x'_2} = \frac{\partial}{\partial x_2}, \quad \frac{\partial}{\partial x'_3} = \frac{\partial}{\partial x_3}. \tag{3.101}$$

In quaternion notation, this can be written very compactly as

$$\Diamond' = i_1 \Diamond^* i_1. \tag{3.102}$$

Now (3.42) and (3.43) cannot depend on our particular choice of coordinates, so they must remain valid in the left-handed coordinate system: that is,

$$\Diamond' \mathcal{F}' = \mu^2 \mathcal{A}', \tag{3.103}$$
$$\bar{\Diamond}' \mathcal{A}' = \mathcal{F}'. \tag{3.104}$$

Substituting (3.102) into (3.103) and (3.104), multiplying by i_1 on the left and on the right, and taking the complex conjugate, we find that

$$\Diamond(-1)i_1\mathcal{F}'^* i_1 = \mu^2 \, i_1 \mathcal{A}'^* i_1, \tag{3.105}$$
$$\bar{\Diamond} i_1 \mathcal{A}'^* i_1 = -i_1 \mathcal{F}'^* i_1. \tag{3.106}$$

To recover (3.42) and (3.43) from (3.105) and (3.106), $\mathcal{F}$ and $\mathcal{A}$ must transform according to[35]

$$\mathcal{F} = (-1)i_1 \mathcal{F}'^* i_1, \tag{3.107}$$
$$\mathcal{A} = i_1 \mathcal{A}'^* i_1. \tag{3.108}$$

Now we notice that the imaginary part of[36] **f** in (3.30) must be an axial vector, for it reverses direction when the coordinate system becomes left-handed (its i_1 component remains the same and both its i_2 and i_3 components change sign; see Figure 3.1). The real part of **f** is a polar vector, for it remains unchanged when the coordinate system becomes left-handed (its i_1 component changes sign and both its i_2 and i_3 components remain the same; see Figure 3.2). As to $\mathcal{A}$, we can infer from its anti-Hermiticity that its time component must be a pure imaginary number while the coefficients of its space components must be real. So the vector part of $\mathcal{A}$ is a polar vector.

---

[35]If the minus sign in (3.105) and (3.106) were grouped with $\mathcal{A}$ and we identified $-i_1 \mathcal{A}'^* i_1$ instead of $i_1 \mathcal{A}'^* i_1$ with $\mathcal{A}$, the time component of $\mathcal{A}$ would change sign when the left-handed coordinates are adopted, which does not make any sense. So the minus sign can only be incorporated in the transformation relation for $\mathcal{F}$ as we do in (3.107) and (3.108).

[36]**f** is the vector whose Cartesian components are the f_n in (3.30).

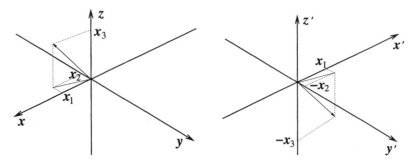

Figure 3.1 Here is a graphical representation of what happens to an axial vector, when the coordinate system becomes left-handed. On the left, we see an axial vector and its Cartesian components in a right-handed coordinate system. On the right, the coordinate system becomes left-handed by reverting the direction of the x-axis. The x-component of the vector remains unchanged, but its y and z components change sign. We can clearly see, on the right, that this makes the vector reverse its direction.

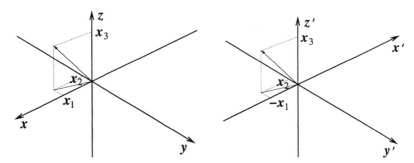

Figure 3.2 Here is a graphical representation of what happens to a polar vector when the coordinate system becomes left-handed. On the left we see an axial vector and its Cartesian components in a right-handed coordinate system. On the right the coordinate system becomes left-handed by reverting the direction of the x-axis. Unlike the axial vector, now it is the x-component that changes sign and both the y and z components remain unchanged. We can clearly see, on the right, that this makes the vector remain unchanged.

The electric field is a polar vector and the magnetic field is an axial vector. So by analogy with them, we will denote by $\mathbf{E}/\sqrt{8\pi\hbar c}$ and $-\mathbf{B}/\sqrt{8\pi\hbar c}$ the real and the imaginary parts, respectively, of $\mathbf{f}$. Also, by analogy of (3.43) with the relations between the fields and their potentials, we denote the vector part of $\mathcal{A}$ by $-\mathbf{A}/\sqrt{8\pi\hbar c}$ (this is also reinforced by the vector potential being a polar vector), and its time component by $i\varphi/\sqrt{8\pi\hbar c}$. The square roots of $8\pi\hbar c$ were included to get the correct dimensions and numerical factors, as $\int d^3r(\mathcal{F}^\dagger\mathcal{F} + \mu^2\mathcal{A}^\dagger\mathcal{A})$ is the four-momentum in units of $\hbar$. Then substituting

$$\mathcal{F} = \sum_{n=1}^{3} i_n \, i \frac{E_n - i B_n}{\sqrt{8\pi\hbar c}}, \tag{3.109}$$

$$\mathcal{A} = i \frac{\varphi}{\sqrt{8\pi\hbar c}} - \sum_{n=1}^{3} i_n \frac{A_n}{\sqrt{8\pi\hbar c}} \tag{3.110}$$

into (3.42), we obtain the following equations in vector notation (see Problem 3.1)

$$\nabla \cdot \mathbf{B} = 0, \tag{3.111}$$

$$\nabla \cdot \mathbf{E} = -\mu^2 \varphi, \tag{3.112}$$

$$\nabla \wedge \mathbf{B} = \frac{1}{c} \frac{\partial \mathbf{E}}{\partial t} - \mu^2 \mathbf{A}, \tag{3.113}$$

$$\nabla \wedge \mathbf{E} = -\frac{1}{c} \frac{\partial \mathbf{B}}{\partial t}. \tag{3.114}$$

Doing the same with (3.43), we find that

$$\mathbf{B} = \nabla \wedge \mathbf{A}, \tag{3.115}$$

$$\mathbf{E} = -\nabla\varphi - \frac{1}{c} \frac{\partial \mathbf{A}}{\partial t}, \tag{3.116}$$

$$\frac{1}{c} \frac{\partial \varphi}{\partial t} + \nabla \cdot \mathbf{A} = 0. \tag{3.117}$$

Substituting (3.109) and (3.110) in the expression for the four-momentum $\hbar \int d^3 r \{ \mathcal{F}^\dagger \mathcal{F} + \mu^2 \mathcal{A}^\dagger \mathcal{A} \}$, we find that the energy density E_D and the three-momentum density $\mathbf{P}_D$ are given by [44, 238]

$$E_D = \frac{1}{8\pi} \left\{ \mathbf{E}^2 + \mathbf{B}^2 + \mu^2 \left(\mathbf{A}^2 + \varphi^2 \right) \right\}, \tag{3.118}$$

$$\mathbf{P}_D = \frac{1}{4\pi c} \left\{ \mathbf{E} \wedge \mathbf{B} + \mu^2 \varphi \mathbf{A} \right\}. \tag{3.119}$$

Equations (3.111) to (3.117) are the Proca equations for a massive spin 1 particle. They look a lot like Maxwell equations except that here the potentials appear in the field equations themselves through the square of the inverse of the Compton wavelength multiplied by the square of 2π (i.e., μ^2). So the potentials here are as physical as the fields and this is emphasized by the lack of gauge invariance, with the Lorentz gauge being compulsory (3.117). Proca equations are the only possible generalization of Maxwell equations for a photon with nonvanishing rest mass which satisfy the following four conditions (see [238]):

1. The Lorentz force expression remains valid.

2. The fields at a point of space in the vacuum are linear in the electric charge and current densities and also in their derivatives evaluated at this point.

3. There are no magnetic monopoles.

4. The transition to Maxwell's theory, when $\mu \to 0$, is smooth.

Could it really be that the photon has a rest mass? The underlying logic of physical theory, particularly relativity, suggests that the rest mass of the photon should vanish exactly. Unfortunately, all that can be hoped from experiment at the moment is an upper bound. But this upper bound is very low. A recent experiment [418] sets the limit at 1.2×10^{-51} g. To have an idea how small this is, even for an elementary particle, consider that the present experimental value for the mass of the electron (a very light particle) is roughly 9.1×10^{-28} g, so that the upper bound on the photon's mass is about 7.6×10^{23} times smaller than the mass of an electron. For a review of the search for the photon's mass and some of the early experiments, see [238, 367]. These experiments basically try to detect some of the consequences of the Proca equations that differ from those of Maxwell equations. For instance, the Proca equations with charges predict deviations from the ordinary electrostatic Coulomb potential. Their solutions lead to the Yukawa potential, which has finite rather than infinite range. This has been exploited by Crandell to devise an undergraduate laboratory experiment to set an upper bound to the photon's mass that can even approach 2×10^{-47} g [128].

3.5 THE LIMIT OF VANISHING REST MASS

For vanishing rest mass, equations (3.111)–(3.114) become Maxwell equations for the free radiation field. The second set of equations (3.115)–(3.117) (i.e., the equations for the potentials) are redundant now because the two sets are no longer coupled. In quaternion notation, this limit is most simply described by a single quaternion equation for the field,

$$\Diamond \mathcal{F} = 0, \tag{3.120}$$

and a single quaternion equation for the wavefunction,

$$\Diamond \Psi = 0. \tag{3.121}$$

All observables are now given only in terms of Ψ, making Φ redundant.

Let us convert our first quantized theory of light from quaternion notation to the more familiar matrix notation. Multiplying (3.121) by $i\hbar$ and writing explicitly in terms of components, the resulting action of the biquaternion gradient operator on the biquaternion wavefunction, we find after some minor rearranging,

$$i\hbar \frac{\partial}{\partial t} \sum_{n=1}^{3} i_n \psi_n = -\hbar c \sum_{n=1}^{3} i_n \epsilon_{nmq} \frac{\partial}{\partial x_m} \psi_q \tag{3.122}$$

for the vector part, and

$$\nabla \cdot \psi = 0 \tag{3.123}$$

for the scalar part. We can write (3.122) in matrix form as a Schrödinger-like equation

$$i\hbar\frac{\partial}{\partial t}\vec{\psi} = H\vec{\psi},$$

(3.124)

where the wavefunction is the column matrix

$$\vec{\psi} = \begin{bmatrix} \psi_1 \\ \psi_2 \\ \psi_3 \end{bmatrix}$$

(3.125)

and the Hamiltonian is the 3×3 Hermitian matrix

$$H = \hbar c \begin{bmatrix} 0 & \partial/\partial x_3 & -\partial/\partial x_2 \\ -\partial/\partial x_3 & 0 & \partial/\partial x_1 \\ \partial/\partial x_2 & -\partial/\partial x_1 & 0 \end{bmatrix}.$$

(3.126)

What about the expectation values. The reader can easily check that the expectation value of the energy [i.e., (3.81) with $\mu \to 0$] is given by

$$\mathcal{E} = \int d^3r \, \vec{\psi}^\dagger \left(i\hbar\frac{\partial}{\partial t}\vec{\psi} \right),$$

(3.127)

where

$$\vec{\psi}^\dagger = \begin{bmatrix} \psi_1^* & \psi_2^* & \psi_3^* \end{bmatrix}.$$

(3.128)

Analogously, for the nth component of momentum [i.e., (3.94) with $\mu \to 0$] we have

$$\mathcal{P}_n = \int d^3r \, \vec{\psi}^\dagger \left(\frac{\hbar}{i}\frac{\partial}{\partial x_n}\vec{\psi} \right).$$

(3.129)

Now for the angular momentum [i.e., (3.95) with $\mu \to 0$] we find immediately that the expectation value of the nth component of the orbital angular momentum is given by

$$L_n = \int d^3r \, \vec{\psi}^\dagger \left(\sum_{mq} \epsilon_{nmq} x_m \frac{\hbar}{i}\frac{\partial}{\partial x_q}\vec{\psi} \right).$$

(3.130)

But the spin contribution to the angular momentum needs a bit of extra thought, as it involves one of the quaternion units sandwiched by the wavefunction.

In terms of components,

$$S\left\{ \Psi^\dagger i_n \Psi \right\} = -\sum_{mq} \epsilon_{nmq} \psi_m^* \psi_q.$$

(3.131)

From (3.131) we see that the spin contribution to (3.95) can be written in the same spirit as (3.127), (3.129), and (3.130) (i.e., as the expectation value of the nth component of a spin operator, $\hat{S}_n$). But unlike the other operators in (3.127), (3.129), and (3.130), this operator must be a matrix. According to (3.131), this matrix must be given by

$$\hat{S}_n = i\hbar \begin{bmatrix} 0 & -\epsilon_{n12} & -\epsilon_{n13} \\ -\epsilon_{n21} & 0 & -\epsilon_{n23} \\ -\epsilon_{n31} & -\epsilon_{n32} & 0 \end{bmatrix} = i\hbar \begin{bmatrix} 0 & -\delta_{n3} & \delta_{n2} \\ \delta_{n3} & 0 & -\delta_{n1} \\ -\delta_{n2} & \delta_{n1} & 0 \end{bmatrix}.$$

(3.132)

Explicitly, the three spin matrices are

$$
\hat{S}_1 = \hbar \begin{bmatrix} 0 & 0 & 0 \\ 0 & 0 & -i \\ 0 & i & 0 \end{bmatrix}, \quad
\hat{S}_2 = \hbar \begin{bmatrix} 0 & 0 & i \\ 0 & 0 & 0 \\ -i & 0 & 0 \end{bmatrix}, \quad
\hat{S}_3 = \hbar \begin{bmatrix} 0 & -i & 0 \\ i & 0 & 0 \\ 0 & 0 & 0 \end{bmatrix}.
\tag{3.133}
$$

You can easily verify that they obey the angular momentum commutation relations (see Problem 3.6)

$$
\left[\hat{S}_n, \hat{S}_m \right] = i\hbar \sum_q \epsilon_{qnm} \hat{S}_q.
\tag{3.134}
$$

You can also easily verify that (see Problem 3.7)

$$
\hat{S}^2 = \sum_n \hat{S}_n = \hbar^2 \begin{bmatrix} 2 & 0 & 0 \\ 0 & 2 & 0 \\ 0 & 0 & 2 \end{bmatrix}.
\tag{3.135}
$$

As the eigenvalues of $\hat{S}^2$ are given by $s(s+1)\hbar^2$, where s must be integer or half-integer, it follows from (3.135) that s can only be 1. So the photon has spin 1.

There is an interesting connection between the spin matrices and the Schrödinger-like equation (3.124). If we define the Pauli-like matrices

$$
\hat{\sigma}_n = \frac{1}{\hbar} \hat{S}_n \quad \text{and} \quad \hat{\boldsymbol{\sigma}} = \mathbf{i}\hat{\sigma}_1 + \mathbf{j}\hat{\sigma}_2 + \mathbf{k}\hat{\sigma}_3,
\tag{3.136}
$$

comparing the pq elements (i.e., pth row and qth column) of $\hat{H}\vec{\psi}$ and $\hat{\sigma}_n \vec{\psi}$:

$$
\left(\hat{H}\vec{\psi} \right)_{pq} = -\hbar c \sum_n \epsilon_{pnq} \frac{\partial}{\partial x_n} \psi_q,
\tag{3.137}
$$

$$
\left(\hat{\sigma}_n \vec{\psi} \right)_{pq} = -i\epsilon_{pnq}\psi_q,
\tag{3.138}
$$

we find that

$$
\left(\hat{H}\vec{\psi} \right)_{pq} = c \left(\sum_n \hat{\sigma}_n \frac{\hbar}{i} \frac{\partial}{\partial x_n} \vec{\psi} \right)_{pq} = c \left(\hat{\boldsymbol{\sigma}} \cdot \hat{\mathbf{p}} \vec{\psi} \right)_{pq}.
\tag{3.139}
$$

Thus the Schrödinger-like equation (3.124) can also be written as

$$
i\hbar \frac{\partial}{\partial t} \vec{\psi} = c \hat{\boldsymbol{\sigma}} \cdot \hat{\mathbf{p}} \vec{\psi}.
\tag{3.140}
$$

This equation was derived before in ways completely different from the present approach by a number of authors, including Oppenheimer [472], Good [241], and Bialynicki-Birula [53–55]. As noted by Bialynicki-Birula [54], this equation has the same form (the $\hat{\boldsymbol{\sigma}}$ matrices are different, of course) as Weyl's equation for neutrinos [638]. Moreover, from it we see that the component of spin that is a good quantum number (i.e., that commutes with the Hamiltonian) is the component along the direction of the momentum of the photon: the helicity. This is another reminder that momentum rather than configuration space is the right place for introducing the wavefunction of a relativistic particle. So we will go to momentum space.

In momentum space, (3.140) reads simply

$$i\hbar \frac{\partial}{\partial t} \vec{\underline{\psi}} = c\hat{\sigma} \cdot \mathbf{p} \vec{\underline{\psi}}, \tag{3.141}$$

where

$$\vec{\underline{\psi}}(\mathbf{p}, t) = \int \frac{d^3 r}{\sqrt{(2\pi)^3 \hbar}} e^{-i\mathbf{p}\cdot\mathbf{r}/\hbar} \vec{\psi}(\mathbf{r}, t). \tag{3.142}$$

The eigenvalue problem for $(\hat{\sigma} \cdot \mathbf{p}/p)$ is

$$\hat{\sigma} \cdot \frac{\mathbf{p}}{p} \vec{\underline{\psi}} = \xi \vec{\underline{\psi}}. \tag{3.143}$$

Then

$$\frac{1}{p} \begin{vmatrix} -p\xi & -ip_3 & ip_2 \\ ip_3 & -p\xi & -ip_1 \\ -ip_2 & ip_1 & -p\xi \end{vmatrix} = 0. \tag{3.144}$$

Calculating the determinant in (3.144), we find that the eigenvalues ξ must satisfy the following secular equation:

$$(\xi^2 - 1)\xi = 0, \tag{3.145}$$

whose solutions are $\xi = -1, 0, 1$. To look at the corresponding eigenvectors, we are better off changing the coordinate system to one where one of the axes coincides with the direction of the momentum. So we will briefly show how we can change to such a coordinate system.

Let $\{e_1, e_2, e_3\}$ be the present right-handed coordinate axes. We want to rewrite the wavefunction and the matrices using the new right-handed set of coordinate axes $\{\varepsilon_1, \varepsilon_2, \varepsilon_3\}$, where $\varepsilon_3 = \mathbf{p}/p$ and ε_1 and ε_2 are two arbitrary directions orthogonal to each other and to $\mathbf{p}$. A given vector can be represented in the present coordinates, $\beta_1 e_1 + \beta_2 e_2 + \beta_3 e_3$, or in the new one, $\alpha_1 \varepsilon_1 + \alpha_2 \varepsilon_2 + \alpha_3 \varepsilon_3$. As it is still the same vector no matter which coordinate system we use, we must have

$$\beta_1 e_1 + \beta_2 e_2 + \beta_3 e_3 = \alpha_1 \varepsilon_1 + \alpha_2 \varepsilon_2 + \alpha_3 \varepsilon_3. \tag{3.146}$$

Taking the scalar product of ε_n with (3.146) and using the orthonormality of $\{\varepsilon_1, \varepsilon_2, \varepsilon_3\}$, we find that

$$\alpha_n = \sum_m (e_m \cdot \varepsilon_n)\beta_m. \tag{3.147}$$

Analogously, taking the scalar product of e_n with (3.146) and using the orthonormality of $\{e_1, e_2, e_3\}$, we obtain

$$\beta_n = \sum_m (\varepsilon_m \cdot e_n)\alpha_m. \tag{3.148}$$

Substituting (3.148) in (3.147), we obtain $\alpha_n = \sum_{m,l}(\mathbf{e}_m \cdot \boldsymbol{\varepsilon}_n)(\boldsymbol{\varepsilon}_l \cdot \mathbf{e}_m)\alpha_l$. As this must hold for any α_l, it follows that

$$\sum_{m,l}(\mathbf{e}_m \cdot \boldsymbol{\varepsilon}_n)(\boldsymbol{\varepsilon}_l \cdot \mathbf{e}_m) = \delta_{ln}. \tag{3.149}$$

In other words, the matrix

$$\mathsf{T} \equiv \begin{bmatrix} \mathbf{e}_1 \cdot \boldsymbol{\varepsilon}_1 & \mathbf{e}_2 \cdot \boldsymbol{\varepsilon}_1 & \mathbf{e}_3 \cdot \boldsymbol{\varepsilon}_1 \\ \mathbf{e}_1 \cdot \boldsymbol{\varepsilon}_2 & \mathbf{e}_2 \cdot \boldsymbol{\varepsilon}_2 & \mathbf{e}_3 \cdot \boldsymbol{\varepsilon}_2 \\ \mathbf{e}_1 \cdot \boldsymbol{\varepsilon}_3 & \mathbf{e}_2 \cdot \boldsymbol{\varepsilon}_3 & \mathbf{e}_3 \cdot \boldsymbol{\varepsilon}_3 \end{bmatrix} \tag{3.150}$$

changes from the present coordinate system to the new one, and its inverse

$$\mathsf{T}^{-1} = \begin{bmatrix} \mathbf{e}_1 \cdot \boldsymbol{\varepsilon}_1 & \mathbf{e}_1 \cdot \boldsymbol{\varepsilon}_2 & \mathbf{e}_1 \cdot \boldsymbol{\varepsilon}_3 \\ \mathbf{e}_2 \cdot \boldsymbol{\varepsilon}_1 & \mathbf{e}_2 \cdot \boldsymbol{\varepsilon}_2 & \mathbf{e}_2 \cdot \boldsymbol{\varepsilon}_3 \\ \mathbf{e}_3 \cdot \boldsymbol{\varepsilon}_1 & \mathbf{e}_3 \cdot \boldsymbol{\varepsilon}_2 & \mathbf{e}_3 \cdot \boldsymbol{\varepsilon}_3 \end{bmatrix} \tag{3.151}$$

changes from the new coordinates back to the old ones.

In the new coordinates, $\hat{\sigma} \cdot (\mathbf{p}/p)\vec{\psi}$ is $\mathsf{T}\hat{\sigma} \cdot (\mathbf{p}/p)\vec{\psi}$. Inserting $\mathsf{T}^{-1}\mathsf{T} = 1$ between $\hat{\sigma} \cdot (\mathbf{p}/p)$ and $\vec{\psi}$, we have

$$\mathsf{T}\hat{\sigma} \cdot \frac{\mathbf{p}}{p}\vec{\psi} = \mathsf{T}\hat{\sigma} \cdot \frac{\mathbf{p}}{p}\mathsf{T}^{-1}\mathsf{T}\vec{\psi}, \tag{3.152}$$

where

$$\vec{\psi}' \equiv \mathsf{T}\vec{\psi} \tag{3.153}$$

is the wavefunction written in the new coordinates, and

$$\hat{\sigma}'_3 \equiv \mathsf{T}\hat{\sigma} \cdot \frac{\mathbf{p}}{p}\mathsf{T}^{-1} \tag{3.154}$$

is $\hat{\sigma} \cdot (\mathbf{p}/p)$ in the new coordinates. Elementwise, (3.154) is given by

$$\begin{aligned}(\hat{\sigma}'_3)_{l'q'} &= \sum_{lq}(\mathbf{e}_l \cdot \boldsymbol{\varepsilon}_{l'})\sum_n (\hat{\sigma}_n)_{lq}\,(\mathbf{e}_n \cdot \boldsymbol{\varepsilon}_3)(\boldsymbol{\varepsilon}_{q'} \cdot \mathbf{e}_q) \\ &= -i\sum_l(\mathbf{e}_l \cdot \boldsymbol{\varepsilon}_{l'})\underbrace{\sum_{qn}\epsilon_{lnq}(\mathbf{e}_n \cdot \boldsymbol{\varepsilon}_3)(\boldsymbol{\varepsilon}_{q'} \cdot \mathbf{e}_q)}_{\mathbf{e}_l \cdot (\boldsymbol{\varepsilon}_3 \wedge \boldsymbol{\varepsilon}_{q'})} \\ &= -i\sum_l(\mathbf{e}_l \cdot \boldsymbol{\varepsilon}_{l'})\mathbf{e}_l \cdot \left\{\sum_m \boldsymbol{\varepsilon}_m \epsilon_{m3q'}\right\} \\ &= -i\sum_{lm}(\mathbf{e}_l \cdot \boldsymbol{\varepsilon}_{l'})(\mathbf{e}_l \cdot \boldsymbol{\varepsilon}_m)\epsilon_{m3q'}, \end{aligned} \tag{3.155}$$

where we have used the fact that $(\hat{\sigma}_n)_{lq} = -i\epsilon_{lnq}$ and that $\boldsymbol{\varepsilon}_3 \wedge \boldsymbol{\varepsilon}_{q'}$ can also be written as $\sum_m \boldsymbol{\varepsilon}_m \epsilon_{m3q'}$. Substituting (3.149) in the last line of (3.155), we finally find that

$$(\hat{\sigma}'_3)_{l'q'} = -i\epsilon_{l'3q'}; \tag{3.156}$$

that is, $\hat{\sigma} \cdot (\mathbf{p}/p)$ in the new coordinate system coincides with the old[37] $\hat{\sigma}_3$:

$$\hat{\sigma}'_3 = \begin{bmatrix} 0 & -i & 0 \\ i & 0 & 0 \\ 0 & 0 & 0 \end{bmatrix}. \tag{3.157}$$

In the new coordinates, the eigenvalue problem is given by

$$\hat{\sigma}'_3 \vec{\psi}' = \xi \vec{\psi}'. \tag{3.158}$$

Now then, for helicity $\xi = -\hbar$, this yields $\psi'_2 = -i\psi'_1$ and $\psi'_3 = 0$, where ψ'_1, ψ'_2, and ψ'_3 are the components of $\vec{\psi}'$. Analogously, for helicity $\xi = \hbar$, (3.158) yields $\psi'_2 = i\psi'_1$ and $\psi'_3 = 0$. Last but not least, for helicity $\xi = 0$, (3.158) yields $\psi'_1 = \psi'_2 = 0$. So for $\xi = \pm\hbar$, $\vec{\psi}'$ is transverse. But for $\xi = 0$, $\vec{\psi}'$ is longitudinal (i.e., parallel to the momentum). This is forbidden by the transversality condition (3.123), which in momentum space reads simply $\psi'_3 = 0$. Thus, due to the transversality condition, the photon can have only two helicities $\xi = \pm\hbar$ despite having spin 1. This is a general consequence of vanishing rest mass, as is proved in the group theory treatment (see the recommended reading at the end of the chapter). In fact, we can see from what was done earlier in this section that when μ does not vanish, the divergence does not either, allowing $\xi = 0$.

3.6 SECOND QUANTIZATION OF THE EXTREME QUANTUM THEORY OF LIGHT

I will not derive the formalism of second quantization from scratch here; for that the reader should look at some of the references mentioned in the recommended reading section at the end of the chapter. I assume a certain familiarity with the many-body second-quantization formalism. This section has two main aims. The first is to clarify the meaning of the negative-energy solutions of the photon's wave equation (i.e., the antiphoton). The second is to establish the connection between this first-quantized theory of light and ordinary quantum electrodynamics derived through canonical quantization.

I would like to stress the importance of obtaining the zero-point energy of the electromagnetic radiation field. There are a number of phenomena that point to the existence of a zero-point energy of the radiation field, such as the Casimir effect (see Chapter 2) and the blackbody radiation spectrum [325, 451]. Bialynicki-Birula [54] deals with this problem using hindsight and the accumulated knowledge on classical electrodynamics (that following Oppenheimer [472], we do not wish to use in our extreme quantum theory of light). Sipe [568] derives a second-quantized Hamiltonian that does not have zero-point energy. In his extension of Cook's theory [119, 120],

---

[37]For a particular example of the general calculation just presented, see Problem 3.8.

Inagaki [310] discusses second quantization but also derives a Hamiltonian with no zero-point energy terms. Akhiezer and Berestetsky [12] (Sec. 15.4) also deal with the second quantization of photons, but they do not discuss the problem of negative energies and the antiphoton, nor do they show how the zero-point energy arises in the second-quantization approach. Hammer and Good [269] do get the zero-point energy term but then throw it away in an ad hoc manner. The approach presented here is similar to theirs, but only in some respects.

We start this discussion by insisting on interpreting the Hamiltonian in (3.140) as the energy. Good [241] also derived the wave equation (3.140) for the photon and called

$$H = c\hat{\sigma} \cdot \hat{\mathbf{p}} \qquad (3.159)$$

the Hamiltonian, but he wrote that it is not to be identified with the energy. He does not explain why. One obvious reason would be that if H is the energy, states with arbitrarily negative energy occur and there is no longer a stable ground state [58]: A positive-energy photon could then tumble into any negative-energy state, releasing an arbitrarily large amount of energy.[38] However, as negative-energy solutions are a well-known feature of relativistic wave equations, we insist on identifying H (3.159) as the energy. In the Dirac equation for the electron, it is the interpretation of the positron (antielectron) as a hole in a sea (the Dirac sea) of filled negative-energy states that leads to the negative zero-point energy of the electron–positron field [451]. Analogously,[39] we will now show how, by interpreting the negative-energy photons as positive-energy antiphotons, the zero-point energy of the photon field also appears.

Following Good [241], let us work out the steady-state plane-wave solutions of (3.140). Substituting $\vec{\psi} = \vec{u} \, \exp(i\{\mathbf{p} \cdot \mathbf{r} - Et\}/\hbar)$ in (3.140), we find three solutions,

$$\vec{\psi}_{\pm}(\mathbf{p}, \mathbf{r}, t) = \vec{u}_{\pm}(\mathbf{p}) \, e^{i(\mathbf{p} \cdot \mathbf{r} \mp cpt)/\hbar} \qquad (3.160)$$

and

$$\vec{\psi}_0(\mathbf{p}, \mathbf{r}, t) = \vec{u}_0(\mathbf{p}) \, e^{i\mathbf{p} \cdot \mathbf{r}/\hbar}, \qquad (3.161)$$

where $\vec{u}_\lambda(\mathbf{p})$ is an eigenstate of the helicity operator corresponding to the eigenvalue λ and momentum $\mathbf{p}$. They are normalized to 1, that is,

$$\vec{u}_\lambda(\mathbf{p})\vec{u}_{\lambda'}(\mathbf{p}) = \delta_{\lambda,\lambda'}. \qquad (3.162)$$

Good [241] writes the explicit forms, which you can easily verify:

$$\vec{u}_{\pm}(\mathbf{p}) = \frac{1}{p\sqrt{2(p_x^2 + p_y^2)}} \begin{bmatrix} \pm i p p_y - p_x p_z \\ \mp i p p_x - p_y p_z \\ p_x^2 + p_y^2 \end{bmatrix} \quad \text{and} \quad \vec{u}_0(\mathbf{p}) = \begin{bmatrix} p_x \\ p_y \\ p_z \end{bmatrix}, \qquad (3.163)$$

---

[38] As Feynman [207] nicely put it: "If the energies were negative we know that we could solve all our energy problems by dumping particles into this pit of negative energy and running the world with the extra energy."

[39] Except that there will be no Dirac sea for photons, as they do not obey the Pauli exclusion principle. This also suggests that, more generally, the solution to the negative-energy-states problem is not connected to the idea of Dirac sea, but rather, to that of zero-point energy.

where $p \equiv \sqrt{p_x^2 + p_y^2 + p_z^2}$. As we have shown in Section 3.5, the $\lambda = 0$ solution is not allowed by the transversality condition (3.123). So we arrive at the interesting conclusion that the transversality condition is equivalent to saying that there is no photon with zero energy. This makes a lot of sense physically, as a photon with zero energy would be a photon at rest (i.e., $p = 0$), which would contradict relativity and experimental evidence. This is also consistent with what happens for massive particles. Relativistic wave equations for particles with rest mass m do not admit solutions with energies between $-mc^2$ and mc^2. For the photon, whose rest mass vanishes, this gap collapses into the single forbidden state $E = 0$.

Note that according to (3.160), only solutions with positive helicity have positive energy. This is because we have derived $\vec{\psi}$ in the preceding section from Ψ_+ only. In order to get a complete description of the photon, we also need to take Ψ_- into account. Using Ψ_- instead of Ψ_+ in the derivation of the preceding section is equivalent to having $\vec{\psi}^*$ rather than $\vec{\psi}$, which satisfies the conjugate wave equation

$$i\hbar \frac{\partial}{\partial t} \vec{\psi}^* = -c\hat{\boldsymbol{\sigma}} \cdot \hat{\mathbf{p}} \vec{\psi}^*. \tag{3.164}$$

This is not equivalent to the description in terms of $\vec{\psi}$ because now a photon of negative helicity has positive energy. This and the analogy with the case of the electron, which can take all the available values of its helicity, show that there must be two positive-energy (negative-energy) photon states for each given momentum: one with positive helicity, the other with negative helicity. To accommodate this, we define the photon wavefunction to be the six-component column matrix (just as Bialynicki-Birula does in [53–55])

$$\psi = \begin{bmatrix} \vec{\psi} \\ \vec{\psi}^* \end{bmatrix} \tag{3.165}$$

and the Hamiltonian to be the 6×6 matrix

$$H = \begin{bmatrix} c\hat{\boldsymbol{\sigma}} \cdot \hat{\mathbf{p}} & 0 \\ 0 & -c\hat{\boldsymbol{\sigma}} \cdot \hat{\mathbf{p}} \end{bmatrix}, \tag{3.166}$$

so that the wave equation is still simply

$$i\hbar \frac{\partial}{\partial t} \psi = H\psi. \tag{3.167}$$

In the standard second-quantization formalism [196, 363], the wavefunction becomes an operator, $\hat{\psi}$. This operator is constructed by writing down a general solution of the wave equation as an arbitrary superposition of the eigenstates, and then replacing the expansion coefficients by annihilation operators. Notice that as this must be an arbitrary superposition, *all* the eigenstates of the wave equation should be present, including the negative-energy states. Thus the second-quantized wavefunction is given by

$$\hat{\psi}(\mathbf{r}) = \frac{1}{\sqrt{2}} \int d^3p \sum_{\lambda} \{\hat{a}_{+,\lambda}(\mathbf{p}) \, \psi_+(\mathbf{p}, \lambda, \mathbf{r}, t) + \hat{a}_{-,\lambda}(\mathbf{p}) \, \psi_-(\mathbf{p}, \lambda, \mathbf{r}, t)\},$$

$$\tag{3.168}$$

where $\psi_+(\mathbf{p}, \lambda, \mathbf{r}, t)$ is the solution of the wave equation for a particle of momentum $\mathbf{p}$, helicity λ, and energy cp, and $\psi_-(\mathbf{p}, \lambda, \mathbf{r}, t)$ is the corresponding negative-energy solution. As you can easily verify, these eigenstate solutions of (3.167) are given by

$$\psi_\epsilon(\mathbf{p}, \lambda, \mathbf{r}, t) = \frac{1}{\sqrt{2}} \left[\begin{array}{c} \delta_{\lambda,\epsilon}\vec{u}_\epsilon(\mathbf{p}) \\ \delta_{\lambda,-\epsilon}\vec{u}_{-\epsilon}(\mathbf{p}) \end{array} \right] \frac{e^{i(\mathbf{p}\cdot\mathbf{r}-\epsilon cpt)/\hbar}}{(2\pi)^{3/2}}, \qquad (3.169)$$

which are normalized as

$$\int d^3r \psi_\epsilon^\dagger(\mathbf{p}, \lambda, \mathbf{r}, t)\psi_{\epsilon'}(\mathbf{p}', \lambda', \mathbf{r}, t) = \delta(\mathbf{p} - \mathbf{p}')\frac{e^{-i(\epsilon'-\epsilon)cpt/\hbar}}{2}$$

$$\times \left\{ \delta_{\lambda,\epsilon} \underbrace{\vec{u}_\epsilon^\dagger \vec{u}_{\epsilon'}}_{\delta_{\epsilon,\epsilon'}} \delta_{\lambda',\epsilon'} + \delta_{\lambda,-\epsilon} \underbrace{\vec{u}_{-\epsilon}^\dagger \vec{u}_{-\epsilon'}}_{\delta_{\epsilon,\epsilon'}} \delta_{\lambda',-\epsilon'} \right\}$$

$$= \delta_{\epsilon,\epsilon'}\delta_{\lambda,\lambda'}\delta(\mathbf{p} - \mathbf{p}'). \qquad (3.170)$$

The operator $\hat{a}_{+,\lambda}(\mathbf{p})$ annihilates a positive-energy particle of momentum $\mathbf{p}$ and helicity λ. The operator $\hat{a}_{-,\lambda}(\mathbf{p})$ annihilates the corresponding negative-energy particle. But if a particle with negative energy does not make any sense, what is the physical meaning of the negative-energy solutions that we need to be able to reconstruct *arbitrary* many-particle states with (3.168)?

We cannot use the idea of Dirac sea here because photons are bosons and do not obey the Pauli exclusion principle. But there is a more general idea of antiparticle where it is not viewed as a hole in a sea of filled negative-energy states [364]. This idea is due to Stückelberg [581], Feynman[40] [197, 198, 203], and others. According to this idea, the negative-energy solutions represent a real positive-energy particle, but instead of moving forward in space-time, this particle moves backward. As Feynman put it [197]: "to someone whose future gradually becomes past through a moving present," this particle moving backward in space-time appears as a particle of the same mass as the ordinary particle that moves forward in space-time but with other properties, such as charge, being the exact opposite. For this reason, these particles that move backward in space-time have been named *antiparticles*. Annihilation (creation) of a particle moving backward in space-time will appear to us as creation (annihilation) of an antiparticle.

So the negative-energy solutions of (3.167) are photons moving backward in space-time. To find out the wavefunction $\bar{\psi}_+$ of a positive-energy antiphoton moving forward in space-time, we must then transform the wavefunction of a negative-energy photon moving forward (i.e., a positive-energy photon moving backward) so that it moves backward. This is done by applying to ψ_- the parity operator $\mathcal{P}$, which changes $\mathbf{r}$ into $-\mathbf{r}$, and the time-inversion operator $\mathcal{T}$, which changes t into $-t$:

$$\bar{\psi}_+(\mathbf{p}, \lambda, \mathbf{r}, t) = \mathcal{P}\mathcal{T}\psi_-(\mathbf{p}, \lambda, \mathbf{r}, t). \qquad (3.171)$$

---

[40]Feynman wrote [197] that it was originally suggested to him by John Archibald Wheeler.

What are these operators for our Schrödinger-like equation (3.167) for the photon?

Under time inversion, $t \to -t$, $\hat{\mathbf{p}} \to -\hat{\mathbf{p}}$, and $\hat{\sigma} \to -\hat{\sigma}$. Let $t' = -t$ and $\psi'(t') = \mathcal{T}\psi(t)$; then the transformed (3.167) reads

$$\mathcal{T}i\mathcal{T}^{-1}\hbar\frac{\partial}{\partial t'}\psi'(t') = -\mathcal{T}H(t)\mathcal{T}^{-1}\psi'(t'). \tag{3.172}$$

For (3.167) to be invariant under time inversion [i.e., for (3.172) to have the same form as (3.167)] we must take $\mathcal{T}i\mathcal{T}^{-1} = -i$ if we take $\mathcal{T}H(t)\mathcal{T}^{-1} = H'(t')$. As the helicity operator does not change under time inversion, $H'(t') = H(t)$. So $\mathcal{T}$ is just the ordinary complex conjugation and, as $(X^*)^* = X$, $\mathcal{T}^{-1}$ is also.

Under space inversion (parity operation), $\mathbf{r} \to -\mathbf{r}$, $\mathbf{p} \to -\mathbf{p}$, and $\hat{\sigma} \to \hat{\sigma}$. Let $\mathbf{r}' = -\mathbf{r}$ and $\psi'(\mathbf{r}') = \mathcal{P}\psi(\mathbf{r})$. The transformed Schrödinger-like equation (3.167) is given by

$$\mathcal{P}i\mathcal{P}^{-1}\hbar\frac{\partial}{\partial t}\psi'(\mathbf{r}') = \mathcal{P}H(\mathbf{r})\mathcal{P}^{-1}\psi'(\mathbf{r}'). \tag{3.173}$$

Then if we take $\mathcal{P}H(\mathbf{r})\mathcal{P}^{-1} = H'(\mathbf{r}')$, we have to take $\mathcal{P}i\mathcal{P}^{-1} = i$ to keep (3.167) invariant. Moreover, as the helicity changes sign under parity, $H'(\mathbf{r}') = -H(\mathbf{r})$. Then we can take $\mathcal{P}$ to be the 4×4 matrix that swaps the pairs of upper and lower rows,

$$\mathcal{P} = \begin{bmatrix} \mathbf{0} & \mathbf{1} \\ \mathbf{1} & \mathbf{0} \end{bmatrix}, \tag{3.174}$$

where $\mathbf{0}$ is the 2×2 null matrix and $\mathbf{1}$ is the 2×2 unit matrix.

Now that we have $\mathcal{P}$ and $\mathcal{T}$, we can find the wavefunction of an antiphoton. Using (3.171), we get

$$\begin{aligned}
\bar{\psi}_+(\mathbf{p}, \lambda, \mathbf{r}, t) &= \mathcal{P}\mathcal{T}\psi_-(\mathbf{p}, \lambda, \mathbf{r}, t) \\
&= \frac{1}{\sqrt{2}}\begin{bmatrix} \mathbf{0} & \mathbf{1} \\ \mathbf{1} & \mathbf{0} \end{bmatrix}\begin{bmatrix} \delta_{\lambda,-1}\vec{u}_-^*(\mathbf{p}) \\ \delta_{\lambda,1}\vec{u}_+^*(\mathbf{p}) \end{bmatrix}\frac{e^{-i(\mathbf{p}\cdot\mathbf{r}+cpt)/\hbar}}{(2\pi)^{3/2}} \\
&= \frac{1}{\sqrt{2}}\begin{bmatrix} \delta_{\lambda,1}\vec{u}_+^*(\mathbf{p}) \\ \delta_{\lambda,-1}\vec{u}_-^*(\mathbf{p}) \end{bmatrix}\frac{e^{-i(\mathbf{p}\cdot\mathbf{r}+cpt)/\hbar}}{(2\pi)^{3/2}}.
\end{aligned} \tag{3.175}$$

As $\vec{u}_\lambda^*(\mathbf{p}) = \vec{u}_\lambda(-\mathbf{p})$, we can rewrite this as

$$\begin{aligned}
\bar{\psi}_+(\mathbf{p}, \lambda, \mathbf{r}, t) &= \frac{1}{\sqrt{2}}\begin{bmatrix} \delta_{\lambda,1}\vec{u}_+(-\mathbf{p}) \\ \delta_{\lambda,-1}\vec{u}_-(-\mathbf{p}) \end{bmatrix}\frac{e^{-i(\mathbf{p}\cdot\mathbf{r}+cpt)/\hbar}}{(2\pi)^{3/2}} \\
&= \psi_+(-\mathbf{p}, \lambda, \mathbf{r}, t).
\end{aligned} \tag{3.176}$$

So a positive-energy antiphoton of momentum $\mathbf{p}$ and helicity λ moving forward in space-time is just a positive-energy photon of momentum $-\mathbf{p}$ and helicity λ moving forward in space-time. The photon is its own[41] antiparticle!

---

[41] A particle that moves backward in space-time can also be seen as one whose proper time evolves in the opposite direction to the coordinate time that we use: say, in the laboratory [125, 197]. The particle's proper time is the time in the reference frame of the particle (i.e., in the reference frame in which the particle is at rest). In the case of the photon, which moves at the speed of light, its proper time is always zero; time does not pass for a photon. So the photon is its own antiparticle.

To annihilate a negative-energy photon moving forward in space-time is the same as to annihilate a positive-energy photon moving backward in space-time. But as it is moving backward in space-time, its annihilation will appear to us as the *creation* of a positive-energy antiphoton moving forward in space-time. Knowing now what an antiphoton is, we can rewrite (3.168) as

$$\hat{\psi}(\mathbf{r}) = \frac{1}{\sqrt{2}} \int d^3p \sum_\lambda \left\{ \hat{a}_\lambda(\mathbf{p}) \, \psi_+(\mathbf{p}, \lambda, \mathbf{r}, t) + \hat{a}_\lambda^\dagger(-\mathbf{p}) \, \psi_-(\mathbf{p}, \lambda, \mathbf{r}, t) \right\},$$

$$(3.177)$$

where the energy index in the creation and annihilation operators has been dropped for convenience, as we will only deal with positive-energy states now.

Now what about the creation and annihilation operators $\hat{a}_\lambda(\mathbf{p})$ and $\hat{a}_\lambda^\dagger(\mathbf{p})$? If they obey the commutation relations

$$[\hat{a}_\lambda(\mathbf{p}), \hat{a}_{\lambda'}^\dagger(\mathbf{p}')] = \delta(\mathbf{p} - \mathbf{p}')\delta_{\lambda'\lambda} \quad \text{and} \quad [\hat{a}_\lambda(\mathbf{p}), \hat{a}_{\lambda'}(\mathbf{p}')] = 0, \qquad (3.178)$$

the particles (our photons) will follow Bose–Einstein statistics. If they obey the anti-commutation relations

$$\{\hat{a}_\lambda(\mathbf{p}), \hat{a}_{\lambda'}^\dagger(\mathbf{p}')\} = \delta(\mathbf{p} - \mathbf{p}')\delta_{\lambda'\lambda} \quad \text{and} \quad \{\hat{a}_\lambda(\mathbf{p}), \hat{a}_{\lambda'}(\mathbf{p}')\} = 0, \qquad (3.179)$$

the particles will follow Fermi–Dirac statistics as electrons, for example, do. In 1940, Pauli showed quite generally that particles of integral spin must follow Bose–Einstein statistics, while those of half-integral spin must follow Fermi–Dirac statistics [478]. This is called the *spin-statistics theorem.* Then, as the photon has spin 1, it follows Bose–Einstein statistics, and $\hat{a}_\lambda(\mathbf{p})$ and $\hat{a}_\lambda^\dagger(\mathbf{p})$ must obey the commutation relations (3.178) rather than the anticommutation relations (3.179).

What are the second-quantized expressions for the various observables? In particular, what is the second-quantized Hamiltonian (i.e., the energy of an arbitrary number of photons)? In ordinary many-body second quantization, as is done for the Dirac equation, for example, the second-quantized wavefunction $\hat{\psi}$ is also the field operator. In that case, the second-quantized version of a first-quantized operator $\mathcal{O}$ is given by

$$\hat{O} = \int d^3r \; \hat{\psi}^\dagger(\mathbf{r}) \mathcal{O} \hat{\psi}(\mathbf{r}). \qquad (3.180)$$

Here we cannot use this expression, because $\hat{\psi}$ is not the field operator. Recall from Section 3.2 that the wavefunction Ψ is related to the field $\mathcal{F}$ by (3.72). Then the field operator $\hat{F}$ is given by

$$\hat{F}(\mathbf{r}) = \frac{1}{\sqrt{2}} \int d^3p \sqrt{|H|} \sum_\lambda \left\{ \hat{a}_\lambda(\mathbf{p}) \, \psi_+(\mathbf{p}, \lambda, \mathbf{r}, t) + \hat{a}_\lambda^\dagger(-\mathbf{p}) \, \psi_-(\mathbf{p}, \lambda, \mathbf{r}, t) \right\},$$

$$(3.181)$$

where H is the first-quantized Hamiltonian (3.166). As we saw in Section 3.2, the energy is the time component of the four-momentum, which is given in terms of the quaternionic fields by (3.71). Translating this into our present context, the second-quantized Hamiltonian (energy) is given in terms of the field operator by

$$
\begin{aligned}
\hat{H} &= \int d^3r \, \hat{F}^\dagger(\mathbf{r}) \hat{F}(\mathbf{r}) \\
&= \frac{1}{2} \int d^3r \int d^3p \sqrt{cp} \sum_\lambda \left\{ \hat{a}_\lambda^\dagger(\mathbf{p}) \psi_+^\dagger(\mathbf{p}, \lambda, \mathbf{r}, t) + \hat{a}_\lambda(-\mathbf{p}) \psi_-^\dagger(\mathbf{p}, \lambda, \mathbf{r}, t) \right\} \\
&\quad \times \int d^3p' \sqrt{cp'} \sum_{\lambda'} \left\{ \hat{a}_{\lambda'}(\mathbf{p}') \psi_+(\mathbf{p}', \lambda', \mathbf{r}, t) + \hat{a}_{\lambda'}^\dagger(-\mathbf{p}') \psi_-(\mathbf{p}', \lambda', \mathbf{r}, t) \right\} \\
&= \frac{1}{2} \int d^3p \, cp \sum_\lambda \left\{ \hat{a}_\lambda^\dagger(\mathbf{p}) \hat{a}_\lambda(\mathbf{p}) + \hat{a}_\lambda(-\mathbf{p}) \hat{a}_\lambda^\dagger(-\mathbf{p}) \right\} \\
&= \int d^3p \, cp \sum_\lambda \left\{ \hat{a}_\lambda^\dagger(\mathbf{p}) \hat{a}_\lambda(\mathbf{p}) + \frac{1}{2} \right\},
\end{aligned}
\tag{3.182}
$$

which is equivalent to (2.148) derived by canonical quantization in Chapter 2, including the zero-point-energy term.

It can be shown that for a given first-quantized operator $\mathcal{O}$, its second-quantized version $\hat{O}$ is given in our formalism by

$$
\hat{O} = \int d^3r \hat{\psi}^\dagger(\mathbf{r}) \frac{H}{|H|} \mathcal{O} \hat{\psi}(\mathbf{r}),
\tag{3.183}
$$

just as in the theory of Hammer and Good [268, 269]. Their theory, which is valid for any spin, reduces to the usual equivalence between the second-quantized wavefunction and the field operator for the case of spin $1/2$. In that case it yields the usual expression (3.180) that is ordinarily used in the second quantization of the Dirac equation.

Let us now find the second-quantized expressions for the electric and magnetic fields introduced in Section 3.4. From (3.109) we see that

$$
\hat{\mathbf{E}}(\mathbf{r} = -i\sqrt{16\pi} \left[\vec{x}^\dagger \quad -\vec{x}^\dagger \right] \hat{F}(\mathbf{r}) \quad \text{and} \quad \hat{B}(\mathbf{r}) = \sqrt{16\pi} \left[\vec{x}^\dagger \quad \vec{x}^\dagger \right] \hat{F}(\mathbf{r}),
\tag{3.184}
$$

where $\vec{x}$ is the three-dimensional column vector whose components are the unit vectors of whatever right-handed Cartesian axes we happen to be using for our three-dimensional space

$$
\vec{x} = \begin{bmatrix} x_1 \\ x_2 \\ x_3 \end{bmatrix}.
\tag{3.185}
$$

This way both $\hat{\mathbf{E}}$ and $\hat{B}$ will be ordinary vectors rather than column-matrix vectors. Substituting (3.181) in (3.184), we find that (see Problem 3.9)

$$\hat{\mathbf{E}}(\mathbf{r}) = i \int d^3 p \sqrt{\frac{cp}{2\pi^2}} \sum_\lambda \left\{ \hat{a}_\lambda(\mathbf{p}) e^{-icpt/\hbar} - \hat{a}_\lambda^\dagger(-\mathbf{p}) e^{icpt/\hbar} \right\} \mathbf{u}_\lambda(\mathbf{p}) e^{i\mathbf{p}\cdot\mathbf{r}/\hbar},$$

$$(3.186)$$

$$\hat{\mathbf{B}}(\mathbf{r}) = i \int d^3 p \sqrt{\frac{cp}{2\pi^2}} \sum_\lambda \left\{ \hat{a}_\lambda(\mathbf{p}) e^{-icpt/\hbar} + \hat{a}_\lambda^\dagger(-\mathbf{p}) e^{icpt/\hbar} \right\} \frac{\mathbf{p}}{p} \wedge \mathbf{u}_\lambda(\mathbf{p}) e^{i\mathbf{p}\cdot\mathbf{r}/\hbar},$$

$$(3.187)$$

where $\mathbf{u}_\lambda(\mathbf{p}) \equiv -\lambda \vec{x}^\dagger \vec{u}_\lambda(\mathbf{p})$ is the ordinary vector form of $-\lambda \vec{u}_\lambda(\mathbf{p})$.

Except for the use of the Schrödinger picture with momentum and helicity rather than the Heisenberg picture with wave vector and linear polarization, (3.186) and (3.187) are completely equivalent to (2.146) and (2.147). The helicity basis $\mathbf{u}_\lambda(\mathbf{p})$ corresponds to circular polarization and the wavefunction of the photon to the mode functions of the field.

RECOMMENDED READING

- Feynman's 1986 Dirac Memorial Lecture at Cambridge University [207] gives an insightful account on why there must be antiparticles and on the spin-statistics theorem.

- Chapter 2 and Appendix 3 of the book by Pike and Sarkar [492] will give you a good idea of what the group theory approach to photon localization is about. Jordan's papers [331–336] on this subject are also quite readable as well as the pedagogical paper by Han and Kim on the little group for photons [270]. The key original references are the 1948 paper by Pryce [501], 1949 paper by Newton (no, this is not Isaac!) and Wigner [464], and Wightman's imprimitivity reformulation [643]. Also related to this group theory approach are [32], [645], [647], [648], and the references mentioned in [464]. For those wishing to study group theory in depth, there are many books on the subject. I find the following books particularly useful: [330], [639], [396], and [650].

- It is worth looking at some of the historical papers on the photon wavefunction and its relativistic wave equation. The paper by Oppenheimer [472], whose philosophy inspired this entire chapter is, of course, a must. The approach presented here has some similarity to that of Good and collaborators [241, 268, 269, 462, 463, 624]. See also Weinberg's approach in [626–628]. If you can read French, have a look at the following papers by Proca: [497–499]. The original paper by Landau and Peierls on the wavefunction of the photon in configuration space [387] is also available in an English translation and is well worth looking at. You might also want to look at Cook's rediscovery of the Landau–Peierls wavefunction [119, 120] as well as Inagaki's [310, 311] extension of Cook's work. For an extensive review of the photon wavefunction, see [54]. See also the nice pedagogical discussion in Sec. 1.5.4 of [552].

- In 1966, Leonard Mandel [428] proposed a configuration-space photon number operator for applications in quantum optics. This operator can be used to introduce a coarse-grained photon wavefunction in configuration space.

- Even though it is impossible to produce a perfectly localized photon state, the possibility of constructing highly localized polychromatic photon states has been demonstrated theoretically [4, 15, 56, 289, 666].

- As in Feynman's 1948 paper [197], the idea of antiparticles can be discussed within classical relativistic theories. The paper by Costella et al. [125] gives a great pedagogical account of how antiparticles are a consequence of relativity alone.

- An interesting topic that we did not cover in this chapter is the connection between the paraxial approximation for light beams and the nonrelativistic dynamics of a massive particle. The photon in the paraxial approximation behaves like a nonrelativistic particle. See the papers by Deutsch and Garrison [151] and by Marte and Stenholm [431].

- Spin was discovered at Leiden by George Uhlenbeck, who was Paul Ehrenfest's postdoc at the time, and Samuel Goudsmit, a Ph.D. student. For an interesting account of this discovery as well as of the history of other discoveries in particle physics, see Pais's book [474]. See also Tomonaga's book on "the story of spin" [601].

- In the literature, *second quantization* is often used to mean field quantization in general, which was introduced in 1927 by Dirac [156]. Here, we are using the term to mean specifically the method for going from a description of a system of particles by a configuration or momentum–space wavefunction to one by an occupation-number wavefunction. This method was developed in 1932 by Vladimir Alexandrovich Fock[42] [216]. As Fock's original paper is in German, you might want to read instead the expository version of Fock's paper in the *Physical Review* which is, of course, in English [123]. Fock wrote [218]: "Without exception, all of the results which Dr. Corson finds are contained in my paper. There is a close parallelism not only between the formulas, but also between the texts of the two papers." Corson forgot to give due credit to Fock, who wrote an angry comment [218] that came short of accusing Corson of plagiarism. The main historical papers on field quantization are mentioned in Chapter 6, page 121, of [543] and in [363]. See also Chapter 3 of [12] which quantizes the electromagnetic field using both canonical quantization and second quantization (in [12], second quantization has the same specific meaning as here). For a nice pedagogical treatment of second quantization, see, for instance, Chapter 10 of Milonni's book [451] and Kobe's paper in the *American Journal of Physics* [363].

---

[42]Hence the name *Fock state* adopted for states of a definite number of particles in a given energy level.

- Oliver Heaviside used to write Maxwell equations in symmetric form, as if magnetic charges (monopoles) existed. He called it the *duplex form* (see the paragraph that starts on page viii of the preface to the first volume of [281], pages 57 and 58 of Volume 3 of [285], and [290]). But even though he liked this symmetrical form of Maxwell equations, Heaviside was fully aware that there was no evidence for magnetic monopoles and would always set the magnetic charges to zero at the end of his calculations. As he wrote on page 452 of the first volume of [281]: "It would then, by a suitable arrangement of matter, be possible to have a unipolar magnet, a quantity of matter round which the magnetic force was everywhere directed outward, or everywhere inward. This being contradicted by universal experience, we must conclude that div $\mathbf{B} = 0$." It was Paul Dirac (who had a first degree in electrical engineering and was an admirer of Heaviside's writings) who took the idea of magnetic monopoles seriously and showed in 1931 [157] that the existence of a single magnetic monopole would explain the quantization of charge (see Chapter 4 of Jackson's book [314] and [545]). The existence of magnetic monopoles has been inferred theoretically by 't Hooft [598] and Polyakov [494], but their predicted energies have so far proved to be too high for experimental confirmation (see e.g., [1]). Recently, Zhong Fang and collaborators [188] have found experimental evidence for effective monopoles at much lower energies in the ground state of crystals (see also [421]). But these are only effective monopoles that have no meaning outside such ground states. For an extensive list of references on magnetic monopoles, see the resource letter published in the *American Journal of Physics* [239].

- Kobe [364] also uses the Stückelberg–Wheeler–Feynman idea to solve the negative-energy-photon problem and interpret the antiphoton. However, he does not do a parity transformation. He considers antiphotons to be negative-energy states which go backward in *time* rather than in *space-time*. As a result, his antiphotons turn out to be photons of the opposite helicity.

- The *American Journal of Physics* has published some excellent pedagogical papers on light polarization and the spin of the photon. You might wish to have a look specially at [46] and [277].

- Recently, the de Broglie wavelength of a two-photon wave packet has been measured [219].

- Ole Keller holds the view that the question of the localizability of the photon cannot be separated from that of how a photon is born, say, out of an atom or molecule. So he considers the interaction with matter and near-fields explicitly. See, for instance, [342–350]. Chan et al. [98, 99] have recently followed a similar idea in which they dealt explicitly with the entanglement between the atom and the emitted photon using a Schmidt decomposition (for a pedagogical account of the use of the Schmidt decomposition for dealing with entangled systems, see [185]). But notice that their use of the dipole approximation in the analysis in principle rules out the possibility of investigating localization of the photon to a region smaller than the typical size of an atom.

- Jauch and Piron [322] dropped Newton's and Wigner's [464] assumption that position measurements should be compatible (commute) with one another and obtained a position operator for the photon. Since then, this idea of dropping the commonsense ideas adopted in [464] of what a position measurement is and what localized states are has lead to a cottage industry of papers on alternative photon position operators. See, for example, [278–280].

- For a review of Proca equations and the experimental search for an upper limit to the photon's mass, see [238, 367]. See also Sec. 12.9 of Jackson's book [314] and the paper by Bass and Schrödinger [44]. Crandall wrote a nice pedagogical paper in the *American Journal of Physics* [128] describing an undergraduate experiment to put an upper limit on the photon's mass.

- Those curious about quaternions should read Appendix C (they are used actively nowadays in computer animation, robotics, and in field theory). The following references should also be quite readable for the average physicist: [2, 67, 95, 96, 146, 179, 180, 244, 256, 257, 263, 264, 266, 291], Chapter 9 of [386], and [506, 566, 567, 588].

- For a discussion on the localization of relativistic particles in the context of the consistent histories interpretation of quantum mechanics, see [471]. The consistent histories approach has been used to resolve [254] some apparent paradoxes related to localization of relativistic particles [286–288]. See also von Zuben's analysis of these paradoxes [621].

Problems

3.1   Substitute

$$\Diamond = \frac{i}{c}\frac{\partial}{\partial t} + \sum_{n=1}^{3} i_n \frac{\partial}{\partial x_n},$$

$$\mathcal{F} = \sum_{n=1}^{3} i_n i (E_n - i B_n),$$

$$\mathcal{A} = i\varphi - \sum_{n=1}^{3} i_n A_n$$

in (3.42) and (3.43). Using (3.28) and property 2 of biquaternions in Section 3.2 (page 77), show that (3.42) and (3.43) lead to (3.111)–(3.117).

3.2   Using (3.98), (3.99), and (3.100), show that if $\mathcal{F}_\alpha$ and $\mathcal{A}_\alpha$ have the properties (3.44) and (3.45), so do Ψ_α and Φ_α [i.e., equations (3.76) hold].

3.3   Derive (3.92).

3.4   Using $i_n \mathcal{K} = -\mathcal{K} i_n - i 2 k_n$, $\sqrt{k^2 + \mu^2} \underline{\Psi}^\dagger = -\underline{\Psi}^\dagger \mathcal{K} + \mu^2 \underline{\Phi}^\dagger$, and $\sqrt{k^2 + \mu^2} \underline{\Phi} = \mathcal{K}\underline{\Phi} + \underline{\Psi}$, show that

$$
J_n = (2\pi)^3 \hbar \int d^3 k \sum_{jk} \epsilon_{njk} S \left\{ \underline{\Psi}^\dagger i_k \frac{\partial}{\partial k_j} \left(\sqrt{k^2 + \mu^2} \underline{\Psi} \right) \right.
$$

$$
\left. + \mu^2 \frac{\partial}{\partial k_j} \left(\sqrt{k^2 + \mu^2} \underline{\Phi}^\dagger \right) i_k \underline{\Phi} \right\}
$$

$$
- (2\pi)^3 \hbar \int d^3 k \sum_{jk} \epsilon_{njk} \frac{k_j}{2\sqrt{k^2 + \mu^2}} S \left\{ \underline{\Psi}^\dagger i_k \underline{\Psi} + \mu^2 \underline{\Phi}^\dagger i_k \underline{\Phi} \right\}.
$$

Then use $\sqrt{k^2 + \mu^2} \underline{\Psi} = -\mathcal{K}\underline{\Psi} + \mu^2 \underline{\Phi}$ and $\sqrt{k^2 + \mu^2} \underline{\Phi}^\dagger = \underline{\Phi}^\dagger \mathcal{K} + \underline{\Psi}^\dagger$ instead of $\sqrt{k^2 + \mu^2} \underline{\Psi}^\dagger = -\underline{\Psi}^\dagger \mathcal{K} + \mu^2 \underline{\Phi}^\dagger$ and $\sqrt{k^2 + \mu^2} \underline{\Phi} = \mathcal{K}\underline{\Phi} + \underline{\Psi}$, and show that

$$
J_n = (2\pi)^3 \hbar \int d^3 k \sum_{jk} \epsilon_{njk} S \left\{ \frac{\partial}{\partial k_j} \left(\sqrt{k^2 + \mu^2} \underline{\Psi}^\dagger \right) \bar{i}_k \underline{\Psi} \right.
$$

$$
\left. + \mu^2 \underline{\Phi}^\dagger \bar{i}_k \frac{\partial}{\partial k_j} \left(\sqrt{k^2 + \mu^2} \underline{\Phi} \right) \right\}
$$

$$
- (2\pi)^3 \hbar \int d^3 k \sum_{jk} \epsilon_{njk} \frac{k_j}{2\sqrt{k^2 + \mu^2}} S \left\{ \underline{\Psi}^\dagger \bar{i}_k \underline{\Psi} + \mu^2 \underline{\Phi}^\dagger \bar{i}_k \underline{\Phi} \right\}.
$$

Now, proceeding in the same way as was done for the momentum, show that J_n is given by (3.95).

3.5   Show that due to the transversality condition $\mathbf{k} \cdot \underline{\psi} = 0$, it is impossible to make a linear combination of plane waves that leads to a delta function in configuration space.[43] Notice that for nonvanishing μ, such a linear combination is possible but it must contain antiparticles (i.e., negative frequencies) as well as particles (i.e., positive frequencies).

3.6   Show that (3.134) holds in two different ways:

1. Use (E.18) of Appendix E to calculate $(\hat{S}_n \hat{S}_m)_{pq} = -\hbar^2 \sum_l \epsilon_{npl} \epsilon_{mlq}$ and $(\hat{S}_m \hat{S}_n)_{pq} = -\hbar^2 \sum_l \epsilon_{mpl} \epsilon_{nlq}$, where the subscript pq stands for the element on row p and column q. Use (E.18) again to calculate $(i\hbar \sum_l \epsilon_{lnm} \hat{S}_l)_{pq} = \hbar^2 \sum_l \epsilon_{lnm} \epsilon_{lpq}$ and show that $(\hat{S}_n \hat{S}_m)_{pq} - (\hat{S}_m \hat{S}_n)_{pq} = (i\hbar \sum_l \epsilon_{lnm} \hat{S}_l)_{pq}$.

2. Check that (3.134) holds by doing the matrix multiplications on the left-hand side of (3.134) for each case.

---

[43] See remark (ii) at the bottom of page 188 of [108].

3.7   Show that (3.135) holds in two different ways:

1. Use (E.18) of Appendix E to calculate $(\hat{S}_n^2)_{pq} = -\hbar^2 \sum_l \epsilon_{npl}\epsilon_{nlq}$ and substitute the result in $(\hat{S}^2)_{pq} = \sum_n (\hat{S}_n^2)_{pq}$.

2. Calculate $\hat{S}_n^2$ for each $n = 1, 2, 3$ by matrix multiplication and add the resulting matrices to get $\hat{S}^2$.

3.8   As a particular example of the coordinate transformation discussed in Section 3.5, take

$$\varepsilon_1 \equiv \frac{p_2}{p_\perp} e_1 - \frac{p_1}{p_\perp} e_2,$$

$$\varepsilon_2 \equiv \frac{p_1 p_3}{p_\perp p} e_1 + \frac{p_2 p_3}{p_\perp p} e_2 - \frac{p_\perp}{p} e_3,$$

$$\varepsilon_3 \equiv \frac{p_1}{p} e_1 + \frac{p_2}{p} e_2 + \frac{p_3}{p} e_3,$$

where $p = \sqrt{p_1^2 + p_2^2 + p_3^2}$ and $p_\perp = \sqrt{p_1^2 + p_2^2}$. Show that

$$\mathsf{T} = \begin{bmatrix} \dfrac{p_2}{p_\perp} & -\dfrac{p_1}{p_\perp} & 0 \\[2mm] \dfrac{p_1 p_3}{p_\perp p} & \dfrac{p_2 p_3}{p_\perp p} & -\dfrac{p_\perp}{p} \\[2mm] \dfrac{p_1}{p} & \dfrac{p_2}{p} & \dfrac{p_3}{p} \end{bmatrix} \quad \text{and} \quad \mathsf{T}^{-1} = \begin{bmatrix} \dfrac{p_2}{p_\perp} & \dfrac{p_1 p_3}{p_\perp p} & \dfrac{p_1}{p} \\[2mm] -\dfrac{p_1}{p_\perp} & \dfrac{p_2 p_3}{p_\perp p} & \dfrac{p_2}{p} \\[2mm] 0 & -\dfrac{p_\perp}{p} & \dfrac{p_3}{p} \end{bmatrix}.$$

Now calculate $\hat{\sigma}_3' = \mathsf{T}\hat{\sigma} \cdot (\mathbf{p}/p)\mathsf{T}^{-1}$ by doing the matrix multiplications and show that

$$\hat{\sigma}_3' = \begin{bmatrix} 0 & -i & 0 \\ i & 0 & 0 \\ 0 & 0 & 0 \end{bmatrix}.$$

3.9   Using (3.184), show that

$$\hat{\mathbf{E}}(\mathbf{r}) = i\int d^3p \sqrt{\frac{cp}{2\pi^2}} \sum_\lambda \left\{ \hat{a}_\lambda(\mathbf{p})e^{-icpt/\hbar} - \hat{a}_\lambda^\dagger(-\mathbf{p})e^{icpt/\hbar} \right\} (-\lambda)\vec{x}^\dagger \vec{u}_\lambda(\mathbf{p})\, e^{i\mathbf{p}\cdot\mathbf{r}/\hbar},$$

$$\hat{\mathbf{B}}(\mathbf{r}) = \int d^3p \sqrt{\frac{cp}{2\pi^2}} \sum_\lambda \left\{ \hat{a}_\lambda(\mathbf{p})e^{-icpt/\hbar} + \hat{a}_\lambda^\dagger(-\mathbf{p})e^{icpt/\hbar} \right\} \vec{x}^\dagger \vec{u}_\lambda(\mathbf{p})\, e^{i\mathbf{p}\cdot\mathbf{r}/\hbar}.$$

Now use (3.185) and (3.163) to show that

$$i\frac{\mathbf{P}}{p} \wedge \mathbf{u}_\lambda(\mathbf{p}) = \vec{x}^\dagger \vec{u}_\lambda(\mathbf{p}),$$

where $\mathbf{u}_\lambda(\mathbf{p})$ is the unit vector

$$\mathbf{u}_\lambda(\mathbf{p}) \equiv -\lambda\vec{x}^\dagger \vec{u}_\lambda(\mathbf{p}).$$

4

A box of photons

Will it ever be possible to perceive the grain of light, the particles of light? Suppose one single photon had been isolated in an experiment, a single grain of light trapped in a box with a peephole, stored in a room sealed from light.

—Thomas Hård af Segerstad [271]

This chapter is about light without matter in a box. Here you will get a quick glimpse of what cavity QED is all about and will be able to appreciate a number of interesting ideas. Having entered the quantum world in the preceding chapters, here we look for its classical limit introducing coherent states, the density matrix, and the P representation. Then we see how nature can be fooled with squeezed states, which have less noise than the standard quantum limit. Among other things, we will see how they can be detected and generated. The latter will lead us to a discussion on the interesting problem of what happens to the quantum radiation field when a cavity changes size suddenly.

4.1 THE CLASSICAL LIMIT OF THE QUANTIZED RADIATION FIELD

One of the first things physicists do with a new theory is try to recover the old theory it surpasses as a limiting case. For quantum and classical mechanics in general, at the time of writing, this is still an open question. For the radiation field, however, we know at least that there are some quantum states that most closely approach the classical realm. In Section 4.1.1 we introduce these *coherent states*. Then we discuss how the idea of quantum state can be generalized to open systems, introducing the density matrix. Finally, in Section 4.1.3 we show how the density matrix of any field can be expressed as a sum of diagonal matrices in a basis of coherent states: the Glauber–Sudarshan diagonal representation. This diagonal representation will provide us with an objective criterion to find out when we are dealing with a nonclassical radiation field.

4.1.1   Coherent states: Quasiclassical states of the radiation field

A classical electromagnetic wave can be written as (2.110), with $\hat{a}$ and $\hat{a}^\dagger$ replaced by complex amplitudes. Are there any quantum states for which the radiation field would essentially behave like such a classical wave? In other words, are there any quantum states for which measurements carried out on the radiation field yield results similar to those expected from a classical wave? To answer this question, we must consider how the radiation field is measured.

Any ordinary photon detector, be it our eyes, a photographic plate, a photodiode, or a photomultiplier, detects photons by destroying them.[1] This means that the observables measured are always in the normal order, where every $\hat{a}$ operator acts on the state before the $\hat{a}^\dagger$ operators. In other words, ordinary measurements measure expectation values of the type $\langle (\hat{a}^\dagger)^m \hat{a}^n \rangle$, where m and n are integers. Now measurements like these will yield results similar to those obtained if the radiation field were a classical wave, when for any m and n,

$$\left\langle \left(\hat{a}^\dagger \right)^m \hat{a}^n \right\rangle = (\alpha^*)^m \alpha^n, \tag{4.1}$$

where α is a complex number. Using (4.1), we can construct the operator

$$\hat{Z} \equiv \hat{a} - \alpha \tag{4.2}$$

such that

$$\left\langle \hat{Z}^\dagger \hat{Z} \right\rangle = 0. \tag{4.3}$$

Rewriting (4.3) as $|\hat{Z}|\alpha\rangle|^2 = 0$, we see that it implies that any state $|\alpha\rangle$ which satisfies (4.1) must also satisfy

$$\hat{Z}|\alpha\rangle = 0. \tag{4.4}$$

So $|\alpha\rangle$ is a right eigenstate of $\hat{a}$ with eigenvalue α, that is,

$$\hat{a}|\alpha\rangle = |\alpha\rangle\alpha. \tag{4.5}$$

---

[1]Recently, a remarkable experiment realized for the first time a quantum nondemolition photon measurement, where a photon was detected without being destroyed [467]. This experiment could, however, detect only a single photon.

Notice that $\hat{a}$ has no left eigenstates (see Problem 4.1), but it has right eigenstates, as can easily be demonstrated by finding the explicit solutions (see Problem 4.2) of (4.5). Physically, (4.5) means that the $|\alpha\rangle$ are states that are insensitive to the annihilation of a photon. This makes such states quasiclassical because then a photo-detection measurement does not disturb them. As condition (4.1) is also the condition for full coherence [235] (i.e., for having maximum fringe contrast in interference experiments), the states $|\alpha\rangle$ are called *coherent states*.

Instead of calculating these coherent states directly from (4.5), as in Problem 4.2, we obtain them here in a more indirect way. This will pay dividends later, as one of the spin-offs of this indirect route will turn out to be a very useful tool.

Consider the unitary operator $\hat{D}(\alpha)$ [i.e., $\hat{D}^\dagger(\alpha)\hat{D}(\alpha) = \hat{1}$], which has the following effect on $\hat{a}$:

$$\hat{D}^\dagger(\alpha)\hat{a}\hat{D}(\alpha) = \hat{a} + \alpha. \tag{4.6}$$

Then the coherent state $|\alpha\rangle$ will be given by

$$|\alpha\rangle = \hat{D}(\alpha)|0\rangle, \tag{4.7}$$

where $|0\rangle$ is the vacuum state (notice that the vacuum state is the only number state that is also a coherent state). Moreover, as long as $|0\rangle$ is normalized, the states $|\alpha\rangle$ obtained from (4.7) will also be because $\hat{D}(\alpha)$ is unitary.

The problem of finding $|\alpha\rangle$ reduces then to that of finding $\hat{D}(\alpha)$. To find $\hat{D}(\alpha)$, let us first make a rotation so that α becomes real. Let $\alpha = r\exp(i\theta)$, then consider the unitary transformation $\hat{R}(\theta)$ that rotates $\hat{a}$:

$$\hat{R}^\dagger(\theta)\hat{a}\hat{R}(\theta) = \hat{a}e^{i\theta}, \tag{4.8}$$

with

$$\hat{R}^\dagger(\theta)\hat{R}(\theta) = \hat{1}. \tag{4.9}$$

You can obtain an explicit expression for $\hat{R}(\theta)$ in Problem 4.3, but here we are not interested in $\hat{R}(\theta)$, only in $\hat{D}(\alpha)$. Sandwiching (4.6) between $\hat{R}^\dagger(\theta)$ and $\hat{R}(\theta)$ and using (4.8), we find that

$$\hat{D}_\theta^\dagger(r)\hat{a}\hat{D}_\theta(r) = \hat{a} + r, \tag{4.10}$$

where

$$\hat{D}_\theta(r) = \hat{R}^\dagger(\theta)\hat{D}(re^{i\theta})\hat{R}(\theta). \tag{4.11}$$

Now consider an infinitesimal $r = \varepsilon$. Up to first order in ε, $\hat{D}_\theta(\varepsilon)$ will be given by the unit operator plus ε times an operator that we still have to determine. As $\hat{D}_\theta(\varepsilon)$ is unitary, this yet undetermined operator must be anti-Hermitian. Let us call it $i\hat{w}$, where $\hat{w}$ is Hermitian. Then

$$\hat{D}_\theta(\varepsilon) = \hat{1} + i\varepsilon\hat{w}, \tag{4.12}$$

and from (4.10) we find that

$$[\hat{a}, \hat{w}] = -i. \tag{4.13}$$

The solution of (4.13) has the form

$$\hat{w} = -i\hat{a}^\dagger + f(\hat{a}), \tag{4.14}$$

where $f(\hat{a})$ is a yet undetermined function of $\hat{a}$. As $\hat{w}$ must be Hermitian, we find that

$$f(\hat{a}) = i\hat{a} + \text{a real number}. \tag{4.15}$$

The real number only adds a global phase and can be ignored. So we obtain

$$\hat{D}_\theta(\varepsilon) = \hat{1} + (\hat{a}^\dagger - \hat{a})\varepsilon. \tag{4.16}$$

Now, $\hat{D}_\theta(r)$ for finite r can be obtained by applying $\hat{D}_\theta(r/N)$ successively N times. Each time we add another r/N to $\hat{a}$, so after N successive applications, we add $(r/N)N = r$ to $\hat{a}$. As $N \to \infty$, r/N will become an infinitesimal ε. Then[2]

$$\hat{D}_\theta(r) = \lim_{N \to \infty} \left[\hat{1} + (\hat{a}^\dagger - \hat{a})\frac{r}{N} \right]^N$$
$$= e^{(\hat{a}^\dagger - \hat{a})r}. \tag{4.17}$$

To obtain $\hat{D}(\alpha)$, we just apply the inverse rotation (see Problem 4.4), that is,

$$\hat{D}(\alpha) = \hat{R}(\theta)\hat{D}_\theta(r)\hat{R}^\dagger(\theta)$$
$$= \exp\left\{ \left[\hat{R}(\theta)\hat{a}^\dagger \hat{R}^\dagger(\theta) - \hat{R}(\theta)\hat{a}\hat{R}^\dagger(\theta) \right] r \right\}$$
$$= \exp\left\{ \left[\hat{a}^\dagger e^{i\theta} - \hat{a}e^{-i\theta} \right] r \right\}$$
$$= e^{\hat{a}^\dagger \alpha - \alpha^* \hat{a}}. \tag{4.18}$$

The operator $\hat{D}(\alpha)$ was introduced by Glauber, and in view of (4.6), it is called the *Glauber displacement operator.* From (4.7) and (4.18), we find that

$$|\alpha\rangle = e^{\hat{a}^\dagger \alpha - \alpha^* \hat{a}}|0\rangle$$

---

[2]To see that $\lim_{N \to \infty} (1 + x/N)^N = \exp(x)$, notice that

$$\left(1 + \frac{x}{N} \right)^N = \sum_{n=0}^{N} \frac{N!}{n!(N-n)!} \frac{x^n}{N^n}.$$

Thus,

$$\lim_{N \to \infty} \left(1 + \frac{x}{N} \right)^N = \sum_{n=0}^{\infty} \frac{x^n}{n!} \lim_{N \to \infty} \frac{N!}{(N-n)!N^n}$$
$$= \sum_{n=0}^{\infty} \frac{x^n}{n!} \lim_{N \to \infty} 1 \left(1 - \frac{1}{N} \right) \cdots \left(1 + \frac{n-1}{N} \right)$$
$$= \sum_{n=0}^{\infty} \frac{x^n}{n!}$$
$$= e^x.$$

$$= e^{-|\alpha|^2/2} e^{\hat{a}^\dagger \alpha} e^{-\hat{a}\alpha^*} |0\rangle$$

$$= e^{-|\alpha|^2/2} e^{\hat{a}^\dagger \alpha} |0\rangle$$

$$= e^{-|\alpha|^2/2} \sum_{n=0}^{\infty} \frac{\alpha^n}{n!} \left(\hat{a}^\dagger\right)^n |0\rangle, \tag{4.19}$$

where we have used the Baker–Hausdorff disentanglement theorem (see Appendix D) to get the second line of (4.19). From (2.122), it follows then that

$$|\alpha\rangle = e^{-|\alpha|^2/2} \sum_{n=0}^{\infty} \frac{\alpha^n}{\sqrt{n!}} |n\rangle. \tag{4.20}$$

Unlike number states, coherent states are not orthogonal[3]; that is,

$$\langle \alpha' | \alpha \rangle = e^{-(|\alpha'|^2 + |\alpha|^2)/2} e^{\alpha'^* \alpha}. \tag{4.21}$$

But they are complete (i.e., any state can be expressed in terms of them). This can be shown by working out the sum of all diagonal coherent-state projectors,

$$\hat{P} = \int d^2\alpha |\alpha\rangle\langle\alpha|. \tag{4.22}$$

Using (4.20),

$$\hat{P} = \sum_{n,m} |n\rangle\langle m| \int d^2\alpha \, e^{-|\alpha|^2} \frac{\alpha^n (\alpha^*)^m}{\sqrt{n!m!}}. \tag{4.23}$$

Now the integral over α in (4.23) can be done easily if we switch to polar coordinates $\alpha = r \exp(i\theta)$.

$$\int d^2\alpha \, e^{-|\alpha|} \alpha^n (\alpha^*)^m = \int_0^\infty dr \, r \int_0^{2\pi} d\theta \, e^{-r^2} r^{2n} r^{m-n} e^{-i(m-n)\theta}$$

$$= \int_0^\infty dr \, r^{m+n+1} e^{-r^2} \underbrace{\int_0^{2\pi} d\theta \, e^{-i(m-n)\theta}}_{2\pi\delta_{m,n}}$$

$$= 2\pi\delta_{m,n} \int_0^\infty dr \, r^{2n+1} e^{-r^2}$$

$$= 2\pi\delta_{m,n} \left\{ (-1)^n \frac{d^n}{d\lambda^n} \int_0^\infty dr \, r \, e^{-\lambda r^2} \right\}_{\lambda=1}$$

$$= 2\pi\delta_{m,n} \left\{ (-1)^n \frac{d^n}{d\lambda^n} \frac{1}{2\lambda} \right\}_{\lambda=1}$$

$$= \pi\delta_{m,n} n!. \tag{4.24}$$

---

[3]Only eigenstates of Hermitian operators have to be orthogonal.

Substituting (4.24) in (4.23), we find that

$$\int d^2\alpha \, |\alpha\rangle\langle\alpha| = \pi \sum_n |n\rangle\langle n|, \tag{4.25}$$

so that

$$\frac{1}{\pi} \int d^2\alpha \, |\alpha\rangle\langle\alpha| = \hat{1}. \tag{4.26}$$

Coherent states have another fascinating property: Any coherent state in an expansion of a given state can itself be expanded in terms of other coherent states and even eliminated. This is called *overcompleteness*. In Section 4.1.3, overcompleteness will be exploited to expand any density matrix as a sum of diagonal coherent-state projectors. But first we must be introduced to the density matrix.

4.1.2 The density matrix

There are certain circumstances when we do not know exactly the state of a quantum system, only the classical probabilities of the system being in any of a certain number of states. For example, suppose that we have a closed system but the initial preparation is not controlled well, so that although the system is in a definite state, we do not know which, we know only the classical probabilities of it being any state in a given set of states. Uncontrollable preparation was very typical in the early days of quantum mechanics. Nowadays, however, this is becoming less and less common in quantum optics. What is far more common is to have an open system. In this case also, we will generally not know the state of the system with certainty. As the case of an open system is more relevant today, we will use it as an example to introduce an idea for dealing with this classical[4] uncertainty in quantum mechanics: the density matrix.

First, let us see how the idea of the density matrix emerges from the basic quantum mechanics formalism. Let $|\psi\rangle$ be the state of a closed system. The measurable quantities that come out of the theory are the expectation values of observables. Let $\hat{A}$ be an operator corresponding to one such observable. Then, as we know from basic quantum mechanics, the expectation value of $\hat{A}$ in the state $|\psi\rangle$ is given by $\langle\psi|\hat{A}|\psi\rangle$. Now, introducing a complete basis $\{|e_i\rangle\}$ for our closed system, where $\sum_i |e_i\rangle\langle e_i| = \hat{1}$, with $\hat{1}$ being the unit operator, we can rewrite any such expectation value as

$$\langle\hat{A}\rangle = \sum_i \langle\psi|\hat{A}|e_i\rangle\langle e_i|\psi\rangle$$

$$= \sum_i \langle e_i|\psi\rangle\langle\psi|\hat{A}|e_i\rangle$$

$$= \mathrm{Tr}\left(\hat{\rho}\hat{A}\right), \tag{4.27}$$

---

[4]Classical because the system is actually in a definite quantum state, although we do not know which. Quantum uncertainty is of a different nature: The system does not even have the uncertain property, only acquiring the property when we measure it.

where

$$\hat{\rho} = |\psi\rangle\langle\psi| \qquad (4.28)$$

is the density matrix and Tr stands for the trace operation, which consists of adding all the diagonal elements of a matrix (operator).

Everything that is ordinarily done in quantum mechanics using quantum states $|\psi\rangle$ can be done using the density matrix $\hat{\rho}$. Even the Schrödinger equation for $|\psi\rangle$ can be rewritten in the form of a master equation to describe the time evolution of $\hat{\rho}$:

$$\frac{\partial}{\partial t}\hat{\rho} = -\frac{i}{\hbar}[\hat{H}, \hat{\rho}], \qquad (4.29)$$

where $\hat{H}$ is the Hamiltonian of the system. But the real usefulness of the density matrix becomes apparent only when we consider open systems.

Any open system can be viewed as part of a very large closed system formed by itself and its environment (see Figure 4.1). For our present purposes, the key feature

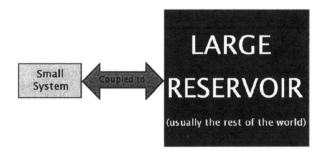

Figure 4.1 The usual way of dealing with a dissipative system in quantum mechanics. An open system together with its environment form a closed system, to which we can apply the ordinary formalism of quantum mechanics. The environment can be thought as a large reservoir to which the small open system is coupled.

of the environment is that it has an immense number of degrees of freedom that we are not able to observe. So the operator $\hat{A}$ appearing in (4.27) is in this case $\hat{A} = \hat{A}_S$, where the subscript S denotes that this operator acts only on the states of the small open system, not on the states of the environment. A complete basis for the large closed system can be formed by taking the tensor product of a complete basis $\{|s_i\rangle\}$ for the small open system with a complete basis $\{|r_j\rangle\}$ for the environment. Then we can write the expectation value of $\hat{A}_S$ as

$$\langle\psi|\hat{A}_S|\psi\rangle = \sum_i \sum_j \langle\psi|\hat{A}_S|s_i\rangle|r_j\rangle\langle r_i|\langle s_i|\psi\rangle$$

$$= \sum_i \langle s_i| \left[\underbrace{\left(\sum_j \langle r_j|\psi\rangle\langle\psi|r_j\rangle \right)}_{\hat{\rho}_S} \hat{A}_S \right] |s_i\rangle$$

$$= \mathrm{Tr}_S \left(\hat{\rho}_S \hat{A}_S \right), \tag{4.30}$$

where

$$\hat{\rho}_S = \mathrm{Tr}_R(\hat{\rho}) \tag{4.31}$$

is the reduced density matrix for the small open system S, with Tr_R and Tr_S being the traces over the environment and the open system, respectively. In other words, instead of using the states $|\psi\rangle$ belonging to the very large space formed by S and R, we can just use the reduced density matrix $\hat{\rho}_S$ belonging to the small open system S.

The density matrix of the entire system, $\hat{\rho}$, is just the projector $|\psi\rangle\langle\psi|$ and leads to the same results that we obtain using the ordinary state vector $|\psi\rangle$. The reduced density matrix, $\hat{\rho}$, on the other hand, opens up a whole new possibility that cannot be described by ordinary state vectors alone: mixed states. If $|\psi\rangle$ is a simple product of a state of the open system by a state of the environment, $\hat{\rho}_S$ will be just a projector as $\hat{\rho}$. So $|\psi\rangle$ must be an entangled state of S and R, if $\hat{\rho}_S$ is to contain any physics that cannot be described by just using the old state vector formalism (i.e., $|\psi\rangle$ must not be factorizable into a product of a state of S and a state of R). The simplest such entangled state has the general form

$$|\psi\rangle = \left(|\phi\rangle|R\rangle + p e^{i\theta} |\phi_\perp\rangle|R'\rangle \right) \frac{1}{\sqrt{1+p^2}}, \tag{4.32}$$

where $|\phi\rangle$ and $|\phi_\perp\rangle$ are two orthogonal states of S, $|R\rangle$ and $|R'\rangle$ are any two states of R, and p and θ are two real numbers. If $|R\rangle$ and $|R'\rangle$ are the same state, $|\psi\rangle$ becomes a product state; otherwise, it will not be factorizable except for the trivial case where $p = 0$.

We obtain the following reduced density matrix from (4.32):

$$\hat{\rho}_S = \frac{1}{1+p^2} \left\{ |\phi\rangle\langle\phi| + p^2 |\phi_\perp\rangle\langle\phi_\perp| + p \left(e^{-i\theta} \langle R'|R\rangle|\phi\rangle\langle\phi_\perp| + \mathrm{H.c.} \right) \right\}, \tag{4.33}$$

where H.c. stands for *Hermitian conjugate*. In this simple case, from the form of (4.33), we can tell whether $\hat{\rho}_S$ represents a pure state or a mixture. When $|\langle R|R'\rangle|^2 = 1$, $|\psi\rangle$ is a product state and $\hat{\rho}_S$ becomes a pure state described by the projector

$$\hat{\rho}_S = \left\{ \frac{1}{\sqrt{1+p^2}} \left(|\phi\rangle + p e^{i\theta} |\phi_\perp\rangle \right) \right\} \left\{ \frac{1}{\sqrt{1+p^2}} \left(\langle\phi| + p e^{-i\theta} \langle\phi_\perp| \right) \right\}. \tag{4.34}$$

When $|\langle R|R'\rangle|^2 = 0$, $|\psi\rangle$ is an entangled state and $\hat{\rho}_S$ becomes a statistical mixture completely devoid of any quantum correlations,

$$\hat{\rho}_S = p_1 |\phi\rangle\langle\phi| + p_2 |\phi_\perp\rangle\langle\phi_\perp|, \tag{4.35}$$

where

$$p_1 = \frac{1}{1+p^2} \tag{4.36}$$

and

$$p_2 = \frac{p^2}{1 + p^2} \tag{4.37}$$

are the classical probabilities (i.e., $p_1 + p_2 = 1$) of finding S in states $|\phi\rangle$ or $|\phi_\perp\rangle$, respectively. When $0 < |\langle R|R'\rangle|^2 < 1$, $|\psi\rangle$ will still be an entangled state, but $\hat{\rho}_S$ will be a mixture with some degree of quantum coherence left in it. This quantum coherence is revealed by the presence of off-diagonal components in (4.33) in this case. The amount of quantum coherence present will depend on how close to 1 $|\langle R|R'\rangle|^2$ is, covering a continuous range from a pure state to a mixture devoid of coherence.

In general, however, it might not be obvious whether a given reduced density matrix is pure or a mixture just by inspecting its expression. There might not even be an expression for $\hat{\rho}_S$. In this case, the trace of $\hat{\rho}_S^2$ can be very helpful. Even though $\mathrm{Tr}(\hat{\rho}_S)$ is always unity, as $\langle \psi|\psi\rangle = 1$, the trace of $\hat{\rho}_S^2$ lets us distinguish reduced density matrices that represent pure states from those that represent mixtures. From (4.33) we find that

$$\mathrm{Tr}\left(\hat{\rho}_S^2\right) = \frac{1}{(1 + p^2)^2} \left(1 + p^4 + 2p^2|\langle R|R'\rangle|^2\right). \tag{4.38}$$

So if the state is a mixture, the trace of $\hat{\rho}_S^2$ will be less than 1. Only when the state is pure is the trace of $\hat{\rho}_S^2$ 1.

4.1.3 The diagonal coherent-state representation

As we have seen in Section 4.1.2, quantum coherence is entirely in the off-diagonal components of the density matrix. For example, in the case of a two-dimensional Hilbert space described by the basis formed by the two orthonormal states $|s_1\rangle$ and $|s_2\rangle$, the most general density matrix has the form

$$\begin{aligned}
\hat{\rho} &= |s_1\rangle\langle s_1|p_1 + |s_2\rangle\langle s_2|p_2 \\
&+ |s_1\rangle\langle s_2|a + |s_2\rangle\langle s_1|a^*,
\end{aligned} \tag{4.39}$$

and unless $a \neq 0$, $\hat{\rho}$ will have some quantum coherence.

However, this identification of quantum coherence with the presence of off-diagonal components is possible only when the basis is orthogonal. In the example above, if we use the states $|\varphi_1\rangle$ and $|\varphi_2\rangle$ as basis instead of $|s_1\rangle$ and $|s_2\rangle$, where

$$|\varphi_1\rangle = \left(|s_1\rangle\sqrt{p_1} + |s_2\rangle\frac{a^*}{\sqrt{p_1}}\right)\sqrt{\frac{p_1}{p_1^2 + |a|^2}} \tag{4.40}$$

and

$$|\varphi_2\rangle = |s_2\rangle, \tag{4.41}$$

then

$$\hat{\rho} = \left(p_1 + \frac{|a|^2}{p_1}\right)|\varphi_1\rangle\langle\varphi_1| + |\varphi_2\rangle\langle\varphi_2|\left(p_2 - \frac{|a|^2}{p_1}\right) \tag{4.42}$$

[i.e., two diagonal components with no off-diagonal components as if it is a statistical mixture even when $\text{Tr}(\hat{\rho}^2) = 1$ and it is a pure state]— all because $|\varphi_1\rangle$ and $|\varphi_2\rangle$ are not orthogonal. The same trick can be played when the dimension of the Hilbert space is infinite, as in the case of a mode of the radiation field (a field oscillator). In this subsection we see that a basis of coherent states allows us to write down any density matrix as a diagonal matrix.

To derive the diagonal coherent-state representation of the density matrix, we first notice that any density matrix $\hat{\rho}$ describing a single mode of the radiation field can be written as

$$\hat{\rho} = \sum_{l=0}^{\infty} \sum_{k=0}^{\infty} r_{k,l} \left(\hat{a}\right)^k \left(\hat{a}^\dagger\right)^l. \tag{4.43}$$

This can be seen by writing the annihilation and creation operators in the photon number basis,

$$\hat{a} = \sum_{n=0}^{\infty} \sqrt{n+1} |n\rangle\langle n+1|, \tag{4.44}$$

$$\hat{a}^\dagger = \sum_{n=0}^{\infty} \sqrt{n+1} |n+1\rangle\langle n|. \tag{4.45}$$

Using (4.44) and (4.45), we can rewrite $\left(\hat{a}\right)^k \left(\hat{a}^\dagger\right)^l$ as

$$\left(\hat{a}\right)^k \left(\hat{a}^\dagger\right)^l = \sum_{n=0}^{\infty} \frac{(n+1+l)!}{\sqrt{n!(n+l-k)!}} |n+l-k\rangle\langle n|. \tag{4.46}$$

Now, substituting (4.46) in (4.43) and changing the summation dummy indices l and k to m and j, where $j \equiv l + k$ and $m \equiv n + l - k$, we find that

$$\sum_{l=0}^{\infty} \sum_{k=0}^{\infty} r_{k,l} \left(\hat{a}\right)^k \left(\hat{a}^\dagger\right)^l = \sum_{n=0}^{\infty} \sum_{m=0}^{\infty} c_{n,m} |m\rangle\langle n|, \tag{4.47}$$

with

$$c_{n,m} = \sum_{j=0}^{\infty} r_{|j-m+n|/2,|j+m-n|/2} \frac{\left(1 + \dfrac{j+m+n}{2}\right)!}{\sqrt{n!m!}} |m\rangle\langle n|. \tag{4.48}$$

As (4.47) is a bona fide photon-number-state expansion of $\hat{\rho}$, which is always possible because the photon number states form a complete basis, we see that (4.43) does hold. Incidentally, a normal order expansion, where all the annihilation operators are to the right of the creation operators, is also possible as well as any other ordering.

With (4.43) established, we can now insert the unit operator (4.26) between $\left(\hat{a}\right)^k$ and $\left(\hat{a}^\dagger\right)^l$, obtaining the diagonal coherent-state representation

$$\hat{\rho} = \int d^2\alpha |\alpha\rangle\langle\alpha| P(\alpha, \alpha^*), \tag{4.49}$$

where

$$P(\alpha, \alpha^*) = \frac{1}{\pi} \sum_{k=0}^{\infty} \sum_{l=0}^{\infty} r_{k,l} \alpha^k (\alpha^*)^l. \tag{4.50}$$

However, as it stands, (4.50) is not very useful, as it is written in terms of $r_{k,l}$ rather than some expression involving $\hat{\rho}$ that we can calculate. To obtain such an expression, we notice first that part of the usefulness of (4.49) is that with it the expectation value of an operator $\hat{O}$ becomes simply

$$\langle \hat{O} \rangle = \mathrm{Tr}\left\{ \hat{\rho}\hat{O} \right\}$$

$$= \sum_{n=0}^{\infty} \langle n | \hat{\rho}\hat{O} | n \rangle$$

$$= \sum_{n=0}^{\infty} \int d^2\alpha \langle n | \alpha \rangle \langle \alpha | \hat{O} | n \rangle P(\alpha, \alpha^*)$$

$$= \int d^2\alpha\, P(\alpha, \alpha^*) \langle \alpha | \hat{O} \sum_{n=0}^{\infty} | n \rangle \langle n | \alpha \rangle$$

$$= \int d^2\alpha\, P(\alpha, \alpha^*) \langle \alpha | \hat{O} | \alpha \rangle. \tag{4.51}$$

So if $\hat{O}$ is in normal order, we can just replace each annihilation operator in $\hat{O}$ by α and each creation operator by α^* in order to obtain $\langle \alpha | \hat{O} | \alpha \rangle$. Now let $\hat{O} = \exp\left(\beta\hat{a}^\dagger\right)\exp\left(-\beta^*\hat{a}\right)$. Then using (4.50), we find that

$$\frac{1}{\pi} \int d^2\alpha' e^{\beta\alpha'^* - \beta^*\alpha'} \sum_{k=0}^{\infty} \sum_{l=0}^{\infty} r_{k,l} \alpha'^k (\alpha'^*)^l = Tr\left\{ \hat{\rho}e^{\beta\hat{a}^\dagger} e^{-\beta^*\hat{a}} \right\} \tag{4.52}$$

To obtain $\sum_{k,l} r_{k,l}\alpha^k (\alpha^*)^l$, we just have to multiply (4.52) by $(1/\pi)\exp(\beta^*\alpha - \beta\alpha^*)$ and integrate (Fourier transform) over β. Using the fact that

$$\frac{1}{\pi^2} \int d^2\beta e^{\beta^*(\alpha - \alpha') - \beta(\alpha - \alpha')^*}$$

$$= \int_{-\infty}^{\infty} \frac{d(2\beta_x)}{2\pi} e^{i2\beta_x(\alpha_y - \alpha'_y)} \int_{-\infty}^{\infty} \frac{d(2\beta_y)}{2\pi} e^{-i2\beta_y(\alpha_x - \alpha'_x)}$$

$$= \delta(\alpha_y - \alpha'_y)\delta(\alpha_x - \alpha'_x)$$

$$= \delta^2(\alpha - \alpha'), \tag{4.53}$$

we finally find an expression relating $P(\alpha, \alpha^*)$ to a quantity we can calculate directly from $\hat{\rho}$:

$$P(\alpha, \alpha^*) = \frac{1}{\pi^2} \int d^2\beta\, e^{\beta^*\alpha - \beta\alpha^*} Tr\left\{ \hat{\rho}e^{\beta\hat{a}^\dagger - \beta^*\hat{a}} \right\}. \tag{4.54}$$

Apart from the practical usefulness of (4.49) for computing expectation values of normal-order operators applying (4.51), (4.49) is also important because it offers us an objective criterion for finding out when a given state (density matrix) of the radiation field is nonclassical.

Classical radiation can be described in the same way as we have been dealing with quantum radiation, using a mode expansion, except that instead of creation and annihilation operators, the expansions involve complex amplitudes for each mode. In general, these complex amplitudes will not be known precisely, but there will be a probability of them having certain values (e.g., as in the case of thermal light). For a single mode, we can call $P_{\text{class}}(\alpha, \alpha^*)$ the probability that the field has the complex amplitude α, with $0 \leq P_{\text{class}}(\alpha, \alpha^*) \leq 1$ and $\int d^2\alpha\, P_{\text{class}}(\alpha, \alpha^*) = 1$. Then the average value of a given measurable function of the radiation field $\mathcal{O}(\alpha, \alpha^*)$ is

$$\langle \mathcal{O} \rangle = \int d^2\alpha\, \mathcal{O}(\alpha, \alpha^*) P_{\text{class}}(\alpha, \alpha^*). \tag{4.55}$$

Comparing (4.55) to (4.51), we see that if $P(\alpha, \alpha^*)$ can be regarded as a probability distribution [i.e., if not only $\int d^2\alpha\, P(\alpha, \alpha^*) = 1$ but also $0 \leq P(\alpha, \alpha^*) \leq 1$], then (4.51) is formally identical to (4.55) and the field described by $\hat{\rho}$ can be regarded as classical. The first condition is always satisfied, but the second one does not always hold. In general, $P(\alpha, \alpha^*)$ can even become negative. This is unacceptable for a probability distribution; for this reason, $P(\alpha, \alpha^*)$ is called a *quasiprobability distribution*. Whenever $P(\alpha, \alpha^*)$ is negative, it is impossible to describe the field in a classical way, and thus we can regard $\hat{\rho}$ as nonclassical. Coherent states and statistical mixtures of them are examples of fields for which $P(\alpha, \alpha^*)$ is always positive, justifying their reputations as classical fields. In the next sections we will see an example of a field where $P(\alpha, \alpha^*)$ can become negative. The problem with this criterion of nonclassicality is that for most fields $P(\alpha, \alpha^*)$ is highly singular. In the case of a coherent state, for example, $P(\alpha, \alpha^*)$ is already a delta function. For this reason, people have sought alternative criteria of nonclassicality sometimes using other quasiprobability distributions, such as the Wigner function [429] and even other methods [155, 616, 617].

4.2 GOING BEYOND THE STANDARD QUANTUM LIMIT: SQUEEZED STATES

From a classical point of view, light is just an electromagnetic wave, as radio waves, except that its frequency is much larger than those of radio. Then we should be able to modulate a light wave as we modulate a radio carrier wave, to carry radio and television channels, for instance. There is a great advantage in using light instead of radio. Because the frequency is several orders of magnitude larger for light than for radio, a light wave can carry many more communication channels than can a radio wave [657].

Unfortunately, this nice idea has a basic flaw: Light is not a wave; it is made up of quantum particles (i.e., photons). As we have seen in Section 4.1, coherent states are the closest that we can get to a wave, and even then, they can only be regarded as waves with amplitude and phase noise.

But there are also radio photons. So why is it that quantum noise is not a problem for radio, where the classical wave description seems to work very well? The answer is that the amount of quantum noise depends on how much energy a photon carries. As this energy is proportional to the frequency, radio photons carry much less energy than do light photons. So a radio coherent state can already be regarded as a classical wave at much lower intensities than it would be possible to regard a light coherent state as a classical wave. In fact, at room temperature, for radio, $\hbar\omega$ is much smaller than kT (i.e., the Boltzmann constant times the absolute temperature), showing that thermal noise is much more of an issue for radio than is quantum noise. For light, the situation is inverted (i.e., $\hbar\omega \gg kT$).

Light can still be used for communication, indeed, it is actually already being used in many fiber optic telephone and data networks. For these engineering applications, it can basically be regarded as a very noisy high-frequency classical radio wave. There is, however, a way to take more out of light: a way to reduce the quantum noise effectively so that we are able to do with it the same things that we can do with low-noise radio waves. This section is about this way of effectively reducing quantum noise. There are applications for this not only in communications, where it increases the channel capacity, but also in optical measurements, where it increases the precision.

The idea is that for radio communications and many interferometric measurements, one does not use the entire complex wave amplitude, only its real or its imaginary part (or any arbitrary quadrature, see Section 4.2.4). For a single mode, these are Hermitian operators proportional to what we call quadratures of $\hat{a}$,

$$\hat{a}_1 \equiv \frac{\hat{a} + \hat{a}^\dagger}{2} \tag{4.56}$$

and

$$\hat{a}_2 \equiv \frac{\hat{a} - \hat{a}^\dagger}{i2}, \tag{4.57}$$

respectively. So all that we have to do is to decrease the noise (i.e., the variance) in the required quadrature. As the uncertainty relation involves the product of the variances in each of two complementary quadratures (just as $\hat{q}$ and $\hat{p}$ in Section 2.4), reducing the variance of only one of them will not violate quantum mechanics, because the variance in the other quadrature is free to increase. In the next subsection we introduce quantum states with such a reduced noise in one of the quadratures. In Section 4.2.2 we discuss a way of generating these states. Then in Section 4.2.3, we present a nice way of visualizing them. Finally, in Section 4.2.4, we discuss how they can be measured.

4.2.1 The squeezing operator

For a single mode, we can define dimensionless position

$$\hat{x} \equiv \sqrt{2}\hat{a}_1 \tag{4.58}$$

and momentum

$$\hat{p} \equiv \sqrt{2}\hat{a}_2 \tag{4.59}$$

operators, such that

$$[\hat{x}, \hat{p}] = i. \tag{4.60}$$

Any coherent state of this single mode can be described in a dimensionless x representation (just like the normal x representation) by a wavefunction $\psi(x)$. Then we can obtain a squeezed state of quadrature $\hat{a}_1$ if we make a scale transformation changing $\psi(x)$ into something proportional to $\psi(\lambda x)$ with λ real and larger than 1. The variance of the transformed function will be $1/\lambda$ of that of the original function, and increasing λ, the new variance can be made as small as we please.

As this transformation only "shrinks" the wavefunction without altering its general shape, the uncertainty product in the $x - p$ uncertainty relation is not affected. In fact, recalling that the x and p representation wavefunctions are related by a Fourier transform, we find that if the variance of the untransformed function in dimensionless p space is Δp, the variance of the transformed function will be $\lambda \, \Delta p$. So we can say that a transformation of this kind only "removes" uncertainty from one quadrature to "add" it to the complementary quadrature, conserving the "total" uncertainty.

Let $\psi_\lambda(x)$ be the squeezed wavefunction; then

$$\psi_\lambda(x) = C\psi(\lambda x), \tag{4.61}$$

where C is a complex quantity. To determine C, we demand that $\psi_\lambda(x)$ be normalized; that is,

$$\begin{aligned}
1 &= \int dx \; \psi_\lambda^*(x)\psi_\lambda(x) \\
&= |C|^2 \int dx \; \psi_\lambda^*(x)\psi_\lambda(x) \\
&= \frac{|C|^2}{\lambda} \int dx \; |\psi(x)|^2 \\
&= \frac{|C|^2}{\lambda},
\end{aligned} \tag{4.62}$$

as the untransformed function $\psi(x)$ is also normalized. Choosing the phase of $\psi_\lambda(x)$ to be the same as that of $\psi(x)$, we find that

$$\psi_\lambda(x) = \sqrt{\lambda}\,\psi(\lambda x). \tag{4.63}$$

As we mentioned in Chapter 2, it is often more convenient to work with quantum states directly, without being limited to any particular basis. Let us then define the squeezing operator $\hat{S}_\lambda$ as the operator that squeezes the state it is applied to, that is,

$$|\psi_\lambda\rangle = \hat{S}_\lambda|\psi\rangle, \tag{4.64}$$

where

$$\langle x|\psi\rangle = \psi(x) \tag{4.65}$$

and

$$\langle x|\psi_\lambda\rangle = \psi_\lambda(x). \tag{4.66}$$

Now let us determine the explicit form of $\hat{S}_\lambda$.

From (4.63) it follows that $\hat{S}_\lambda$ is unitary, so we can always write it down as

$$\hat{S}_\lambda = e^{-i\hat{M}(\lambda)}, \tag{4.67}$$

where $\hat{M}(\lambda)$ is a Hermitian operator that is a function of λ. For $\lambda = 1$ we know that $\hat{S}_1 = \hat{1}$. If $\lambda = 1 + \epsilon$ with $|\epsilon| \ll 1$, we can expand (4.67) as a power series in ϵ and neglect terms higher than first order in ϵ:

$$\hat{S}_{1+\epsilon} = \sum_{n=0}^\infty \frac{\epsilon^n}{n!} \frac{d^n}{d\epsilon^n} \left[e^{-i\hat{M}(1+\epsilon)} \right]_{\epsilon=0}$$
$$\approx \hat{1} - i\epsilon\hat{Z}, \tag{4.68}$$

where

$$\hat{Z} = \left(\frac{d\hat{M}}{d\lambda} \right)_{\lambda=1} \tag{4.69}$$

is the generator[5] of the unitary transformation $\hat{S}_\lambda$. To obtain $\hat{Z}$, we notice that (4.63) implies that

$$\hat{S}_\lambda^\dagger \hat{x} \hat{S}_\lambda = \frac{\hat{x}}{\lambda}, \tag{4.70}$$

which, to first order in ϵ, reduces to

$$[\hat{Z}, \hat{x}] = i\hat{x}. \tag{4.71}$$

Now, we know that $\hat{p}$ is the generator of translations.[6] Our scale transformation (i.e., squeezing) is a translation proportional to $\hat{x}$. Thus, we might think that $\hat{Z}$ is just something like $\hat{x}\hat{p}$. But $\hat{x}\hat{p}$ is not Hermitian! So, instead of $\hat{x}\hat{p}$, we try the Anzats $(\hat{x}\hat{p} + \hat{p}\hat{x})/2$, which is Hermitian.[7] By direct substitution in (4.71), we can see that

$$\hat{Z} = -\frac{1}{2} (\hat{x}\hat{p} + \hat{p}\hat{x}) + f(\hat{x}) \tag{4.72}$$

is a solution of (4.71).

As any function of $\hat{x}$ commutes with $\hat{x}$, $f(\hat{x})$ would be completely arbitrary if all we had was (4.71). There is, however, another key equation that $\hat{Z}$ must satisfy.

---

[5]See Chapter VI of [243].
[6]The reader who is not familiar with this should have a look at Chapter VI of [243].
[7]Notice that the coefficients of $\hat{x}\hat{p}$ and $\hat{p}\hat{x}$ in the sum have to be the same; otherwise, the sum is not Hermitian. This is the reason for adopting the symmetric product $(\hat{x}\hat{p} + \hat{p}\hat{x})/2$ rather than just any arbitrary linear combination of $\hat{x}\hat{p}$ and $\hat{p}\hat{x}$.

As $\hat{S}_\lambda$ is unitary, it must preserve the commutation relations between operators; in particular, it cannot alter (4.60). In order not to alter (4.60), we must have

$$\hat{S}_\lambda^\dagger \hat{p} \hat{S}_\lambda = \lambda \hat{p}. \tag{4.73}$$

Then

$$[\hat{Z}, \hat{p}] = -i\hat{p}. \tag{4.74}$$

The only way to satisfy both (4.71) and (4.74) is to take $f(\hat{x})$ as a complex number. From (4.67) we see that a complex $f(\hat{x})$ just leads to a trivial phase factor in $\hat{S}_\lambda$. But we have already chosen the phase of $\psi_\lambda(x)$ when we wrote (4.63), and to satisfy (4.63), $f(\hat{x}) = 0$, so

$$\hat{Z} = -\frac{1}{2}\left(\hat{x}\hat{p} + \hat{p}\hat{x}\right). \tag{4.75}$$

We have determined the operator that performs an infinitesimal squeezing operation. Finite squeezing can be achieved by a succession of several infinitesimal squeezing operations. In other words, as $\hat{S}_\lambda$ has the group property

$$\hat{S}_{\lambda_1}\hat{S}_{\lambda_2} = \hat{S}_{\lambda_1 \lambda_2}, \tag{4.76}$$

we can obtain $\hat{S}_\lambda$ from the product of N infinitesimal operators $\hat{S}_{1+\epsilon}$ if we choose N large enough so that $(1 + \epsilon)^N = \lambda$. Then let us take $\epsilon = (\ln \lambda)/N$, so that as $N \to \infty$, $\lim_{N \to \infty}(1 + \epsilon)^N = \lim_{N \to \infty}[1 + (\ln \lambda)/N]^N = \exp(\ln \lambda) = \lambda$. Using (4.76), we find that

$$\begin{aligned}
\hat{S}_\lambda &= \lim_{N \to \infty}\left(\hat{1} - i\frac{\ln \lambda}{N}\hat{Z}\right)^N \\
&= \exp\left(-i\hat{Z}\ln \lambda\right) \\
&= \exp\left[\frac{i}{2}\left(\hat{x}\hat{p} + \hat{p}\hat{x}\right)\ln \lambda\right] \\
&= \exp\left[\frac{1}{2}\left(\hat{a}^2 - \hat{a}^{\dagger 2}\right)\ln \lambda\right].
\end{aligned} \tag{4.77}$$

Let us examine now a particular type of squeezed state that is also a minimum uncertainty state. This type of squeezed state can be seen as a generalization of coherent states where the product of uncertainties in the two canonically conjugate quadratures is still minimum, but with each quadrature having a different uncertainty[8] (i.e., variance). In fact, as the squeezing operation does not alter the uncertainty product, we can obtain such a state in a straightforward way just by applying $\hat{S}_\lambda$ to a coherent state. But we can get more insight if instead of doing so, we calculate $\hat{S}_\lambda \hat{a} \hat{S}_\lambda^\dagger$.

From (4.58), (4.59), (4.70), and (4.73), taking $\lambda \to 1/\lambda$, we find that

$$\hat{S}_\lambda \hat{a} \hat{S}_\lambda^\dagger = \hat{b}_\lambda, \tag{4.78}$$

---

[8]Coherent states have the same variance in both quadratures.

where

$$\hat{b}_\lambda = \mu_\lambda \hat{a} + \nu_\lambda \hat{a}^\dagger, \tag{4.79}$$

with

$$\mu_\lambda = \cosh r, \tag{4.80}$$
$$\nu_\lambda = \sinh r, \tag{4.81}$$

and

$$r = \ln \lambda. \tag{4.82}$$

Now, we notice that the minimum uncertainty squeezed state $|\beta\rangle_\lambda$, which we obtain when $\hat{S}_\lambda$ is applied to a coherent state $|\beta\rangle$, is a right-hand-side eigenstate of the operator $\hat{b}_\lambda$; that is,

$$\begin{aligned}
\hat{b}_\lambda |\beta\rangle_\lambda &= \hat{S}_\lambda \hat{a} \hat{S}_\lambda^\dagger \left(\hat{S}_\lambda |\beta\rangle \right) \\
&= \hat{S}_\lambda \hat{a} |\beta\rangle \\
&= \beta \hat{S}_\lambda |\beta\rangle \\
&= |\beta\rangle_\lambda \, \beta.
\end{aligned} \tag{4.83}$$

This analogy with coherent states can be taken even further.

Consider $\hat{b}_\lambda^\dagger \hat{b}_\lambda$. As $[\hat{b}_\lambda, \hat{b}_\lambda^\dagger] = 1$, $\hat{b}_\lambda^\dagger \hat{b}_\lambda$ behaves as a generalized number operator and $\hat{b}_\lambda$, $\hat{b}_\lambda^\dagger$ as generalized annihilation and creation operators. The eigenstates of $\hat{b}_\lambda^\dagger \hat{b}_\lambda$ are generalized number states, and $|\beta\rangle_\lambda$ has an expansion in terms of these generalized number states which is formally identical to that of coherent states in terms of number states, with a Poissonian distribution of generalized quanta.

This particular type of squeezed state, which is also a minimum uncertainty state, has been named a *two-photon coherent state* by Yuen [660]. The coherent-state analogy mentioned above explains why this type of squeezed state can also be seen as a sort of generalized coherent state. The reason that this generalization can be regarded as a two-photon generalization, with ordinary coherent states being one-photon coherent states, has to do with the form of the squeezing operator. If we compare $\hat{S}_\lambda$ and $\hat{D}(\alpha)$, we see that $\hat{S}_\lambda$ is $\hat{D}(\alpha)$ with $\alpha \to (1/2) \ln \lambda$ and $\hat{a} \to \hat{a}^2$. So while the argument of the exponential in $\hat{D}(\alpha)$ is linear in $\hat{a}$ and $\hat{a}^\dagger$, in $\hat{S}_\lambda$ it is quadratic. In Chapter 5 when we treat the interaction with matter, you will realize that $\hat{S}_\lambda$ and $\hat{D}(\alpha)$ can be seen as time evolution operators associated with interaction Hamiltonians between a single mode of the radiation field and a classical current.[9] In the case of $\hat{D}(\alpha)$, the interaction is linear, describing a one-photon transition. In the case of $\hat{S}_\lambda$, the interaction is nonlinear (quadratic), describing a two-photon transition.

---

[9]This might seem a bit contradictory at first sight, but such a "classical" two-photon current can actually be realized using the nonlinear dielectric permittivity of certain crystals. This method was in fact used in the first demonstration of squeezing in a parametric oscillator [569].

4.2.2 A curious way of generating squeezed states

As the interaction with matter is dealt with only in Chapter 5, we will not describe here the way of generating squeezed states mentioned above. Instead, we will describe a less practical but more dramatic and theoretically interesting alternative way that will also help illustrate some of the points discussed earlier.

As a single-cavity mode is equivalent to a harmonic oscillator, let us start this discussion by considering a harmonic oscillator of unit mass whose Hamiltonian is given by

$$\hat{H} = \frac{\hat{p}^2}{2} + \omega^2 \frac{\hat{q}^2}{2}, \tag{4.84}$$

where $\hat{p}$ is the particle's momentum operator and $\hat{q}$ its position operator. The oscillator energy quanta annihilation operator is defined by (see Section 2.4.1)

$$\hat{b} \equiv \sqrt{\frac{\omega}{2\hbar}} \, \hat{q} + i \frac{\hat{p}}{\sqrt{2\omega\hbar}}. \tag{4.85}$$

Now what happens if the frequency of this oscillator were to change quite suddenly[10] from ω to ω'? As we are working in the Schödinger picture, the particle's position and momentum operators remain unchanged, but the energy quanta annihilation operator will change to

$$\hat{a} \equiv \sqrt{\frac{\omega'}{2\hbar}} \, \hat{q} + i \frac{\hat{p}}{\sqrt{2\omega'\hbar}}. \tag{4.86}$$

From (4.85) and (4.86) we find that the energy quanta annihilation operator after the sudden frequency change is related to the annihilation and creation operators before by [319]

$$\hat{a} = \frac{1}{2\sqrt{\omega'\omega}} \left\{ (\omega' + \omega)\hat{b} + (\omega' - \omega)\hat{b}^\dagger \right\}. \tag{4.87}$$

As during a sudden change of Hamiltonian, the quantum state remains unchanged [441], (4.87) is entirely equivalent to (4.79). Thus, the sudden change in the oscillator's frequency leads to squeezing [246, 319].

But how can we suddenly change the frequency of an oscillator? Janszky and Yushin [319] suggested using a Franck–Condon transition. This is a molecular transition between two vibrational states belonging to different electronic states (see Secs. 14.1 and 16.4 of [262]). Now we will look into a way to achieve this frequency jump in cavity QED.

In a cavity, the frequency of the oscillator corresponding to a mode can be changed by changing the size of the cavity. As the size that matters is the effective optical size, it can be changed by actually moving the cavity walls or by changing the dielectric permittivity inside the cavity [656]. But either way, the analogy between a single cavity-mode and a harmonic oscillator will no longer hold, as the change in cavity size

---

[10]*Quite suddenly* means in a time much shorter than the period of oscillation.

exposes the true multimode nature of the cavity (i.e., there is a major rearrangement of modes during the size change that breaks down the formal analogy between a single-mode cavity and an oscillator). Can we generate squeezed light from the vacuum by suddenly changing the size of a cavity? What happens to the field when the cavity changes size suddenly?

To answer these questions, we consider a simple one-dimensional model where a cavity is formed by a pair of perfect mirrors placed at $x = 0$ and $x = L_i$. The mirror at $x = L_i$ is suddenly[11] moved from its original position to a new position at $x = L_f$ (see Figure 4.2).

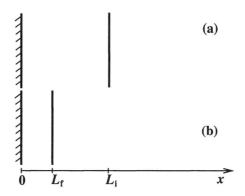

Figure 4.2   Can we squeeze the vacuum by physically squeezing a cavity very rapidly? This diagram represents our one-dimensional cavity model. A cavity is formed by a pair of perfect mirrors in the vacuum initially located at $x = 0$ and $x = L_i$ as in (a). Then suddenly the mirror on the right is pushed from $x = L_i$ to $x = L_f$, as in (b). What happens to the quantum field inside the cavity?

The electric and magnetic fields can be written in terms of the normal modes for the perfect cavity $\sqrt{2/L}\sin(n\pi x/L)$ and the external continuum modes $\sqrt{2/\pi}\sin([x - L]k)$, where L is L_i before the sudden change in cavity size and becomes L_f after. The discrete cavity modes are normalized to 1, and the external continuum modes are delta-function normalized. Before the sudden change, the fields are given by

$$\hat{E}(x) = \begin{cases} i2 \displaystyle\sum_{n=1}^{\infty} \sqrt{\dfrac{\pi\hbar c k_n}{L_i}}\,(\hat{b}_n - \hat{b}_n^\dagger)\sin k_n x & \text{for } 0 < x < L_i \\[18pt] i2 \displaystyle\int_0^\infty dk\,\sqrt{\hbar c k}\,\left\{\hat{b}(k) - \hat{b}^\dagger(k)\right\}\sin([x - L_i]k) & \text{for } x > L_i, \end{cases}$$

$$\text{(4.88)}$$

---

[11] *Suddenly* here means this: Any mirror becomes transparent for high-enough frequencies. Let ω_c be the frequency around which the mirror (moving) starts becoming transparent. The sudden change of cavity size we are considering here is such that it happens in a time τ_c much shorter than $1/\omega_c$. Relativity, of course, requires that during the sudden change the mirror never exceed the speed of light. This results in a restriction on how much L_f can differ from L_i: $|L_f - L_i| \ll c/\omega_c$.

where $k_n \equiv n\pi/L_i$. After the change they are given by

$$\hat{E}(x) = \begin{cases} i2 \displaystyle\sum_{n=1}^{\infty} \sqrt{\dfrac{\pi\hbar c k'_n}{L_f}}(\hat{a}_n - \hat{a}_n^\dagger)\sin k'_n x & \text{for } 0 < x < L_f \\[4mm] i2 \displaystyle\int_0^\infty dk \sqrt{\hbar c k}\left\{\hat{a}(k) - \hat{a}^\dagger(k)\right\}\sin([x - L_f]k) & \text{for } x > L_f, \end{cases}$$

(4.89)

where $k'_n \equiv n\pi/L_f$.

Even though the expression for the fields in terms of annihilation and creation operators is different before and after the sudden change in cavity length, the field operators, like $\hat{q}$ and $\hat{p}$ for the harmonic oscillator, remain unchanged.[12] Using this, we can work out how the annihilation and creation operators after the sudden change relate to the annihilation and creation operators before.

Assuming that $L_f \leq L_i$, we multiply the electric field expressions in (4.88) and (4.89) by $\sqrt{2/L_f}\sin(n\pi x/L_f)$ and integrate over x from 0 to L_f. Equating the results, we find that

$$\hat{a}_n - \hat{a}_n^\dagger = -\frac{1}{\pi L_i}\sum_{m=1}^{\infty}\sqrt{\frac{m}{n}}(\hat{b}_m - \hat{b}_m^\dagger)$$
$$\times \left\{\frac{\sin\left(\left[\dfrac{n}{L_f} - \dfrac{m}{L_i}\right]\pi L_f\right)}{\dfrac{n}{L_f} - \dfrac{m}{L_i}} + \frac{\sin\left(\left[\dfrac{n}{L_f} + \dfrac{m}{L_i}\right]\pi L_f\right)}{\dfrac{n}{L_f} + \dfrac{m}{L_i}}\right\}. \quad (4.90)$$

Analogously, multiplying the magnetic field expressions in (4.88) and (4.89) by $\sqrt{2/L_f}\cos(n\pi x/L_f)$, integrating over x from 0 to L_f, and equating the results, we obtain

$$\hat{a}_n - \hat{a}_n^\dagger = -\frac{1}{\pi L_i}\sum_{m=1}^{\infty}\sqrt{\frac{m}{n}}(\hat{b}_m + \hat{b}_m^\dagger)$$
$$\times \left\{\frac{\sin\left(\left[\dfrac{n}{L_f} - \dfrac{m}{L_i}\right]\pi L_f\right)}{\dfrac{n}{L_f} - \dfrac{m}{L_i}} - \frac{\sin\left(\left[\dfrac{n}{L_f} + \dfrac{m}{L_i}\right]\pi L_f\right)}{\dfrac{n}{L_f} + \dfrac{m}{L_i}}\right\}. \quad (4.91)$$

---

[12]We are working in the Schrödinger picture.

Adding (4.90) and (4.91), we get[13]

$$
\hat{a}_n = -\frac{L_f}{\pi L_i} \sum_{m=1}^{\infty} \sqrt{\frac{m}{n}} \left\{ \frac{\sin\left(\left[n - m\frac{L_f}{L_i}\right]\pi\right)}{n - (L_f/L_i)m} \hat{b}_m - \frac{\sin\left(\left[n + m\frac{L_f}{L_i}\right]\pi\right)}{n + (L_f/L_i)m} \hat{b}_m^\dagger \right\}.
$$

(4.92)

As the quantum state of the field remains unchanged for a sudden change in the Hamiltonian, (4.92) implies that the initial vacuum becomes squeezed. But as we had already foreseen, the new annihilation operators are related not only to the annihilation and creation operators of the corresponding oscillator (mode) before the sudden change, but also to those of every other oscillator (mode) as well.

4.2.3 A geometrical picture for squeezed states

Any reader who has ever browsed through the literature on squeezing has probably come across pictures of noise ellipses and circles. In this subsection we explain how this geometrical picture of quantum noise works.

In his pioneering paper on squeezed states, Yuen [660] defined his two-photon coherent states as in (4.79):

$$
\hat{b}_{\mu,\nu} = \mu\hat{a} + \nu\hat{a}^\dagger,
$$

(4.93)

except that rather than demanding real μ and ν, his only requirement on them was

$$
|\mu|^2 - |\nu|^2 = 1,
$$

(4.94)

otherwise leaving them free to assume any complex value.

Apart from a global arbitrary phase multiplying $\hat{b}_{\mu,\nu}$, Yuen's requirement on μ and ν allows us to write

$$
\mu = e^{-i\theta} \cosh r,
$$

(4.95)

$$
\nu = e^{i\theta} \sinh r,
$$

(4.96)

where θ is any angle and r is any real number, no longer necessarily related to λ through $\lambda = \ln r$. Inverting (4.93) gives

$$
\hat{a} = \mu^* \hat{b}_{\mu,\nu} - \nu\hat{b}_{\mu,\nu}^\dagger,
$$

(4.97)

---

[13]Notice that

$$
\lim_{L_f \to L_i} \frac{\sin\left(\left[n - m\frac{L_f}{L_i}\right]\pi\right)}{n - (L_f/L_i)m} = \pi\delta_{nm} \qquad \lim_{L_f \to L_i} \frac{\sin\left(\left[n + m\frac{L_f}{L_i}\right]\pi\right)}{n + (L_f/L_i)m} = 0,
$$

so that $\lim_{L_f \to L_i} \hat{a}_n = \hat{b}_n$.

and using (4.56) and (4.57), we obtain the following expression for the product of the quadrature uncertainties (variances) in eigenstates of $\hat{b}_\lambda$:

$$\langle(\Delta\hat{a}_1)^2\rangle\langle(\Delta\hat{a}_2)^2\rangle = \frac{1}{16}|\mu_\lambda^2 - \nu_\lambda^2|^2$$

$$= \frac{1}{16}(\cosh^4 r - 2\cos 4\theta \, \cosh^2 r \, \sinh^2 r + \sinh^4 r).$$

$$(4.98)$$

Equation (4.98) yields the minimum uncertainty product (i.e., $\langle(\Delta\hat{a}_1)^2\rangle \times \langle(\Delta\hat{a}_2)^2\rangle = 1/16$), when $\theta = 0$, $\pi/2$, π, and $3\pi/2$ for θ defined in the interval $[0, 2\pi)$. Apart from a global phase, the operator $\hat{b}_{\mu,\nu}$ obtained from (4.93) with (4.95) and (4.96) for these values of θ corresponds to either $\hat{b}_{\ln r}$ or $\hat{b}_{1/\ln r}$ obtained from (4.79) with (4.80) and (4.81). For instance, with $\theta = 0$, Yuen's $\hat{b}_\lambda$ correspond to our old $\hat{b}_{\ln r}$, but with $\theta = \pi$, it corresponds to our old $\hat{b}_{1/\ln r}$. For other values of θ, $\hat{b}_{\mu,\nu}$ does not correspond to any $\hat{b}_\lambda$, and its right-hand-side eigenstates are not minimum uncertainty states. Having said that, how can we insist on claiming that the eigenstates of $\hat{b}_{\mu,\nu}$ are minimum-uncertainty states? We will show now that in a certain sense, these eigenstates are not very different from the eigenstates of $\hat{b}_\lambda$. A geometrical representation of these squeezed states will then naturally emerge as a by-product of this discussion.

In classical mechanics, we can talk about phase space and realize, for example, a rotation in phase space. In quantum mechanics, momentum and position cannot both be known simultaneously with absolute certainty, but it is still possible to think about an analogue of rotation in phase space. This is the operation performed by $\hat{R}(\theta)$ defined by (4.8). Using (4.58) and (4.59) in (4.8), we see that $\hat{R}(\theta)$ performs the following transformation on our dimensionless position and momentum operators:

$$\hat{x} \to \hat{x}\cos\theta + \hat{p}\sin\theta, \qquad (4.99)$$
$$\hat{p} \to -\hat{x}\sin\theta + \hat{p}\cos\theta, \qquad (4.100)$$

just as a rotation in classical phase space even though we cannot construct a phase space with $\hat{x}$ and $\hat{p}$.

Now, we notice that $\hat{b}_{\mu,\nu}$ is just $\hat{b}_\lambda$ rotated by θ; that is,

$$\hat{R}(\theta)\hat{b}_\lambda\hat{R}^\dagger(\theta) = \hat{a}e^{-i\theta}\cosh r + \hat{a}^\dagger e^{i\theta}\sinh r$$
$$= \hat{b}_{\mu,\nu}. \qquad (4.101)$$

So if we "look" in the right direction, the eigenstates of $\hat{b}_{\mu,\nu}$ become minimum uncertainty states, just as those of $\hat{b}_\lambda$. The right direction is that given by the rotated quadratures $\hat{a}_1(\theta) = \hat{R}(\theta)\hat{a}_1\hat{R}^\dagger(\theta)$ and $\hat{a}_2(\theta) = \hat{R}(\theta)\hat{a}_2\hat{R}^\dagger(\theta)$. For these rotated quadratures the right-hand-side eigenstates of $\hat{b}_{\mu,\nu}$ are minimum uncertainty states because, according to (4.101), $\hat{R}(\theta)$ applied to such an eigenstate yields a right-hand-side eigenstate of $\hat{b}_\lambda$, which we have already seen to be a minimum uncertainty state of

$\hat{a}_1$ and $\hat{a}_2$. Let us examine in more detail the dependence of the quadrature variances on the rotation angle θ.

For an eigenstate of $\hat{b}_{\mu,\nu}$, the variances of $\hat{a}_1$ and $\hat{a}_2$ are given by

$$\langle(\Delta\hat{a}_1)\rangle = \frac{1}{4}|\mu - \nu|^2$$
$$= a^2 - 2ab\cos 2\theta + b^2, \tag{4.102}$$
$$\langle(\Delta\hat{a}_2)\rangle = \frac{1}{4}|\mu + \nu|^2$$
$$= a^2 + 2ab\cos 2\theta + b^2, \tag{4.103}$$

where

$$a = \frac{1}{2}\cosh r, \tag{4.104}$$

$$b = \frac{1}{2}\sinh r. \tag{4.105}$$

Now we will show that these variances behave as if they are projections in Cartesian axes of an ellipse rotated by an angle θ, whose minimum and maximum width (i.e., the principal axes of the ellipse) correspond to $\langle\Delta\hat{a}_1(\theta)^2\rangle$ and $\langle\Delta\hat{a}_2(\theta)^2\rangle$.

Consider an ellipse whose principal axes have the lengths[14] $a - b$ and $a + b$. For Cartesian coordinate axes x and y aligned with the principal axes, the equation describing this ellipse is the familiar

$$\frac{x^2}{(a-b)^2} + \frac{y^2}{(a+b)^2} = \frac{1}{4}. \tag{4.106}$$

Let us now rotate the ellipse by θ [i.e., the classical version of (4.99) and (4.100)],

$$x \rightarrow x\cos\theta + y\sin\theta, \tag{4.107}$$
$$y \rightarrow -x\sin\theta + y\cos\theta. \tag{4.108}$$

The rotated ellipse is shown in Figure 4.3. Its equation is given by

$$(a^2 + 2ab\cos 2\theta + b^2)x^2 + 4abxy\sin 2\theta + (a^2 - 2ab\cos 2\theta + b^2)y^2 = \frac{(a^2 - b^2)^2}{4}. \tag{4.109}$$

To prove our point, we must obtain the projections Δx and Δy of the ellipse on the Cartesian axes. We see from Figure 4.3 that Δx is the distance between the two values of x for which dx/dy vanishes, and Δy is the distance between the two values of y for which dy/dx vanishes. Calculating these distances, we find that

$$(\Delta x)^2 = a^2 - 2ab\cos 2\theta + b^2, \tag{4.110}$$
$$(\Delta y)^2 = a^2 + 2ab\cos 2\theta + b^2. \tag{4.111}$$

---

[14]We take a and b as real, of course, with a positive and greater than b, as in (4.104) and (4.105).

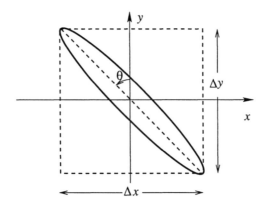

Figure 4.3 An ellipse rotated by θ. The projections on the x and y axes are denoted by Δx and Δy, respectively.

As you can see, these equations are equivalent to (4.102) and (4.103).

This sort of classical behavior of the quadrature variances of right-hand-side eigenstates of $\hat{b}_{\mu,\nu}$ is not a general rule. In fact, for an arbitrary quantum state, a simple calculation reveals that the rotated quadratures $\langle\Delta\hat{a}_1(\theta)^2\rangle$ and $\langle\Delta\hat{a}_2(\theta)^2\rangle$ are related to the "old" ones, $\langle(\Delta\hat{a}_1)^2\rangle$ and $\langle(\Delta\hat{a}_1)^2\rangle$, by

$$\langle\Delta\hat{a}_1(\theta)^2\rangle = \langle(\Delta\hat{a}_1)^2\rangle \cos^2\theta + \langle(\Delta\hat{a}_1)^2\rangle \sin^2\theta$$
$$+ \langle\Delta\hat{a}_1\,\Delta\hat{a}_2 + \Delta\hat{a}_2\,\Delta\hat{a}_1\rangle \cos\theta\sin\theta, \qquad (4.112)$$
$$\langle\Delta\hat{a}_2(\theta)^2\rangle = \langle(\Delta\hat{a}_1)^2\rangle \sin^2\theta + \langle(\Delta\hat{a}_1)^2\rangle \cos^2\theta$$
$$+ \langle\Delta\hat{a}_1\,\Delta\hat{a}_2 + \Delta\hat{a}_2\,\Delta\hat{a}_1\rangle \cos\theta\sin\theta \qquad (4.113)$$

For a right-hand-side eigenstate of $\hat{b}_{\mu,\nu}$ the factor $\langle\Delta\hat{a}_1\Delta\hat{a}_2 + \Delta\hat{a}_2\Delta\hat{a}_1\rangle$ vanishes because $\langle\Delta\hat{a}_1\Delta\hat{a}_2\rangle = i/4$ and $\langle\Delta\hat{a}_2\Delta\hat{a}_1\rangle = -i/4$. Taking $\langle(\Delta\hat{a}_1)^2\rangle = (a-b)^2$ and $\langle(\Delta\hat{a}_1)^2\rangle = (a+b)^2$, we recover (4.102) and (4.103). Notice that $\langle\Delta\hat{a}_1\Delta\hat{a}_2 + \Delta\hat{a}_2\Delta\hat{a}_1\rangle$ is a sort of symmetrical operator for the correlations between fluctuations in each of these two quadratures. For a state where such symmetrized correlations exist, it is no longer possible to express the rotated quadratures as a function of the "old" quadratures only. It is if they can no longer be treated as independent coordinates, and the analogy with the ellipse fails. However, whenever this symmetrized correlations can be neglected, we will be able to adopt a very useful heuristic description based on this convenient transformation property of the rotated variances.

Caves [94] developed a method to represent the quantized radiation field that, within the limits mentioned above, manages to deal with both its wave and particle properties. This is the now widely used (and sometimes abused) complex phase diagram; an attempt to extend the classical electrical engineering concept of phasor to quantum optics. The idea is to express the annihilation operator $\hat{a}$ as a phasor $Ae^{i\delta}$ corresponding to its average value in the quantum state we are dealing with, exactly as in classical mechanics, except for a *noise ellipse* centered at the end of the phasor describing the quantum fluctuations around this average. This is shown in Figure 4.4. Whenever the symmetrized correlations mentioned in the preceding paragraph vanish

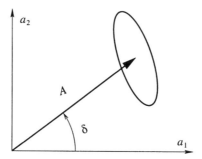

Figure 4.4 Schematic representation of a squeezed state. The phasor describes the expectation value of $\hat{a}$ and the ellipse, the noise. The variance of a quadrature, say a_1, is given by the projection of the ellipse on the corresponding axis.

or can be neglected, the fluctuations of $\hat{a}$ in any direction can be obtained simply by projecting the noise ellipse on that particular direction. The complex phase diagrams perfectly account for this convenient transformation property of the quantum noise of these states. However, as the projections of the noise ellipse are complex numbers corresponding to expectation values of operators rather than the operators themselves, problems appear when we attempt to use products of these projections to describe products of the respective operators.

To illustrate the usefulness of these diagrams as well as the problem mentioned above, let us consider a specific example. The simplest observable involving a product of $\hat{a}$ and $\hat{a}^\dagger$ is the number operator. Let us draw Caves's noise ellipse diagram for the photon number fluctuations of a coherent state.

A simple calculation shows that the rotated quadrature variances, $\langle \Delta \hat{a}_1(\theta)^2 \rangle$ and $\langle \Delta \hat{a}_2(\theta)^2 \rangle$, of a coherent state $|\alpha\rangle$ are independent of θ and always equal to $1/4$. In other words, the noise ellipse of a coherent state is a circle of radius[15] $1/4$. Thus, taking $\alpha = Ae^{i\delta}$, we represent this coherent state by Figure 4.5.

Now we notice that the number operator remains unchanged under a rotation $\hat{R}(\theta)$. So we can write the number operator as a product of $\hat{a}_\parallel^\dagger$ and $\hat{a}_\parallel$, where $\hat{a}_\parallel$ is the annihilation operator rotated by δ (i.e., $\langle \hat{a}_\parallel \rangle = A$). In terms of Caves's diagram, this means that to describe $\hat{N}$, only the amplitude of the phasor and the projection of the noise ellipse along the direction of the phasor have to be taken into account. As the noise ellipse is in this case a circle of diameter $1/2$, its projection in any direction

---

[15]So that the square of its diameter is $1/4$.

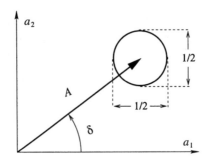

Figure 4.5 Schematic representation of a coherent state. The noise is represented by a circle, because *any* quadrature has the same variance of $1/4$.

is always $1/2$. Then

$$\Delta A^2 = \left(A + \frac{1}{2} + \frac{i}{2}\right)^* \left(A + \frac{1}{2} + \frac{i}{2}\right) - A^2$$

$$= \left(\frac{1}{2} + \frac{i}{2}\right)A + \left(\frac{1}{2} - \frac{i}{2}\right)A + \left(\frac{1}{2} - \frac{i}{2}\right)\left(\frac{1}{2} + \frac{i}{2}\right)$$

$$= A + \frac{1}{2}. \tag{4.114}$$

Before we can say that this is the variance of $\hat{n}$, let us think about what we cannot trust and should not take seriously in such a calculation. There is a very good reason why terms involving products of noise ellipse projections should not be taken seriously. As $\hat{a}_{\|}$ and $\hat{a}_{\|}^{\dagger}$ do not commute, the operators associated with their variances do not commute either, so that the expectation values of mixed products of them will generally be different from the products of their expectation values. Thus, we should throw away the $1/2$ in (4.114) and write

$$\sqrt{\langle(\Delta\hat{n})^2\rangle} = A. \tag{4.115}$$

Surprisingly enough, this is the exact result! Let us see why.

For a state where $\langle\hat{a}_{\|}\rangle = A$, we can always write

$$\hat{a}_{\|} = A + \Delta\hat{a}_{\|}, \tag{4.116}$$

where

$$\Delta\hat{a}_{\|} = \hat{a}_{\|} - \langle\hat{a}_{\|}\rangle. \tag{4.117}$$

Writing $\hat{a}_{\|}$ in terms of $\hat{a}_1(\delta)$ and $\hat{a}_2(\delta)$, we find that

$$\hat{a}_{\|} = A + \Delta\hat{a}_1(\delta) + i\,\Delta\hat{a}_2(\delta). \tag{4.118}$$

In this example, we can obtain the correct values for $\sqrt{\langle\Delta\hat{a}_1(\delta)^2\rangle}$ and $\sqrt{\langle\Delta\hat{a}_2(\delta)^2\rangle}$ from the projections of the noise ellipse parallel and perpendicular to the phasor, respectively, because $\langle \Delta \hat{a}_1 \, \Delta \hat{a}_2 + \Delta \hat{a}_2 \, \Delta \hat{a}_1 \rangle$ vanishes. This, however, does not explain why we can obtain an exact result for $\langle (\Delta \hat{n})^2 \rangle$ by discarding terms involving products of fluctuations. To understand this point, let us look into the quantum term corresponding to the $1/2$ in the last line of (4.114).

Using (4.116), we write the number operator as

$$\hat{n} = A^2 + 2A \, \Delta \hat{a}_1(\delta) + \Delta \hat{a}_{\parallel}^{\dagger} \, \Delta \hat{a}_{\parallel}. \tag{4.119}$$

The last two terms in (4.119) correspond to the operator $\Delta \hat{n}$ describing the fluctuations in the photon number. The reader can easily verify that this operator can also be written as

$$\begin{aligned} \Delta \hat{n} = {} & 2A \, \Delta \hat{a}_1(\delta) \\ & + [\Delta \hat{a}_1(\delta) - i \, \Delta \hat{a}_2(\delta)] \, [\Delta \hat{a}_1(\delta) + i \, \Delta \hat{a}_2(\delta)]. \end{aligned} \tag{4.120}$$

The term $[\Delta \hat{a}_1(\delta) - i \, \Delta \hat{a}_2(\delta)] \, [\Delta \hat{a}_1(\delta) + i \, \Delta \hat{a}_2(\delta)]$ corresponds to the product of the two noise ellipse projections, $(1/2 - i/2)(1/2 + i/2)$, that gave rise to the $1/2$ in the last line of (4.114). Caves's diagrams cannot reproduce any quantum correlations between these two noise operators. However, if the average amplitude A is large enough, this term will be negligible in comparison with the first term on the right-hand side of (4.120). Then the calculation in terms of Caves's diagrams will be a good approximation to the exact result.

The coherent-state example we have considered is the best of both worlds, because for a coherent state the last term in (4.120) does not contribute to the final result. The variance of $\hat{n}$ is given by

$$\begin{aligned} \langle (\Delta \hat{n})^2 \rangle = {} & 4A^2 \langle \Delta \hat{a}_1(\delta)^2 \rangle + \langle \Delta \hat{a}_{\parallel}^{\dagger} \, \Delta \hat{a}_{\parallel} \, \Delta \hat{a}_{\parallel}^{\dagger} \, \Delta \hat{a}_{\parallel} \\ & + 2A \{ \Delta \hat{a}_1(\delta) \, \Delta \hat{a}_{\parallel}^{\dagger} \, \Delta \hat{a}_{\parallel} + \Delta \hat{a}_{\parallel}^{\dagger} \, \Delta \hat{a}_{\parallel} \, \Delta \hat{a}_1(\delta) \} \rangle, \end{aligned} \tag{4.121}$$

and for a coherent state the last term on the right-hand side of (4.121) vanishes, yielding

$$\langle (\Delta \hat{n})^2 \rangle = 4A^2 \langle \Delta \hat{a}_1(\delta)^2 \rangle, \tag{4.122}$$

which coincides with (4.115).

The variance $\langle \Delta \hat{a}_1(\delta)^2 \rangle$ is related to amplitude fluctuations, whereas $\langle \Delta \hat{a}_2(\delta)^2 \rangle$ is related to phase fluctuations. Equation (4.122) shows that for a coherent state, photon number fluctuations are connected only to amplitude fluctuations. So the Caves's diagram technique has given us some insight into the nature of coherent states. In the next subsection it will help us to understand a method of detecting squeezed states.

4.2.4 Observing squeezed states: Homodyne detection

Homodyne detection was first developed for radio [460] but is now widely used in optics. In this section we follow closely the treatment of homodyne detection presented in [416].

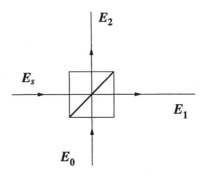

Figure 4.6 The four ports of a beam splitter.

Consider a beam splitter with two plane waves E_s and E_0 incident on each of its faces, respectively, as in Figure 4.6. If the complex amplitude reflection and transmission coefficients are r and t, respectively, we can write

$$E_1 = rE_0 + tE_s \tag{4.123}$$

and

$$E_2 = rE_s + tE_0. \tag{4.124}$$

Or in a more compact form,

$$\begin{bmatrix} E_1 \\ E_2 \end{bmatrix} = \begin{bmatrix} r & t \\ t & r \end{bmatrix} \begin{bmatrix} E_0 \\ E_s \end{bmatrix}. \tag{4.125}$$

For a lossless beam splitter, energy conservation implies that the 2×2 matrix in (4.125) is unitary:

$$|r|^2 + |t|^2 = 1 \tag{4.126}$$

and

$$t^*r + r^*t = 0. \tag{4.127}$$

From (4.127) it follows that $\arg(r) - \arg(t) = \pi/2$. For a 50:50 beam splitter, $|r| = |t| = 1/\sqrt{2}$. Choosing r as real, we find that $r = 1/\sqrt{2}$ and $t = i/\sqrt{2}$.

In the case of a quantized field, the plane wave complex amplitudes are annihilation operators. So the fields at the output ports of a 50:50 beam splitter can be described by the annihilation operators $\hat{d}$ and $\hat{c}$, given by

$$\begin{bmatrix} \hat{d} \\ \hat{c} \end{bmatrix} = \begin{bmatrix} 1/\sqrt{2} & i/\sqrt{2} \\ i/\sqrt{2} & 1/\sqrt{2} \end{bmatrix} \begin{bmatrix} \hat{a} \\ \hat{e} \end{bmatrix}. \tag{4.128}$$

But if we make $\hat{e} \rightarrow i\hat{e}$ and $\hat{c} \rightarrow i\hat{c}$, we obtain

$$\hat{d} = \frac{1}{\sqrt{2}}(\hat{a} - \hat{e}) \tag{4.129}$$

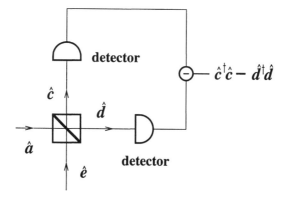

Figure 4.7 Diagram of a balanced homodyne detector. The local oscillator field (intense coherent state) is injected in $\hat{a}$. The field to be measured, with the same frequency of the local oscillator, is injected in $\hat{e}$. The outputs $\hat{c}$ and $\hat{d}$ are measured by two different photodetectors whose difference in photocounts is monitored.

and

$$\hat{c} = \frac{1}{\sqrt{2}}(\hat{a} + \hat{e}). \tag{4.130}$$

Suppose now that in the port corresponding to $\hat{e}$, we inject squeezed vacuum, and in that corresponding to $\hat{a}$, we inject a coherent state $|\alpha(t)\rangle$ with the same frequency but with a phase difference θ, so that $\alpha(t) = B \exp(i[\theta - \omega t])$. The balanced homodyne detection scheme consists of measuring the difference in the photon numbers at ports 1 and 2 (see Figure 4.7). Writing the operators $\hat{a}$ and $\hat{e}$ as before, that is,

$$\hat{a} = \alpha(t) + \Delta\hat{a} \tag{4.131}$$

and

$$\hat{e} = \Delta\hat{e}, \tag{4.132}$$

we find that the difference in photon number is given by

$$\hat{c}^{\dagger}\hat{c} - \hat{d}^{\dagger}\hat{d} = 2B\,\Delta\hat{e}_1(\theta - \omega t) + [\Delta\hat{e}^{\dagger}\,\Delta\hat{a} + \Delta\hat{a}^{\dagger}\,\Delta\hat{e}]. \tag{4.133}$$

As the coherent state $|\alpha(t)\rangle$ is very intense (a laser beam), we can neglect the second term square brackets on the right-hand side of (4.133). Now $\Delta\hat{e}_1(\theta - \omega t)$ is the variance of the quadrature $\hat{e}_1$ in a reference frame rotated by $\theta - \omega t$ in relation to the old reference frame. The rotation by $-\omega t$ allows it to follow the movement of $\hat{e}$ in phase space due to its free time evolution, so that $\Delta\hat{e}_1(\theta - \omega t)$ does not change in time. The dependence on θ allows us to choose the direction in which we want to measure the field fluctuations (i.e., it allows us to choose which quadrature we want to look at).

RECOMMENDED READING

- In Section 4.1 we saw how an open system becomes entangled to the immense number of degrees of freedom of the environment. This entanglement leads to a mixed reduced-density matrix for the open system. Controlled entanglement, rather than this ordinary entanglement with the inaccessible degrees of freedom of the environment, has many interesting applications, including teleportation, quantum cryptography, and quantum computing. For more on these applications that are now part of the new field of quantum information, see [71].

- The Glauber–Sudarshan quasiprobability distribution seen in Section 4.1 is not the only one used in quantum optics. Another example of quasiprobability distribution is the Wigner function [644], which is associated with symmetrical ordering rather than normal ordering. It introduces the idea of phase-space in quantum mechanics (for a phase space approach to quantum optics, see Schleich's book [537]). Some of the advantages of the Wigner function are that unlike P, it is not singular for coherent states and for physical quantum states, and that it makes important features such as quantum interference easy to identify visually.

- In the 1990s, the Wigner function of a light beam was determined experimentally for the first time using the technique of quantum tomography. For an excellent account of this as well as a list of some of the key references in that period, see [399]. Unfortunately, this technique does not apply to quantum fields in high-Q microwave cavities, as those fields cannot be taken out of the cavity and as there are no photon detectors for those frequencies. For this typical cavity QED scenario, a different technique was proposed in which atoms are used to probe the field: atomic homodyne detection [168, 170, 652].

- Outside the former Iron Curtain, interest in moving mirrors and time-dependent boundary conditions was sparked by the possibility of particle generation due to the expansion of the universe [57, 475]. The first known paper on the problem of the quantum radiation field in a cavity with moving walls was published by Gerald T. Moore in 1970 [457]. There are, however, older papers on the classical theory of time-dependent boundary conditions that were published in the former Soviet Union. For a comprehensive review of the moving mirrors problem (also known as the dynamic Casimir effect), see [161]. More recently, Law found an effective Hamiltonian that explicitly describes the physical processes that happen during a change of cavity size [394, 395]. Also of interest are a recent calculation of the decoherence produced by moving boundaries [141] and the proposal by Schwinger [546–550] to explain the phenomenon of sonoluminescence[16] in terms of the dynamic Casimir effect, but there are alternative theories, such as [292].

- For more on the use of squeezed states for optical communications, see [558, 657, 661, 662].

- Although they only became popular in the 1960s with the work of Roy Glauber, coherent states were actually introduced in 1926 by Schrödinger [540]. Squeezed states were first introduced by Kennard in 1927 [351], long before Horace Yuen's seminal paper on two-photon coherent states of the radiation field [660]. For a historical account of these discoveries, see [465].

Problems

4.1   Show that $\hat{a}$ cannot have any left eigenstate.
Hint: Find a recursive relation for the Fock state components of such a hypothetical eigenstate. Then show that the vacuum component must vanish so that all the other components will also vanish.

4.2   Derive the explicit expression (4.20) for the normalized coherent state $|n\rangle$ directly from (4.5).
Hint: Multiply (4.5) by a Fock state $\langle n|$ from the left and work out a recursive relation for $\langle n|\alpha\rangle$.

4.3   Show that $\hat{R}(\theta) = \exp\left(i\theta\hat{a}^\dagger\hat{a}\right)$.
Hint: Start with an infinitesimal rotation $\delta\theta$.

4.4   Show that if $\hat{U}^\dagger\hat{U} = \hat{1}$, then $\hat{U}^\dagger \exp\left(\hat{C}\right)\hat{U} = \exp\left(\hat{U}^\dagger\hat{C}\hat{U}\right)$.
Hint: Use a Taylor series expansion of $\exp\left(\hat{C}\right)$.

4.5   Draw Caves's diagrams for the fields in each of the ports of the beam splitter used in homodyne detection. Use these diagrams to write down an expression for the difference in the fluctuations of the average photon number of c and d.

---

[16]Sonoluminescence is the emission of intense bursts of light by ultrasonically driven gas bubbles in liquids, when they collapse suddenly. The cause of this phenomenon is still an open question. For a brief history of sonoluminescence, see [22].

5

Let matter be!

Matter has always been and will always be one of the main objects of physics.
—Wolfgang Pauli [479]

When there are source charges, the quantization of the radiation field is a bit more subtle than what we saw in Chapter 2. For the free field, the classical dynamics is described completely by the sourceless Maxwell equations in the vacuum

$$\nabla \cdot \mathbf{E} = 0, \tag{5.1}$$

$$\nabla \cdot \mathbf{B} = 0, \tag{5.2}$$

$$\nabla \wedge \mathbf{E} = -\frac{1}{c}\frac{\partial}{\partial t}\mathbf{B} \tag{5.3}$$

$$\nabla \wedge \mathbf{B} = \frac{1}{c}\frac{\partial}{\partial t}\mathbf{E}. \tag{5.4}$$

Then both $\mathbf{B}$ and $\mathbf{E}$ are transverse. We have seen in Chapter 2 that in reciprocal space (mode space) we can easily identify a pair of canonically conjugate variables[1] and construct a Hamiltonian with them.

The situation becomes more complicated as soon as charges are introduced. To start with, as the fields are described by Maxwell equations in the vacuum with sources

$$\nabla \cdot \mathbf{E} = 4\pi\rho, \tag{5.5}$$

---

[1]They are proportional to the components of the fields in reciprocal space.

145

$$\nabla \cdot \mathbf{B} = 0, \tag{5.6}$$

$$\nabla \wedge \mathbf{E} = -\frac{1}{c}\frac{\partial}{\partial t}\mathbf{B}, \tag{5.7}$$

$$\nabla \wedge \mathbf{B} = \frac{4\pi}{c}\mathbf{J} + \frac{1}{c}\frac{\partial}{\partial t}\mathbf{E}, \tag{5.8}$$

the electric field is no longer transverse. But even more important, however, the force on a charge $\bar{e}_\alpha$ due to its interaction with the electromagnetic field,

$$\mathbf{F} = \bar{e}_\alpha \mathbf{E}(\mathbf{r}_\alpha) + \bar{e}_\alpha \frac{\dot{\mathbf{r}}_\alpha}{c} \wedge \mathbf{B}(\mathbf{r}_\alpha), \tag{5.9}$$

depends not only on the charge's position $\mathbf{r}_\alpha$, but also on its velocity $\dot{\mathbf{r}}_\alpha$. This means that this force cannot be obtained from a position-dependent potential as is ordinarily done in a Lagrangian or Hamiltonian description. Fortunately, as we will see in Section 5.2, it is still possible to define a potential dependent on both position and velocity that allows us to use the Lagrangian or the Hamiltonian formalism with the Lorentz force (5.9).

This is not possible for every velocity-dependent force. Viscous friction for example does not allow a Lagrangian or a Hamiltonian description [240, 386]. The price that we have to pay, however, is that we can no longer deal with the fields directly, but instead, are forced to use electromagnetic potentials in a Lagrangian or Hamiltonian description with charges. An important consequence of this is that the vector potential acquires a real physical significance in quantum mechanics which it did not have in classical electrodynamics [9].

It is possible, though, to introduce an approximate coupling between the radiation field and matter without having to use electromagnetic potentials, avoiding all the problems mentioned above. We can do this when the atoms or molecules can be, except for their internal dynamics, approximated by electric dipoles interacting with the radiation field. In most of quantum optics and cavity QED, this approximation holds extremely well and is often used. In the next section we discuss this approach. Then in Section 5.2, we present the complete quantization procedure (i.e., containing all multipolar interactions) starting from a fully classical theory for both field and matter. In Section 5.3 we show that the general interaction term derived in Section 5.2 can also be interpreted as a consequence of the principle of gauge invariance.

5.1 A SINGLE POINT DIPOLE

Consider a single point dipole sitting at the origin. This is described classically by (5.5)–(5.8), where, as there are no free charges, the source charge density and current are given by

$$\rho = -\nabla \cdot \mathbf{P}, \tag{5.10}$$

$$\mathbf{J} = \frac{\partial}{\partial t}\mathbf{P}, \tag{5.11}$$

with the polarization due to the single dipole being

$$\mathbf{P}(\mathbf{r}) = \mathbf{p}\delta(\mathbf{r}). \tag{5.12}$$

In Chapter 4 when we considered the free radiation field, both $\mathbf{E}$ and $\mathbf{B}$ were transverse. Here we have to be more careful because although $\mathbf{B}$ is still transverse, (5.5) shows that $\mathbf{E}$ is no longer transverse. The longitudinal part of $\mathbf{E}$ decouples from $\mathbf{B}$ in the equations of motion and is, in fact, a function of $\mathbf{P}$ only:[2]

$$\mathbf{E}_{||}(\mathbf{r}) = -\boldsymbol{\nabla} \left[\boldsymbol{\nabla} \cdot \left(\frac{\mathbf{P}}{r} \right) \right]. \tag{5.13}$$

So $\mathbf{E}_{||}$ does not not have any independent degrees of freedom and should not be quantized. The displacement vector $\mathbf{D} \equiv \mathbf{E} + 4\pi\mathbf{P}$, however, is transverse. Thus, in the present case, $\mathbf{D}$ and $\mathbf{B}$, rather than $\mathbf{E}$ and $\mathbf{B}$, are the right pair of field vectors that contain only the truly independent degrees of freedom of the field, and they are what we should use in the canonical quantization scheme.

Rewriting Maxwell equations in terms of $\mathbf{D}$ and $\mathbf{B}$, we find that

$$\boldsymbol{\nabla} \cdot \mathbf{D} = 0, \tag{5.14}$$

$$\boldsymbol{\nabla} \cdot \mathbf{B} = 0, \tag{5.15}$$

$$\boldsymbol{\nabla} \wedge \mathbf{D} = -\frac{1}{c}\frac{\partial}{\partial t}\mathbf{B} - 4\pi(\mathbf{p} \wedge \boldsymbol{\nabla})\delta(\mathbf{r}), \tag{5.16}$$

$$\boldsymbol{\nabla} \wedge \mathbf{B} = \frac{1}{c}\frac{\partial}{\partial t}\mathbf{D}. \tag{5.17}$$

---

[2]To obtain (5.13) from (5.5), just solve the reciprocal space version of (5.5), which yields $\mathcal{E}_{||}(\mathbf{k}) = -4\pi(\mathbf{k} \cdot \mathbf{p})/k$ and invert:

$$\mathbf{E}_{||}(\mathbf{r}) = -\frac{4\pi}{(2\pi)^3} \int d^3k \frac{(\mathbf{k} \cdot \mathbf{p})\mathbf{k}}{k^2} e^{i\mathbf{k}\cdot\mathbf{r}}.$$

Using the identity $(\mathbf{k} \cdot \mathbf{p})\mathbf{k} = \mathbf{k} \wedge (\mathbf{k} \wedge \mathbf{p}) + k^2\mathbf{p}$, we find that

$$\mathbf{E}_{||}(\mathbf{r}) = -4\pi\mathbf{P}(\mathbf{r}) - \frac{1}{2\pi^2} \int d^3k \frac{\mathbf{k} \wedge (\mathbf{k} \wedge \mathbf{p})}{k^2} e^{i\mathbf{k}\cdot\mathbf{r}}$$

$$= -4\pi\mathbf{P}(\mathbf{r}) - \frac{1}{2\pi^2} \boldsymbol{\nabla} \wedge \left\{ \boldsymbol{\nabla} \wedge \mathbf{p} \int d^3k \frac{1}{k^2} e^{i\mathbf{k}\cdot\mathbf{r}} \right\}.$$

Now

$$\int d^3k \frac{e^{i\mathbf{k}\cdot\mathbf{r}}}{k^2} = \frac{2\pi}{ir} \int_{-\infty}^{\infty} d\eta \frac{e^{i\eta}}{\eta} = \frac{2\pi^2}{r}.$$

Then

$$\mathbf{E}_{||}(\mathbf{r}) = -4\pi\mathbf{P}(\mathbf{r}) - \boldsymbol{\nabla} \wedge \left\{ \boldsymbol{\nabla} \wedge \left(\frac{\mathbf{P}}{r} \right) \right\}.$$

Finally, using the identity

$$\boldsymbol{\nabla} \wedge \left\{ \boldsymbol{\nabla} \wedge \left(\frac{\mathbf{P}}{r} \right) \right\} = \boldsymbol{\nabla} \left[\boldsymbol{\nabla} \cdot \left(\frac{\mathbf{P}}{r} \right) \right] - \boldsymbol{\nabla}^2 \left(\frac{1}{r} \right) \mathbf{p},$$

we obtain (5.13).

These equations are very similar to the free field equations we saw in Chapter 2 except for the source term on the left-hand side of (5.16). This term describes the interaction between the point dipole and the radiation field.

As both $\mathbf{D}$ and $\mathbf{B}$ are transverse, we can describe their reciprocal space counterparts using a pair of unit transverse vectors $\epsilon_j(\mathbf{k})$, with $j = 1, 2$, which are orthogonal to each other and to k:

$$\mathbf{D}(\mathbf{r}) = \int d^3k \sum_j \epsilon_j(\mathbf{k}) \mathcal{D}_j(\mathbf{k}) \, e^{i\mathbf{k}\cdot\mathbf{r}}, \tag{5.18}$$

$$\mathbf{B}(\mathbf{r}) = \int d^3k \sum_j (-i)\frac{\mathbf{k}}{k} \wedge \epsilon_j(\mathbf{k}) \mathcal{B}_j(\mathbf{k}) e^{i\mathbf{k}\cdot\mathbf{r}}. \tag{5.19}$$

The number of degrees of freedom is doubled when we go over to reciprocal space, because $\mathcal{D}_j$ and $\mathcal{B}_j$ are complex rather than real. These extra degrees of freedom cannot be independent. In fact, as $\mathbf{D}$ and $\mathbf{B}$ are real, if we take the complex conjugate of (5.18) and (5.19), make the change of variable $\mathbf{k} \to -\mathbf{k}$ in the integrals, and define $\epsilon_j(-\mathbf{k})$ to be equal to $\epsilon_j(\mathbf{k})$, we find that

$$\mathcal{D}_j(-\mathbf{k}) = \mathcal{D}_j^*(\mathbf{k}) \quad \text{and} \quad \mathcal{B}_j(-\mathbf{k}) = \mathcal{B}_j^*(\mathbf{k}). \tag{5.20}$$

To deal only with truly independent degrees of freedom, we consider only $\mathbf{k}$ in the half space defined by, say, positive k_z.

Now using

$$\delta(\mathbf{r}) = \frac{1}{(2\pi)^3} \int d^3k \, e^{i\mathbf{k}\cdot\mathbf{r}}, \tag{5.21}$$

we can rewrite Maxwell equations (5.14)–(5.17) as

$$\dot{\mathcal{B}}_j(\mathbf{k}) = ck\mathcal{D}_j(\mathbf{k}) - \frac{1}{2\pi^2}ck\mathbf{p} \cdot \epsilon_j(\mathbf{k}), \tag{5.22}$$

$$\dot{\mathcal{D}}_j(\mathbf{k}) = -ck\mathcal{B}_j(\mathbf{k}). \tag{5.23}$$

There are only two equations now rather than four because we have already used the fact that both $\mathbf{D}$ and $\mathbf{B}$ are transverse and therefore (5.18) and (5.19) automatically satisfy the two vanishing divergency Maxwell equations (5.14) and (5.15).

Equations (5.22) and (5.23) look very much like the equations of motion for the canonical position and momentum of a harmonic oscillator with a driving term. Let us follow Lord Kelvin's and Maxwell's method of looking for analogies in nature [87] and develop this idea further.

The Hamiltonian of a one-dimensional harmonic oscillator of frequency ω and mass m under the effect of an external potential U is given by

$$H = \frac{p^2}{2m} + \frac{m\omega^2}{2}q^2 + U(q), \tag{5.24}$$

where q is its canonical position and p (not to be confused with the norm of the single dipole $\mathbf{p}$) is its canonical momentum.

The equations of motion are obtained from $\dot{p} = -\partial H/\partial q$ and $\dot{q} = \partial H/\partial p$, which yield

$$\dot{p} = -m\omega^2 q - \frac{d}{dq}U, \tag{5.25}$$

$$\dot{q} = \frac{p}{m}. \tag{5.26}$$

They are very similar to (5.22) and (5.23). There is one important difference, though. Unlike p and q in (5.25) and (5.26), $\mathcal{B}_j$ and $\mathcal{D}_j$ in (5.22) and (5.23) are complex. This is not a big problem, however, because as we have shown in Chapter 2, with a few minor adjustments we can use the Hamiltonian formalism with complex canonical coordinates. Let us discuss this in a bit more detail here than we did there.

Consider a complex version of the equations of motion (5.25) and (5.26):

$$\dot{\mathcal{Q}} = \frac{\mathcal{P}}{m}, \tag{5.27}$$

$$\dot{\mathcal{P}} = -m\omega^2\mathcal{Q} - \mathcal{F}, \tag{5.28}$$

where $\mathcal{F}$ is a driving force (independent of $\mathcal{Q}$, $\mathcal{P}$). These two complex equations actually contain four equations of motion for four real variables. Let us define these four real variables in the following way:

$$\mathcal{Q} \equiv \frac{q_r + iq_i}{\sqrt{2}}, \tag{5.29}$$

$$\mathcal{P} \equiv \frac{p_r + ip_i}{\sqrt{2}}, \tag{5.30}$$

so that

$$q_r = \frac{\mathcal{Q} + \mathcal{Q}^*}{\sqrt{2}}, \quad q_i = \frac{\mathcal{Q} - \mathcal{Q}^*}{i\sqrt{2}}, \quad p_r = \frac{\mathcal{P} + \mathcal{P}^*}{\sqrt{2}}, \quad \text{and} \quad p_i = \frac{\mathcal{P} - \mathcal{P}^*}{i\sqrt{2}}. \tag{5.31}$$

The reason for the $\sqrt{2}$ factor will become evident soon.

For each of the two pairs of real variables (q_r, p_r) and (q_i, p_i) in (5.31), we have a pair of driven harmonic oscillator equations

$$\dot{q}_r = \frac{p_r}{m}, \quad \dot{p}_r = -m\omega^2 q_r - F_r \tag{5.32}$$

and

$$\dot{q}_i = \frac{p_i}{m}, \quad \dot{p}_i = -m\omega^2 q_i - F_i, \tag{5.33}$$

where

$$\mathcal{F} \equiv \frac{F_r + iF_i}{\sqrt{2}}. \tag{5.34}$$

The equations of motion (5.32) and (5.33) can be obtained from a Hamiltonian such as (5.24). As they are all independent variables, the four equations of motion can be obtained from a Hamiltonian given by the sum of these two separate Hamiltonians; that is,

$$\mathcal{H} = \frac{p_r^2}{2m} + m\frac{\omega^2}{2}q_r^2 + q_r F_r + \frac{p_i^2}{2m} + m\frac{\omega^2}{2}q_i^2 + q_i F_i \qquad (5.35)$$

with the equations of motion given by

$$\dot{q}_r = \frac{\partial \mathcal{H}}{\partial p_r}, \qquad \dot{p}_r = -\frac{\partial \mathcal{H}}{\partial q_r}, \qquad (5.36)$$

and

$$\dot{q}_i = \frac{\partial \mathcal{H}}{\partial p_i}, \qquad \dot{p}_i = -\frac{\partial \mathcal{H}}{\partial q_i}. \qquad (5.37)$$

In terms of the complex variables $\mathcal{Q}$ and $\mathcal{P}$, (5.35) is given simply by

$$\mathcal{H} = \frac{\mathcal{P}^*\mathcal{P}}{2m} + m\frac{\omega^2}{2}\mathcal{Q}^*\mathcal{Q} + \mathcal{Q}^*\mathcal{F} + \text{c.c.}, \qquad (5.38)$$

where c.c. stands for *complex conjugate,* showing that $\mathcal{H}$ is real as a Hamiltonian should be.

Now, from (5.31), we find that

$$\frac{\partial}{\partial p_r} = \frac{1}{\sqrt{2}}\frac{\partial}{\partial \mathcal{P}} + \frac{1}{\sqrt{2}}\frac{\partial}{\partial \mathcal{P}^*}, \qquad \frac{\partial}{\partial p_i} = \frac{i}{\sqrt{2}}\frac{\partial}{\partial \mathcal{P}} - \frac{i}{\sqrt{2}}\frac{\partial}{\partial \mathcal{P}^*} \qquad (5.39)$$

and

$$\frac{\partial}{\partial q_r} = \frac{1}{\sqrt{2}}\frac{\partial}{\partial \mathcal{Q}} + \frac{1}{\sqrt{2}}\frac{\partial}{\partial \mathcal{Q}^*}, \qquad \frac{\partial}{\partial q_i} = \frac{i}{\sqrt{2}}\frac{\partial}{\partial \mathcal{Q}} - \frac{i}{\sqrt{2}}\frac{\partial}{\partial \mathcal{Q}^*}. \qquad (5.40)$$

Substituting (5.39) and (5.40) in (5.36) and (5.37), we obtain

$$\frac{\dot{q}_r}{\sqrt{2}} = \frac{1}{2}\frac{\partial \mathcal{H}}{\partial \mathcal{P}} + \frac{1}{2}\frac{\partial \mathcal{H}}{\partial \mathcal{P}^*}, \qquad \frac{\dot{p}_r}{\sqrt{2}} = -\frac{1}{2}\frac{\partial \mathcal{H}}{\partial \mathcal{Q}} - \frac{1}{2}\frac{\partial \mathcal{H}}{\partial \mathcal{Q}^*} \qquad (5.41)$$

and

$$i\frac{\dot{q}_i}{\sqrt{2}} = -\frac{1}{2}\frac{\partial \mathcal{H}}{\partial \mathcal{P}} + \frac{1}{2}\frac{\partial \mathcal{H}}{\partial \mathcal{P}^*}, \qquad i\frac{\dot{p}_i}{\sqrt{2}} = \frac{1}{2}\frac{\partial \mathcal{H}}{\partial \mathcal{Q}} - \frac{1}{2}\frac{\partial \mathcal{H}}{\partial \mathcal{Q}^*}. \qquad (5.42)$$

Differentiating (5.29) and (5.30) with respect to time and substituting in (5.41) and (5.42), we find the complex version of Hamilton equations of motion:

$$\dot{\mathcal{Q}} = \frac{\partial \mathcal{H}}{\partial \mathcal{P}^*}, \qquad \dot{\mathcal{P}} = -\frac{\partial \mathcal{H}}{\partial \mathcal{Q}^*}. \qquad (5.43)$$

From the Poisson brackets of the real canonical variables, we can work out those of the complex ones:

$$[\mathcal{Q}, \mathcal{P}^*]_{PB} = \frac{1}{2}[q_r + iq_i, p_r + ip_i]_{PB}$$

$$= \frac{1}{2}\left(\underbrace{[q_r, p_r]_{PB}}_{1} - i\underbrace{[q_r, p_i]_{PB}}_{0} + i\underbrace{[q_i, p_r]_{PB}}_{0} + \underbrace{[q_i, p_i]_{PB}}_{1}\right)$$

$$= 1. \qquad (5.44)$$

Analogously,

$$[\mathcal{Q}, \mathcal{P}]_{PB} = 0, \qquad [\mathcal{Q}, \mathcal{Q}^*]_{PB} = 0, \qquad [\mathcal{P}, \mathcal{P}^*]_{PB} = 0. \tag{5.45}$$

Now comparing (5.27) and (5.28) with (5.22) and (5.23), we find that we can make the following: correspondences

$$\omega = ck, \tag{5.46}$$

$$m = 1, \tag{5.47}$$

$$\mathcal{Q}_j(\mathbf{k}) = -\frac{\mathcal{N}}{ck}\mathcal{D}_j, \tag{5.48}$$

$$\mathcal{P}_j(\mathbf{k}) = \mathcal{N}\,\mathcal{B}_j, \tag{5.49}$$

$$\mathcal{F}_j(\mathbf{k}) = \frac{\mathcal{N}}{2\pi^2}ck\mathbf{p}\cdot\boldsymbol{\epsilon}_j(\mathbf{k}), \tag{5.50}$$

where $\mathcal{N}$ is an arbitrary constant, and

$$[\mathcal{Q}_j(\mathbf{k}), \mathcal{P}_{j'}^*(\mathbf{k}')] = \delta_{jj'}\delta(\mathbf{k} - \mathbf{k}'), \tag{5.51}$$

as these are continuous rather than discrete variables.

So this interesting analogy tells us that each pair of field variables $\mathcal{D}_j(\mathbf{k})$ and $\mathcal{B}_j(\mathbf{k})$ corresponds to a Hamiltonian of the sort (5.38) by way of (5.46)–(5.50). Now, provided that we restrict $\mathbf{k}$ to a half space, pairs corresponding to different values of j and $\mathbf{k}$ are independent. Thus, the Hamiltonian for our single electric dipole interacting with the radiation field is just the sum (integral) of each of these individual Hamiltonians for each pair $\mathcal{D}_j(\mathbf{k})$ and $\mathcal{B}_j(\mathbf{k})$:

$$H = \frac{\mathcal{N}^2}{2}\int' d^3k \sum_j \left\{ \mathcal{B}_j^2(\mathbf{k}) + \mathcal{D}_j^2(\mathbf{k}) - \frac{1}{\pi^2}\mathbf{p}\cdot\boldsymbol{\epsilon}_j(\mathbf{k})\,\mathcal{D}_j(\mathbf{k}) + \text{c.c.}\right\} + H_{\text{atm}},$$

$$\tag{5.52}$$

where we have added the Hamiltonian H_{atm} describing the free evolution of the atom (dipole) and the prime on the integral sign is to remind us that this integral is over half of the space rather than over the entire space, as usual. Choosing the arbitrary constant so that $\mathcal{N}^2 = 2\pi^2$, we make the Hamiltonian coincide with the total energy; that is, the Hamiltonian in real space is given by

$$H = \frac{1}{8\pi}\int d^3r\,(\mathbf{B}^2 + \mathbf{D}^2) - \int d^3r\,\mathbf{P}\cdot\mathbf{D} + H_{\text{atm}}. \tag{5.53}$$

Up to now everything is classical. We have shown that $-\pi\sqrt{2}\mathcal{D}_j/ck$ and $\pi\sqrt{2}\mathcal{B}_j$ are canonical variables and we have found their Hamiltonian. Then to quantize the radiation field, we can just apply the ordinary canonical quantization procedure as in Chapter 4. We take the commutator of the quantized variables as their classical Poisson bracket times $i\hbar$. Then, defining

$$\hat{a}_j(\mathbf{k}) \equiv \frac{1}{\sqrt{2\hbar ck}}\left\{\mathcal{P}_j(\mathbf{k}) - ick\mathcal{Q}_j(\mathbf{k})\right\}, \tag{5.54}$$

so that

$$[\hat{a}_j(\mathbf{k}), \hat{a}_{j'}^\dagger(\mathbf{k}')] = \delta_{jj'}\delta(\mathbf{k} - \mathbf{k}') \tag{5.55}$$

and

$$\frac{\mathcal{N}^2}{2} \int d^3k \; ck \sum_j \left\{ \boldsymbol{\mathcal{B}}_j^\dagger(\mathbf{k}) \cdot \boldsymbol{\mathcal{B}}_j(\mathbf{k}) + \boldsymbol{\mathcal{D}}_j^\dagger(\mathbf{k}) \cdot \boldsymbol{\mathcal{D}}_j(\mathbf{k}) + \text{H.c.} \right\}$$

$$= \frac{\hbar}{2} \int d^3k \; ck \sum_j \left\{ \hat{a}_j^\dagger(\mathbf{k})\hat{a}_j(\mathbf{k}) + \hat{a}_j(\mathbf{k})\hat{a}_j^\dagger(\mathbf{k}) \right\}, \tag{5.56}$$

as for a continuum of harmonic oscillators, we find that

$$\mathbf{D}(\mathbf{r}) = -i \int d^3k \; \sqrt{\frac{ck\hbar}{4\pi^2}} \sum_j \boldsymbol{\epsilon}_j(\mathbf{k}) \left\{ \hat{a}_j(\mathbf{k}) - \hat{a}_j^\dagger(-\mathbf{k}) \right\} e^{i\mathbf{k}\cdot\mathbf{r}} \tag{5.57}$$

and

$$\mathbf{B}(\mathbf{r}) = -i \int d^3k \; \sqrt{\frac{ck\hbar}{4\pi^2}} \sum_j \frac{\mathbf{k}}{k} \wedge \boldsymbol{\epsilon}_j(\mathbf{k}) \left\{ \hat{a}_j(\mathbf{k}) + \hat{a}_j^\dagger(-\mathbf{k}) \right\} e^{i\mathbf{k}\cdot\mathbf{r}}. \tag{5.58}$$

We have quantized the electromagnetic field, but what about matter? Here matter is represented by a single atom that is approximated by a point electric dipole $\hat{\mathbf{p}}$. How can we quantize this dipole? What is its free evolution[3] Hamiltonian $\hat{H}_{\text{atm}}$? To address these questions, we must first understand how the dipole approximation arises. Suppose for simplicity that the atom is hydrogen.[4] Assume that it is at rest. As the nucleus is much heavier than the electron, we can think of the nucleus as fixed at the origin. The electron, on the other hand, will move about and after quantization its position will be described by the position operator $\hat{\mathbf{r}}$. Then the charge density associated with this atom is given by

$$\rho_{\text{atm}}(\mathbf{r}) = \bar{e}\,\delta(\mathbf{r}) - \bar{e}\,\delta(\mathbf{r} - \hat{\mathbf{r}}), \tag{5.59}$$

where $-\bar{e}$ is the charge of the electron.

Now if $r = |\mathbf{r}|$ is much larger[5] than the atomic size, we can approximate $\delta(\mathbf{r} - \hat{\mathbf{r}})$ by

$$\delta(\mathbf{r} - \hat{\mathbf{r}}) = \int_{-\infty}^{\infty} \frac{d^3k}{(2\pi)^3} \; e^{i r \mathbf{k}\cdot([\mathbf{r}/r] - [\hat{\mathbf{r}}/r])}$$

---

[3]That is, uncoupled to the field.

[4]This is not a very restrictive assumption. Rydberg atoms used in many cavity QED experiments can be very well approximated by a scaled model of hydrogen. In a Rydberg atom one of the electrons is in a very high energy level. As the lower-energy electrons are much closer to the nucleus, they partially shield the nucleus charge so that the outer electron experiences an effective Coulomb potential similar to that in a hydrogen atom.

[5]This is the case when we look at the atom's charge distribution from a macroscopic distance, then we cannot see all the details of its charge distribution (i.e., the several multipole terms in its multipole expansion). But more important here, this is also the case when the wavelength λ of the electromagnetic radiation interacting with the atom is much larger than the size of the atom (e.g., the Bohr radius for a hydrogen atom). For we can easily see, then, using a Fourier argument, that only $r \approx \lambda$ will contribute significantly to the interaction Hamiltonian with the radiation field.

$$\approx \int_{-\infty}^{\infty} d^3k \, \frac{e^{i\mathbf{k}\cdot\mathbf{r}}}{(2\pi)^3} \{1 - i\mathbf{k}\cdot\hat{\mathbf{r}}\}$$
$$= \delta(\mathbf{r}) - \hat{\mathbf{r}}\cdot\boldsymbol{\nabla}\delta(\mathbf{r}). \tag{5.60}$$

Then

$$\rho_{\text{atm}}(\mathbf{r}) = \bar{e}\hat{\mathbf{r}}\cdot\boldsymbol{\nabla}\delta(\mathbf{r})$$
$$= \boldsymbol{\nabla}\cdot\{\bar{e}\hat{\mathbf{r}}\delta(\mathbf{r})\}. \tag{5.61}$$

So the atom produces the polarization $\mathbf{P}(\mathbf{r}) = \bar{e}\hat{\mathbf{r}}\delta(\mathbf{r})$, which means that the quantized point dipole we wanted to determine is given by

$$\hat{\mathbf{p}} = \bar{e}\hat{\mathbf{r}}. \tag{5.62}$$

A great simplification is possible when the conditions are such that only a pair of atomic levels is significantly involved in the interaction with the radiation field. This is the *two-level atom approximation,* and it is a very good approximation in many cavity QED experiments involving Rydberg atoms. Let us call $|\uparrow\rangle$ the upper atomic state and $|\downarrow\rangle$ the lower atomic state. The essence of the two-level atom approximation is that these two states span the whole of the relevant part of atomic space; that is, for all practical purposes, we can take

$$\hat{1} = |\downarrow\rangle\langle\downarrow| + |\uparrow\rangle\langle\uparrow|, \tag{5.63}$$

where $\hat{1}$ is the atomic unity operator. Then inserting $\hat{1}$ on both sides of (5.62), we find that

$$\hat{\mathbf{p}} = \hat{1}\hat{\mathbf{p}}\hat{1}$$
$$= (|\downarrow\rangle\langle\downarrow| + |\uparrow\rangle\langle\uparrow|)\bar{e}\hat{\mathbf{r}}(|\downarrow\rangle\langle\downarrow| + |\uparrow\rangle\langle\uparrow|)$$
$$= |\uparrow\rangle\langle\downarrow|\chi + |\downarrow\rangle\langle\uparrow|\chi^*, \tag{5.64}$$

where $\chi \equiv \langle\uparrow|\hat{\mathbf{r}}|\downarrow\rangle$ and we have assumed that $\langle\downarrow|\hat{\mathbf{r}}|\downarrow\rangle = \langle\uparrow|\hat{\mathbf{r}}|\uparrow\rangle = 0$ (which is often the case for symmetry reasons).

The atomic Hamiltonian also simplifies in this case. Setting the zero of energy halfway between the two atomic levels for convenience, the atomic Hamiltonian becomes

$$\hat{H}_{\text{atm}} = \hbar\frac{\omega}{2}(|\uparrow\rangle\langle\uparrow| - |\downarrow\rangle\langle\downarrow|), \tag{5.65}$$

where ω is the frequency of the atomic transition between $|\uparrow\rangle$ and $|\downarrow\rangle$.

When adopting the two-level atom approximation, one often represents the atomic operators by Pauli matrices. These matrices are defined by

$$\hat{\sigma} \equiv |\downarrow\rangle\langle\uparrow|, \tag{5.66}$$
$$\hat{\sigma}_x \equiv \hat{\sigma} + \hat{\sigma}^\dagger, \tag{5.67}$$
$$\hat{\sigma}_y \equiv i\hat{\sigma} - i\hat{\sigma}^\dagger, \tag{5.68}$$
$$\hat{\sigma}_z \equiv \hat{\sigma}^\dagger\hat{\sigma} - \hat{\sigma}\hat{\sigma}^\dagger. \tag{5.69}$$

Pauli matrices are very convenient to use in calculations because they have commutators that resemble vector multiplication of Cartesian vectors:

$$[\hat{\sigma}_i, \hat{\sigma}_j] = i2 \sum_k \varepsilon_{ijk} \hat{\sigma}_k. \qquad (5.70)$$

5.2   AN ARBITRARY CHARGE DISTRIBUTION[6]

Consider a set of charged particles q_α, of mass m_α, located at $\mathbf{r}_\alpha$, interacting with the electromagnetic field in free space. We will assume that we can neglect relativistic effects in the motion of these charges. Therefore, this motion will be described by Newton's law,

$$m_\alpha \frac{d^2 \mathbf{r}_\alpha}{dt^2} = \mathbf{F}_\alpha, \qquad (5.71)$$

where $\mathbf{F}_\alpha$ is the Lorentz force, given by

$$\mathbf{F}_\alpha = q_\alpha \left\{ \mathbf{E}\left(\mathbf{r}_\alpha, t\right) + \frac{\dot{\mathbf{r}}_\alpha}{c} \wedge \mathbf{B}\left(\mathbf{r}_\alpha, t\right) \right\}. \qquad (5.72)$$

The Lorentz force accounts for the influence of the field upon the charges, but the charges also affect the field. They are themselves sources of field. The field dynamics is described by Maxwell equations (5.5)–(5.8), where

$$\rho\left(\mathbf{r}, t\right) = \sum_\alpha q_\alpha \delta\left(\mathbf{r} - \mathbf{r}_\alpha(t)\right) \qquad (5.73)$$

is the density of charge associated with the point charges and

$$\mathbf{J}\left(\mathbf{r}, t\right) = \sum_\alpha \dot{\mathbf{r}}_\alpha(t) q_\alpha \delta\left(\mathbf{r} - \mathbf{r}_\alpha(t)\right) \qquad (5.74)$$

is the corresponding current density.

Since it is the independent degrees of freedom of the field that have to be quantized, we first have to identify these degrees of freedom. As we saw in Chapter 2, this is greatly simplified if we work with Maxwell equations in reciprocal instead of in real space [108].

Let $\mathcal{F}(\mathbf{k}, t)$ be the reciprocal space counterpart of a vector field $\mathbf{F}(\mathbf{r}, t)$ given by

$$\mathcal{F}(\mathbf{k}, t) = \frac{1}{(2\pi)^{3/2}} \int d^3 r \, e^{-i\mathbf{k}\cdot\mathbf{r}} \mathbf{F}(\mathbf{r}, t). \qquad (5.75)$$

As we are dealing with real fields, $\mathbf{F}(\mathbf{r}, t)$, $\mathcal{F}(\mathbf{k}, t)$ is not completely independent. Its values for negative $\mathbf{k}$ are determined by those for positive $\mathbf{k}$, as in Section 5.1, through the relation

$$\mathcal{F}(-\mathbf{k}, t) = \mathcal{F}^*(\mathbf{k}, t). \qquad (5.76)$$

---

[6]This section is based on [165] and follows closely the approach of [108].

So to avoid introducing redundant degrees of freedom, we will always work in half reciprocal space. Then the vector field $\mathbf{F}(\mathbf{r}, t)$ is given in terms of $\mathcal{F}(\mathbf{k}, t)$ by

$$\mathbf{F}(\mathbf{r}, t) = \frac{1}{(2\pi)^{3/2}} \int' d^3k \, e^{i\mathbf{k}\cdot\mathbf{r}} \mathcal{F}(\mathbf{k}, t) + \text{c.c.} \tag{5.77}$$

Equation (5.6) only states that the magnetic field is transverse. The remaining Maxwell equations yield the following equations in reciprocal space:

$$\mathcal{E}_{\parallel}(\mathbf{k}, t) = -i4\pi \frac{\mathbf{k}}{k^2}\sigma(\mathbf{k}, t), \tag{5.78}$$

$$\dot{\mathcal{B}}(\mathbf{k}, t) = -ic\mathbf{k} \wedge \mathcal{E}_{\perp}(\mathbf{k}, t), \tag{5.79}$$

$$\dot{\mathcal{E}}_{\perp}(\mathbf{k}, t) = ic\mathbf{k} \wedge \mathcal{B}(\mathbf{k}, t) - 4\pi\mathcal{J}_{\perp}(\mathbf{k}, t), \tag{5.80}$$

where $\sigma(\mathbf{k}, t)$ is $\rho(\mathbf{k}, t)$ in reciprocal space, $\mathcal{E}_{\parallel}(\mathbf{k}, t)$ is the longitudinal component of the electric field in reciprocal space, and $\mathcal{E}_{\perp}(\mathbf{k}, t)$, $\mathcal{B}(\mathbf{k}, t)$, and $\mathcal{J}_{\perp}(\mathbf{k}, t)$ are the transverse components of $\mathbf{E}(\mathbf{r}, t)$, $\mathbf{B}(\mathbf{r}, t)$, and $\mathbf{J}(\mathbf{r}, t)$ in reciprocal space. The longitudinal component of the current density satisfies the equation of conservation of charge:

$$\dot{\sigma}(\mathbf{k}, t) + ik\mathcal{J}_{\parallel}(\mathbf{k}, t) = 0. \tag{5.81}$$

We will adopt the canonical quantization procedure described in Chapter 2. To do so, we have to obtain the canonical variables associated with the Hamiltonian description of this system. These are variables such that when the Hamiltonian of the system is expressed in terms of them, Maxwell equations and the Newton–Lorentz equations for the charged particles are recovered from Hamilton equations of motion.

The Hamiltonian of the system is just the total energy, given by

$$H = \frac{1}{2} \sum_{\alpha} m_{\alpha} \dot{\mathbf{r}}_{\alpha}^2 + V_{\text{Coul}} + H_{\text{trans}}, \tag{5.82}$$

where H_{trans} is the energy stored in the transverse fields,

$$H_{\text{trans}} = \frac{1}{8\pi} \int d^3k \left[|\mathcal{E}_{\perp}(\mathbf{k}, t)|^2 + |\mathcal{B}(\mathbf{k}, t)|^2 \right], \tag{5.83}$$

and V_{Coul} is the Coulomb energy of the charged particles,

$$V_{\text{Coul}} = \sum_{\alpha} \frac{q_{\alpha}^2}{64\pi^4} \int \frac{d^3k}{k^2} + \frac{1}{2} \sum_{\alpha \neq \beta} \frac{q_{\alpha}q_{\beta}}{|\mathbf{r}_{\alpha} - \mathbf{r}_{\beta}|}. \tag{5.84}$$

The first term in (5.84) is the sum of the Coulomb self-energies of each particle α, and the second term is the Coulomb interaction between pairs of charges. The former diverges. This is in part due to the fact that it is inconsistent to try to account, in a nonrelativistic theory, for the interaction of the particles with the high-frequency modes of the field. An immediate solution is to introduce a cutoff k_{α} in all integrals over k in reciprocal space. For each particle α the cutoff must be on the order of

$m_\alpha c/\hbar$ so as to leave out interactions with modes whose energy is comparable to $m_\alpha c^2$. Such a cutoff gives a finite value for the Coulomb self-energy on the order of

$$\frac{1}{2} \sum_\alpha \frac{q_\alpha^2}{(2\pi)^3} k_\alpha. \tag{5.85}$$

A proper solution to the problem of the divergency of the self-energy requires a relativistic theory. When a relativistic theory of quantum electrodynamics was first developed, however, this divergency associated with the Coulomb self-energy and others associated with the infinite number of degrees of freedom of the field remained unsolved. These divergencies were dealt with only by renormalization theory [202]. We do not discuss this problem here and refer the reader to [203, 313], for example.

The Hamiltonian (5.82) is meaningless unless the associated canonical variables are defined. It turns out that the canonical variables do not involve the fields directly, but the electromagnetic potentials. This gives the potentials a physical significance in quantum mechanics that they do not possess in classical mechanics[7] [9, 318]. A complete discussion should provide a derivation of the Hamiltonian and Canonical variables from a Lagrangian. Such an approach is adopted by Cohen-Tannoudji et al. [108]. In this section I merely show that a given choice of canonical variables is possible.

The potentials $\mathbf{A}$ and U are defined such that

$$\mathbf{E}(\mathbf{r}, t) = -\nabla U(\mathbf{r}, t) - \frac{1}{c}\dot{\mathbf{A}}(\mathbf{r}, t), \tag{5.86}$$

$$\mathbf{B}(\mathbf{r}, t) = \nabla \wedge \mathbf{A}(\mathbf{r}, t). \tag{5.87}$$

In reciprocal space, this becomes

$$\mathcal{E}(\mathbf{k}, t) = -\frac{1}{c}\dot{\mathcal{A}}(\mathbf{k}, t) - i\mathbf{k}\mathcal{U}(\mathbf{k}, t), \tag{5.88}$$

$$\mathcal{B}(\mathbf{k}, t) = i\mathbf{k} \wedge \mathcal{A}(\mathbf{k}, t). \tag{5.89}$$

A description of the electromagnetic field in terms of potentials introduces extra degrees of freedom that cannot possibly be independent. This is reflected in the fact that the fields remain unchanged under the following gauge transformations of the potentials:

$$\mathbf{A}(\mathbf{r}, t) \rightarrow \mathbf{A}(\mathbf{r}, t) + \nabla F(\mathbf{r}, t), \tag{5.90}$$

$$U(\mathbf{r}, t) \rightarrow U(\mathbf{r}, t) - \frac{1}{c}\frac{\partial}{\partial t}F(\mathbf{r}, t), \tag{5.91}$$

where $F(\mathbf{r}, t)$ is an arbitrary function of $\mathbf{r}$ and t. Indeed, when a particular gauge is chosen, the redundant degrees of freedom can be eliminated using the constraint

---

[7]Here we are referring to the transverse component of the vector potential, which is gauge invariant. This component can give rise to an observable physical effect in a region where there are no fields (Bohm–Aharonov effect [9]).

relations introduced by the choice of gauge. Depending on the choice of gauge, however, the separation of the truly independent degrees of freedom of the fields from the rest might not be so straightforward as when we were dealing with the fields themselves. In nonrelativistic quantum electrodynamics, such a separation is possible by the adoption of the Coulomb gauge.

The Coulomb gauge is the gauge where the vector potential **A** is transverse:

$$\mathcal{A}(\mathbf{k}, t) = \mathcal{A}_{\perp}(\mathbf{k}, t). \tag{5.92}$$

The transverse vector potential, however, has only two independent components in reciprocal space. These are components along two orthogonal directions on the plane perpendicular to **k**. We will denote a given choice of two such directions by the normalized vectors $\epsilon_1(\mathbf{k})$ and $\epsilon_2(\mathbf{k})$. The components will be indicated by the subscripts ϵ_1 and ϵ_2.

In the Coulomb gauge, the transverse fields depend on the vector potential only:

$$\mathcal{E}_{\perp}(\mathbf{k}, t) = -\frac{1}{c} \dot{\mathcal{A}}(\mathbf{k}, t), \tag{5.93}$$

$$\mathcal{B}(\mathbf{k}, t) = i\mathbf{k} \wedge \mathcal{A}(\mathbf{k}, t), \tag{5.94}$$

and the longitudinal component of the electric field depends only on the scalar potential:

$$\mathcal{E}_{\|}(\mathbf{k}, t) = -ik\,\mathcal{U}(\mathbf{k}, t). \tag{5.95}$$

So the vector potential is rid of all redundant degrees of freedom, with the scalar potential being determined completely by the dynamic variables of the charged particles. This is the great advantage of the Coulomb gauge.

There are, however, some disadvantages. The Coulomb gauge is not so well suited to the relativistic theory because it is not manifestly covariant[8] [108, 314]. Moreover, the instantaneous Coulomb scalar potential raises a causality question [108, 314]. We do not address this problem here. I refer the reader to Brill and Goodman [75] or Cohen-Tannoudji et al. [108] for a detailed discussion.

Now we show that the canonical variables of our system are $\mathbf{r}_\alpha$, $\mathbf{p}_\alpha$, $\mathcal{A}(\mathbf{k}, t)$, and $\mathbf{\Pi}(\mathbf{k}, t)$, where $\mathbf{p}_\alpha$ and $\mathbf{\Pi}(\mathbf{k}, t)$, the conjugate momenta of $\mathbf{r}_\alpha$ and $\mathcal{A}(\mathbf{k}, t)$, respectively, are given by

$$\mathbf{p}_\alpha = m_\alpha \dot{\mathbf{r}}_\alpha + \frac{q_\alpha}{c} \mathbf{A}(\mathbf{r}_\alpha, t), \tag{5.96}$$

$$\mathbf{\Pi}(\mathbf{k}, t) = \frac{1}{c} \dot{\mathcal{A}}(\mathbf{k}, t). \tag{5.97}$$

The Hamiltonian (5.82), expressed in terms of these variables, becomes

$$H = \sum_\alpha \frac{1}{2m_\alpha} \left\{ \mathbf{p}_\alpha - \frac{q_\alpha}{c} \left[\int d^3k \, \mathcal{A}(\mathbf{k}, t) \frac{e^{i\mathbf{k}\cdot\mathbf{r}_\alpha}}{(2\pi)^{3/2}} + \text{c.c.} \right] \right\}^2 + V_{\text{Coul}}$$

$$+ \frac{1}{4\pi} \int d^3k \left\{ \mathbf{\Pi}^*(\mathbf{k}, t) \cdot \mathbf{\Pi}(\mathbf{k}, t) + c^2 k^2 \mathcal{A}^*(\mathbf{k}, t) \cdot \mathcal{A}(\mathbf{k}, t) \right\}, \tag{5.98}$$

---

[8]The Coulomb gauge refers to a particular Lorentz frame of reference.

where the first term on the right-hand side of (5.98) is the *minimal coupling* term describing the interaction with matter.

Hamilton equations of motion for the discrete variables $\{\mathbf{r}_\alpha, \mathbf{p}_\alpha\}$ are obtained from the Hamiltonian H by [108]

$$\dot{r}_{\alpha,j} = \frac{\partial H}{\partial p_{\alpha,j}}, \tag{5.99}$$

$$\dot{p}_{\alpha,j} = -\frac{\partial H}{\partial r_{\alpha,j}}, \tag{5.100}$$

where j stands for each of the three independent Cartesian components of these vectors.

Hamilton equations of motion for the continuous complex variables $\{\mathcal{A}(\mathbf{k}, t), \Pi(\mathbf{r}, t)\}$ are obtained from the Hamiltonian H by [108]

$$\dot{\mathcal{A}}_\epsilon^*(\mathbf{k}, t) = \frac{\partial H}{\partial \Pi_\epsilon(\mathbf{k}, t)}, \tag{5.101}$$

$$\dot{\Pi}_\epsilon^*(\mathbf{k}, t) = -\frac{\partial H}{\partial \mathcal{A}_\epsilon(\mathbf{k}, t)}, \tag{5.102}$$

where ϵ denotes each of the two independent components of the transverse dynamical variables, and the functional derivative of a functional ϕ of $u(\mathbf{k})$ is defined by [108]

$$\frac{\partial \phi}{\partial u(\mathbf{k})} = D(\mathbf{k}), \tag{5.103}$$

with $D(\mathbf{k})$ such that

$$\delta\phi = \phi(u + \delta u) - \phi(u)$$
$$= \int d^3k\, D(\mathbf{k})\delta u(\mathbf{k}) + O\left(\delta u\right)^2. \tag{5.104}$$

So to obtain the functional derivative in equation (5.102), we have to calculate the change in the Hamiltonian (5.98) when $\mathcal{A}_\epsilon$ varies by $\delta\mathcal{A}_\epsilon$

$$\delta H = \sum_\alpha \frac{1}{2m_\alpha} \left\{ \mathbf{p}_\alpha - \frac{q_\alpha}{c} \left[\int d^3k\, (\mathcal{A} + \epsilon\delta\mathcal{A}_\epsilon) \frac{e^{i\mathbf{k}\cdot\mathbf{r}_\alpha}}{(2\pi)^{3/2}} + \text{c.c.} \right] \right\}^2 + V_{\text{Coul}}$$

$$+ \frac{1}{4\pi} \int d^3k \left\{ \Pi^* \cdot \Pi + c^2k^2\mathcal{A}^* \cdot (\mathcal{A} + \epsilon\delta\mathcal{A}_\epsilon) \right\} - H$$

$$= -\sum_\alpha \frac{q_\alpha}{m_\alpha c} p_{\alpha,\epsilon} \int d^3k\, \delta\mathcal{A}_\epsilon \frac{e^{i\mathbf{k}\cdot\mathbf{r}_\alpha}}{(2\pi)^{3/2}}$$

$$+ \sum_\alpha \frac{q_\alpha^2}{m_\alpha c^2} \mathcal{A}_\epsilon(\mathbf{r}_\alpha) \int d^3k\, \delta\mathcal{A}_\epsilon \frac{e^{i\mathbf{k}\cdot\mathbf{r}_\alpha}}{(2\pi)^{3/2}}$$

$$+ \frac{1}{4\pi} \int d^3k\, c^2k^2\, \mathcal{A}_\epsilon^*\delta\mathcal{A}_\epsilon + O\left(\delta\mathcal{A}_\epsilon\right)^2. \tag{5.105}$$

Then we find that

$$
\begin{aligned}
\frac{\partial H}{\partial \mathcal{A}_\epsilon} &= \sum_\alpha \frac{q_\alpha^2}{m_\alpha c^2} A_\epsilon(\mathbf{r}_\alpha) \frac{e^{i\mathbf{k}\cdot\mathbf{r}_\alpha}}{(2\pi)^{3/2}} - \sum_\alpha \frac{q_\alpha}{m_\alpha c} p_{\alpha,\epsilon} \frac{e^{i\mathbf{k}\cdot\mathbf{r}_\alpha}}{(2\pi)^{3/2}} + \frac{c^2 k^2}{4\pi} \mathcal{A}_\epsilon^*(\mathbf{k}) \\
&= -\underbrace{\sum_\alpha \frac{q_\alpha}{c} \dot{r}_{\alpha,\epsilon} \frac{e^{i\mathbf{k}\cdot\mathbf{r}_\alpha}}{(2\pi)^{3/2}}}_{\mathcal{J}_{\perp,\epsilon}^*(\mathbf{k})} + \frac{c^2 k^2}{4\pi} \mathcal{A}_\epsilon^*(\mathbf{k}),
\end{aligned}
\tag{5.106}
$$

thus, Hamilton's equation of motion for Π_ϵ^* (5.102) yields

$$
\dot{\Pi}_\epsilon^*(\mathbf{k}, t) = \mathcal{J}_{\perp,\epsilon}^*(\mathbf{k}, t) - \frac{c^2 k^2}{4\pi} \mathcal{A}_\epsilon^*(\mathbf{k}, t).
\tag{5.107}
$$

Now, using the definition of Π (5.97), (5.94), (5.94) and

$$
\mathbf{k} \wedge [\mathbf{k} \wedge \mathcal{A}(\mathbf{k}, t)] = -k^2 \mathcal{A}(\mathbf{k}, t),
\tag{5.108}
$$

we find that equation (5.107) is just the complex conjugate of the Maxwell equation for the time derivative of the transverse component of the electric field in reciprocal space (5.80).

   The other Hamilton equations yield the following: the equations of motion of $\mathbf{r}_\alpha$ (5.99) and of $\mathcal{A}$ (5.102) only recover the definitions (5.96) and (5.97) of the conjugate momenta $\mathbf{p}_\alpha$ and Π, respectively. The remaining Hamilton equation (5.100), however, yields

$$
\dot{p}_{\alpha,j} = \frac{q_\alpha}{m_\alpha c} \left[\mathbf{p}_\alpha - \frac{q_\alpha}{c} \mathbf{A}(\mathbf{r}_\alpha, t) \right] \cdot \frac{\partial}{\partial r_{\alpha,j}} \mathbf{A}(\mathbf{r}_\alpha) + q_\alpha E_{\parallel,j}(\mathbf{r}_\alpha).
\tag{5.109}
$$

Substituting equation (5.99) or (5.96) and its total time derivative in equation (5.109), we obtain

$$
m_\alpha \ddot{r}_{\alpha,j} = q_\alpha E_{\parallel,j}(\mathbf{r}_\alpha, t)
$$
$$
+ \underbrace{\frac{q_\alpha}{c} \dot{\mathbf{r}}_\alpha \cdot \frac{\partial}{\partial r_{\alpha,j}} \mathbf{A}(\mathbf{r}_\alpha, t) - \frac{q_\alpha}{c} (\dot{\mathbf{r}}_\alpha \cdot \boldsymbol{\nabla}) A_j(\mathbf{r}_\alpha, t)}_{(q_\alpha/c)\,[\dot{\mathbf{r}}_\alpha \wedge \mathbf{B}(\mathbf{r}_\alpha, t)]_j} \underbrace{- \frac{q_\alpha}{c} \frac{\partial}{\partial t} A_j(\mathbf{r}_\alpha, t)}_{q_\alpha E_{\perp,j}(\mathbf{r}_\alpha, t)},
$$
$$
\tag{5.110}
$$

where we have used the vector identity

$$
[\mathbf{C} \wedge (\boldsymbol{\nabla} \wedge \mathbf{G})]_j = \mathbf{C} \cdot \partial_j \mathbf{G} - (\mathbf{C} \cdot \boldsymbol{\nabla}) G_j
\tag{5.111}
$$

to recognize the first term on the right-hand side of equation (5.110) as the j component of magnetic force on charge q_α [i.e., $(q_\alpha/c)\,[\dot{\mathbf{r}}_\alpha \wedge \mathbf{B}(\mathbf{r}_\alpha, t)]_j$]. So equation (5.100) allows us to recover the Newton–Lorentz equations (5.71) and (5.72).

We have shown that $\{\mathbf{r}_\alpha, \mathbf{p}_\alpha\}$ and $\{\mathcal{A}(\mathbf{k}, t), \boldsymbol{\Pi}(\mathbf{k}, t)\}$ are canonical variables. Therefore, their fundamental Poisson brackets[9] are given by [240]

$$(r_{\alpha,i}, p_{\beta,j}) = \delta_{\alpha\beta}\, \delta_{ij}, \tag{5.112}$$

$$(\mathcal{A}_\epsilon(\mathbf{k}, t), \Pi_{\epsilon'}^*(\mathbf{k}', t)) = \delta_{\epsilon\epsilon'}\, \delta(\mathbf{k} - \mathbf{k}'), \tag{5.113}$$

with all other Poisson brackets involving pairs of these variables vanishing. We can quantize the theory making the correspondence discussed in Chapter 2 between the Poisson brackets and the quantum commutators, that is,

$$(u, v) \rightarrow \frac{1}{i\hbar}[u, v]. \tag{5.114}$$

This yields the following commutators for our pairs of conjugate canonical variables

$$[r_{\alpha,i}, p_{\beta,j}] = i\hbar\delta_{\alpha\beta}\, \delta_{ij}, \tag{5.115}$$

$$\left[\mathcal{A}_\epsilon(\mathbf{k}), \Pi_{\epsilon'}^\dagger(\mathbf{k}')\right] = i\hbar\delta_{\epsilon\epsilon'}\, \delta(\mathbf{k} - \mathbf{k}'). \tag{5.116}$$

From the commutators (5.115) and (5.116) we can derive the whole of nonrelativistic quantum electrodynamics. It is more convenient, however, to express the quantized field in terms of annihilation and creation operators instead of the canonical variables $\mathcal{A}_\epsilon(\mathbf{k})$ and $\Pi_\epsilon(\mathbf{k})$. The annihilation operator is given by

$$a_\epsilon(\mathbf{k}) = \frac{1}{\sqrt{8\pi\hbar ck}}\left[ck\mathcal{A}_\epsilon(\mathbf{k}) + \frac{i}{c}\Pi_\epsilon(\mathbf{k})\right]. \tag{5.117}$$

The operator $a_\epsilon(\mathbf{k})$ and its Hermitian conjugate obey the following commutation relations deduced from (5.116) and (5.117):

$$[a_\epsilon(\mathbf{k}), a_{\epsilon'}(\mathbf{k}')] = 0, \tag{5.118}$$

$$[a_\epsilon^\dagger(\mathbf{k}), a_{\epsilon'}^\dagger(\mathbf{k}')] = 0, \tag{5.119}$$

$$[a_\epsilon(\mathbf{k}), a_{\epsilon'}^\dagger(\mathbf{k}')] = \delta_{\epsilon\epsilon'}\, \delta(\mathbf{k} - \mathbf{k}'). \tag{5.120}$$

In terms of $a_\epsilon(\mathbf{k})$ and $a_\epsilon^\dagger(\mathbf{k})$, the expression for H_{trans} takes the familiar form

$$H_{\text{trans}} = \int d^3k \sum_{l=1,2}\left[a_{\epsilon_l}^\dagger(\mathbf{k})a_{\epsilon_l}(\mathbf{k}) + \frac{1}{2}\right]\hbar ck. \tag{5.121}$$

---

[9]We notice that because the canonical variables of the field have only two independent components, the Poisson brackets associated with arbitrary Cartesian components of these variables do not have the form given by (5.113). In fact, a straightforward calculation of such Poisson brackets in terms of (5.113) yields $\left[\mathcal{A}_i(\mathbf{k}, t), \Pi_j^*(\mathbf{k}', t)\right] = \left\{\delta_{ij} - k_i k_j / k^2\right\}\delta(\mathbf{k} - \mathbf{k}').$

5.3 MATTER–RADIATION COUPLING AS A CONSEQUENCE OF GAUGE INVARIANCE

There is an elegant and beautiful way of deriving the form of the coupling between matter and radiation: to use the basic principle of gauge invariance. This principle says that the physics cannot depend on our choice of gauge for the electromagnetic potentials; that is, when we make the gauge change

$$\mathbf{A}'(\mathbf{r}, t) = \mathbf{A}(\mathbf{r}, t) + \nabla \chi(\mathbf{r}, t),$$
$$\phi'(\mathbf{r}, t) = \phi(\mathbf{r}, t) - \frac{1}{c}\frac{\partial}{\partial t}\chi(\mathbf{r}, t), \tag{5.122}$$

where χ is any given function of $\mathbf{r}$ and t, the observable quantities should not change. To get any further, we must make this physical requirement more precise by translating it into equations.

Let us denote by unprimed quantities the quantum-mechanical description in the old gauge and, by primed quantities, the description in the new gauge. If the two descriptions are equivalent, they must be related by a unitary transformation $\hat{T}$. This way, both descriptions will yield the same values for any observable quantity. Suppose that $\hat{O}$ is an observable; then its expectation value in the new gauge is given by

$$\langle \psi' | \hat{O}' | \psi' \rangle = \langle \psi | \hat{T}^\dagger \hat{T} \hat{O} \hat{T}^\dagger \hat{T} | \psi \rangle = \langle \psi | \hat{O} | \psi \rangle, \tag{5.123}$$

as $\hat{T}^\dagger \hat{T} = 1$. The unitarity of $\hat{T}$ also allows us to write $\hat{T}$ as

$$\hat{T} = \exp\left(\frac{i}{\hbar}\hat{G}\right), \tag{5.124}$$

where $\hat{G}$ is a Hermitian operator.

To find out how the Hamiltonian in the new gauge $\hat{H}'$ is related to the Hamiltonian in the old gauge $\hat{H}$, we write down the Schrödinger equation in the new gauge and use the fact that $|\psi'\rangle = \hat{T}|\psi\rangle$. Then, as

$$i\hbar \frac{\partial}{\partial t}\left(\hat{T}|\psi\rangle\right) = i\hbar \hat{T}\frac{\partial}{\partial t}|\psi\rangle - \left(\frac{\partial \hat{G}}{\partial t}\right)\hat{T}|\psi\rangle, \tag{5.125}$$

multiplying the Schrödinger equation in the new gauge by $\hat{T}^\dagger$ from the left, we find that

$$i\hbar \frac{\partial}{\partial t}|\psi\rangle = \hat{T}^\dagger \left(\hat{H}' + \frac{\partial \hat{G}}{\partial t}\right)\hat{T} |\psi\rangle. \tag{5.126}$$

Comparing (5.126) with the Schrödinger equation in the old gauge,

$$i\hbar \frac{\partial}{\partial t}|\psi\rangle = \hat{H}|\psi\rangle, \tag{5.127}$$

we see that the Hamiltonian in the new gauge, $\hat{H}'$, must be related to that in the old gauge, $\hat{H}$, by

$$\hat{H}' = \hat{T}\hat{H}\hat{T}^{\dagger} - \frac{\partial\hat{G}}{\partial t}. \tag{5.128}$$

So the Hamiltonian in the new gauge will only be an observable when $\hat{G}$ is not an explicit function of time.

Now consider the typical Hamiltonian for an atom or molecule:

$$\hat{H}_A = \sum_{\alpha} \frac{\mathbf{p}_{\alpha}^2}{2m_{\alpha}} + \sum_{\alpha} q_{\alpha}\phi(\mathbf{r}_{\alpha}, t). \tag{5.129}$$

For instance, the Hamiltonian for the hydrogen atom that is studied in elementary quantum mechanics has this form. What happens when we make the general gauge change (5.122)? Substituting ϕ by ϕ' given by (5.122) in the Hamiltonian (5.129), we find the following expression for the atomic Hamiltonian in the new gauge:

$$\hat{H}'_A = \hat{H}_A - \sum_{\alpha} \frac{q_{\alpha}}{c} \frac{\partial}{\partial t}\chi(\mathbf{r}_{\alpha}, t). \tag{5.130}$$

Comparing (5.128) and (5.130), we see that we must take

$$\hat{G} = \sum_{\alpha} \frac{q_{\alpha}}{c}\chi(\mathbf{r}_{\alpha}, t). \tag{5.131}$$

Unfortunately, this does not work, as the resulting $\hat{T}$ yields

$$\hat{T}\hat{H}_A\hat{T}^{\dagger} = \sum_{\alpha} \frac{1}{2m_{\alpha}}\left(\mathbf{p}_{\alpha} - \nabla_{\alpha}\hat{G}\right)^2 + \sum_{\alpha} q_{\alpha}\phi(\mathbf{r}_{\alpha}, t), \tag{5.132}$$

which can be rewritten as

$$\hat{H}'_A = \hat{T}\hat{H}_A\hat{T}^{\dagger} - \sum_{\alpha} \frac{q_{\alpha}}{c}\frac{\partial}{\partial t}\chi(\mathbf{r}_{\alpha}, t) + \sum_{\alpha} \frac{1}{m_{\alpha}}(\nabla_{\alpha}\hat{G}) \cdot \mathbf{p}_{\alpha} - \sum_{\alpha} \frac{(\nabla_{\alpha}\hat{G})^2}{2m_{\alpha}}. \tag{5.133}$$

Looking at (5.130), we see that the last two terms in (5.133) should not have been present. The presence of these two terms shows that $\hat{H}_A$ is not gauge invariant. But this is not very surprising, as $\hat{H}_A$ does not include the radiation field. Any system of charged particles subject to acceleration would also involve electromagnetic radiation.

Let us then complement $\hat{H}_A$ adding the Hamiltonian of the radiation field, $\hat{H}_F$, which was derived in Chapter 2, and a yet unknown coupling term $\hat{H}_I$. The complete Hamiltonian is given by $\hat{H} = \hat{H}_A + \hat{H}_I + \hat{H}_F$. As $\hat{H}'_F = \hat{H}_F$, $\hat{H}_I$ must be such that when transformed, it generates terms that cancel the last two terms on the right-hand side of (5.133). In other words, writing H_I as a functional of the electromagnetic potentials, $H_I[\mathbf{A}, \phi]$, it must be such that

$$\hat{T}\underbrace{\hat{H}_I[\mathbf{A}, \phi]\hat{T}^{\dagger}}_{\hat{H}'_I} = \hat{H}_I[\mathbf{A}', \phi'] + \sum_{\alpha} \frac{1}{m_{\alpha}}(\nabla_{\alpha}\hat{G}) \cdot \mathbf{p}_{\alpha} - \sum_{\alpha} \frac{(\nabla_{\alpha}\hat{G})^2}{2m_{\alpha}}. \tag{5.134}$$

As you can easily verify, the minimal-coupling interaction derived in Section 5.2 is the solution of this equation. Without the minimal-coupling interaction, the Hamiltonian is not gauge invariant.

Historically, the form of the coupling between matter and radiation was known long before the discovery of the symmetry under gauge transformations of the quantum-mechanical system of charged particles interacting with the electromagnetic field [315]. This symmetry was discovered by Fock in 1926 [215]. But the idea of gauge invariance as a basic unifying principle predates quantum mechanics and can be traced back [413, 414] to Weyl's [636] failed attempt to unify electromagnetism and gravitation in 1919 [315]. Much later, in 1928, Weyl wrote [637]: "But I now believe that this gauge invariance does not tie together electricity and gravitation, but rather electricity and matter in the manner described above."

The quantum nature of the electromagnetic field gives rise to many physical phenomena, some of which we mentioned earlier. In Chapter 6 we use the formalism we have presented here to study spontaneous emission. First, we examine spontaneous emission in free space and show how it is triggered by the zero-point fluctuations in the field. Then we describe how a cavity can alter this zero-point field and the consequences it has for spontaneous emission.

RECOMMENDED READING

- For a discussion of minimal coupling as a consequence of gauge invariance using a field description for matter as well as for the field, see Sec. 3.5 of [116].

- For a gauge-invariant formulation of perturbation theory in QED, see [366].

- The minimal coupling can also be seen as a consequence of Lorentz invariance [406]. For a nonrelativistic particle, a rephrasing of the principle of restricted Galilean invariance also leads to the minimal coupling Hamiltonian [320, 321].

- For a fully relativistic derivation of the interaction between electric and *magnetic* dipoles with an external electromagnetic field, see [20]. For a discussion of the Aharonov–Bohm phase in this context, see also [397, 572].

- See the book by Allen and Eberly [13] for a detailed discussion of the experimental requirements for the validity of the two-level atom approximation.

- Richard P. Feynman once derived the Lorentz force and two of Maxwell equations from the assumption that a nonrelativistic point particle's position and momentum satisfy the well-known quantum-mechanical commutation relation and that the particle obeys Newton's second law of motion [176]. This crazy derivation can be reinterpreted as a proof that the most general velocity-dependent force that can be described in a Lagrangian or Hamiltonian formalism is one that has the same general form of the Lorentz force [302]. This is related to the question of whether knowing the equations of motion of a physical system is enough to allow us to construct a Lagrangian or Hamiltonian formalism describing that system. See [301, 470, 646].

6

Spontaneous emission: From irreversible decay to Rabi oscillations[1]

Humpty Dumpty sat on a wall,
Humpty Dumpty had a great fall.
All the king's horses,
And all the king's men,
Couldn't put Humpty together again!

—Old nursery rhyme[2]

This chapter is divided into two sections. In the first section we use the formalism developed in Chapter 5 to discuss spontaneous emission in free space. We address the problem of atomic stability when the coupling with the dynamic electromagnetic field is included and the atom is allowed to radiate. We examine the roles of vacuum fluctuations and radiation reaction in the spontaneous emission process and the stability of the ground state of the atom [3, 129, 131, 449, 450, 453, 456, 557]. Then

---

[1]This chapter is based on [165].

[2]The popular image of Humpty Dumpty today is that of the egglike creature invented by Lewis Carroll. In fifteenth-century England, however, "Humpty Dumpty" was a colloquial term describing an obese person. The "Humpty Dumpty" in this rhyme might have referred to a big fat cannon used by the Royalists during the English Civil War (1642–1649). It was mounted on top of the St. Mary's at the Wall Church in Colchester defending the city against siege in the summer of 1648. A shot from a Parliamentary cannon damaged the wall beneath and "Humpty Dumpty had a great fall." The Royalists, "all the King's men," attempted to raise Humpty Dumpty onto another part of the wall. However, the cannon was so heavy that "All the King's horses and all the King's men couldn't put Humpty together again!"

we show that spontaneous emission in free space leads to an exponential decay of the atomic energy until it reaches the ground state.

In the second section we address the problem of spontaneous emission in a cavity. We examine two limits: the weak and strong coupling regimes. We show that in the weak-coupling regime, there is still an exponential decay, but the rate is changed by the cavity. In the strong-coupling regime, however, spontaneous emission becomes reversible with the atom performing Rabi oscillations. We also discuss what happens for long interaction times compared to the Rabi frequency, and we show that the oscillation collapses to revive only several periods later. We give an interpretation for these collapses and revivals where we find that the revivals are a unique feature of the quantum nature of the field [559].

6.1 SPONTANEOUS EMISSION IN FREE SPACE

When Schrödinger [541] first solved the hydrogen atom problem in quantum mechanics, he found the energy levels corresponding to the observed spectral lines that classical theory had failed to account for. These levels, however, were eigenstates of the Hamiltonian. So once the atom was found to be in one of them, it would stay there forever in the absence of external perturbations. This is in disagreement with experiment, which shows that only the ground state of an atom is stable; excited states eventually decaying by spontaneous emission, or other incoherent processes, to the ground state.

Schrödinger could not have obtained the spontaneous emission decay of excited states because his Hamiltonian did not include the interaction with the electromagnetic field. It incorporated only the instantaneous Coulomb potential. So there was nothing to disturb the atom and make it lose energy, bringing it to lower energy states. When the interaction with the electromagnetic field is included, the decay of atomic energy can be understood even in a crude classical model. The atom can simply be regarded as an oscillating classical dipole [274]. Then the radiation reaction force [314], which is the force on the atom due to its own field, damps the oscillation decreasing the atomic energy.

This, however, does not explain why there is a stable ground state because such decay would not stop at a minimum energy state but would continue until the collapse of the atom. What prevents the collapse? We will show in this section that it is the vacuum fluctuations of the electromagnetic field which are responsible for the stability of the ground state. This will give us insight into the nature of spontaneous emission and will provide us with a picture of the roles of radiation reaction (a classical effect) and vacuum fluctuations (a quantum effect) in this process. Our discussion will be based on the approach of Dalibard et al. [131].

Although the existence of spontaneous emission was pointed out by Einstein [182], the first calculation of the spontaneous emission rate from first principles using quantum electrodynamics was done by Weisskopf and Wigner [631]. Our approach in this section retains the crucial assumption of the original Weisskopf–Wigner calculation:

the Markovian assumption [223]. In the next section we study an example where this assumption breaks down.

The hydrogen atom consists of an electron of charge $-q$ and a proton of charge q bound together by their Coulomb interaction. Because the mass of the proton is much larger than the mass of the electron, we can neglect the motion of the former. The problem then reduces to an electron moving under the action of a Coulomb potential. We include the interaction with the quantized electromagnetic field as given earlier:

$$H = \frac{1}{2m} \left[\mathbf{p} + \frac{q}{c} \mathbf{A}(\mathbf{r}) \right]^2 + V_{\text{Coul}} + H_{\text{trans}}, \tag{6.1}$$

where

$$H_{\text{trans}} = \frac{1}{8\pi} \int d^3k \left\{ \boldsymbol{\Pi}^\dagger(\mathbf{k}) \cdot \boldsymbol{\Pi}(\mathbf{k}) + c^2 k^2 \boldsymbol{\mathcal{A}}^\dagger(\mathbf{k}) \cdot \boldsymbol{\mathcal{A}}(\mathbf{k}) \right\} \tag{6.2}$$

and

$$V_{\text{Coul}} = -\frac{q^2}{r}. \tag{6.3}$$

As can be seen from (6.3), we have neglected the constant self-energy contribution to the Coulomb energy. This term leads to a shift in the mass of the bare particle. It is the same mass renormalization as that present in classical theory [131].

The linear size of a typical hydrogen atom is on the order of 1Å. Only x-rays have wavelengths of that magnitude. The wavelengths of electromagnetic radiation emitted and absorbed by electronic transitions are usually much longer than this. When such radiation interacts with the atom, the field experienced by the electron is, to a very good approximation, the same experienced by the nucleus. So in equation (6.1) we can replace the position operator of the electron, $\mathbf{r}$, in the argument of $\mathbf{A}$ by the position of the proton, which we assume fixed at the origin. This is called the *dipole approximation* (see Sec. 5.1).

In the electric dipole approximation, the electron is insensitive to the magnetic field and the Heisenberg equation for the Cartesian momentum $\boldsymbol{\pi} = \mathbf{p} + (q/c)\mathbf{A}(0)$ together with the Heisenberg equation for the position of the electron (i.e., $\dot{\mathbf{r}} = \boldsymbol{\pi}/m$) yields the Newton–Lorentz equation without the magnetic force term

$$m\ddot{\mathbf{r}} = -\boldsymbol{\nabla} V_{\text{Coul}} - q\mathbf{E}_\perp(0). \tag{6.4}$$

The Heisenberg equation for the annihilation operator, $a_\epsilon(\mathbf{k})$, is

$$\dot{a}_\epsilon(\mathbf{k}) = -i\omega a_\epsilon(\mathbf{k}) - i\frac{q}{mc\hbar} \mathcal{A}_0(k)\boldsymbol{\epsilon}(\hat{\mathbf{k}}) \cdot \boldsymbol{\pi}, \tag{6.5}$$

where

$$\mathcal{A}_0(k) = \sqrt{\frac{hc^2}{(2\pi)^3\omega}}. \tag{6.6}$$

We can integrate equation (6.5),

$$a_\epsilon(\mathbf{k}, t) = a_\epsilon(\mathbf{k}, 0)e^{-i\omega t} - \frac{q}{mc\hbar} \mathcal{A}_0(k) \int_0^t dt'\, e^{-i\omega(t-t')} \boldsymbol{\epsilon}(\hat{\mathbf{k}}) \cdot \boldsymbol{\pi}(t'). \tag{6.7}$$

Let us write the vector potential as a sum of two terms:

$$\mathbf{A}(0, t) = \mathbf{A}_{\text{vac}}(0, t) + \mathbf{A}_{\text{s}}(0, t), \tag{6.8}$$

where the first term is the vacuum part:

$$\mathbf{A}_{\text{vac}}(0, t) = \int d^3k \, \mathcal{A}_0(k) \sum_{\epsilon} \left[a_\epsilon(\mathbf{k}, 0)e^{-i\omega t} + a_\epsilon^\dagger(\mathbf{k}, 0)e^{i\omega t} \right] \boldsymbol{\epsilon}(\hat{\mathbf{k}}), \tag{6.9}$$

and the second term in (6.8) is due to the radiation produced by the atom:

$$\mathbf{A}_{\text{s}}(0, t) = -\frac{iq}{mc\hbar} \int_0^t dt' \int d^3k \, \mathcal{A}_0^2(k) \sum_{\epsilon} \left[e^{-i\omega(t-t')} - e^{i\omega(t-t')} \right] \boldsymbol{\epsilon}(\hat{\mathbf{k}}) \cdot \boldsymbol{\pi}(t')\boldsymbol{\epsilon}(\hat{\mathbf{k}}). \tag{6.10}$$

We can do the integral in equation (6.10) in the following way:

$$
\begin{aligned}
\mathbf{A}_{\text{s}}(0, t) &= -\frac{q}{(2\pi)^2 mc^2} \int_0^t dt' \int d\Omega \sum_{\epsilon} \boldsymbol{\epsilon}(\hat{\mathbf{k}}) \cdot \boldsymbol{\pi}(t')\boldsymbol{\epsilon}(\hat{\mathbf{k}}) \\
&\quad \times \frac{d}{dt'} \int_{-\omega_M}^{\omega_M} d\omega \, e^{-i\omega(t-t')} \\
&= -\frac{4q}{3mc^2} \int_0^t dt' \, \boldsymbol{\pi}(t') \frac{d}{dt'} \delta_M(t - t') \\
&= -\frac{4q\delta_M(0)}{3mc^2} \boldsymbol{\pi}(t) + \frac{2q}{3mc^2} \dot{\boldsymbol{\pi}}(t),
\end{aligned}
\tag{6.11}
$$

where we have introduced the cutoff frequency ω_M, on the order of $mc^2/\hbar$, mentioned in Section 5.2, with $\delta_M(\tau)$ being a truncated delta function, given by

$$\delta_M(\tau) = \frac{1}{2\pi} \int_{-\omega_M}^{\omega_M} d\omega \, e^{-i\omega\tau}. \tag{6.12}$$

The first term on the last line of equation (6.11) will lead to the same classical mass renormalization as the sum of the Coulomb self-energies of the particles. Nonetheless, because this is not a relativistic theory, we cannot include contributions from the interactions with the high frequency-modes, so this mass correction turns out to be only $4/3$ of the mass correction obtained from the Coulomb self-energy [131].

If we now substitute equations (6.11) and (6.8) in equation (6.4), we obtain the quantum version of the classical Abraham–Lorentz equation [314]:

$$\left[m + \frac{4q^2\delta_M(0)}{3mc^3} \right] \ddot{\mathbf{r}} = -\nabla V_{\text{Coul}} + \frac{2q^2}{3c^3} \dddot{\mathbf{r}} - q\mathbf{E}_{\perp,\text{vac}}(0). \tag{6.13}$$

The coefficient multiplying the term on the left is the renormalized classical mass of the electron. The first term on the right-hand side is the instantaneous Coulomb force of the proton on the electron. The next term is the radiation reaction force that damps the motion of the electron as it radiates. The last term is absent in the classical theory when there is no incident external radiation on the atom. In the quantum theory, however, it is the electric field operator of the vacuum. The fact that this last term cannot be put equal to zero is the source of most of the differences between the classical and the quantum theory. Next, we will see that the vacuum fluctuations introduced by this term are in fact responsible for the existence of a stable ground state. As an approach based on the quantum Lorentz equation (6.13) presents many difficulties [131], we will proceed in a different way.

The rate of decay of atomic energy is given by

$$\frac{d}{dt} H_{\text{atom}} = -\frac{i}{\hbar} [H_{\text{atom}}, H], \tag{6.14}$$

where

$$H_{\text{atom}} = \frac{p^2}{2m} + V_{\text{Coul}}. \tag{6.15}$$

We want to calculate the expectation value of (6.14) when the field is initially in the vacuum state and the atom in one of the atomic energy levels. Then we want to identify a contribution from radiation reaction and another from vacuum fluctuations. To do so, we have to split the vector potential operator into two terms as in equation (6.8). However, because the commutator that appears in equation (6.14) involves a product of a field operator and an atomic operator:

$$[H_{\text{atom}}, H] = i\hbar \frac{q^3}{2mc} \frac{\mathbf{r} \cdot \mathbf{A}(0)}{r^3}, \tag{6.16}$$

we have to be careful when interpreting individual terms as radiation reaction and vacuum contributions. For although the total vector potential operator commutes with atomic operators at all times, the same cannot be said of each of the two terms on the right-hand side of equation (6.8) taken separately.

Apparently, any ordering of atomic and split field operators is possible in (6.14) as long as it is used in both the term involving only $\mathbf{A}_s$ and in that involving $\mathbf{A}_{\text{vac}}$ only [3, 453]. Nevertheless, Dalibard et al. [131] pointed out that to be able to assign a physical meaning to such terms in (6.14), each of them has to be Hermitian; otherwise, they cannot be associated with observables. The Hermiticity requirement allows only one ordering [131]: the totally symmetric ordering. So we can write

$$\frac{d}{dt} H_{\text{atom}} = \left(\frac{d}{dt} H_{\text{atom}}\right)_s + \left(\frac{d}{dt} H_{\text{atom}}\right)_{\text{vac}}, \tag{6.17}$$

where the first term on the right-hand side of (6.17) is the radiation reaction contribution,

$$\left(\frac{d}{dt} H_{\text{atom}}\right)_s = \frac{q^3}{4mc} \left\{ \frac{\mathbf{r}}{r^3} \cdot \mathbf{A}_s(0) + \mathbf{A}_s(0) \cdot \frac{\mathbf{r}}{r^3} \right\}, \tag{6.18}$$

and the second term is the vacuum fluctuation contribution,

$$\left(\frac{d}{dt} H_{\text{atom}}\right)_{\text{vac}} = \frac{q^3}{4mc} \left\{ \frac{\mathbf{r}}{r^3} \cdot \mathbf{A}_{\text{vac}}(0) + \mathbf{A}_{\text{vac}}(0) \cdot \frac{\mathbf{r}}{r^3} \right\}. \tag{6.19}$$

Now, we are going to calculate the expectation values of (6.18) and (6.19) up to first order in the fine structure constant. We start with the radiation reaction contribution (6.18). Because $\mathbf{A}_s$ is already proportional to q, we can approximate all atomic operators by the corresponding unperturbed atomic operators. The unperturbed atomic operators are those of the uncoupled atom. We will denote them by the subscript a. They satisfy the following Heisenberg equations of motion:

$$\dot{\mathbf{r}}_a = -\frac{i}{\hbar}[\mathbf{r}_a, H_{\text{atom}}] = \frac{\mathbf{P}_a}{m}, \tag{6.20}$$

$$\dot{\mathbf{p}}_a = -\frac{i}{\hbar}[\mathbf{p}_a, H_{\text{atom}}] = -\frac{q^2}{2}\frac{\mathbf{r}_a}{r_a^3}. \tag{6.21}$$

From these two equations, we can write the first-order expansion in the fine structure constant of equation (6.18) as

$$\left(\frac{d}{dt}H_{\text{atom}}\right)_s = -\frac{q}{2c}\{\ddot{\mathbf{r}}_a \cdot \mathbf{A}_{s,a}(0) + \mathbf{A}_{s,a}(0) \cdot \ddot{\mathbf{r}}_a\}, \tag{6.22}$$

where $\mathbf{A}_{s,a}$ is given by

$$\mathbf{A}_{s,a}(0) = -i\frac{qc}{(2\pi)^2 m}$$
$$\times \int_0^t dt' \int \frac{d^3k}{\omega} \sum_\epsilon \epsilon(\hat{\mathbf{k}}) \cdot \mathbf{p}_a(t')\left[e^{-i\omega(t-t')} - e^{i\omega(t-t')}\right]\epsilon(\hat{\mathbf{k}})$$
$$= -\frac{q}{(2\pi)^2 mc^2}\int_0^t dt' \int d\Omega$$
$$\times \int_0^\infty d\omega \sum_\epsilon \epsilon(\hat{\mathbf{k}}) \cdot \mathbf{p}_a(t')\frac{d}{dt'}\left[e^{-i\omega(t-t')} + e^{i\omega(t-t')}\right]\epsilon(\hat{\mathbf{k}})$$
$$= \frac{q}{2(2\pi)^2 mc^2}\int d\Omega \sum_\epsilon \epsilon(\hat{\mathbf{k}}) \cdot \ddot{\mathbf{r}}_a\epsilon(\hat{\mathbf{k}}). \tag{6.23}$$

In the last step in (6.23), we have used equation (6.20) and neglected the contribution to mass renormalization in (6.23).

Substituting (6.23) in (6.22) and taking the expectation value for the case where initially the atom is in an arbitrary atomic energy level $|a\rangle$ and the field in the vacuum state, we obtain

$$\left\langle \frac{d}{dt}H_{\text{atom}}\right\rangle_s = -\frac{2q^2}{3c^3}\langle a|(\ddot{\mathbf{r}}_a)^2|a\rangle. \tag{6.24}$$

In the case of the vacuum contribution (6.19), $\mathbf{A}_{\text{vac}}$ is of zero order in the coupling constant, so the atomic operators have to be computed up to first order in the coupling constant. It is convenient to write the perturbation expansion of (6.19), up to first order in the fine structure constant, as

$$\left(\frac{d}{dt}H_{\text{atom}}\right)_{\text{vac}} = -\frac{iq}{2mc\hbar}\{[V_{\text{Coul}}^I, \mathbf{p}_a] \cdot \mathbf{A}_{\text{vac}}(0) + \mathbf{A}_{\text{vac}}(0) \cdot [V_{\text{Coul}}^I, \mathbf{p}_a]\}, \tag{6.25}$$

where V_{Coul}^I is the first perturbation correction to the Coulomb potential, which obeys the equation[3]

$$\frac{d}{dt}V_{\text{Coul}}^I = -\frac{i}{\hbar}\left[V_{\text{Coul,a}}, \frac{q}{mc}\mathbf{p}_a \cdot \mathbf{A}_{\text{vac}}(0)\right]$$
$$= -\frac{q}{mc}\dot{\mathbf{p}}_a \cdot \mathbf{A}_{\text{vac}}(0). \tag{6.26}$$

Integrating (6.26),

$$V_{\text{Coul}}^I(t) = \frac{q}{mc}\int_0^t d\tau\, \dot{\mathbf{p}}_a \cdot \mathbf{A}_{\text{vac}}(0, t - \tau), \tag{6.27}$$

and substituting this expression for V_{Coul}^I in (6.25), we obtain

$$\left(\frac{d}{dt}H_{\text{atom}}\right)_{\text{vac}} = -i\frac{q^2}{2\hbar m^2 c^2}\int_0^t d\tau \sum_{i,j}[\dot{p}_{a,i}(t-\tau), p_{a,j}(t)]$$
$$\times\{A_{\text{vac},i}(t-\tau)A_{\text{vac},j}(t) + A_{\text{vac},j}(t)A_{\text{vac},i}(t+\tau)\}. \tag{6.28}$$

The expectation value of (6.28) is given by

$$\left\langle\frac{d}{dt}H_{\text{atom}}\right\rangle_{\text{vac}} = -i\frac{q^2}{2(2\pi)^2 m^2 c^3}\int_0^t d\tau \int d\Omega \int_{-\infty}^{\infty}|\omega|\,d\omega\, e^{i\omega\tau}$$
$$\times\sum_\epsilon\langle a\,|[\dot{p}_{a,\epsilon}(t-\tau), p_{a,\epsilon}(t)]|\,a\rangle$$

$$= -i\frac{q^2}{2(2\pi)^2 m^2 c^3}\int_0^t d\tau \int d\Omega \int_{-\infty}^{\infty}|\omega|\,d\omega\, e^{i\omega\tau}$$
$$\times\sum_{\epsilon,b}\Big\{\langle a\,|\dot{p}_{a,\epsilon}(t)|\,b\rangle\,\langle b\,|p_{a,\epsilon}(t)|\,a\rangle\, e^{-i\omega_{ab}\tau}$$
$$- \langle a\,|p_{a,\epsilon}(t)|\,b\rangle\,\langle b\,|\dot{p}_{a,\epsilon}(t)|\,a\rangle\, e^{i\omega_{ab}\tau}\Big\}$$

$$\approx -i\frac{q^2}{2(2\pi)^2 m^2 c^3}\int d\Omega \int_0^t d\tau\, \delta(\tau)\sum_{\epsilon,b}|\omega_{ab}|$$
$$\times\Big\{\langle a\,|\dot{p}_{a,\epsilon}(t)|\,b\rangle\,\langle b\,|p_{a,\epsilon}(t)|\,a\rangle - \langle a\,|p_{a,\epsilon}(t)|\,b\rangle\,\langle b\,|\dot{p}_{a,\epsilon}(t)|\,a\rangle\Big\}, \tag{6.29}$$

where we have inserted a complete set of unperturbed atomic eigenstates between $\dot{p}_{a,\epsilon}$ and $p_{a,\epsilon}$. The angular frequencies $\omega_{ab} \equiv \omega_a - \omega_b$ correspond to the atomic transitions from $|a\rangle$ to $|b\rangle$.

---

[3]Equation (6.25) only involves the unperturbed momentum, $\mathbf{p}_a$, because the perturbation does not affect momentum. The perturbation only affects functions of the canonically conjugate variable $\mathbf{r}$ such as the Coulomb potential.

Now, if we use the relation

$$\frac{\omega_{ab}}{m} \langle a | p_{a,\epsilon}(t) | b \rangle = -i \langle a | \ddot{r}_{a,\epsilon}(t) | b \rangle, \tag{6.30}$$

we can rewrite (6.29) as

$$\left\langle \frac{d}{dt} H_{\text{atom}} \right\rangle_{\text{vac}} = -\frac{2q^2}{3c^3} \left\{ \sum_{\omega_{ab}>0} \langle a | \ddot{\mathbf{r}}_a(t) | b \rangle \cdot \langle b | \ddot{\mathbf{r}}_a(t) | a \rangle \right.$$
$$\left. - \sum_{\omega_{ab}<0} \langle a | \ddot{\mathbf{r}}_a(t) | b \rangle \cdot \langle b | \ddot{\mathbf{r}}_a(t) | a \rangle \right\}. \tag{6.31}$$

The total average rate of variation of atomic energy is the sum of the individual rates (6.24) and (6.31). This sum becomes evident if we introduce a complete set of unperturbed atomic eigenstates in (6.24) and rewrite it as

$$\left\langle \frac{d}{dt} H_{\text{atom}} \right\rangle_{\text{s}} = -\frac{2q^2}{3c^3} \left\{ \sum_{\omega_{ab}>0} \langle a | \ddot{\mathbf{r}}_a(t) | b \rangle \cdot \langle b | \ddot{\mathbf{r}}_a(t) | a \rangle \right.$$
$$\left. + \sum_{\omega_{ab}<0} \langle a | \ddot{\mathbf{r}}_a(t) | b \rangle \cdot \langle b | \ddot{\mathbf{r}}_a(t) | a \rangle \right\}. \tag{6.32}$$

Then the total rate is simply

$$\left\langle \frac{d}{dt} H_{\text{atom}} \right\rangle_{\text{total}} = -\frac{4q^2}{3c^3} \sum_{\omega_{ab}>0} \langle a | \ddot{\mathbf{r}}_a(t) | b \rangle \cdot \langle b | \ddot{\mathbf{r}}_a(t) | a \rangle. \tag{6.33}$$

As you can see from (6.33), when $|a\rangle$ is the ground state, ω_{ab} will never be positive so that the summation will vanish. In other words, in the ground state the contribution of the vacuum fluctuations cancels the radiation reaction contribution exactly and the atom does not decay. This shows that the ground state of the atom is stable only because of the vacuum fluctuations of the field. The self-reaction term alone would lead to the collapse of the ground state as classical electrodynamics predicts and the atomic commutation relations $[r_i, p_j] = i\hbar\delta_{ij}$ would not hold.[4]

So far, our treatment of spontaneous emission has been perturbative. We will now go beyond the perturbative approach and we will show that an initially excited atom decays exponentially to the ground state in free space. Before we do so, however, it is convenient to simplify the problem further.

---

[4] According to classical theory, an atom cannot have a minimum-energy state even when only the Coulomb interaction is included in the Hamiltonian. It is the quantum commutation relations of the atomic operators that prevent the electron from colliding with the proton by giving it enough kinetic energy to overcome the Coulomb attraction every time the electron is localized within a certain neighborhood of the proton [409]. When the coupling with the radiation field is included to account for atomic emission, the atomic commutation relations are violated unless the field commutation relations are included [131, 556].

The canonical variables we have used to describe this system are not unique. They are related to other possible sets of canonical variables by unitary transformations; the quantum analog of classical contact transformations. We will perform such a transformation on these canonical variables to obtain a more convenient set of variables. When written in terms of this new set, the Hamiltonian will acquire a form more amenable to physical interpretation. We will then be able to use physical insight to make some approximations that will greatly simplify the problem.

Let us consider the unitary transformation T that takes $\mathbf{p}$ into $\mathbf{p} - (q/c)\mathbf{A}(0)$, that is,

$$T\mathbf{p}T^{\dagger} = \mathbf{p} - \frac{q}{c}\mathbf{A}(0). \tag{6.34}$$

This is a translation in momentum space of $-(q/c)\mathbf{A}(0)$. We will now show that the generator of infinitesimal translations in one dynamical variable is the canonical conjugate variable [243]. By applying the infinitesimal translations successively, we will obtain an expression for a general finite translation that will yield T as a particular case.

Let $T_u(\alpha)$ be a unitary transformation that produces a translation in the space of a given dynamical variable u:

$$T_u(\alpha)uT_u^{\dagger}(\alpha) = u + \alpha. \tag{6.35}$$

We can write the unitary transformation $T_u(\delta)$ corresponding to an infinitesimal translation δ as

$$T_u(\delta) = 1 + i\delta X, \tag{6.36}$$

where X is a Hermitian operator which is the generator of infinitesimal translations. From

$$T_u(\delta)uT_u^{\dagger}(\delta) = u + \delta \tag{6.37}$$

we obtain the following commutation relation between X and u

$$[X, u] = -i. \tag{6.38}$$

Equation (6.38) implies that

$$X = -\frac{1}{\hbar}v, \tag{6.39}$$

where v is the canonical conjugate variable corresponding to u:

$$[v, u] = i\hbar. \tag{6.40}$$

A finite translation $T_u(\alpha)$ is given by

$$T_u(\alpha) = \lim_{N \to \infty} \left(1 - \frac{1}{\hbar}\frac{\alpha}{N}v\right)^N$$

$$= \exp\left(-\frac{i}{\hbar}\alpha v\right). \tag{6.41}$$

In the case where $u = p_j$ and $\alpha = -(q/c)A_j(0)$, we obtain

$$T_{p_j}\left[-\frac{q}{c}A_j(0)\right] = \exp\left[\frac{iq}{c\hbar}A_j(0)r_j\right]. \tag{6.42}$$

The complete translation T can be achieved by translating each component of $\mathbf{p}$ successively:

$$T = \prod_j T_{p_j}\left[-\frac{q}{c}A_j(0)\right] = \exp\left[\frac{iq}{c\hbar}\mathbf{A}(0)\cdot\mathbf{r}\right]. \tag{6.43}$$

The Hamiltonian (6.1) acquires the following form[5] when we apply the unitary transformation T:

$$THT^\dagger = \frac{\mathbf{p}^2}{2m} + V_{\text{Coul}} + H_{\text{trans}} - \int d^3k\left(\left[\boldsymbol{\Pi}(\mathbf{k}) + \boldsymbol{\Pi}^\dagger(\mathbf{k})\right]\cdot\mathbf{r}q\right.$$
$$\left. + q^2\left\{[\mathbf{r}\cdot\boldsymbol{\epsilon}_1(\mathbf{k})]^2 + [\mathbf{r}\cdot\boldsymbol{\epsilon}_2(\mathbf{k})]\right\}\right). \tag{6.44}$$

We notice that $\boldsymbol{\Pi}$ in the new representation is no longer associated with the electric field but to the displacement field in this representation:

$$\mathcal{D}'(\mathbf{k}) = T\mathcal{D}(\mathbf{k})T^\dagger$$
$$= T\left(-\boldsymbol{\Pi}(\mathbf{k}) - \frac{q}{2\pi}\left\{\boldsymbol{\epsilon}_1(\mathbf{k})\left[\boldsymbol{\epsilon}_1(\mathbf{k})\cdot\mathbf{r}\right] + \boldsymbol{\epsilon}_2(\mathbf{k})\left[\boldsymbol{\epsilon}_2(\mathbf{k})]\right\}\right)T^\dagger$$
$$= -\boldsymbol{\Pi}(\mathbf{k}). \tag{6.45}$$

So equation (6.44) can be written as

$$H' = \frac{\mathbf{p}^2}{2m} + V_{\text{Coul}} + H_{\text{trans}} - \mathbf{D}'(0)\cdot\mathbf{d} + \varepsilon_{\text{dip}}, \tag{6.46}$$

where $\mathbf{d}$ is the atomic dipole $-q\mathbf{r}$,

$$H_{\text{trans}} = \frac{1}{8\pi}\int d^3k\left\{\mathcal{D}'^\dagger(\mathbf{k})\cdot\mathcal{D}'(\mathbf{k}) + \mathcal{B}^\dagger(\mathbf{k})\cdot\mathcal{B}(\mathbf{k})\right\}, \tag{6.47}$$

and ε_{dip} is the dipole self-energy,

$$\varepsilon_{\text{dip}} = \int d^3k\, q^2\left\{[\mathbf{r}\cdot\boldsymbol{\epsilon}_1(\mathbf{k})]^2 + [\mathbf{r}\cdot\boldsymbol{\epsilon}_2(\mathbf{k})]^2\right\}. \tag{6.48}$$

The self-energy apparently diverges, but in fact, we must restrict the integral in (6.48) to long wavelengths compared to an angstrom in order to be consistent with the dipole approximation. This term leads to another sort of mass renormalization. Because we are not interested in discussing renormalization, we simply ignore (6.48).

---

[5]From the fact that T can be written as $\exp\left[(iq/c\hbar)\int d^3k\,\mathbf{r}\cdot\mathcal{A}(\mathbf{k}) - \text{h.c.}\right]$ and the fact that $\mathcal{A}_\epsilon$ and Π_ϵ are canonical conjugate variables, a straightforward generalization of the results presented in the preceding paragraph yields $T\Pi_\epsilon T^\dagger = \Pi_\epsilon - qr_\epsilon$. We also notice that T affects neither $\mathcal{A}$ nor $\mathbf{r}$.

The problem can be simplified even further if we assume that there are only two atomic levels involved. This is the famous two-level atom approximation. Within this approximation, the Hamiltonian (6.46) becomes

$$H = \underbrace{\hbar \frac{\omega_0}{2} \sigma_z}_{\text{Atomic Energy}} + \underbrace{\int d^3k \, \hbar\omega \sum_{\epsilon} \left[a_\epsilon^\dagger(\mathbf{k}) a_\epsilon^\dagger(\mathbf{k}) + \frac{1}{2} \right]}_{\text{Field Energy}}$$

$$+ \underbrace{\hbar \left(\sigma^\dagger - \sigma \right) \int d^3k \sum_{\epsilon} \left[a_\epsilon(\mathbf{k}) - a_\epsilon^\dagger(\mathbf{k}) \right] g_\epsilon(\mathbf{k})}_{\text{Interaction}}, \qquad (6.49)$$

where we have dropped the primes and expressed the displacement field in terms of the creation and annihilation operators defined by equation (5.117). The coupling constant $g_\epsilon(\mathbf{k})$ is obtained from equation (6.46) when the atomic dipole is expressed in terms of the two-level atom operators [i.e., $\mathbf{d} = -i\hbar\mu \left(\sigma^\dagger - \sigma \right) \mathbf{x}$]. We have shifted the zero of energy to the average of the energies of the two atomic levels as in Chapter 5.

Now we want to obtain a nonperturbative expression for the expectation value of the atomic energy $\sigma_z \hbar\omega_0/2$ as a function of time for the case where the field is initially in the vacuum state. We can solve for the field operators as we have done before:

$$a_\epsilon(\mathbf{k}, t) = a_\epsilon(\mathbf{k}, 0)e^{-i\omega t} + ig_\epsilon(\mathbf{k}) \int_0^t dt' \left[\sigma^\dagger(t') - \sigma(t') \right] e^{-i\omega(t-t')}. \qquad (6.50)$$

The Heisenberg equation for σ_z is

$$\dot{\sigma}_z(t) = -i2 \left[\sigma^\dagger(t) - \sigma(t) \right] \int d^3k \sum_{\epsilon} \left[a_\epsilon(\mathbf{k}, t) - a_\epsilon^\dagger(\mathbf{k}, t) \right] g_\epsilon(\mathbf{k}). \qquad (6.51)$$

Because we are interested in the total rate of change of atomic energy, we can choose normal ordering[6] when we substitute equation (6.50) in (6.51). This yields

$$\langle \dot{\sigma}_z(t) \rangle = 2 \left\langle \sigma^\dagger(t) \int_0^t dt' \int d^3k \sum_{\epsilon} g_\epsilon^2(\mathbf{k}) \left[\sigma^\dagger(t') - \sigma(t') \right] e^{-i\omega(t-t')} \right\rangle + \text{c.c.} \qquad (6.52)$$

We notice that the time evolution of the atomic creation and annihilation operators $\sigma^\dagger$ and σ consists of a fast oscillation of angular frequency ω_0 and a slowly varying envelope [13]

$$\sigma(t) = \tilde{\sigma}(t)e^{-i\omega_0 t}. \qquad (6.53)$$

---

[6] The choice of normal ordering simplifies the calculation of the expectation value of (6.51) for the case where the field is initially in the vacuum state because the first term on the right-hand side of equation (6.50), the free field, will not contribute.

Using (6.53) in (6.52), we obtain

$$\langle \dot{\sigma}_z(t) \rangle = -2 \int_0^t dt' \int d^3k \sum_\epsilon g_\epsilon^2(\mathbf{k})$$

$$\times \left\langle \tilde{\sigma}^\dagger(t) \left[\tilde{\sigma}^\dagger(t') e^{i\omega_0(t+t')-i\omega(t-t')} - \tilde{\sigma}(t') e^{-i(\omega-\omega_0)(t-t')} \right] \right\rangle + \text{c.c.}$$

$$\approx - \langle \sigma^\dagger(t)\sigma(t) \rangle \, 4\pi \underbrace{\int d^3k \sum_\epsilon g_\epsilon^2(\mathbf{k})\delta(\omega - \omega_0)}_{\gamma}$$

$$= -\gamma \left[1 + \langle \sigma_z(t) \rangle \right]. \tag{6.54}$$

When we integrate equation (6.54), we find that the atomic energy decays exponentially with the rate $-\gamma$:

$$\langle \sigma_z(t) \rangle = -1 + \left[1 + \langle \sigma_z(0) \rangle \right] e^{-\gamma t}. \tag{6.55}$$

The exponential decay in (6.55) is typical of irreversible processes. The atom has an infinite continuum of modes to emit the photon, and once emitted, it is extremely unlikely that the photon will find its way back to the atom and be reabsorbed. In the next section we will see that a cavity can change this decay rate, and in the strong coupling regime, even make spontaneous emission a reversible process.

6.2 SPONTANEOUS EMISSION IN A CAVITY

Let us assume that the atomic transition is resonant with the mode of lowest frequency supported by the cavity and very far off resonance from any other modes. In this case, we can neglect all cavity modes except for the lowest one, and the Hamiltonian of the entire system becomes

$$H = \hbar\frac{\omega}{2}\sigma_z + \hbar\omega \left(a^\dagger a + \frac{1}{2} \right) + \hbar g \left(\sigma^\dagger - \sigma \right) \left(a - a^\dagger \right), \tag{6.56}$$

where

$$[a, a^\dagger] = 1. \tag{6.57}$$

The terms σa and $\sigma^\dagger a^\dagger$ in the Hamiltonian (6.56) do not conserve energy and can only contribute to the dynamics through virtual transitions. This contribution is much smaller than that of the real energy-conserving transitions associated with the terms $\sigma a^\dagger$ and $\sigma^\dagger a$. So we will make the *rotating-wave approximation* and keep only the energy-conserving terms in (6.56):

$$H = \hbar\frac{\omega}{2}\sigma_z + \hbar\omega \left(a^\dagger a + \frac{1}{2} \right) + \hbar g \left(\sigma^\dagger a + a^\dagger \sigma \right). \tag{6.58}$$

Real cavities are not made of perfectly reflective mirrors. In a real cavity, radiation eventually decays because of losses. The usual approach is to introduce losses by coupling the cavity to a large reservoir. When the reservoir variables are traced over, the evolution of the atom + lossy cavity system is given by a master equation for the reduced density matrix[7] of such a system, ρ. As this is a well-known procedure (see, e.g., [533]), we will not repeat such calculations here. We will, however, derive the same result from a physical argument for the case where the environment is at zero temperature.

Let us consider for a moment a hypothetical one-dimensional cavity composed of two mirrors a distance l apart. After a time $t = 2Nl/c$ corresponding to N round trips in the cavity,[8] the electric field of a classical electromagnetic wave at position x, $E(x)$, will decrease from its initial value at $t = 0$, due to the partial reflections at the mirrors, according to

$$E(x,t) = R^N E(x,0), \tag{6.59}$$

where R is the reflectivity of the mirrors; assumed the same for both, with no change of phase on reflection. Substituting $N = tc/2l$ in (6.59), we can rewrite (6.59) as

$$E(x,t) = E(x,0)e^{-\kappa t/2}, \tag{6.60}$$

where $\kappa = -c \ln(R)/4l$.

Equation (6.60) is purely formal, because the field does not decay continuously in time but only at those discrete times when it hits one of the mirrors. If, however, the mirrors are sufficiently good reflectors so that electromagnetic radiation can complete many round trips before it loses a very small fraction of its energy,[9] we can neglect the fact that the losses are localized at the mirrors and assume them to be distributed homogeneously in the cavity and occur continuously in time.[10] In this case, equation (6.60) is a good approximation for any time $t > 0$, even for times that are not integral multiples of a round-trip time.

In our crude argument above, we have omitted an important detail. When the mean energy of the field in the cavity becomes comparable to the mean energy of blackbody radiation, equation (6.60) breaks down. This is because the energy leaking out of the cavity will be balanced by energy being fed into the cavity by blackbody radiation. At zero temperature, however, there is no blackbody radiation and we would expect (6.60) to be valid all the way down to the quantum zero-point energy. The quantum states that recover classical radiation in the limit of high intensity are coherent states. So if (6.60) can be carried out to the quantum limit, a coherent state $|\alpha\rangle$ at time t decays to

$$|\alpha\rangle \rightarrow \left|e^{-\kappa t/2}\alpha\right\rangle. \tag{6.61}$$

We will use this result to derive the master equation describing the decay of any cavity field.

---

[7] An alternative approach keeps the state vector description but replaces Schrödinger's equation by a stochastic equation [88, 131, 590].

[8] Here c is the speed of light in the cavity.

[9] That is, if the decay time is much longer than the time of flight of a light signal through the cavity.

[10] This implies coarse graining at a time scale on the order of the time of flight of a light signal through the cavity (see [362]).

Any field density matrix $\rho(t)$ can be written as a linear combination of diagonal coherent-state density matrices[11] [236, 583]

$$\rho = \int d^2\alpha \, P(\alpha) \, |\alpha\rangle\langle\alpha|, \tag{6.62}$$

where

$$P(\alpha) = \frac{1}{\pi^2} \int d^2\xi \, \text{Tr} \left(\rho e^{\xi a^\dagger - \xi^* a} \right) e^{\alpha\xi^* - \alpha^*\xi}. \tag{6.63}$$

So for an infinitesimal time interval δt,

$$\rho(t + \delta t) = \int d^2\alpha \, P(\alpha) \, \left| e^{-\kappa\delta t/2}\alpha \right\rangle \left\langle \alpha e^{-\kappa\delta t/2} \right|. \tag{6.64}$$

Let us calculate $|\alpha \exp(-\kappa\, \delta t/2)\rangle\langle\alpha \exp(-\kappa\, \delta t/2)|$:

$$
\begin{aligned}
\left| e^{-\kappa\delta t/2}\alpha \right\rangle &= \exp\left(-\frac{|\alpha|^2}{2} e^{-\kappa\delta t} \right) \sum_{n=0}^{\infty} \frac{\left(e^{-\kappa\delta t/2}\alpha \right)^n}{\sqrt{n!}} |n\rangle \\
&= \exp\left(-\frac{|\alpha|^2}{2} e^{-\kappa\delta t} \right) e^{-(\kappa/2)\delta t\, a^\dagger a} \sum_{n=0}^{\infty} \frac{\alpha^n}{\sqrt{n!}} |n\rangle \\
&= \left(1 + \frac{\kappa}{2} |\alpha|^2 \, \delta t \right) \left(1 - \frac{\kappa}{2} \delta t\, a^\dagger a \right) |\alpha\rangle + \text{higher-order terms} \\
&= \left(1 + \frac{\kappa}{2} |\alpha|^2 \, \delta t - \frac{\kappa}{2} a^\dagger a\, \delta t \right) |\alpha\rangle + O\left(\delta t^2 \right), \tag{6.65}
\end{aligned}
$$

so

$$
\begin{aligned}
\left| e^{-\kappa\delta t/2}\alpha \right\rangle \left\langle e^{-\kappa\delta t/2}\alpha \right| &= |\alpha\rangle\langle\alpha| \left(-\frac{\kappa}{2} a^\dagger a \right) |\alpha\rangle\langle\alpha| \delta t \\
&\quad - \frac{\kappa}{2} |\alpha\rangle\langle\alpha| a^\dagger a\, \delta t + \kappa|\alpha|^2 |\alpha\rangle\langle\alpha| \delta t + O\left(\delta t^2 \right) \\
&= |\alpha\rangle\langle\alpha| - \frac{\kappa}{2} \left\{ a^\dagger a, |\alpha\rangle\langle\alpha| \right\} \delta t + \kappa a|\alpha\rangle\langle\alpha| a^\dagger \, \delta t \\
&\quad + O\left(\delta t^2 \right). \tag{6.66}
\end{aligned}
$$

Substituting (6.66) in (6.64), we obtain

$$\rho(t + \delta t) = \rho(t) - \frac{\kappa}{2} \left\{ a^\dagger a, \rho(t) \right\} \delta t + \kappa a\rho(t) a^\dagger \, \delta t + O\left(\delta t^2 \right), \tag{6.67}$$

where the braces stand for the anticommutator. The master equation is given by [411]

$$
\begin{aligned}
\frac{d}{dt}\rho(t) &= \lim_{\delta \to 0} \frac{\rho(t + \delta t) - \rho(t)}{\delta t} \\
&= -\frac{\kappa}{2} \left\{ a^\dagger a, \rho(t) \right\} + \kappa a\rho(t) a^\dagger. \tag{6.68}
\end{aligned}
$$

---

[11]This is a consequence of the overcompleteness of the basis of coherent states; see Chapter 4.

When the two-level atom is in the cavity, the master equation is given by

$$\frac{d}{dt}\rho = \frac{1}{i\hbar}[H, \rho] - \frac{\kappa}{2}\{a^\dagger a, \rho\} + \kappa a\rho a^\dagger. \tag{6.69}$$

We are now ready to look into how the cavity changes spontaneous emission. The system starts from the product state $|\uparrow\rangle|0\rangle$ with the atom in the excited state and the cavity field in the vacuum. Then there are only three states [34, 273] involved in the time evolution described by (6.69):

$$|1\rangle = |\uparrow\rangle|0\rangle, \tag{6.70}$$
$$|2\rangle = |\downarrow\rangle|1\rangle, \tag{6.71}$$
$$|3\rangle = |\downarrow\rangle|0\rangle. \tag{6.72}$$

The corresponding components of the density matrix obey the following differential equations derived from (6.69):

$$\frac{d}{dt}\rho_{11} = ig\left(\rho_{12} - \rho_{21}\right), \tag{6.73}$$

$$\frac{d}{dt}\rho_{22} = -\kappa\rho_{22} - ig\left(\rho_{12} - \rho_{21}\right), \tag{6.74}$$

$$\frac{d}{dt}\left(\rho_{12} - \rho_{21}\right) = -\frac{\kappa}{2}\left(\rho_{12} - \rho_{21}\right) + i2g\left(\rho_{11} - \rho_{22}\right), \tag{6.75}$$

$$\frac{d}{dt}\rho_{33} = \kappa\rho_{22}. \tag{6.76}$$

Equation (6.76) is not independent from the other equations, because $\rho_{11} + \rho_{22} + \rho_{33} = 1$. The remaining equations can be put in matrix form:

$$\frac{d}{dt}\begin{bmatrix} \rho_{11} \\ \rho_{22} \\ \rho_{12} - \rho_{21} \end{bmatrix} = \begin{bmatrix} 0 & 0 & ig \\ 0 & -\kappa & -ig \\ i2g & -i2g & -\kappa/2 \end{bmatrix}\begin{bmatrix} \rho_{11} \\ \rho_{22} \\ \rho_{12} - \rho_{21} \end{bmatrix}. \tag{6.77}$$

The eigenvalues of the matrix in (6.77) are the three roots of the following secular equation:

$$\left(\lambda + \frac{\kappa}{2}\right)\left[(\lambda + \kappa)\lambda + 4g^2\right] = 0. \tag{6.78}$$

These roots are given by

$$\lambda_0 = -\frac{\kappa}{2}, \tag{6.79}$$

$$\lambda_\pm = -\frac{\kappa}{2} \pm \frac{\kappa}{2}\sqrt{1 - \frac{16g^2}{\kappa^2}}. \tag{6.80}$$

We examine two limiting cases. The weak-coupling regime, where the cavity decay rate is much larger than the coupling between the atom and the cavity mode, $\kappa \gg g$, and the strong-coupling regime, where the coupling is much larger than the cavity decay rate, $g \gg \kappa$.

In the weak-coupling regime, we can approximate (6.80) by

$$\lambda_{\pm} = -\frac{\kappa}{2} \pm \frac{\kappa}{2} \left(1 - \frac{8g^2}{\kappa^2} \right). \tag{6.81}$$

Then we can write the probability of finding the atom in the upper state as

$$\rho_{11}(t) = A e^{-\kappa t/2} + B e^{-4g^2 t/\kappa} + D e^{-\kappa t - 4g^2 t/\kappa}, \tag{6.82}$$

where the constants A, B, and D are determined by the initial conditions. The condition $\rho_{11}(0) = 1$ yields

$$A + B + D = 1. \tag{6.83}$$

The conditions

$$\frac{d}{dt}\rho_{11}(0) = 0, \tag{6.84}$$

$$\frac{d^2}{dt^2}\rho_{11}(0) = -2g^2 \tag{6.85}$$

yield

$$\frac{\kappa}{2}A + \frac{4g^2}{\kappa}B + \left(\kappa - \frac{4g^2}{\kappa} \right) D = 0, \tag{6.86}$$

$$\left(\frac{\kappa}{2} \right)^2 A + \left(\frac{4g^2}{\kappa} \right)^2 B + \left(\kappa - \frac{4g^2}{\kappa} \right)^2 D = -2g^2. \tag{6.87}$$

Solving these three equations for A, B, and D and using the fact that $g/\kappa \ll 1$, we obtain

$$\rho_{11}(t) \approx \exp\left(-\frac{4g^2}{\kappa}t \right). \tag{6.88}$$

So in the weak-coupling regime, the atomic energy still decays exponentially as in free space but with a different decay rate [502]. Spontaneous emission remains an irreversible process in this regime; however, the cavity modifies the mode density, leading to a different decay rate as can be seen from a simple argument using Fermi's golden rule [273].

Now, let us look at the strong-coupling regime. Equation (6.80) can be approximated by

$$\lambda_{\pm} = -\frac{\kappa}{2} \pm i2g. \tag{6.89}$$

Then the probability of finding the atom in the upper state can be written as

$$\rho_{11}(t) = R e^{-\kappa t/2} + C e^{-\kappa t/2} e^{i2gt} + C^* e^{-\kappa t/2} e^{-i2gt}, \tag{6.90}$$

where R is a real constant and C is an imaginary one. After using the initial conditions in (6.90), we find that

$$\rho_{11}(t) \approx \frac{e^{-\kappa t/2}}{2} (1 + \cos 2gt). \tag{6.91}$$

Equation (6.91) shows that ρ_{11} undergoes the *Rabi oscillations* (named after Isidor Isaac Rabi). This means that the atom emits a photon, then reabsorbs it and remits it again and again, until the photon leaks out of the cavity eventually. So, in the strong-coupling regime, spontaneous emission becomes almost reversible. This is a consequence of two things. First, there is only a single mode in the cavity where the photon can go. Second, the photon survives for long enough in the cavity to allow the atom to reabsorb it. As the physics in this regime is so surprisingly different from ordinary spontaneous emission, it is interesting to look into this matter a bit further.

Let us examine what happens when the cavity decay rate is so much smaller than the coupling constant that the atom can perform several Rabi oscillations before the photon has any chance of leaving the cavity.[12] In this case we can neglect cavity decay. This leads us to the *Jaynes–Cummings model* [323], which is just the Hamiltonian evolution part of (6.69).

Now suppose that we prepare a coherent state $|\alpha\rangle$ in the cavity at $t = 0$ with the atom in the lower state. The probability of detecting the atom in the excited state at an arbitrary future time t is given by

$$\rho_{11} = \frac{1}{2}\left[1 + \sum_{n=0}^{\infty} \frac{e^{-|\alpha|^2}|\alpha|^{2n}}{n!} \cos\left(2gt\sqrt{n+1}\right)\right]. \tag{6.92}$$

This is a superposition of Rabi oscillations of different frequencies weighted by a Poissonian distribution. The total sum will oscillate initially, but the oscillation will eventually collapse when the cosine terms become out of phase. We can verify that by expanding $\sqrt{n+1}$ around $|\alpha|^2$ for $|\alpha|^2 \gg 1$ in equation (6.92):

$$\rho_{11}(t) = \frac{1}{2}\left(\frac{1}{2} + \frac{e^{-|\alpha|^2}}{2}\sum_{n=0}^{\infty} \frac{|\alpha|^{2n}}{n!}e^{igt|\alpha|}e^{i(gt/|\alpha|)n} + \text{c.c.}\right)$$

$$= \frac{1}{2}\left[\frac{1}{2} + \frac{e^{-|\alpha|^2}}{2}e^{igt|\alpha|}\exp\left(|\alpha|^2 e^{igt/|\alpha|}\right) + \text{c.c.}\right]. \tag{6.93}$$

Now if $gt/|\alpha| \ll 1$, we can expand the exponential in (6.93):

$$e^{igt/|\alpha|} \approx 1 + \frac{igt}{|\alpha|} - \frac{g^2 t^2}{2|\alpha|^2}. \tag{6.94}$$

Then equation (6.93) becomes [130]

$$\rho_{11}(t) \approx \frac{1}{2}\left(1 + e^{-g^2 t^2/2}\cos 2gt|\alpha|\right). \tag{6.95}$$

So the collapse happens at a time on the order of $1/g$ [34, 178]. They occur because of the noise in the field, which leads to the finite variance of the photon

---

[12]It will become evident later that for this to be true, the cavity decay constant κ has to be at least much smaller than $g/\sqrt{\bar{n}}$, where $\bar{n}$ is the average number of photons in the cavity mode.

number, causing the eventual dephasing of the various cosine terms in (6.92). This noise is of quantum origin here, but classical noise would be equally effective.

Expression (6.95) is valid only for times much smaller than $|\alpha|/g$. In fact, we notice by inspecting (6.93) that the destructive interference disappears when the oscillating complex exponential argument of the exponential function in (6.93) becomes unit again. So a revival of the oscillations should occur at

$$t_{\mathrm{R}} \approx 2\pi \frac{|\alpha|}{g}. \tag{6.96}$$

Although it is possible to recover the collapse of the oscillations in a calculation using a classical field with classical noise, the revival is a genuine quantum effect. Revivals can only take place because the field energy is *discrete* rather than continuous, so that the various cosine components of the oscillations can eventually come back in phase with each other. The recent observation of revivals [153, 517] provided more evidence of the quantum nature of the electromagnetic interaction.

In this chapter we have applied QED to the problem of spontaneous emission by a single atom in the *vacuum*. We have seen that a cavity can make spontaneous emission become reversible, leading to Rabi oscillations. In Chapter 7 we study how QED can be modified to account for material media similar to the process in classical electrodynamics: through a dielectric permittivity.

RECOMMENDED READING

- For a discussion of atomic stability where coupling with the radiation field is neglected and only the Coulomb potential is considered, see [175, 177, 409, 575].

- There are several approximations and assumptions in the Weisskopf–Wigner approach. For a detailed study of these and alternative improvements on the original Weisskopf–Wigner theory, see [7, 144, 360, 555].

- For a discussion of renormalization in the context of atomic physics, I refer the reader to Weisskopf [630], Welton [632], Dalibard et al. [131], and Haroche [274].

- For a thorough discussion of the two-level atom approximation, see Allen and Eberly [13].

7

Macroscopic QED: Quantum electrodynamics in material media

... zu jener Zeit ... elektrische bezw. magnetische „Feldstärken" und „Verschiebungen" wurden als gleich elementare Grössen behandelt, der leere Raum als Spezialfall eines dielektrischen Körpers. Die Materie erschien als Träger des Feldes, nicht der Raum. ... Es war das grosse Verdienst von H. A. Lorentz, dass er hier in überzeugender Weise Wandel schuf. Im Prinzip gibt es nach ihm ein Feld nur im leeren Raume. Die atomistisch gedachte Materie ist einziger Sitz der elektrischen Ladungen; zwischen den materiellen Teilchen ist leerer Raum, der Sitz des elektromagnetischen Feldes ... Dielektrizität, Leitungsfähigkeit, etc. sind ausschliesslich durch die Art der mechanischen Bindung der Teilchen bedingt, aus welchen die Körper bestehen.[1]

—Albert Einstein [184]

There are two main versions of classical electrodynamics. The first was originally created by Maxwell and is now referred to as *macroscopic electrodynamics*. It involves four vectorial field quantities in a material medium: the electric field $\mathbf{E}$, the magnetic field $\mathbf{M}$, the electric displacement $\mathbf{D}$, and the magnetic induction $\mathbf{B}$. The

---

[1]Translation: ... at that time ... electric or magnetic "field intensities" and "displacements" were treated as equally elementary variables, empty space as a special instance of a dielectric body. *Matter* appeared as the bearer of the field, not *space*. ... It was the great merit of H. A. Lorentz that he brought about a change here in a convincing fashion. In principle a field exists, according to him, only in empty space. Matter—considered as atoms—is the only seat of electric charges; between the material particles there is empty space, the seat of the electromagnetic field. ... Dielectricity, conductivity, etc. are determined exclusively by the type of mechanical tie connecting the particles, of which the bodies consist.

second version was introduced by Lorentz. It is simpler and more fundamental. It assumes an atomistic point of view, describing every situation in electromagnetism in terms of charged particles in the vacuum interacting with only two field vectors: the electric field **E** and the magnetic field **B**. Lorentz showed that Maxwell's original theory could be obtained from his simpler version by averaging the electromagnetic quantities in such a way as to ignore microscopic detail. This kind of averaging became known as *macroscopic averaging,* and as we shall see later in the chapter, is a somewhat subtle idea.[2]

Quantum electrodynamics was developed using the more fundamental version of classical electrodynamics (i.e., charged particles in the vacuum). After all, it is supposed to be the most fundamental theory for the interaction between charged particles. Just as in classical electrodynamics, however, it is often too complicated to describe everything in terms of their constituent charged particles. So far in this book, we have looked only at simple systems composed most of the time by a single atom interacting with a single mode of the radiation field in a cavity. In Chapters 2 to 4 we even did away with the atom altogether. But consider, for example, a semiconductor laser. It has a macroscopic number of charged particles composing the material medium that forms the chunk of semiconductor where the laser was grown. Each atom has a number of electrons and protons, and so on. To keep track of all these charged particles in a theory is an enormous task. So in this chapter we will see how we can develop a quantum counterpart of Maxwell's original theory of classical electrodynamics in material media.

There are two main approaches to this problem. The first is to start from Maxwell's macroscopic equations and attempt to quantize them as their microscopic counterparts were quantized in Chapter 2. This approach has a number of problems, especially when the media are dispersive and absorptive [306–308]. Moreover, it is a phenomenological rather than a first-principles approach and has already been presented in a nice pedagogical fashion by Vogel and Welsch in their textbook [618]. So in this chapter we discuss the first-principles approach to QED in dielectrics. We start from ordinary quantum electrodynamics for atoms or molecules in the vacuum and we show how we can use the method of macroscopic averaging introduced by Lorentz to obtain an *approximate* theory, where the effect of all the numerous atoms or molecules is taken into account by a single function: the dielectric permittivity.

In the next Section we introduce a simple cavity-QED model of a dielectric medium. In Section 7.2 we diagonalize the Hamiltonian using dressed states that mix photons with excitations of the medium (polaritons). Section 7.3 introduces the idea of macroscopic averaging and discusses the medium contribution to the quantum noise of the radiation field. Section 7.4 shows how, in many cases, the noise contribution of the medium can be neglected, allowing us to account for the medium using only a permittivity function in QED. The remaining sections describe how to

---

[2]The idea of macroscopic averaging has a long and tortuous history (for more on this, see the recommended reading at the end of the chapter). It was only relatively recently that a proper kind of macroscopic averaging was introduced by Robinson [525]. Many textbooks still give confusing and mistaken discussions of what a macroscopic average should be.

include absorption in this macroscopic version of QED. In Section 7.5 we present the Kramers–Kronig relation, which shows that whenever there is dispersion, somewhere else (in frequency) you must find absorption. Section 7.6 shows how we can modify our simple cavity QED model of a material medium to account for absorption. In Section 7.7 we use classical electrodynamics to determine the dielectric permittivity in this model. In the last section we discuss the quantum theory and show how the dielectric permittivity, calculated classically, emerges in expressions for the quantized macroscopic–electromagnetic fields.

7.1 A SIMPLE MODEL FOR QED IN MATERIAL MEDIA: THE DIELECTRIC JCM

The Jaynes–Cummings model is the simplest and most fundamental case of interaction between atoms and radiation. With a minor modification it can also retain the essential features of material media without burdening us with nonessential features such as local field effects and the nontrivial mode structure that a dielectric gives rise to[3] in free space [237]. So we adopt what we believe to be the simplest one-dimensional microscopic model of lossless material media: N two-level atoms having the same resonance frequency ω_0 in a single-mode cavity of resonance frequency ω. Although we cannot propagate a wave packet in a single-mode cavity to see the effect of dispersion in the packet width, we are able to describe the essential cause of dispersion with this model: the frequency dependence of the dielectric permittivity.

For simplicity, we deal only with linear media. Fano has shown [189] that in the linear-medium approximation, the atoms constituting the medium can be described by harmonic oscillators, just as in the classical Lorentz model [415, 619]. The linear-medium approximation works because we assume that these atoms couple only weakly to the cavity mode. The essence of Fano's result is, in a nutshell, this: A harmonic oscillator has a ladder of energy levels, but when it is coupled weakly to the field, the upper levels are almost never populated, and it is effectively equivalent to a two-level system.

To probe the atom–field interaction within this medium, we consider a single guest two-level atom of resonance frequency ω_a immersed in the material medium and coupled strongly to the field. This guest atom will not, of course, be approximated by a harmonic oscillator (see Figure 7.1). Knoester and Mukamel [361] adopted a similar model to study impurity molecules in a dielectric host crystal, but in their model the dielectric medium is not placed inside a single-mode cavity.

Recalling Section 5.1, the displacement field is given by

$$D(x) = \sqrt{\frac{2}{L}\hbar\omega}(\hat{a} + \hat{a}^\dagger)\sin\left(\frac{\omega}{c}x\right), \tag{7.1}$$

---

[3]Unless the permittivity is not position dependent.

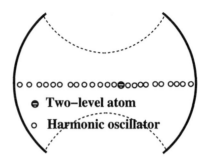

Figure 7.1 Diagram representation of our dielectric Jaynes–Cummings model. A dielectric medium composed of a large number of atoms fills a single-mode cavity of resonance frequency ω (the dotted line represents the single mode). The atoms of the medium have a single resonance frequency relevant to the problem; as they couple only weakly to the cavity mode, they are approximated by harmonic oscillators (represented by unfilled circles in the figure). A single two-level atom (represented by a circle with a double bar inside) of resonance frequency ω_a and strongly coupled to the cavity mode is embedded in the medium.

and the polarization of the medium by

$$P(x) = \sqrt{\frac{\hbar}{2\omega_0}} \sum_{j=1}^{N} (\hat{b}_j + \hat{b}_j^\dagger) q_j \delta(x - x_j), \tag{7.2}$$

where L is the length of the cavity and $\sqrt{\hbar/2\omega_0}(\hat{b}_j + \hat{b}_j^\dagger)$ is the position operator of an oscillator of effective charge q_j; the product $\sqrt{\hbar/2\omega_0} q_j (\hat{b}_j + \hat{b}_j^\dagger)$ is the electric dipole moment operator of the atom of the medium that is located at x_j, which we are approximating by a harmonic oscillator.

The operators $\hat{a}$, $\hat{a}^\dagger$, $\hat{b}_j$, and $\hat{b}_j^\dagger$ satisfy the following commutation relations:

$$[\hat{a}, \hat{a}^\dagger] = 1, \tag{7.3}$$

$$[\hat{b}_j, \hat{b}_{j'}^\dagger] = \delta_{jj'}, \tag{7.4}$$

$$[\hat{b}_j, \hat{b}_{j'}] = 0, \tag{7.5}$$

and $\hat{a}$, $\hat{a}^\dagger$ commute with $\hat{b}_j$, $\hat{b}_j^\dagger$. The Hamiltonian is given by

$$\hat{H} = \hbar\omega\, \hat{a}^\dagger \hat{a} + \hbar\omega_0 \sum_{j=1}^{N} \hat{b}_j^\dagger \hat{b}_j - \int dx\, D(x) P(x) + \frac{\hbar\omega_a}{2}\hat{\sigma}_z + \hbar\Omega(\hat{a} + \hat{a}^\dagger)(\hat{\sigma} + \hat{\sigma}^\dagger). \tag{7.6}$$

The first term on the right-hand side of (7.6) is the energy stored in the single-mode cavity (i.e., the cavity Hamiltonian). The second term is the energy in the medium (i.e., the medium Hamiltonian). The third is the coupling (in the dipole approximation) of the atoms of the medium to the single mode of the cavity. The fourth term describes the energy in the guest atom (i.e., it is the Hamiltonian of the guest atom) (we are using the Pauli matrix notation introduced in Chapter 5). The last term is the coupling between the guest atom and the single mode of the cavity (notice that we are not making the rotating-wave approximation yet), where

$$\Omega = -d\sqrt{\frac{4\pi\omega}{L\hbar}} \sin\left(\frac{\omega}{c}x_a\right) \qquad (7.7)$$

is the Rabi frequency of the guest atom located at x_a whose electric dipole moment strength is d.

This model is much simpler than it seems on a first impression. To see how simple it really is, let us first use the delta function in (7.2) to do the integral in (7.6) explicitly. Then the Hamiltonian becomes

$$\hat{H} = \hbar\omega\,\hat{a}^\dagger\hat{a} + \hbar\omega_0 \sum_{j=1}^{N} \hat{b}_j^\dagger\hat{b}_j + \hbar \sum_{j=1}^{N} g_j(\hat{a}+\hat{a}^\dagger)(\hat{b}_j+\hat{b}_j^\dagger) + \frac{\hbar\omega_a}{2}\hat{\sigma}_z + \hbar\Omega(\hat{a}+\hat{a}^\dagger)(\hat{\sigma}+\hat{\sigma}^\dagger),$$

$$(7.8)$$

where g_j is given by

$$g_j = -q_j\sqrt{\frac{2\pi\omega}{L\omega_0}} \sin\left(\frac{\omega}{c}x_j\right). \qquad (7.9)$$

As you can see, the interaction of the atoms of the material medium with the cavity mode depends only on the weighted sum of all the atomic creation and annihilation operators, $\sum_{j=1}^{N} g_j(\hat{b}_j + \hat{b}_j^\dagger)$. Now it is possible to define a collective atomic annihilation operator composed of the ordinary atomic annihilation operators weighted by g_j,

$$\hat{B}_N = \frac{1}{\sqrt{\sum_{m=1}^{N} g_m^2}} \sum_{j=1}^{N} g_j\hat{b}_j, \qquad (7.10)$$

where the normalizing factor $1/\sqrt{\sum_{m=1}^{N} g_m^2}$ is there to make

$$[\hat{B}_N, \hat{B}_N^\dagger] = 1, \qquad (7.11)$$

whatever the values of the g_j's.

This suggests that if we can find a complete set of collective annihilation and creation operators for the atoms of the medium with $\hat{B}_N\,\hat{B}_N^\dagger$ being one of its elements, we might be able to rewrite (7.8) in an exact yet much simpler form: a form where only $\hat{B}_N$ and $\hat{B}_N^\dagger$ couple to the cavity mode, with all the other $N-1$ collective atomic operators decoupled. This would effectively reduce the problem of describing the atomic medium to the single pair of creation and annihilation operators $\hat{B}_N$ and $\hat{B}_N^\dagger$ rather than N pairs, as the dynamics of the uncoupled $N-1$ pairs would be irrelevant then.

To make this decoupling come true, our set of collective atomic operators should not only be complete but also be such that the Hamiltonian of the medium $\sum_j \hat{b}_j^\dagger\hat{b}_j$ remain diagonal when expressed in terms of the collective operators in the set. Such a set of N pairs of collective atomic-medium creation and annihilation operators containing $\hat{B}_1$ and $\hat{B}_1^\dagger$ can be constructed in the following way. Let

$$G_n = \sqrt{\sum_{m=1}^{n} g_m^2}. \tag{7.12}$$

Then (7.10) can be rewritten as

$$\hat{B}_N = \frac{1}{G_N} \sum_{j=1}^{N} g_j \hat{b}_j. \tag{7.13}$$

We choose the other $N - 1$ annihilation operators to be given by

$$\hat{B}_k = \frac{1}{G_{k+1} G_k} \sum_{j=1}^{k} \left(g_j^2 \hat{b}_{k+1} - g_{k+1} g_j \hat{b}_j \right), \tag{7.14}$$

where $k = 1, 2, 3, \ldots, N - 1$. You can easily check that they all commute with each other and with $\hat{B}_1$ and that together with their Hermitian conjugates, they satisfy the commutation relations of annihilation and creation operators,

$$[\hat{B}_j, \hat{B}_{j'}^\dagger] = \delta_{jj'} \quad \text{and} \quad [\hat{B}_j, \hat{B}_{j'}] = 0. \tag{7.15}$$

We will now show that the Hamiltonian of the medium remains diagonal (i.e., that $\sum_{j=1}^{N} \hat{b}_j^\dagger \hat{b}_j = \sum_{k=1}^{N} \hat{B}_k^\dagger \hat{B}_k$). This is more conveniently demonstrated by induction.

Induction proofs are suited to hypotheses that depend on an arbitrary integer parameter, such as our number N of atoms in the material media. One first has to show that the hypothesis holds for $N = 1$. Then, if it can also be shown that when it holds for $N - 1$, it will hold for N, the hypothesis will have been proved for arbitrary N. In our case, it is trivial to see that it holds for $N = 1$, as in that case there is only one "collective" annihilation operator, $\hat{B}_1$, and according to (7.13), $\hat{B}_1 = \hat{b}_1$. The other collective operators given by (7.14) do not even come into play in the $N = 1$ case. So let us consider the next case, $N = 2$.

For $N = 2$, (7.13) and (7.14) yield

$$\hat{B}_1 = \frac{g_1 \hat{b}_2 - g_2 \hat{b}_1}{\sqrt{g_1^2 + g_2^2}} \quad \text{and} \quad \hat{B}_2 = \frac{g_1 \hat{b}_1 + g_2 \hat{b}_2}{\sqrt{g_1^2 + g_2^2}}. \tag{7.16}$$

Then

$$\hat{B}_1^\dagger \hat{B}_1 = \frac{1}{g_1^2 + g_2^2} \left(g_1^2 \hat{b}_2^\dagger \hat{b}_2 + g_2^2 \hat{b}_1^\dagger \hat{b}_1 - g_1 g_2 \hat{b}_2^\dagger \hat{b}_1 - g_1 g_2 \hat{b}_1^\dagger \hat{b}_2 \right), \tag{7.17}$$

$$\hat{B}_2^\dagger \hat{B}_2 = \frac{1}{g_1^2 + g_2^2} \left(g_2^2 \hat{b}_2^\dagger \hat{b}_2 + g_1^2 \hat{b}_1^\dagger \hat{b}_1 + g_1 g_2 \hat{b}_2^\dagger \hat{b}_1 + g_1 g_2 \hat{b}_1^\dagger \hat{b}_2 \right), \tag{7.18}$$

so that

$$\sum_{m=1}^{2} \hat{B}_m^\dagger \hat{B}_m = \sum_{j=1}^{2} \hat{b}_j^\dagger \hat{b}_j. \tag{7.19}$$

Now suppose that our hypothesis holds for $N = n$, that is,

$$\sum_{m=1}^{n} \hat{B}_m^\dagger \hat{B}_m = \sum_{j=1}^{n} \hat{b}_j^\dagger \hat{b}_j. \tag{7.20}$$

For $N = n + 1$, we denote the collective atomic operators with a prime to remind us that even though the index of such an operator might coincide with that of one of the collective operators for the $N = n$ case, they do not have to be defined in terms of the $\hat{b}_j$'s in the same way. We can see from (7.14), however, that the first $n - 1$ collective operators will be defined in the same way for the $N = n + 1$ case as they were for the $N = n$ case. So

$$\sum_{k=1}^{n+1} \hat{B}_k'^\dagger \hat{B}_k' = \sum_{k=1}^{n-1} \hat{B}_k^\dagger \hat{B}_k + \hat{B}_n'^\dagger \hat{B}_n' + \hat{B}_{n+1}'^\dagger \hat{B}_{n+1}'. \tag{7.21}$$

From (7.14) we also see that we can write $\hat{B}_n'$ as

$$\hat{B}_n' = \frac{1}{\sqrt{G_n^2 + g_{n+1}^2}} \left(G_n \hat{b}_{n+1} - g_{n+1} \hat{B}_n \right), \tag{7.22}$$

and from (7.13), we see that we can write $\hat{B}_{n+1}'$ as

$$\hat{B}_{n+1}' = \frac{1}{\sqrt{G_n^2 + g_{n+1}^2}} \left(G_n \hat{B}_n + g_{n+1} \hat{b}_{n+1} \right). \tag{7.23}$$

Notice the similarity between these last two expressions and those for the $N = 2$ case. As in that case, we obtain

$$\hat{B}_n'^\dagger \hat{B}_n' = \frac{1}{G_n^2 + g_{n+1}^2} \left(G_n^2 \hat{b}_{n+1}^\dagger \hat{b}_{n+1} - G_n g_{n+1} \hat{b}_{n+1}^\dagger \hat{B}_n \right.$$
$$\left. + g_{n+1}^2 \hat{B}_n^\dagger \hat{B}_n - g_{n+1} G_n \hat{B}_n^\dagger \hat{b}_{n+1} \right), \tag{7.24}$$

$$\hat{B}_{n+1}'^\dagger \hat{B}_{n+1} = \frac{1}{G_n^2 + g_{n+1}^2} \left(g_{n+1}^2 \hat{b}_{n+1}^\dagger \hat{b}_{n+1} + G_n g_{n+1} \hat{b}_{n+1}^\dagger \hat{B}_n \right.$$
$$\left. + G_n^2 \hat{B}_n^\dagger \hat{B}_n + g_{n+1} G_n \hat{B}_n^\dagger \hat{b}_{n+1} \right). \tag{7.25}$$

Adding (7.24) and (7.25), we get

$$\hat{B}_n'^\dagger \hat{B}_n' + \hat{B}_{n+1}'^\dagger \hat{B}_{n+1} = \hat{B}_n^\dagger \hat{B}_n + \hat{b}_{n+1}^\dagger \hat{b}_{n+1}. \tag{7.26}$$

Substituting (7.26) in (7.21), we get

$$\sum_{k=1}^{n+1} \hat{B}_k'^\dagger \hat{B}_k' = \sum_{k=1}^{n} \hat{B}_k^\dagger \hat{B}_k + \hat{b}_{n+1}^\dagger \hat{b}_{n+1}. \tag{7.27}$$

Using (7.20) in (7.27), we find that

$$\sum_{k=1}^{n+1} \hat{B}_k'^{\dagger} \hat{B}_k' = \sum_{j=1}^{n+1} \hat{b}_j^{\dagger} \hat{b}_j, \tag{7.28}$$

which completes our proof by induction.

As promised, the Hamiltonian (7.8) becomes much simpler when written in terms of this special set of collective atomic operators. It becomes the sum of two uncoupled Hamiltonians,

$$\hat{H} = \hat{H}_b + \hat{H}_m, \tag{7.29}$$

where

$$\hat{H}_m = \hbar\omega \, \hat{a}^{\dagger}\hat{a} + \hbar\omega_0 \hat{B}_1^{\dagger}\hat{B}_1 + \hbar G_N(\hat{a}+\hat{a}^{\dagger})(\hat{B}_1+\hat{B}_1^{\dagger}) + \frac{\hbar\omega_a}{2}\hat{\sigma}_z + \hbar\Omega(\hat{a}+\hat{a}^{\dagger})(\hat{\sigma}+\hat{\sigma}^{\dagger}) \tag{7.30}$$

is the relevant Hamiltonian for us here and

$$\hat{H}_b = \hbar\omega_0 \sum_{k=2}^{N} \hat{B}_k^{\dagger}\hat{B}_k \tag{7.31}$$

is the part that does not matter.

The Hamiltonian (7.31) is irrelevant for our problem because it just describes the free evolution of the atomic operators $\hat{B}_k$, $\hat{B}_k^{\dagger}$ with $k = 1, \ldots, N-1$, which is simply given by

$$\hat{B}_k(t) = \hat{B}_k(0) \exp(-i\omega_0 t), \tag{7.32}$$

and goes on forever undisturbed by the rest of the system described by $\hat{H}_m$. These $N - 1$ pairs of operators represent collective excitations that cannot be excited by the single field mode $\hat{a}$. The collective excitation associated with $\hat{B}_N$ is the only one that can be excited by the field mode, and the strength of this interaction is dependent on the effective coupling constant G_N. The physical meaning of $\hat{B}_N$ will become clear in Section 7.3.

As $\hat{H}_m$ is all that matters for us here, we will simplify the notation, dropping the index N from G_N and $\hat{B}_N$. So from now on, we will be concerned with the following Hamiltonian:

$$\hat{H} = \hat{H}_0 + \hat{H}_A, \tag{7.33}$$

where

$$\hat{H}_0 = \hbar\omega \, \hat{a}^{\dagger}\hat{a} + \hbar\omega_0 \hat{B}^{\dagger}\hat{B} + \hbar G(\hat{a} + \hat{a}^{\dagger})(\hat{B} + \hat{B}^{\dagger}), \tag{7.34}$$

$$\hat{H}_A = \frac{\hbar\omega_a}{2}\hat{\sigma}_z + \hbar\Omega(\hat{a} + \hat{a}^{\dagger})(\hat{\sigma} + \hat{\sigma}^{\dagger}), \tag{7.35}$$

$$\hat{B} = \frac{1}{G}\sum_{j=1}^{N} g_j \hat{b}_j \quad \text{and} \quad G = \sqrt{\sum_{j'=1}^{N} g_{j'}^2}. \tag{7.36}$$

7.2 HOPFIELD'S POLARITON–PHOTON DRESSED EXCITATIONS

It was Hopfield who first studied a quantum-mechanical model of a dielectric medium
[297, 298]. He named the atomic medium quanta (i.e., that of the bare operators
$\hbar\omega_0 \hat{b}_j^\dagger \hat{b}_j$) *polaritons* because they are energy quanta of the polarization field, as photons are energy quanta of the light field. Here we will call polaritons the quanta of
the collective operator $\hbar\omega_0 \hat{B}^\dagger \hat{B}$, for reasons that will become apparent in Section
7.3. Hopfield also worked out the normal modes (i.e., the polariton–photon dressed
operators). Let us do this now.

We define two pairs of dressed annihilation and creation operators given by

$$\hat{c}_k = x_1^k \hat{a} + y_1^k \hat{a}^\dagger + x_2^k \hat{B} + y_2^k \hat{B}^\dagger, \tag{7.37}$$

with $k = 1, 2$, satisfying the usual commutation relations,

$$[\hat{c}_k, \hat{c}_{k'}^\dagger] = \delta_{kk'}, \tag{7.38}$$

$$[\hat{c}_k, \hat{c}_{k'}] = 0, \tag{7.39}$$

and diagonalizing $\hat{H}_0$,

$$\left[\hat{c}_k, \hat{H}_0\right] = \hbar\Omega_k \hat{c}_k. \tag{7.40}$$

From (7.3), (7.15), (7.37), (7.38), and (7.40), we find that[4]

$$x_j^k = \frac{1}{2}\left(v_j^k + u_j^k\right), \tag{7.41}$$

$$y_j^k = \frac{1}{2}\left(v_j^k - u_j^k\right), \tag{7.42}$$

where

$$u_1^k = \frac{\omega}{\Omega_k} v_1^k, \tag{7.43}$$

$$u_2^k = \frac{\Omega_k^2 - \omega^2}{2G\Omega_k} v_1^k, \tag{7.44}$$

$$v_2^k = \frac{\Omega_k^2 - \omega^2}{2G\omega_0} v_1^k, \tag{7.45}$$

$$v_1^k = \sqrt{\frac{4G^2\Omega_k\omega_0}{(\Omega_k^2 - \omega^2)^2 + 4\omega_0\omega G^2}}, \tag{7.46}$$

and

$$\Omega_k^2 = \frac{1}{2}\left\{\omega_0^2 + \omega^2 + (-1)^k\Lambda\right\}, \tag{7.47}$$

---

[4]Notice that in order for Ω_1 not to become imaginary, we must have $4G'^2 \leq \omega_0^2$, where Ω' is given by
(7.49) (see [166]).

$$\Lambda = (\omega_0^2 - \omega^2)\sqrt{1 + \frac{16\omega_0\omega G^2}{(\omega_0^2 - \omega^2)^2}}. \tag{7.48}$$

As in Hopfield's treatment [298], (7.47) shows that there are no dressed modes in the frequency interval between $\sqrt{\omega_0^2 - 4G'^2}$ and ω_0, where

$$G' = \sqrt{\frac{\omega_0}{\omega}}G \tag{7.49}$$

is the effective oscillator strength[5] that we hold constant. But unlike Hopfield, we have chosen the sign of Λ in (7.47) and (7.48) accordingly for Ω_1 to approach the field frequency ω, and Ω_2 approaches the atomic frequency ω_0, as $G' \to 0$. This way of defining the dressed modes makes Ω_1 and Ω_2 discontinuous functions of ω at $\omega = \omega_0$ (see Figure 7.2), but it will prove useful when we discuss the role of the dielectric permittivity in this full quantum theory.

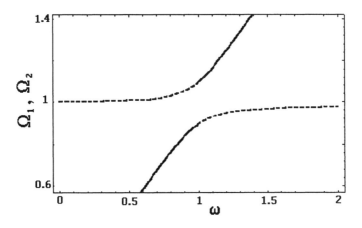

Figure 7.2 This is a plot of the frequencies of the polariton modes in units of ω_0 as a function of ω/ω_0 for $G' = 0.1\omega_0$. The thick line shows Ω_1 and the dotted line shows Ω_2. Our definition of Ω_1 and Ω_2 makes them discontinuous at $\omega = \omega_0$. In the usual definition, Ω_1 is the upper continuous curve and ω_2 the lower continuous curve. (Reproduced from [166], ©1997 The American Physical Society.)

We can invert (7.37) in the following simple way. Let

$$\hat{a} \equiv u_1^1\hat{c}_1 + v_1^1\hat{c}_1^\dagger + u_2^1\hat{c}_2 + v_2^1\hat{c}_2^\dagger \quad \text{and} \quad \hat{B} \equiv u_1^2\hat{c}_1 + v_1^2\hat{c}_1^\dagger + u_2^2\hat{c}_2 + v_2^2\hat{c}_2^\dagger. \tag{7.50}$$

Taking the commutator of these equations with $\hat{c}_k$ and $\hat{c}_k^\dagger$, we find that

$$u_k^1 = [\hat{a}, \hat{c}_k^\dagger], \quad v_k^1 = [\hat{c}_k, \hat{a}], \quad u_k^2 = [\hat{B}, \hat{c}_k^\dagger], \quad \text{and} \quad v_k^2 = [\hat{c}_k, \hat{B}]. \tag{7.51}$$

---

[5]In units of ω_0.

Substituting (7.37) in (7.51), we get

$$\hat{a} = x_1^{1*}\hat{c}_1 - y_1^1\hat{c}_1^\dagger + x_1^{2*}\hat{c}_2 - y_1^2\hat{c}_2^\dagger \quad \text{and} \quad \hat{B} = x_2^{1*}\hat{c}_1 - y_2^1\hat{c}_1^\dagger + x_2^{2*}\hat{c}_2 - y_2^2\hat{c}_2^\dagger. \tag{7.52}$$

Using (7.52) in (7.35), we can reduce the original problem to the case of a single atom coupled to two polariton modes,

$$\hat{H} = \hbar\Omega_1\hat{c}_1^\dagger\hat{c}_1 + \hbar\Omega_2\hat{c}_2^\dagger\hat{c}_2 + \frac{\hbar\omega_a}{2}\hat{\sigma}_z + \hbar\Omega u_1^1(\hat{c}_1 + \hat{c}_1^\dagger)(\hat{\sigma} + \hat{\sigma}^\dagger) + \hbar\Omega u_1^2(\hat{c}_2 + \hat{c}_2^\dagger)(\hat{\sigma} + \hat{\sigma}^\dagger). \tag{7.53}$$

7.3 QUANTUM NOISE GENERATED BY MATTER AND MACROSCOPIC AVERAGES: IS A DIELECTRIC PERMITTIVITY ENOUGH?

Is it possible to describe the effect of the atoms of material media on the field only in terms of a dielectric constant? When atoms and field are described quantum mechanically, the atoms can contribute to the quantum noise in the field. As a dielectric permittivity function cannot account for this atomic contribution to the quantum noise in the field, this could undermine any attempt to describe a material medium only by a permittivity function. Let us calculate the noise coming from the medium.

Consider the simpler case where there is no quest atom and the total system is in the ground state of the dressed operators $\hat{c}_1$ and $\hat{c}_2$ (i.e., in the state $|g\rangle$ such that $\hat{c}_1|g\rangle = 0$ and $\hat{c}_2|g\rangle = 0$). In this state, the variance of the quadrature associated with the electric displacement field operator (7.1) is given by

$$\langle[\Delta(\hat{a} + \hat{a}^\dagger)]^2\rangle \equiv \langle(\hat{a} + \hat{a}^\dagger)^2\rangle - \langle\hat{a} + \hat{a}^\dagger\rangle^2 = \sum_{k=1}^{2} \frac{\omega}{\Omega_k} \frac{\Omega_k^2 - \omega_0^2}{2\Omega_k^2 - \omega_0^2 - \omega^2}. \tag{7.54}$$

This expression contains the effect of dispersion. Let us look at the even simpler case where dispersion is negligible (i.e., when $\omega \ll \omega_0$). Moreover, if there is a situation where we would expect to be able to neglect the noise from the medium, this situation is when the atoms of the medium are only weakly coupled to the field (i.e., $G', \omega \ll \omega_0$). Then up to second order in ω/ω_0 and G'/ω_0, (7.54) reduces to

$$\langle[\Delta(\hat{a} + \hat{a}^\dagger)]^2\rangle \approx 1 + 2\frac{G'^2}{\omega_0^2}. \tag{7.55}$$

Introducing the frequency-independent dielectric permittivity ε_r in terms of the change in the cavity resonance frequency due to the presence of the medium

$$\Omega_1 = \frac{\omega}{\sqrt{\varepsilon_r}}, \tag{7.56}$$

we find that it is given by

$$\varepsilon_r \approx 1 + 4\frac{G'^2}{\omega_0^2}. \tag{7.57}$$

So (7.55) can be rewritten as

$$\langle [\Delta(\hat{a} + \hat{a}^\dagger)]^2 \rangle \approx \sqrt{\varepsilon_r}. \tag{7.58}$$

This is in agreement with what Glauber and Lewenstein obtained by directly quantizing the macroscopic Maxwell equations for the case of a nondispersive and nonabsorptive dielectric medium [237].

So there is no extra noise in the *displacement field* coming from the atoms of the medium in this case. Rosewarne, however, claimed that there is such noise in the *electric field* [528]. He calculated the variance of the electric field in a scalar version of the Hopfield model and found that even in the case of weak coupling ($G', \omega \ll \omega_0$), there is a noise contribution from the atoms on the order of $G'^2/\omega_0\omega$. So it is even *larger* than the $4G'^2/\omega_0^2$ contribution of the medium to the dielectric permittivity. This extra noise is *not* accounted for when the medium is described only in terms of a dielectric permittivity. Then how can we use the idea of dielectric permittivity in quantum electrodynamics without being inconsistent? Let us examine the electric field in more detail.

In terms of the displacement field, the electric field is given by

$$E = D - P. \tag{7.59}$$

As D has no noise contribution from the medium up to second order, the contribution Rosewarne found can come only from the polarization P. But the polarization (7.2) vanishes whenever one is not at one of the positions occupied by an atom of the medium. So except for these positions, there should be no noise contribution from the medium just as there is not for D. Rosewarne's E had no such discontinuities, because he assumes a continuous distribution of atoms. This means that the dynamical variables he uses are supposed to be macroscopic averages from the beginning. How can we define a macroscopic average?

Lorentz introduced macroscopic averages to show how Maxwell's equations in material media can be obtained from more fundamental equations for the electromagnetic fields in the vacuum and the constituent charged particles that make up the media. This average was over "physically infinitesimal" volume elements, which were yet large enough to contain many atoms. There are some problems with Lorentz's definition of macroscopic average. Perhaps the most important for us is that in many physical situations there is no physically infinitesimal volume that is both macroscopically small and yet contains a large number of atoms. In optics, for example, the volume element cannot exceed 10^9 Å^3; otherwise, optical oscillations would be averaged to zero. But the addition of a single extra electron to such a volume would cause an enormous change in the charge density: about 160 C/m^3—by no means an infinitesimal change!

Many alternative definitions of macroscopic averaging have been proposed in the literature.[6] A well-established trend is to regard macroscopic averaging as just a

---

[6]See the recommended reading for some of the main ones.

sort of ensemble averaging. Robinson [524, 525] points out, however, that the aim of doing macroscopic averages is to discard unwanted microscopic detail. This is different from the *ensemble averages* of statistical physics, for there we do an average only because our knowledge of the macroscopic state of the system is incomplete. Robinson showed that the best way to get rid of the unwanted microscopic detail from a given dynamical variable is to eliminate high-frequency components from its spatial Fourier expansion. We will adopt his idea here, as it seems to have "hit the nail on the head" as far as macroscopic averaging is concerned.

As the cavity has a finite length, we can expand the delta functions in (7.2) into a Fourier series. Moreover, as the atoms of the medium can be everywhere inside the cavity except at the positions of the cavity mirrors, this Fourier series will be a sine series; that is,

$$\delta(x - x_j) = \frac{2}{L} \sum_{n=1}^{\infty} \sin\left(\frac{n\pi}{L}x_j\right) \sin\left(\frac{n\pi}{L}x\right). \tag{7.60}$$

In our single-mode model, the high spatial frequencies describing the unwanted microscopic detail are, of course, all those above ω/c. Discarding them from (7.60) and using the resulting expression in (7.2), we find that the macroscopic polarization is given by

$$\bar{P}(x) = \frac{2}{L}\sqrt{\frac{\hbar}{2\omega_0}} \sin\left(\frac{\omega}{c}x\right) \sum_{j=1}^{N} (\hat{b}_j + \hat{b}_j^\dagger) q_j \sin\left(\frac{\omega}{c}x_j\right). \tag{7.61}$$

Using the definitions of g_j, $\hat{B}$, G and G', we can express (7.61) in terms of the collective atomic operators $\hat{B}$ and $\hat{B}^\dagger$:

$$\bar{P}(x) = -\frac{2G'}{\sqrt{\omega_0\omega}}\sqrt{\frac{2}{L}\hbar\omega}(\hat{B}^\dagger + \hat{B}) \sin\left(\frac{\omega}{c}x\right). \tag{7.62}$$

We can now see that $\hat{B}$ and $\hat{B}^\dagger$ are the annihilation and creation operators of macroscopic polarization quanta.

Substituting (7.62) in (7.59), we find the following expression for the macroscopic electric field

$$\bar{E}(x) = \sqrt{\frac{2}{L}\hbar\omega}\left\{\hat{a}^\dagger + \hat{a} + \frac{2G'}{\sqrt{\omega_0\omega}}(\hat{B}^\dagger + \hat{B})\right\} \sin\left(\frac{\omega}{c}x\right). \tag{7.63}$$

The ground-state variance of the term inside the braces is given by

$$\left\langle \left\{\Delta\left[\hat{a}^\dagger + \hat{a} + \frac{2G'}{\sqrt{\omega_0\omega}}(\hat{B}^\dagger + \hat{B})\right]\right\}^2 \right\rangle$$

$$= \frac{4G'^2}{\omega_0\omega}\left\langle \left[\Delta(\hat{B}^\dagger + \hat{B})\right]^2 \right\rangle + \left\langle (\hat{a}^\dagger + \hat{a})^2 \right\rangle.$$

$$+ \frac{2G'}{\sqrt{\omega_0\omega}}\left\langle (\hat{B}^\dagger + \hat{B})(\hat{a}^\dagger + \hat{a}) + (\hat{a}^\dagger + \hat{a})(\hat{B}^\dagger + \hat{B}) \right\rangle \tag{7.64}$$

To calculate the expectation values in (7.64), we write $\hat{a}$, $\hat{a}^\dagger$, $\hat{B}$, and $\hat{B}^\dagger$ in terms of $\hat{c}_1$, $\hat{c}_1^\dagger$, $\hat{c}_2$, and $\hat{c}_2^\dagger$ using (7.52). Then

$$\left\langle (\hat{B}^\dagger + \hat{B})(\hat{a}^\dagger + \hat{a}) + (\hat{a}^\dagger + \hat{a})(\hat{B}^\dagger + \hat{B}) \right\rangle = \sum_{k=1}^{2} \frac{4\omega^2 G'^2 (\Omega_k^2 - \omega^2)\sqrt{\omega_0/\omega}}{[(\Omega_k^2 - \omega^2)^2 + 4\omega^2 G'^2]\,\Omega_k}$$
(7.65)

and

$$\left\langle \left[\Delta(\hat{B}^\dagger + \hat{B}) \right]^2 \right\rangle = \sum_{k=1}^{2} \frac{\omega_0}{\Omega_k} \frac{(\Omega_k^2 - \omega^2)^2}{(\Omega_k^2 - \omega^2)^2 + 4\omega^2 G'^2},$$
(7.66)

so that up to second order in G'/ω_0, we obtain

$$\left\langle \left\{ \Delta \left[\hat{a}^\dagger + \hat{a} + \frac{2G'}{\sqrt{\omega_0\omega}}(\hat{B}^\dagger + \hat{B}) \right] \right\}^2 \right\rangle \approx 1 - \underbrace{6\frac{G'^2}{\omega_0^2}}_{\varepsilon_r^{-3/2}} + 4\frac{G'^2}{\omega_0\omega}.$$
(7.67)

The $\varepsilon_r^{-3/2}$ contribution to the variance is the only one yielded by a theory where the macroscopic Maxwell equations for a nondispersive and nonabsorptive dielectric medium are quantized directly [237]. The last term on the right-hand side of (7.67) is the noise from the medium derived by Rosewarne [528], which cannot exist in a theory where the medium is only described by a dielectric permittivity. Moreover, this noise is larger than the change $4G'^2/\omega_0$ caused by the medium in the unit permittivity of the vacuum.

Does this result mean that all macroscopic versions of QED based only on a dielectric–permittivity description of media are doomed to fail? No. In the next section we show that the extra noise coming from the medium is in a very different frequency than that of the original field in the absence of a medium. So as long as the physical systems affected by the macroscopic field fluctuations are reasonably selective in frequency, they will not "feel" the noise of the medium if they are detuned from it.

7.4 HOW A MACROSCOPIC DESCRIPTION IS POSSIBLE

In our simple model, the physical system that probes the macroscopic fields in our medium is the guest two-level atom. The noise from the medium calculated in Section 7.3 comes from the variance of the dressed operators $\hat{c}_2$ and $\hat{c}_2^\dagger$. These are the dressed operators associated with the dressed quanta that are more polaritonlike than photonlike. The resonance frequency Ω_2 of these dressed excitations is never very far from the resonance frequency of the medium. If guest atom's resonance frequency ω_a is very detuned from Ω_2, the guest atom will not feel these dressed excitations. To find out how large the detuning must be for this to be so, we must require that the probability of these dressed quanta inducing transitions in the guest atom be much less than 1. Then we find that the detuning must satisfy the condition

$$|\Omega_2 - \omega_a| \gg |u_2^2 \Omega|.$$
(7.68)

Condition (7.68) is similar to that for neglecting the counterrotating terms in the Jaynes–Cummings model [323, 559], and when it is not met, the nonresonant dressed excitations will induce a Bloch–Siegert shift in the guest atom [60]. When the coupling between the atoms of the medium and the field is weak, (7.68) becomes simply $| \omega_0 - \omega | \gg | \Omega |$; in other words, the field frequency ω must be detuned from the resonance of the medium ω_0 by an amount much larger than the Rabi frequency of the guest atom.

Assuming that (7.68) is satisfied, we can drop the second dressed mode and write the following macroscopic Hamiltonian

$$\hat{H}_{\mathrm{mac}} = \hbar\Omega_1 \hat{c}_1^\dagger \hat{c}_1 + \frac{\hbar\omega_a}{2}\hat{\sigma}_z + \hbar\Omega u_1^1(\hat{c}_1 + \hat{c}_1^\dagger)(\hat{\sigma} + \hat{\sigma}^\dagger). \qquad (7.69)$$

Using (7.7), (7.43), and (7.46), we can rewrite this Hamiltonian in the more suggestive form

$$\hat{H}_{\mathrm{mac}} = \hbar\Omega_1 \hat{c}_1^\dagger \hat{c}_1 + \frac{\hbar\omega_a}{2}\hat{\sigma}_z - \frac{d}{\varepsilon_0} D_{\mathrm{mac}}(x_a)(\hat{\sigma} + \hat{\sigma}^\dagger), \qquad (7.70)$$

where x_a is the position of the guest atom, d is its electric dipole strength, and

$$D_{\mathrm{mac}}(x) = \left[\frac{2\hbar\Omega_1 \varepsilon_r \sqrt{\varepsilon_r}}{L\gamma} \right]^{1/2} (\hat{c}_1 + \hat{c}_1^\dagger) \sin\left(\sqrt{\varepsilon_r}\frac{\Omega_1}{c}x \right) \qquad (7.71)$$

is the macroscopic displacement field. The frequency-dependent permittivity ε_r in (7.71) obtained from (7.47) is given by

$$\varepsilon_r = 1 + \frac{4G'^2}{\omega_0'^2 - \Omega_1^2}. \qquad (7.72)$$

This expression justifies the physical interpretation of G'^2 as the effective oscillator strength of the medium. The correction on the resonance frequency of the medium due to the interaction with the field [619] is given by

$$\omega_0'^2 = \omega_0^2 - 4G'^2. \qquad (7.73)$$

In (7.71),

$$\gamma = \left(\frac{1}{v_1^1} \right)^2 \qquad (7.74)$$

and using (7.46) and (7.72), we see that it is given by

$$\gamma = \frac{d}{d\Omega_1}(\Omega_1 \sqrt{\varepsilon_r}) \qquad (7.75)$$

(i.e., it is the ratio between the speed of light in vacuum and the group velocity in the medium).

The macroscopic polarization experienced by our guest atom is (7.62) without its $\hat{c}_2$ and $\hat{c}_2^\dagger$ components. Using this polarization and (7.71) in (7.59), we find the following expression for the macroscopic electric field:

$$E_{\mathrm{mac}}(x) = \left(\frac{2\hbar\Omega_1}{\sqrt{\varepsilon_r}L\gamma} \right)^{1/2} (\hat{c}_1 + \hat{c}_1^\dagger) \sin\left(\sqrt{\varepsilon_r}\frac{\Omega_1}{c}x \right). \qquad (7.76)$$

Milonni quantized the macroscopic Maxwell equations for a dispersive (but non-absorptive) medium [452]. Even though his approach is completely different from the first-principles approach adopted here, the expressions he derived for the fields are equivalent to ours. He starts from the phenomenological macroscopic-Maxwell equations and shows that for a narrow range of frequencies where absorption is negligible, it is possible to define a Hamiltonian local in time which allows the field to be quantized in the usual way. In the remainder of this chapter, we show how medium absorption can be included in our theory.

7.5 IF THERE IS DISPERSION, THERE MUST ALSO BE ABSORPTION SOMEWHERE: THE KRAMERS–KRONIG DISPERSION RELATION

In 1926, Ralph de Laer Kronig and Hendrik Anton Kramers discovered independently the first dispersion relation in physics. The Kramers–Kronig dispersion relation, as it is now known in physics, says that if the dielectric permittivity ε depends on the frequency ω, it must not only have a real part $\varepsilon_R(\omega)$ but also an imaginary part $\varepsilon_I(\omega)$, and the two must be related by[7]

$$\epsilon_R(\omega) = 1 + \frac{1}{\pi}\mathcal{P}\int_{-\infty}^{\infty} d\omega' \, \frac{\omega'\varepsilon_I(\omega')}{\omega'^2 - \omega^2}. \tag{7.77}$$

This relation is a consequence of causality (see, e.g., Chapter 1 of [468]). A striking counterexample that n and β cannot be arbitrary was given by Toll [600]. He showed that if a medium could act as a perfect filter for some frequency, causality would be violated: An incoming well-localized wave packet would generate an output field before the arrival of the incident packet!

Our dielectric permittivity (7.72) does not obey the Kramers–Kronig relation because it is not the dielectric permittivity for all frequencies. It is only valid for frequencies that are quite detuned from the resonance of the medium. In most actual dielectric media, if the frequency of the light approaches a medium resonance, the energy is transferred to phonons and other nonradiative forms and does not get back to the field. It is absorbed. To account for such absorption, we will change our simple model to allow the atoms of the medium to couple to a reservoir with infinite degrees of freedom representing these nonradiative mechanisms.

7.6 INCLUDING ABSORPTION IN THE DIELECTRIC JCM

Our dielectric JCM is a simplified version of Hopfield's model of dielectrics. The first to modify the Hopfield model to account for absorption were Huttner and Barnett

---

[7]There is another Kramers–Kronig dispersion relation that we do not mention explicitly here, where ε_I is given as an integral of ε_R over frequency.

[33, 306, 307]. Following their idea, we will now introduce in the dielectric JCM a coupling with a reservoir to model absorption [167].

Let the oscillators of the medium be coupled to a reservoir consisting of a continuum of harmonic oscillators. We will assume that this coupling strength $V(\omega)$ has a maximum modulus that is much smaller than ω_0. This allows us to make the rotating-wave approximation in the interaction term between the oscillators of the medium and those of the reservoir.

We will also make a white-noise assumption that amounts to neglecting the frequency dependency of $V(\omega)$ within the range over which it broadens the linewidth of the medium. This frequency range is given by $\pi|V(\omega_0)|^2$. So we are assuming that $V(\omega)$ is approximately flat over a frequency interval of length $\pi|V(\omega_0)|^2$ centered at ω_0. That is nothing more than the regime where Fermi's golden rule holds. In this regime, the final result is independent of the particular form of $V(\omega)$. Thus, for simplicity, we will adopt a Lorentzian shape for $|V(\omega)|^2$, with $V(\omega)$ given by

$$V(\omega) = \sqrt{\frac{\kappa}{\pi}}\frac{i\Delta}{\omega - \omega_0 + i\Delta}, \qquad (7.78)$$

where $\omega_0 \gg \Delta \gg \kappa$. Figure 7.3 shows a plot of $|V(\omega)|^2$ in the golden rule regime we are considering.

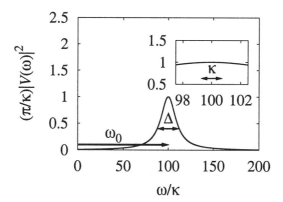

Figure 7.3 We are considering the regime where Fermi's golden rule holds. In this regime the actual shape of $V(\omega)$ is irrelevant as long as it satisfies the broad requirements mentioned in the text. For simplicity, we take $V(\omega)$ given by (7.78) so that $|V(\omega)|^2$ is a Lorentzian. Then the golden rule regime is $\omega_0 \gg \Delta \gg \kappa$, exemplified in this plot. All frequencies in the plot are in units of κ. We used $\omega_0 = 100\kappa$, $\Delta = 10\kappa$. The inset shows an enlargement of the central part of the Lorentzian peak, where you can clearly see that $V(\omega)$ is reasonably flat over the medium linewidth $\pi|V(\omega_0)|^2$.

The new Hamiltonian is given by

$$\hat{H} = \hbar\omega_c \hat{a}^\dagger \hat{a} + \hbar\omega_0 \sum_{j=1}^{N} \hat{b}_j^\dagger \hat{b}_j - \int_0^L dx \, \hat{D}(x)\hat{P}(x) + \int_0^\infty d\eta \sum_{k=1}^{N} \hat{W}_k^\dagger(\eta)\hat{W}_k(\eta)\,\eta$$

$$+ \hbar \int_0^\infty d\eta \sum_{k=1}^{N} \left\{ V(\eta)\hat{b}_k^\dagger \hat{W}_k(\eta) + V^*(\eta)\hat{W}_k^\dagger(\eta)\hat{b}_k \right\}, \tag{7.79}$$

where $\hat{W}_k^\dagger(\eta)$ and $\hat{W}_k(\eta)$ are the reservoir creation and annihilation operators. They obey the usual commutation rules for the continuum

$$[\hat{W}_k(\eta), \hat{W}_{k'}^\dagger(\eta')] = \delta_{kk'}\delta(\eta - \eta') \quad \text{and} \quad [\hat{W}_k(\eta), \hat{W}_{k'}(\eta')] = 0, \tag{7.80}$$

and commute with $\hat{a}^\dagger$, $\hat{a}$, $\hat{b}_j^\dagger$, and $\hat{b}_j$. It is important to stress that there is another assumption implied in (7.79). This is the assumption that each atom has its own reservoir[8] and that the differences in their coupling strengths is negligible. We will see more about that in chapter 9 in the context of cavity damping.

Substituting (7.1) and (7.2) in (7.79), we find that

$$\hat{H} = \hbar\omega_c \hat{a}^\dagger \hat{a} + \hbar\omega_0 \sum_{j=1}^{N} \hat{b}_j^\dagger \hat{b}_j + \hbar \sum_{j=1}^{N} g_j \left(\hat{a}^\dagger + \hat{a}\right)\left(\hat{b}_j^\dagger + \hat{b}_j\right)$$

$$+ \hbar \int_0^\infty d\eta \sum_{k=1}^{N} \hat{W}_k^\dagger(\eta)\hat{W}_k(\eta)\eta$$

$$+ \hbar \int_0^\infty d\eta \sum_{k=1}^{N} \left\{ V(\eta)\hat{b}_k^\dagger \hat{W}_k(\eta) + V^*(\eta)\hat{W}_k^\dagger(\eta)\hat{b}_k \right\}, \tag{7.81}$$

where

$$g_j = -q_j \sqrt{\frac{\omega_c}{2L\omega_0}} \sin\left(\frac{\omega_c}{c}x_j\right). \tag{7.82}$$

As we have done in Section 7.1, we introduce the collective operators

$$\hat{B}_k = \sum_{j=1}^{N} \phi_{kj}\hat{b}_j, \tag{7.83}$$

where

$$\phi_{Nj} = \frac{g_j}{G_N}, \tag{7.84}$$

$$\phi_{kj} = \delta_{k+1,j}\frac{G_{k-1}}{G_k} - (1 - \delta_{k+1,j})\frac{g_k g_j}{G_{k-1}G_k} \quad \text{for } j \leq k+1 \text{ and } k < N, \tag{7.85}$$

---

[8]Phonons or any other incoherent process in the dielectric medium play the role of reservoir.

$$\phi_{kj} = 0 \qquad \text{for } j > k+1 \text{ and } k \neq N, \tag{7.86}$$

$$G_n = \sqrt{\sum_{j'=1}^{n} g_{j'}^2}. \tag{7.87}$$

Like the $\hat{b}_j$, they are annihilation operators obeying the commutation relations

$$[\hat{B}_j, \hat{B}_{j'}^\dagger] = \delta_{jj'} \quad \text{and} \quad [\hat{B}_j, \hat{B}_{j'}] = 0. \tag{7.88}$$

In terms of them, the "lossless" part of the Hamiltonian [i.e., the first three terms on the right-hand side of (7.81)] becomes

$$\hbar\omega_c \hat{a}^\dagger \hat{a} + \hbar\omega_0 \sum_{j=1}^{N} \hat{B}_j^\dagger \hat{B}_j + \hbar G_N \left(\hat{a}^\dagger + \hat{a} \right) \left(\hat{B}_N^\dagger + \hat{B}_N \right). \tag{7.89}$$

The part of the Hamiltonian involving the reservoirs and the coupling with each atom of the medium [i.e., the last two terms on the right-hand side of (7.81)] can be rewritten in terms of the collective operators substituting the $\hat{b}_k$ by

$$\hat{b}_k = \sum_{j=1}^{N} \phi_{jk}^* \hat{B}_j. \tag{7.90}$$

Then

$$\sum_{k=1}^{N} \hat{b}_k^\dagger \hat{W}_k(\eta) = \sum_{k=1}^{N} \sum_{j=1}^{N} \phi_{jk} \hat{B}_j^\dagger \hat{W}_k(\eta) = \sum_{j=1}^{N} \hat{B}_j^\dagger \hat{Y}_j(\eta), \tag{7.91}$$

where

$$\hat{Y}_j(\eta) = \sum_{k=1}^{N} \phi_{jk} \hat{W}_k(\eta). \tag{7.92}$$

As you can easily verify, the $\hat{Y}_j(\eta)$ are annihilation operators; that is,

$$[\hat{Y}_j(\eta), \hat{Y}_{j'}^\dagger(\eta')] = \delta_{jj'}\delta(\eta - \eta') \quad \text{and} \quad [\hat{Y}_j(\eta), \hat{Y}_{j'}(\eta')] = 0. \tag{7.93}$$

Moreover, just as we have shown in Section 7.1 for the $\hat{B}_j$ and the $\hat{b}_k$,

$$\sum_{k=1}^{N} \hat{W}_k^\dagger(\eta) \hat{W}_k(\eta) = \sum_{k=1}^{N} \hat{Y}_k^\dagger(\eta) \hat{Y}_k(\eta). \tag{7.94}$$

So each collective operator $\hat{B}_k$ has its own reservoir associated with the collective annihilation and creation operators $\hat{Y}_k(\eta)$ and $\hat{Y}_k^\dagger(\eta)$. Therefore, the total Hamiltonian is given in terms of these collective operators by

$$\hat{H} = \hbar\omega_c \hat{a}^\dagger \hat{a} + \hbar\omega_0 \sum_{j=1}^{N} \hat{B}_j^\dagger \hat{B}_j + \hbar G_N \left(\hat{a}^\dagger + \hat{a} \right) \left(\hat{B}_N^\dagger + \hat{B}_N \right)$$

$$+ \hbar \int_0^\infty d\eta \sum_{k=1}^N \hat{Y}_k^\dagger(\eta) \hat{Y}_k(\eta) \eta$$

$$+ \hbar \int_0^\infty d\eta \sum_{k=1}^N \left\{ V(\eta) \hat{B}_k^\dagger \hat{Y}_k(\eta) + V^*(\eta) \hat{Y}_k^\dagger(\eta) \hat{B}_k \right\}.$$

(7.95)

As in Section 7.1, we can separate (7.95) into one single damped oscillator, represented by the Nth collective atomic mode, that alone couples to the cavity mode, and $N - 1$ uncoupled damped oscillators. The latter are irrelevant here, for they do not affect the field dynamics. Just as we have done earlier in this chapter, we drop the subscript N from $\hat{B}_N$, $\hat{Y}_N(\eta)$, and G_N to simplify the notation. So we will be concerned with the following Hamiltonian:

$$\hat{H} = \hat{H}_{\text{rad}} + \hat{H}_{\text{mat}} + \hat{H}_{\text{int}},$$

(7.96)

where

$$\hat{H}_{\text{rad}} = \hbar \omega_c \hat{a}^\dagger \hat{a},$$

(7.97)

$$\hat{H}_{\text{mat}} = \hbar \omega_0 \hat{B}^\dagger \hat{B} + \hbar \int_0^\infty d\eta \, \hat{Y}^\dagger(\eta) \hat{Y}(\eta) \, \eta$$

$$+ \hbar \int_0^\infty d\eta \left\{ V(\eta) \hat{B}^\dagger \hat{Y}(\eta) + V^*(\eta) \hat{Y}^\dagger(\eta) \hat{B} \right\},$$

(7.98)

$$\hat{H}_{\text{int}} = \hbar G \left(\hat{a}^\dagger + \hat{a} \right) \left(\hat{B}^\dagger + \hat{B} \right).$$

(7.99)

Unlike Huttner and Barnett [306, 307], we have a strictly microscopic model for the medium, where each constituent atom sits in the vacuum a given distance away from the others. The medium is discrete rather than continuous. From a practical point of view, however, there is no need to account for all this microscopic detail. So we take the macroscopic average of the physical variables in our theory. As we have done in Section 7.4, we expand the Dirac delta functions in (7.2) into Fourier sine series inside the cavity. Robinson's irrelevant spatial Fourier components are here all those with a spatial frequency higher than ω_c/c.

7.7 DIELECTRIC PERMITTIVITY

Dielectric permittivity is a classical concept. Thus, the permittivity in our quantum theory should be the same that we obtain in classical electrodynamics. In this section we temporarily disregard the quantum nature of the radiation field to calculate the dielectric permittivity in our model. Later in this chapter this will be compared with the dielectric permittivity that appears in the Huttner–Barnett theory.

We will denote the classical counterparts of $\hat{a}$, $\hat{B}$, and $\hat{Y}(\eta)$ by $\alpha/\sqrt{\hbar}$, $\beta/\sqrt{\hbar}$, and $\mathcal{Y}(\eta)/\sqrt{\hbar}$, respectively. Making these replacements in (7.96) and using Poisson brackets rather than commutators (see Chapter 2), we obtain the following equations of motion:

$$\frac{d}{dt}\alpha = -i\omega_c\alpha - iG\left(\beta + \beta^*\right), \tag{7.100}$$

$$\frac{d}{dt}\beta = -i\omega_0\beta - iG\left(\alpha + \alpha^*\right) - i\int_0^\infty d\eta\, V(\eta)\mathcal{Y}(\eta), \tag{7.101}$$

$$\frac{d}{dt}\mathcal{Y}(\eta) = -i\eta\,\mathcal{Y}(\eta) - iV^*(\eta)\,\beta. \tag{7.102}$$

Assuming that the reservoir is not initially excited, we may take $\mathcal{Y}(\eta)$ as vanishing at $t = 0$. Then integrating (7.102), we find that

$$\mathcal{Y}(\eta, t) = -iV^*(\eta)\int_0^\infty dt'\, e^{i(t'-t)\eta}\beta(t'). \tag{7.103}$$

Substituting this formal result in (7.101), we get

$$\frac{d}{dt}\beta = -i\omega_0\beta - iG\left(\alpha + \alpha^*\right) - \int_0^\infty d\eta\, |V(\eta)|^2 \int_0^t dt'\, e^{i(t'-t)\eta}\beta(t'). \tag{7.104}$$

In the golden rule regime ($\omega_0 \gg \Delta \gg \kappa$) considered here, the integrals on last term on the right-hand side of (7.104) can be done (Problem 7.1) and we find[9] that

$$\frac{d}{dt}\beta = -\left(\kappa + i\omega_0\right)\beta - iG\left(\alpha + \alpha^*\right). \tag{7.105}$$

The electric displacement D, magnetic field B, and polarization P are given by

$$D = \sqrt{\frac{\omega_c}{L}}\mathcal{D}\sin\left(\frac{\omega_c}{c}x\right), \tag{7.106}$$

$$B = -\sqrt{\frac{\omega_c}{L}}\mathcal{B}\cos\left(\frac{\omega_c}{c}x\right), \tag{7.107}$$

$$P = -\frac{2G'}{\sqrt{\omega_0\omega_c}}\sqrt{\frac{\omega_c}{L}}\chi\sin\left(\frac{\omega_c}{c}x\right), \tag{7.108}$$

where

$$\mathcal{D} = \alpha + \alpha^*, \quad \mathcal{B} = \frac{\alpha - \alpha^*}{i}, \quad \text{and } \chi = \beta + \beta^*. \tag{7.109}$$

To write down a closed system of differential equations for the real variables in (7.109) from the equations of motion (7.100), (7.102), and (7.105) for the complex variables α, β, and $\mathcal{Y}(\eta)$, we must enlarge (7.109) by including the imaginary part of β,

$$\varpi \equiv \frac{\beta - \beta^*}{i}. \tag{7.110}$$

---

[9]This is the famous Markovian approximation.

The physical meaning of ϖ is this. If we think the polarization P is produced by an oscillator of effective charge $G'\sqrt{L}$ in a region of size L, ϖ is proportional to the momentum of this oscillator.

From (7.100), (7.102), (7.105), (7.109), and (7.110), we find that

$$\frac{d}{dt}\mathcal{D} = \omega_c \mathcal{B}, \tag{7.111}$$

$$\frac{d}{dt}\mathcal{B} = -\omega_c \mathcal{D} - 2G\chi, \tag{7.112}$$

$$\frac{d}{dt}\chi = -\kappa\chi + \omega_0\varpi, \tag{7.113}$$

$$\frac{d}{dt}\varpi = -\kappa\varpi - \omega_0\chi - 2G\mathcal{D}. \tag{7.114}$$

Differentiating (7.113) and using (7.114) to eliminate $d\varpi/dt$, we obtain

$$\frac{d^2}{dt^2}\chi = -\kappa\frac{d}{dt}\chi + \omega_0\left\{-\kappa\varpi - \omega_0\chi - 2G\mathcal{D}\right\}. \tag{7.115}$$

Using (7.113) to eliminate ϖ from equation (7.115), we get

$$\frac{d^2}{dt^2}\chi + 2\kappa\frac{d}{dt}\chi + \left(\omega_0^2 + \kappa^2\right)\chi = -2\omega_0 G\mathcal{D}. \tag{7.116}$$

From (7.111) and (7.112), we obtain

$$\frac{d^2}{dt^2}\mathcal{D} = -\omega_c^2\mathcal{D} - 2\omega_c G\chi. \tag{7.117}$$

Now, assuming that the field oscillates in a definite frequency ω, so that

$$\mathcal{D}(t) = \mathcal{D}_0\cos\left(\omega t + \theta\right), \tag{7.118}$$

and that the polarization is induced by the field (i.e., χ does not oscillate by itself), (7.116) has the following solution (Problem 7.2):

$$\chi = -2\omega_0 G\mathcal{D}_0\,\text{Re}\left\{\frac{e^{i(\omega t + \theta)}}{\omega_0^2 + \kappa^2 - \omega^2 + i2\kappa\omega}\right\}. \tag{7.119}$$

Substituting (7.119) in (7.117), we obtain this equation for ω:

$$-\omega^2 = -\omega_c^2 + \frac{4\omega_0\omega_c G^2}{\omega_0^2 + \kappa^2 - \omega^2 - i2\kappa\omega}. \tag{7.120}$$

It is the single-mode lossless cavity that constrains the field inside it to oscillate at its resonance frequency. The dielectric medium, however, changes this resonance from ω_c to ω given by (7.120). The square of the resonance frequency is given by

$$\omega^2 = \frac{\omega_c}{\varepsilon}, \tag{7.121}$$

where ε is the dielectric permittivity of the medium. From (7.120) and (7.121), we find the following expression for ε:

$$\varepsilon(\omega) = 1 + \frac{4G'^2}{\omega_0'^2 - \omega^2 - i2\kappa\omega},\tag{7.122}$$

where

$$\omega_0'^2 \equiv \omega_0^2 + \kappa^2 - 4G'^2\tag{7.123}$$

is the shifted resonance of the medium. The κ^2 contribution is due to the interaction with the reservoir. The $-4G'^2$ contribution is due to the local field, as can be calculated from a one-dimensional Clausius–Mossotti approach (see, e.g., page 42 of [186]). The reader can easily check (Problem 7.3) that (7.122) satisfies the Kramers–Kronig relation (7.77).

The real part of the dielectric permittivity is the square root of the refractive index. The imaginary part describes absorption (the extinction coefficient). We plot both the real and imaginary parts of (7.122) in Figure 7.4.

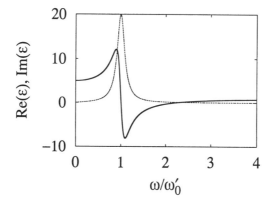

Figure 7.4 The solid line shows the real part of (7.122), and the dotted line, the imaginary part. All parameters are in units of the shifted resonance of the medium ω_0' given by (7.123). We used $G' = \omega_0'$ and $\kappa = 0.1\omega_0'$. This complex dielectric permittivity exhibits all the typical features we expect. There is an region of anomalous dispersion near the resonance of the medium, where the refractive index decreases with frequency rather than increasing. For large frequencies, the real part of ε tends to 1, making the refractive index of the medium the same as that of the vacuum.

7.8 THE FULL QUANTUM THEORY: HUTTNER AND BARNETT'S IMPROVEMENT OF THE HOPFIELD MODEL

Huttner and Barnett's approach consists in finding the set of annihilation and creation operators that diagonalizes the total Hamiltonian (7.96), then expressing the electromagnetic field operators in terms of them. This way they were able to develop a fully Canonical quantum theory of electrodynamics in a linear dielectric, including dispersion and absorption.

The diagonalization is done using a technique developed by Fano [190] in 1961, which they adapted to deal directly with operators rather than wavefunctions [38]. First, $\hat{H}_{\mathrm{mat}}$ is diagonalized, then the rest of the Hamiltonian. We will follow their approach with a few technical changes required by our simplifying assumptions of small losses and white noise. These assumptions not only make the discussion more pedagogical, but also allow us to calculate explicitly things for which they could only derive formal expressions. One of these things is the permittivity. In their general approach, Huttner and Barnett have *interpreted* a certain function as the dielectric permittivity. They have not shown that it is the permittivity that one finds in the classical theory. Here we will be able to calculate this function and show that it coincides[10] with (7.122).

7.8.1   Diagonalization of the matter Hamiltonian

To obtain tractable expressions, we will use our small coupling, white noise assumption with $\omega_0 \gg \Delta \gg \kappa$. These assumptions will allow us to extend integrals over η to $-\infty$, taking $V(\eta)$ as constant and equal to $V \equiv V(\omega_0) = \sqrt{\kappa/\pi}$.

Let us define the dressed operator

$$\hat{B}(\nu) \equiv \alpha(\nu)\hat{B} + \int_0^\infty d\eta \, \beta(\nu, \eta)\hat{Y}(\eta), \tag{7.124}$$

in terms of which the matter Hamiltonian is diagonal:

$$[\hat{B}(\nu), \hat{H}_{\mathrm{mat}}] = \hbar\nu\hat{B}(\nu), \tag{7.125}$$

and such that $\hat{B}(\nu)$ be an annihilation operator:

$$[\hat{B}(\nu), \hat{B}^\dagger(\nu')] = \delta(\nu - \nu'). \tag{7.126}$$

There is no need to impose the other commutation rule, that is,

$$[\hat{B}(\nu), \hat{B}(\nu')] = 0, \tag{7.127}$$

as it is trivially satisfied by definition with $\hat{B}(\nu)$ given by (7.124). So (7.124), (7.126), and (7.126) are enough to define $\hat{B}(\nu)$ uniquely except for a global phase factor. We must now determine $\alpha(\nu)$ and $\beta(\nu, \eta)$ and show that the diagonalization is consistent.

From (7.124) and (7.125), we obtain the following system of coupled equations (Problem 7.4):

$$\nu\alpha(\nu) = \omega_0\alpha(\nu) + \int_0^\infty d\eta \, V(\eta)\beta(\nu, \eta), \tag{7.128}$$

$$\nu\beta(\nu, \eta) = \eta\beta(\nu, \eta) + V(\eta)\alpha(\nu). \tag{7.129}$$

---

[10]For a general demonstration that the formal expression derived by Huttner and Barnett is indeed the same dielectric permittivity that appears in the classical theory, see [167].

Solving (7.129) formally for $\beta(\nu, \eta)$, we find that

$$\beta(\nu, \eta) = V(\eta)\alpha(\nu)\mathcal{P}\frac{1}{\nu - \eta} + z(\nu)\delta(\nu - \eta). \tag{7.130}$$

To determine the function $z(\eta)$, we substitute (7.130) into (7.128). We get

$$z(\nu) = \frac{\nu - \omega_0}{V}\alpha(\nu); \tag{7.131}$$

thus,

$$\beta(\nu, \eta) = V\alpha(\nu)\mathcal{P}\frac{1}{\nu - \eta} + \frac{\nu - \omega_0}{V}\alpha(\nu)\delta(\nu - \eta). \tag{7.132}$$

To determine $\alpha(\nu)$, we substitute (7.132) in (7.126):

$$\alpha^*(\nu')\alpha(\nu) + V^2\alpha^*(\nu')\alpha(\nu)\int_{-\infty}^{\infty} d\eta\, \mathcal{P}\frac{1}{\nu' - \eta}\mathcal{P}\frac{1}{\nu - \eta} + \alpha^*(\nu')\alpha(\nu)\frac{\nu - \omega_0}{\nu' - \nu}$$

$$+ \alpha^*(\nu')\alpha(\nu)\frac{\nu' - \omega_0}{\nu - \nu'} + \left(\frac{\nu - \omega_0}{V}\right)^2\alpha^*(\nu')\alpha(\nu)\delta(\nu - \nu')$$

$$= \delta(\nu - \nu'). \tag{7.133}$$

We show in Appendix F that

$$\int_{-\infty}^{\infty} d\eta\, \mathcal{P}\frac{1}{\nu' - \eta}\mathcal{P}\frac{1}{\nu - \eta} = \pi^2\delta(\nu - \nu'). \tag{7.134}$$

Then (7.133) becomes

$$\left\{\pi^2 V^2\alpha^*(\nu')\alpha(\nu) + \left(\frac{\nu - \omega_0}{V}\right)^2\alpha^*(\nu')\alpha(\nu)\right\}\delta(\nu - \nu') = \delta(\nu - \nu'). \tag{7.135}$$

From this equation we find that

$$|\alpha(\nu)|^2 = \frac{V^2}{(\nu - \omega_0)^2 + \pi^2 V^4}. \tag{7.136}$$

Choosing the arbitrary phase, we can write

$$\alpha(\nu) = \frac{\sqrt{\kappa/\pi}}{\nu - \omega_0 + i\kappa}. \tag{7.137}$$

To check consistency, let us calculate $[\hat{B}, \hat{B}^\dagger]$. If the $\hat{B}(\nu)$ form a complete set, we can express $\hat{B}$ as a linear combination (Problem 7.5) of the $\hat{B}(\nu)$, that is,

$$\hat{B} = \int_0^{\infty} d\nu\, \alpha^*(\nu)\hat{B}(\nu). \tag{7.138}$$

Then

$$[\hat{B}, \hat{B}^\dagger] = \int_{-\infty}^{\infty} d\nu \, |\alpha(\nu)|^2 = 1, \tag{7.139}$$

as it should.

We can also check that $[\hat{Y}(\eta), \hat{Y}^\dagger(\eta')] = \delta(\eta - \eta')$. Writing $\hat{Y}(\eta)$ as a linear combination of $\hat{B}(\nu)$,

$$\hat{Y}(\eta) = \int_0^{\infty} d\nu \, \beta^*(\nu, \eta) \, \hat{B}(\nu), \tag{7.140}$$

we find that

$$[\hat{Y}(\eta), \hat{Y}^\dagger(\eta')] = \int_0^{\infty} d\nu \, \beta^*(\nu, \eta) \beta(\nu, \eta')$$

$$= \frac{\kappa}{\pi} \int_{-\infty}^{\infty} d\nu \, |\alpha(\nu)|^2 \mathcal{P} \frac{1}{\nu - \eta} \mathcal{P} \frac{1}{\nu - \eta'} + |\alpha(\eta)|^2 \frac{\eta - \omega_0}{\eta - \eta'}$$

$$+ |\alpha(\eta')|^2 \frac{\eta' - \omega_0}{\eta' - \eta} + \left(\frac{\eta - \omega_0}{V} \right)^2 |\alpha(\eta)|^2 \delta(\eta - \eta'). \tag{7.141}$$

The product of the two principal parts above can be separated in partial fractions in the following way (see Appendix F or Fano's paper [190]):

$$\mathcal{P} \frac{1}{\nu - \eta} \mathcal{P} \frac{1}{\nu - \eta'} = \frac{1}{\eta' - \eta} \left\{ \mathcal{P} \frac{1}{\nu - \eta'} - \mathcal{P} \frac{1}{\nu - \eta} \right\} + \pi^2 \delta(\eta - \eta') \delta \left(\nu - \frac{\eta + \eta'}{2} \right). \tag{7.142}$$

Then we find that (Problem 7.6)

$$[\hat{Y}(\eta), \hat{Y}^\dagger(\eta')] = \delta(\eta - \eta'), \tag{7.143}$$

as it should.

7.8.2 Diagonalization of the total Hamiltonian

When written in terms of the operators $\hat{B}(\nu)$, the total Hamiltonian is given by

$$\hat{H} = \hbar \omega_c \hat{a}^\dagger \hat{a} + \hbar \int_0^{\infty} d\nu \, \hat{B}^\dagger(\nu) \hat{B}(\nu) \nu$$

$$+ \hbar G \left(\hat{a}^\dagger + \hat{a} \right) \int_0^{\infty} d\nu \left\{ \alpha(\nu) \hat{B}^\dagger(\nu) + \alpha^*(\nu) \hat{B}(\nu) \right\}. \tag{7.144}$$

Let us introduce the operators

$$\hat{C}(\omega) = \alpha_1(\omega) \hat{a} + \alpha_2(\omega) \hat{a}^\dagger + \int_0^{\infty} d\nu \left\{ \beta_1(\omega, \nu) \hat{B}(\nu) + \beta_2(\omega, \nu) \hat{B}^\dagger(\nu) \right\}, \tag{7.145}$$

defined such that they diagonalize the Hamiltonian (7.144), that is,

$$[\hat{C}(\omega), \hat{H}] = \hbar \omega \hat{C}(\omega), \tag{7.146}$$

and are annihilation operators, with their Hermitian conjugates being creation operators, satisfying the usual commutation relations,

$$[\hat{C}(\omega), \hat{C}^\dagger(\omega')] = \delta(\omega - \omega') \tag{7.147}$$

and

$$[\hat{C}(\omega), \hat{C}(\omega')] = 0. \tag{7.148}$$

As in the matter case, the reader can easily check later that (7.148) is satisfied automatically and does not have to be imposed to derive the solution.

We must now determine $\alpha_1(\omega)$, $\alpha_2(\omega)$, $\beta_1(\omega, \nu)$, and $\beta_2(\omega, \nu)$ using (7.144), (7.145), (7.146), (7.147), and (7.148). From (7.144), (7.145), and (7.148), we obtain

$$\omega_c \alpha_1(\omega) + G \int_0^\infty d\nu \ \{\alpha(\nu)\beta_1(\omega, \nu) - \alpha^*(\omega)\beta_2(\omega, \nu)\} = \omega \ \alpha_1(\omega), \tag{7.149}$$

$$-\omega_c \alpha_2(\omega) + G \int_0^\infty d\nu \ \{\alpha(\nu)\beta(\omega, \nu) - \alpha^*(\nu)\beta_2(\omega, \nu)\} = \omega \ \alpha_2(\omega), \tag{7.150}$$

$$\nu\beta_1(\omega, \nu) + G\alpha^*(\nu) \{\alpha_1(\omega) - \alpha_2(\omega)\} = \omega \ \beta_1(\omega, \nu), \tag{7.151}$$

$$-\nu\beta_2(\omega, \nu) + G\alpha(\nu) \{\alpha_1(\omega) - \alpha_2(\omega)\} = \omega \ \beta_2(\omega, \nu). \tag{7.152}$$

From (7.149) and (7.150), we obtain

$$\alpha_2(\omega) = \frac{\omega - \omega_c}{\omega + \omega_c} \alpha_1(\omega). \tag{7.153}$$

Solving (7.151) and (7.152) formally and then using (7.153), we find

$$\beta_1(\omega, \nu) = \frac{2G\omega_c}{\omega + \omega_c} \alpha^*(\nu)\alpha_1(\omega)\mathcal{P}\frac{1}{\omega - \nu} + z_1(\omega)\delta(\omega - \nu), \tag{7.154}$$

$$\beta_2(\omega, \nu) = \frac{2G\omega_c}{\omega + \omega_c} \alpha(\nu)\alpha_1(\omega)\mathcal{P}\frac{1}{\omega + \nu}. \tag{7.155}$$

Substituting (7.154) and (7.155) in (7.149), we get the following equation involving $z_1(\omega)$:

$$(\omega - \omega_c) \alpha_1(\omega) = \frac{2\omega_c G^2}{\omega + \omega_c} \alpha_1(\omega) \int_{-\infty}^\infty d\nu \ |\alpha(\nu)|^2 \left\{ \mathcal{P}\frac{1}{\omega - \nu} - \mathcal{P}\frac{1}{\omega + \nu} \right\}$$
$$+ G\alpha(\omega)z_1(\omega). \tag{7.156}$$

Noticing that

$$\int_{-\infty}^\infty d\nu |\alpha(\nu)|^2 \mathcal{P}\frac{1}{\omega - \nu} = \frac{\pi}{\kappa} (\omega - \omega_0) |\alpha(\omega)|^2, \tag{7.157}$$

$$\int_{-\infty}^\infty d\nu |\alpha(\nu)|^2 \mathcal{P}\frac{1}{\omega + \nu} = \frac{\pi}{\kappa} (\omega + \omega_0) |\alpha(-\omega)|^2, \tag{7.158}$$

after some algebraic manipulation, we can rewrite (7.156) as

$$\alpha(\omega)z_1(\omega) = \frac{\omega - \omega_c}{G} \alpha_1(\omega) + \frac{2\omega_c G^2}{\omega + \omega_c} \omega_0 \frac{\pi}{\kappa} [\alpha(\omega)\alpha^*(-\omega) + \alpha(-\omega)\alpha^*(\omega)] \alpha_1(\omega).$$
$$\tag{7.159}$$

The rest of the calculation is very analogous to what we have done for the matter Hamiltonian. From (7.147) we find an equation for $|\alpha_1(\omega)|^2$. As the phase is arbitrary (we can include any phase in the definition of the operator $\hat{a}$), we choose it conveniently so that $\alpha_1(\omega)$ is given simply by

$$\alpha_1(\omega) = (\omega + \omega_c) G' \sqrt{\frac{\omega_c}{\omega_0}} \alpha(\omega) \frac{\varepsilon^*(\omega)}{\omega^2 \varepsilon^*(\omega) - \omega_c^2}, \tag{7.160}$$

where $\varepsilon^*(\omega)$ is the complex conjugate of the classical dielectric permittivity (7.122) derived in Section 7.7.

As in Problem 7.4, we can use (7.145) to write $\hat{a}$ in terms of $\hat{C}(\omega)$ and $\hat{C}^\dagger(\omega)$,

$$\hat{a} = \int_0^\infty d\omega \left\{ \alpha_1^*(\omega)\hat{C}(\omega) - \alpha_2(\omega)\hat{C}^\dagger(\omega) \right\}. \tag{7.161}$$

Using this equation in (7.59), with $\alpha_1(\omega)$ given by (7.160) and $\alpha_2(\omega)$ given by (7.153), we see that in terms of the operators $\hat{C}(\omega)$ and $\hat{C}^\dagger(\omega)$, for which the total Hamiltonian is a continuous sum (integral) of uncoupled harmonic oscillators, the electric displacement is given by

$$\hat{D}(x) = \sqrt{\frac{2\hbar\omega_c^2}{\omega_0 L}} \int_0^\infty d\omega \, 2\omega_c G' \alpha^*(\omega) \frac{\varepsilon(\omega)}{\omega^2 \varepsilon(\omega) - \omega_c^2} \hat{C}(\omega) \, \sin\left(\frac{\omega_c}{c} x\right) + \text{H.c.} \tag{7.162}$$

To obtain the electric field, we need to find the polarization and use it in (7.59). The polarization is given by (7.2) in terms of the operators $\hat{B}$ and $\hat{B}^\dagger$. So we have to find $\hat{B}$ in terms of $\hat{C}(\omega)$ and $\hat{C}^\dagger(\omega)$. Inverting (7.145) as in Problem 7.4, we find that

$$\hat{B}(\nu) = \int_0^\infty d\omega \left\{ \beta_1^*(\omega, \nu)\hat{C}(\omega) - \beta_2(\omega, \nu)\hat{C}^\dagger(\omega) \right\}. \tag{7.163}$$

Substituting this equation in (7.138), we obtain

$$\hat{B} = \int_0^\infty d\omega \int_0^\infty d\nu \left\{ \beta_1^*(\omega, \nu)\hat{C}(\omega) - \beta_2(\omega, \nu)\hat{C}^\dagger(\omega) \right\} \alpha^*(\nu). \tag{7.164}$$

Thus, after some algebra,

$$\hat{E}(x) = \sqrt{\frac{2\hbar\omega_c^2}{\omega_0 L}} \int_0^\infty d\omega \left\{ \frac{2\omega_c G' \alpha^*(\omega)}{\omega^2 \varepsilon(\omega) - \omega_c^2} + \frac{2G' \alpha^*(\omega)}{\omega_c} \right\} \hat{C}(\omega) \sin\left(\frac{\omega_c}{c} x\right) + \text{H.c.} \tag{7.165}$$

RECOMMENDED READING

- For a review of the foundations of macroscopic electrodynamics, see Robinson's book [525] and van Kranendonk and Sipe's review paper [615]. The statistical ensemble approach to the derivation of the macroscopic Maxwell equations is exemplified by de Groot and Vlieger's paper [148]. See also [539].

- For a review on the Huttner–Barnett theory, see [33].

- By treating the cavity walls from first principles as a collection of atoms in the vacuum, Saito and Hyuga derived the dynamic Casimir effect (see Chapter 4) without having to deal with time-dependent boundary conditions [531]. Another first-principles approach to reflecting walls is Cook and Milonni's derivation of modified spontaneous emission (see Chapter 6), Rabi oscillations, and the Ewald–Oseen extinction theorem [121].

- It is possible to cancel dispersion using nonlocal quantum effects, see [324, 578, 579].

- It is possible to use quantum coherence between atomic states of atoms prepared in a quantum superposition to enhance the index of refraction and at the same time, cancel absorption. See [551] for the theory and [665] for the experimental demonstration. A pedagogical discussion of this can be found in [552].

- For a pedagogical discussion of the macroscopic quantization approach, where one attempts to quantize the macroscopic Maxwell equations in media directly, see Vogel and Welsch's book [618].

- Macroscopic QED has a long history. See [166] for a list of the main historical papers on this subject.

Problems

7.1   In the regime where $\omega_0 \gg \Delta \gg \kappa$, do the integrals on the last term on the right-hand side of (7.104) in two ways:

1. Argue that as $|V(\eta)|^2$ varies very slowly with η (it is basically flat) over the bandwidth of β and $\omega_0 \gg 0$,

$$\int_0^\infty d\eta \, |V(\eta)|^2 \int_0^t dt' \, e^{i(t'-t)\eta} \beta(t') \approx |V(\omega_0)|^2 \int_{-\infty}^\infty d\eta \int_0^t dt' \, e^{i(t'-t)\eta} \beta(t').$$

Now do these integrals and derive (7.105).

2. As $\omega_0 \gg \Delta$, the integral over η can be extended to $-\infty$. Do this integral by contour integration and show that

$$\int_0^\infty d\eta \, |V(\eta)|^2 \int_0^t dt' \, e^{i(t'-t)\eta} \beta(t') \approx -\kappa\Delta \int_0^t dt' \, e^{(t'-t)\Delta} e^{i(t'-t)\omega_0} \beta(t').$$

Now use that $\Delta \gg \kappa$ to do the integral on the right-hand side above by the method of steepest descent and derive (7.105).

7.2   Use the method of Heaviside's operational calculus described in Chapter 2 to show that with $\mathcal{D}$ given by (7.118), the particular solution of (7.116) is given by (7.119).

7.3   Do the integral in (7.77) explicitly by contour integration using the dielectric permittivity (7.122).

7.4   Using (7.98) and (7.124), calculate the commutator of $\hat{B}(\nu)$ with $\hat{H}_{\mathrm{mat}}$. Substitute the result in the left-hand side of (7.125). Substitute $\hat{B}(\nu)$ in the right-hand side of (7.125) by its expression in (7.124). You have now an equation of the form $C_1(\nu)\hat{B} + \int_0^\infty d\eta\, C_2(\nu, \eta)\hat{Y}(\eta) = 0$. Notice that if you take the commutator of both sides of this equation with $\hat{B}^\dagger$, only the term proportional to $\hat{B}$ survives. Analogously, if you take the commutator of this equation with $\hat{Y}^\dagger(\eta)$, only the term proportional to $\hat{Y}(\eta)$ survives. We can make an analogy between the set of operators $\{\hat{B}, \hat{Y}(\eta)\}$ and a set of orthogonal vectors and between the commutator and the inner product of these vectors. In other words, $C_1(\nu)\hat{B} + \int_0^\infty d\eta\, C_2(\nu, \eta)\hat{Y}(\eta) = 0$ implies the equations $C_1(\nu) = 0$ and $C_2(\nu, \eta) = 0$. Show that the latter equations are given by (7.128) and (7.129).

7.5   Assume that $\hat{B}$ can be expressed as a linear combination of the $\hat{B}(\nu)$,

$$\hat{B} = \int_0^\infty d\nu\, C(\nu)\hat{B}(\nu).$$

Now, calculate the commutator $[\hat{B}, \hat{B}^\dagger(\nu)]$ in two ways. First, substitute the linear combination of the $\hat{B}(\nu)$ above for $\hat{B}$. Second, substitute the Hermitian conjugate of (7.124) for $\hat{B}^\dagger(\nu)$. Conclude that $C(\nu) = \alpha^*(\nu)$, as in (7.138).

7.6   Substitute (7.142) in (7.141) and show that (7.143) holds.

8

The maser, the laser, and their cavity QED cousins

On that morning in Franklin Park, the goal of boosting energy gave me an incentive to think more deeply about stimulated emission than I had before. How could one get such a nonequilibrium set up? ... Rabi, right at Columbia, had been working with molecular and atomic beams that he manipulated by deflecting atoms in excited states from those of lower energies. The result could be a beam enriched in excited atoms. ... I took an envelope from my pocket to try to figure out how many molecules it would take to make an oscillator able to produce and amplify millimeter or submillimeter waves. ... Any resonator has losses, so we would need a certain threshold number of molecules in the flow to keep the wave from dying out. Beyond that threshold, a wave would not only sustain itself bouncing back and forth, but it would gain energy with each pass. The power would be limited only by the rate at which molecules carried energy into the cavity.

—Charles H. Townes [602]

Radio transmitters generate electromagnetic radiation that, to a good approximation, can be regarded as coherent waves. This coherence allows us to modulate the wave's amplitude (AM) or its frequency (FM) to carry sound to far-away radio receivers. Before the invention of the laser, there was no such coherent source for electromagnetic radiation in the optical region of the spectrum.

The laser is often regarded as a direct by-product of the modern ideas of the "quantum revolution." But it is also the natural generalization of the idea of radio transmitters to optical frequencies. This chapter starts by showing the similarities between lasers and radio transmitters. Then we move on to discuss how the nonnegligible quantum nature of lasers brings about a number of phenomena that have no counterparts in classical radio transmitters. This departure from well-known classical behavior becomes even more dramatic as we go from traditional lasers and masers to their cavity QED cousins: the micromaser and the microlaser.

In the next section we introduce the basic idea behind the maser and the laser: amplification by stimulated emission of radiation (ASER).[1] Unlike in radio transmitters, noise is usually very important in lasers. So in Section 8.2 we have an interlude where we learn how to add noise to the usual deterministic differential equations of motion encountered in classical physics. In Section 8.3 we show the effect of spontaneous emission noise and introduce the idea of laser threshold. In Section 8.4 we begin our quantum description of ASERs by examining another simple model of a laser that allows us to calculate the quantum state of light generated in a laser. Then we go directly to the other extreme, the quantum regime, and examine a truly microscopic device: the micromaser. The micromaser is the closest experimental realization of the fundamental interaction between a two-level atom and a single mode of the quantized radiation field described by the celebrated Jaynes–Cummings model presented in Chapter 6. In Section 8.6 we sketch the connection between this simple microscopic device and complicated macroscopic lasers and masers, showing how dissipation can transform an optical micromaser into an ordinary laser. On the way we see how strong coupling can lead to the disappearance of the laser threshold. Then in the final section of the chapter we discuss briefly a recent breakthrough in cavity QED: the realization of the one-and-the-same atom laser in the strong-coupling regime.

8.1 THE ASER IDEA: AMPLIFICATION BY STIMULATED EMISSION OF RADIATION

In the 1950s there was a lot of excitement about radar and, on the basic research side, about spectroscopy. Radar was crucial for the allied war effort. Some say that it was radar which in fact won the war for the allies. Spectroscopy helped to bring us quantum mechanics and important advances in chemistry. Developments in both radar and spectroscopy called for an oscillator for microwaves and beyond.

Radar uses radio waves to locate distant objects, as a bat uses sound to find its way through the night. The resolution with which it can locate things depends on the wavelength of these waves. The shorter the wavelength, the higher the resolution, allowing it to locate smaller objects.[2] In spectroscopy, the resolution with which one can resolve an atomic absorption line, say, depends on the bandwidth of the radiation used to probe the line. For atomic resonances in the visible region, one had to use a light bulb or some other thermal source of light. To get as close to a monochromatic signal as possible, the light was fed through a series of frequency filters. The problem

---

[1] We also use this acronym to mean a device rather than a process: that is, an amplifier by stimulated emission of radiation.

[2] Unfortunately, the water vapor in the atmosphere absorbs certain microwave frequencies (that is why microwave ovens work). When this was discovered, attempts to increase the resolution of radar by using microwaves instead of radio were dropped.

is that the light gets weaker and weaker as you filter out its constituent frequencies. What was needed was a sustained oscillator that would produce radiation of a single frequency, such as the ones used in radar, but at the much higher frequencies of microwaves and visible light.

As early as 1883, Lord Rayleigh considered several ways in which an oscillator could keep going despite damping [510]. One of the cases he considered was that of Melde's experiment, where a fine string is maintained in transverse vibration by attaching one of its ends to the vibrating prong of a massive tuning fork, such that the direction of motion of the point of attachment is parallel to the length of the string. This is the first example of a parametric oscillator (Problem 8.1). A more familiar example is a child on a playground swing, as we considered in Chapter 2. Children soon learn that they can pump up the swing amplitude by lowering their center of gravity on the downswing and raising it on the upswing. They also learn that this pumping must be done at twice the swing resonance frequency. When they find this out, you do not need to push their swing any more: They have realized a sustained oscillator. But the sort of sustained oscillator that led to the maser and laser is different.

Consider a typical LC oscillator. It was the Irish physicist George Francis FitzGerald who first suggested using this oscillator to generate radio waves [213]. His idea was later adopted by Hertz in his famous experimental confirmation of Maxwell's theory of electrodynamics. As any actual circuit always has some resistance, the oscillations eventually die out. To make a sustained oscillator out of a good old LC circuit, what we need is to couple it to some amplifying element to effectively cancel the resistance in the circuit. In the early years of radio, this amplifying element was the negative resistance voltaic arc [460]. The first mathematical analysis of this sustained LC oscillator, however, was carried out by van der Pol for a vacuum tube, which soon replaced the arc as a practical amplifying element.

Figure 8.1 shows the diagram of the circuit considered by van der Pol. He derived the following differential equation for the deviation v from the mean of the anode voltage:

$$\frac{d^2v}{dt^2} - \frac{d}{dt}\left(\alpha v - \eta v^3\right) + \omega_0^2 v = 0, \tag{8.1}$$

where $\omega_0 = 1/\sqrt{LC}$, α is the linear gain, and η is the saturation coefficient.[3] In the case where $\omega_0 \gg \alpha, \eta$, it is useful to write v as the product of a slowly varying amplitude $V(t)$ and a rapidly oscillating ω_0 frequency component (see Sec. 4-1 of [533]):

$$v(t) = V(t)\,e^{-i\omega_0 t} + V^*(t)e^{i\omega_0 t}. \tag{8.2}$$

Then it is a good approximation to neglect the second derivative of V and the products of $\dot{V}$ and α and η, when (8.2) is substituted in (8.1). This is the slowly varying envelope approximation. Moreover, the effect on, say, the $V\exp(-i\omega_0 t)$ component of v from

---

[3]α and η are related to the parameters of the saturating triode characteristic; see the review paper by van der Pol [609].

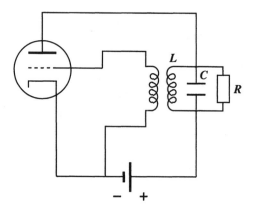

Figure 8.1 This is a circuit diagram of a van der Pol oscillator. The triode compensates for the ohmic losses, keeping the oscillations going without attenuation. Having a pure sinusoidal wave rather than a damped one is vital for the transmission of human voice and music. The triode made commercial radio a reality.

terms in the equation of motion oscillating with the different frequencies $-3\omega_0$, $-\omega_0$, and $3\omega_0$ will also be negligible. This is called the *rotating-wave approximation*. With these two approximations, (8.1) yields the following first-order differential equation for V:

$$\frac{dV}{dt} = \frac{1}{2}\left(\alpha - \eta'|V|^2\right)\, V, \tag{8.3}$$

where $\eta' \equiv 3\eta/4$.

For simplicity, let us regard V as real. Equation (8.3) has two steady-state solutions, obtained by setting $\dot{V} = 0$. One is $V_s = 0$ (i.e., the oscillator does not oscillate at all). The other is $V_s = \sqrt{\alpha/\eta'}$. Let us see what happens when V deviates only a little bit from a steady-state solution. Let us take $V = V_s + \delta$ and substitute it in (8.3), only keeping terms up to first order in δ. For the first steady-state solution, $V_s = 0$, we find that

$$\dot{\delta} = \frac{\alpha}{2}\delta. \tag{8.4}$$

As $\dot{\delta}$ is proportional to δ with a *positive* proportionality constant, if $V_s = 0$ is given even a tiny perturbation δ, be it positive or negative, the tendency will be to deviate even more from $V_s = 0$ with time. In other words, the steady-state solution $V_s = 0$ is unstable.

For the second steady-state solution, $V_s = \sqrt{\alpha/\eta'}$, we find that

$$\dot{\delta} = -\alpha\, \delta. \tag{8.5}$$

We can see from this equation that as the time derivative of δ is proportional to δ with a negative constant of proportionality (because α is positive), it is always of opposite sign to δ; the tendency is to restore the oscillator to its $V_s = \sqrt{\alpha/\eta'}$ steady state.

Thus we may conclude that given a small perturbation, such as the random voltage fluctuations that are present in any electronic circuit, (8.3) will predict the buildup of oscillations from $V = 0$, settling at the stable steady state $V_s = \sqrt{\alpha/\eta'}$ after an initial transient period.

As the frequency increases, some problems begin to appear with the basic idea of the electronic circuit in Figure 8.1. First, we can no longer find a pure inductor, capacitor, and so on. There are not really any truly pure electronic components anyway. A capacitor is always a bit of an inductor and of a resistor, but at low frequencies (such as radio), most of the capacity in the circuit will be lumped at the capacitors, most of the inductance at the inductors, and so on. As the frequency gets higher and higher, however, this approximation breaks down and inductors acquire as much capacity as capacitors, and there is inductance, capacity, and resistance (impedance) all over the place instead of only in tidy little lumped elements (see, e.g., Chapter 23 of [206]). So what can we use at higher frequencies instead of our old electronic LC circuit? If you have read the previous chapters of this book, you should have the answer at the tip of your tongue: a cavity.

The second problem is what to use instead of the vacuum tube to amplify these higher-frequency electromagnetic fields. Charles Townes's great idea was to use atomic or molecular electronic transitions to do that. He imagined sending excited atoms or molecules through a small hole across the cavity, each of which could then decay to their ground state emitting a photon into the cavity.

Although a cavity mode is equivalent to a harmonic oscillator as an LC circuit is, a cavity has many modes. How can we make sure that the photons will go all (or at least mostly) to a single mode? Conservation of energy will provide selectivity by allowing emission only in the mode that lies within the linewidth of the atomic or molecular transition. But how can this radiation generated by many independent atomic/molecular emissions be as coherent as that of a radio transmitter, which is generated by a single-dipole antenna?

It was Einstein who in 1905 discovered that excited atoms or molecules can emit light in two qualitatively different ways: spontaneous and stimulated emission. In spontaneous emission, the photon is emitted randomly without particular preference for any mode (see Chapter 6). In stimulated emission, however, the photon is preferentially emitted into a mode that is already occupied by other photon(s), preserving the coherence of the radiation field. Stimulated emission is a consequence of photons being bosons, as we shall see briefly next.

Emission of photons into a given mode k is represented by the term proportional to $\hat{a}_k^\dagger$ in the coupling part of the Hamiltonian, as discussed in Chapter 5. A simple Fermi golden rule argument then shows that the probability of emission into a given mode k is proportional to $|\hat{a}_k^\dagger|\psi\rangle|^2$. We can rewrite $|\hat{a}_k^\dagger|\psi\rangle|^2$ as $\langle\psi|\hat{a}_k\hat{a}_k^\dagger|\psi\rangle$. Using the commutator between $\hat{a}_k$ and $\hat{a}_k^\dagger$, $[\hat{a}_k, \hat{a}_k^\dagger] = 1$, which is a consequence of the photon being a boson, we find that $|\hat{a}_k^\dagger|\psi\rangle|^2 = 1 + \bar{n}_k$, where $\bar{n}_k = \langle\psi|\hat{n}_k|\psi\rangle$ is the average number of photons in mode k. Thus, the probability of emission into a mode that is already filled with $\bar{n}_k$ photons on average is $1 + \bar{n}_k$ greater than that of emitting into an unpopulated mode.[4]

With these ideas, we are already in a position to write down a "back of the envelope" description of a laser. In this "hand-waving" description, we will regard the various quantum-mechanical observables as if they were classical.

Laser cavities are usually open, so some of the photons emitted can actually go into free space. Let us call β the fraction of emission into the lasing mode. Then if we call n the number of photons in the lasing mode, N the population of the upper lasing level, and $\gamma_\parallel$ the decay rate of the upper lasing level to the lower lasing level, the rate at which stimulated emission photons will increase the population of the lasing mode is[5] $\beta\gamma_\parallel Nn$. Cavity damping will tend to decrease this population at a rate of $-\kappa n$, where κ^{-1} is the photon lifetime in the lasing mode. Absorption of photons by the population in the lower lasing level (making then a transition to the upper level) will also tend to decrease n. But if the lower lasing level is rapidly emptied into lower-lying levels that do not take part in the laser, there will be no absorption. For simplicity, let us assume that this is the case. Then it is only the population N of the upper lasing level that appears in the laser equations. Last but not least, there is the spontaneous emission contribution to the population of the lasing mode, $R_{\mathrm{sp}} = \beta\gamma_\parallel N$. Putting all this together, we have

$$\dot{n} = -\kappa n + \beta\gamma_\parallel Nn + R_{\mathrm{sp}}. \tag{8.6}$$

To complete our description, we must write down the rate equation for N. The spontaneous (into all modes now) and stimulated emission terms will, of course, tend to decrease N. To keep the laser going, we must feed energy in it. This is done by pumping the population in the upper lasing level with some rate Λ. Thus,

$$\dot{N} = \Lambda - \gamma_\parallel N - \beta\gamma_\parallel Nn. \tag{8.7}$$

The lifetime of molecular and atomic energy levels involved in optical transitions is often so short that even a moderate cavity Q value is already large enough to make κ much smaller than $\gamma_\parallel$. In that case, the relaxation dynamics of N will be much faster than that of n and N will "follow n adiabatically." This is called *adiabatic elimination*. It can be derived in the following way. If we multiply (8.7) by $\exp(\gamma_\parallel t)$ and notice that $\gamma_\parallel N \exp(\gamma_\parallel t) = N d \exp(\gamma_\parallel t)/dt$, we can rewrite (8.7) as

$$\frac{d}{dt}\left(N e^{\gamma_\parallel t}\right) = \left(\Lambda - \gamma_\parallel \beta Nn\right)e^{\gamma_\parallel t}. \tag{8.8}$$

---

[4]For fermions we have the anticommutation relation $\{\hat{b}_k, \hat{b}_k^\dagger\} \equiv \hat{b}_k\hat{b}_k^\dagger + \hat{b}_k^\dagger\hat{b}_k = 1$ rather than the commutation relation above. As a consequence, the probability of emission into a given mode k turns out to be proportional to $|\hat{b}_k^\dagger|\psi\rangle|^2 = 1 - \bar{n}_k$, so that the probability of emission actually decreases if the mode is not absolutely unpopulated. Notice that the average number of fermions on mode k, $\bar{n}_k$, cannot be larger than 1; otherwise, the square modulus of $\hat{b}_k^\dagger|\psi\rangle$ becomes negative.

[5]The stimulated emission rate is that part of the emission rate that exists only when the mode is already populated. Hence, it is the $\bar{n}_k$ part of $\bar{n}_k + 1$ derived in the preceding paragraphs. The "1" part is the spontaneous emission contribution, which is always present.

Integrating this equation from the initial time $-\infty$, when the upper level was still unpopulated, to t, we get

$$N(t) = \int_{-\infty}^{t} dt' \left\{ \Lambda - \gamma_{\parallel} \beta N(t') n(t') \right\} e^{-(t-t')\gamma_{\parallel}}. \tag{8.9}$$

As $\gamma_{\parallel} \gg \kappa$, the integrand above is dominated by the exponential. Using the steepest descent method, we can approximate this integral by

$$N(t) \approx \left\{ \Lambda - \gamma_{\parallel} \beta N(t) n(t) \right\} \int_{-\infty}^{t} dt' \, e^{-(t-t')\gamma_{\parallel}} = \frac{\Lambda - \gamma_{\parallel} \beta N(t) n(t)}{\gamma_{\parallel}}. \tag{8.10}$$

Solving for N yields

$$N = \frac{\Lambda / \gamma_{\parallel}}{1 + \beta n}, \tag{8.11}$$

which shows that adiabatic elimination of N is equivalent to setting $\dot{N}$ to zero in (8.7).

In ordinary lasers, β is often much smaller than 1. A HeNe 633-nm laser, for example, has $\beta \approx 10^{-8}$, and an index-guided edge-emitting semiconductor laser typically has $\beta \approx 10^{-5}$ [654]. Thus, for $n \ll 1/\beta$, we can expand the denominator on the right-hand side of (8.11) to first order in βn. Substituting the result in (8.6), and neglecting the spontaneous emission contribution R_{sp}, we get the following differential equation for n:

$$\dot{n} = \left(\Lambda \beta - \kappa - \Lambda \beta^2 n \right) n. \tag{8.12}$$

Calling $\Lambda \beta - \kappa$ and $\Lambda \beta^2$ on the right-hand side of (8.12) $\alpha/2$ and $\eta'/2$, respectively, we can recognize here the van der Pol oscillator in the slowly varying envelope and rotating-wave approximations. But this is a van der Pol oscillator that oscillates at optical rather than radio frequencies, an increase of seven orders of magnitude: from hundreds of megahertz to hundreds of terahertz.

We mentioned earlier that it is noise that knocks the van der Pol oscillator out of the unstable $V_s = 0$ steady state when we switch on the circuit. Without noise, it would never start oscillating unless we perturbed it out of $V_s = 0$ some other way. But we have otherwise neglected noise in our analysis so far. Although in electronics neglecting noise is not so bad, in lasers (quantum electronics) it is a very poor approximation. This is because at radio frequencies, noise is of classical origin. In that regime, the energy of an electromagnetic quantum (photon) is much smaller than the energy of thermal fluctuations (kT) at room temperature. As classical noise can be reduced, we can get it to a level where it is negligible. At optical frequencies, however, quantum noise is no longer negligible. As quantum noise cannot be eliminated,[6] we just have to live with it. In the laser it plays an important

---

[6]Quantum noise can be reduced for one observable only at the expense of increasing it for the complementary observable, as in squeezed light.

role. To add noise to this simple description of a laser, we need to make a small interlude, where we talk a little bit about botany and also about a mathematical technique that finds applications not only in laser theory but also in modeling the stock market and other diverse fields.

8.2 HOW TO ADD NOISE: BROWNIAN MOTION; THE LANGEVIN EQUATION; ITO'S AND STRATONOVICH'S STOCHASTIC CALCULUS

The prototype for describing noise mathematically is a phenomenon that was discovered in 1827 by the Scottish botanist Robert Brown. He observed under a microscope a suspension of pollen grains in water and noticed that the grains were moving, as if they were alive (Figure 8.2). But as he later observed the same "swarming" motion

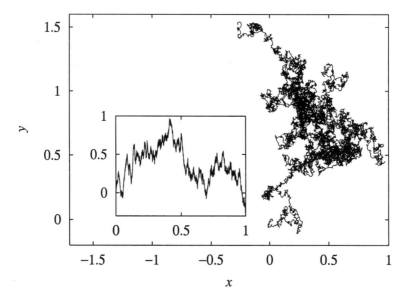

Figure 8.2 This is the result of a numerical simulation of the two dimensional Brownian motion of a single particle (e.g., a grain of pollen). The large plot shows the actual trajectory of the particle. The position the particle had when we started to follow its path was chosen as the origin of coordinates (i.e., $x = 0$, $y = 0$). On the inset, the x-coordinate of the particle is plotted against the time. The units are arbitrary and the diffusion constant was taken as 1.

on suspensions of fine inorganic particles, Brown soon concluded that it was not a manifestation of life.

8.2.1   Einstein's approach to Brownian motion

The first satisfactory explanation of the riddle of *Brownian motion* was given by Albert Einstein in 1905, his famous *anno mirabilis* [181]. What follows is an account of his theory slightly adapted to fit our discussion here.

Einstein started his analysis from the hypothesis that the swarming motion of a grain of pollen was caused by the frequent pushes it received from collisions with the thermally agitated molecules of the liquid. Now if the viscosity of the liquid is high enough, the movement imparted by one such impact will quickly dampen, so that the motion caused by further impacts will be completely independent of the preceding one (i.e., there is no memory of the previous motion). Assuming that this was the case, Einstein introduced a time interval τ much shorter than observable time intervals, yet large enough for the viscous friction to do its job of making the motions in two successive time intervals τ be independent events.

Einstein examined the motion in one dimension, which we can call the x-axis. For $\mathcal{N}$ suspended particles in a viscous liquid, the x-coordinate of each particle will increase by a certain amount in a time interval τ. As we cannot keep track of the movement of every molecule in the liquid, all we can hope for is to describe this amount by a statistical probability distribution. Let $d\mathcal{N}$ be the number of particles whose increase in x is between Δ and $\Delta + d\Delta$. Then the probability distribution $\phi(\Delta)$ of increase in x is independent of time and defined by

$$d\mathcal{N} = \mathcal{N}\,\phi(\Delta)\,d\Delta, \tag{8.13}$$

where, as any probability distribution, the sum (integral) of all $\phi(\Delta)$ must yield 1:

$$\int_{-\infty}^{\infty} d\Delta\,\phi(\Delta) = 1. \tag{8.14}$$

Moreover, $\phi(\Delta)$ is nonvanishing only for small Δ, and if the kicks are random enough, we can assume that $\phi(-\Delta) = \phi(\Delta)$.

The number of particles $d\mathcal{N}(x,t)$ with their x-coordinates between x and $x + dx$ at time t will also be described by a probability distribution, which we call $p(x,t)$, defined by

$$d\mathcal{N}(x,t) = \mathcal{N}p(x,t)\,dx, \tag{8.15}$$

where

$$\int_{-\infty}^{\infty} dx\,p(x,t) = 1. \tag{8.16}$$

The number of particles $d\mathcal{N}(x, t + \tau)$ with their x-coordinates between x and $x + dx$ at time $t + \tau$ is given by the number of particles with their x-coordinates between $x - \Delta$ and $x - \Delta + dx$ at time t that will receive an increment Δ after time τ:

$$d\mathcal{N}(x, t + \tau) = \mathcal{N}\,dx \int_{-\infty}^{\infty} d\Delta\,p(x - \Delta, t)\phi(\Delta). \tag{8.17}$$

As τ is very small and $\phi(\Delta)$ differs from zero only for small Δ, we can expand both $p(x, t + \tau)$ and $p(x - \Delta, t)$ in power series of τ and Δ, respectively:

$$p(x, t + \tau) = p(x, t) + \tau \frac{\partial}{\partial t} p(x, t) + \frac{\tau^2}{2} \frac{\partial^2}{\partial t^2} p(x, t) + \cdots \tag{8.18}$$

$$p(x - \Delta, t) = p(x, t) - \Delta \frac{\partial}{\partial x} p(x, t) + \frac{\Delta^2}{2} \frac{\partial^2}{\partial x^2} p(x, t) + \cdots \tag{8.19}$$

Substituting these expansions in (8.17) and keeping only the lowest nonvanishing order after zeroth order, Einstein found that p obeyed the diffusion equation

$$\frac{\partial p}{\partial t} - \frac{D}{2} \frac{\partial^2 p}{\partial x^2} = 0, \tag{8.20}$$

where

$$D \equiv \frac{1}{\tau} \int_{-\infty}^{\infty} d\Delta \, \Delta^2 \phi(\Delta) \tag{8.21}$$

is the diffusion coefficient. So Brownian motion is the mechanism behind diffusion.

8.2.2 Langevin's approach to Brownian motion

Three years after Einstein's paper, Paul Langevin revisited the problem of Brownian motion, adopting a more direct approach [392]. Assuming viscous friction, he wrote down the following equation of motion for a particle of pollen of mass m:

$$m \frac{d^2 x}{dt^2} = -\gamma \frac{dx}{dt} + \mathcal{F}, \tag{8.22}$$

where γ is the coefficient of friction and $\mathcal{F}$ is the random fluctuating force that the particle experiences due to the impact of the thermally agitated molecules of the liquid.

Multiplying (8.22) by x and noticing that

$$x \frac{d^2 x}{dt^2} = \frac{1}{2} \frac{d^2 x^2}{dt^2} - \left(\frac{dx}{dt} \right)^2, \tag{8.23}$$

we obtain

$$\frac{m}{2} \frac{d^2 x^2}{dt^2} - m \left(\frac{dx}{dt} \right)^2 + \frac{\gamma}{2} \frac{dx^2}{dt} = x\mathcal{F}. \tag{8.24}$$

Then Langevin takes the average of (8.24) over a large number of particles. He assumes that $\langle x\mathcal{F} \rangle$ vanishes because the random force is uncorrelated to x and fluctuates equally between positive and negative values.

He also notices that in thermal equilibrium,[7]

$$\left\langle \frac{m}{2} \left(\frac{dx}{dt} \right)^2 \right\rangle = \frac{1}{2} kT, \tag{8.25}$$

---

[7]This follows from the equipartition theorem (see, e.g., [515]).

where k is Boltzmann's constant and T is the absolute temperature. Thus, the average of (8.24) yields

$$\frac{d^2}{dt^2}\langle x^2\rangle + \frac{\gamma}{m}\frac{d}{dt}\langle x^2\rangle = \frac{kT}{m}. \tag{8.26}$$

Calling $d\langle x^2\rangle/dt$ X, we have a first-order nonhomogeneous ordinary differential equation for X. The general solution of this equation is an arbitrary linear combination of the particular solution with the solution of the associated homogeneous equation (the equation where the right-hand side of this equation is set to zero):

$$\frac{d}{dt}\langle x^2\rangle = \frac{kT}{\gamma} + Ce^{-\gamma t/m}, \tag{8.27}$$

where C is an arbitrary constant.

Now notice that for high-enough viscosity such that $\gamma/m \gg 1/\tau$, the exponential in the solution above will be negligible in any observable time interval. Integrating once more, Langevin then arrived at the following result:

$$\langle x^2(t)\rangle - \langle x^2(0)\rangle = \frac{kT}{\gamma}t. \tag{8.28}$$

This is the same result that one obtains starting with the diffusion equation (8.20), provided that $D = kT/\gamma$. So Langevin determined Einstein's diffusion coefficient in terms of the temperature, the coefficient of viscous friction, and Boltzmann's constant.

8.2.3 The modern form of Langevin's equation used in laser physics

You can often get more insight by deriving a well-known result in a different way. Let us reflect a bit more about Langevin's approach. What happens if we use the assumption of quite high viscosity ($\gamma/m \gg 1/\tau$) right from the beginning? Then neglecting $m\,d^2x/dt^2$ compared with $\gamma\,dx/dt$, instead of (8.22) we can write

$$\frac{dx}{dt} = F, \tag{8.29}$$

where $F \equiv \gamma\mathcal{F}$. Now let us try to obtain the properties of F without using the thermal equilibrium result (8.25) explicitly. The question we ask is: What properties must F have for this simple Langevin equation to be equivalent to the diffusion equation (8.20) deduced by Einstein?

The ensemble average of F can be given in terms of that of x by averaging (8.29):

$$\langle F(t)\rangle = \frac{d}{dt}\langle x\rangle. \tag{8.30}$$

We can calculate $\langle x\rangle$ if we determine the probability distribution $p(x,t)$ that obeys Einstein's diffusion equation (8.20). Assuming that the particle starts at the origin at $t = 0$, you can show (Problem 8.2) that the solution of (8.20) is given by

$$p(x,t) = \frac{1}{\sqrt{2\pi Dt}}\exp\left(-\frac{x^2}{2Dt}\right). \tag{8.31}$$

Then $\langle x \rangle = 0$ and we find that

$$\langle F(t) \rangle = 0, \tag{8.32}$$

as we would expect from a random force that pushes as much as it pulls.

What about the self-correlation of F [i.e., $\langle F(t)F(t') \rangle$]? Using (8.29) again, we can write this as

$$\langle F(t)F(t') \rangle = \frac{\partial^2}{\partial t\, \partial t'} \langle x(t)x(t') \rangle. \tag{8.33}$$

Let us calculate $\langle x(t)x(t') \rangle$.

For $t > t'$, we can write this average as

$$\langle x(t)x(t') \rangle = \int_{-\infty}^{\infty} dx'\int_{-\infty}^{\infty} dx\, p(x',t')\, x'\, p(x,t|x',t')\, x, \tag{8.34}$$

where $p(x,t|x',t')$ is the *conditional* probability of finding the particle at position x at time t given that it was at position x' at the earlier time t'. This conditional probability is just the probability of the particle diffusing to x at time t having started at x' at time t'. You can easily show that this probability is given by[8]

$$p(x,t|x',t') = \frac{1}{\sqrt{2\pi(t-t')D}} \exp\left(-\frac{x^2}{2\,(t-t')D}\right). \tag{8.35}$$

Substituting (8.31) and (8.35) in (8.34) and doing the integrals (Problem 8.3), we find that

$$\langle x(t)x(t') \rangle = Dt' \quad \text{for } t > t'. \tag{8.36}$$

Swapping t and t', we find that

$$\langle x(t)x(t') \rangle = Dt \quad \text{for } t' > t. \tag{8.37}$$

Equations (8.36) and (8.37) are equivalent to the single equation

$$\langle x(t)x(t') \rangle = Dt'\Theta(t-t') + Dt\Theta(t'-t), \tag{8.38}$$

where $\Theta(u)$ is the Heaviside step function, which is unity for positive u and vanishes when u is negative. Substituting (8.38) in (8.33), we obtain

$$\langle F(t)F(t') \rangle = D\frac{\partial}{\partial t}\Big\{\Theta(t-t') + \underbrace{t\delta(t'-t) - t'\delta(t-t')}_{(t-t')\delta(t-t')=0}\Big\}$$

$$= D\,\delta(t-t'). \tag{8.39}$$

In other words, F is a very special force: it fluctuates so much that it is only instantaneously correlated to itself.

This is all that we need for introducing noise in our simple laser model. If, however, you wish to know a bit more about Langevin equations, read the next subsection.

---

[8] In fact, the probability $p(x,t)$ can also be viewed as a conditional probability of finding the particle at x at time t *given* that it started at the origin at time 0. To see this, just take $x' = 0$ and $t' = 0$ in (8.35) and compare with (8.31).

8.2.4 Ito's and Stratonovich's stochastic calculus

Langevin's equation was the first example of what mathematicians now call a *stochastic differential equation*. As it stands, however, (8.29) can lead to ambiguities. Let us try, for example, to derive Langevin's assumption that $\langle x(t)F(t)\rangle$ vanishes using only (8.29), and (8.39). Integrating (8.29), and taking $x(0) = 0$, we can write

$$\langle x(t)F(t)\rangle = \int_0^t dt' \, \langle F(t)F(t')\rangle. \tag{8.40}$$

Using (8.39), we obtain

$$\langle x(t)F(t)\rangle = D \int_0^t dt' \, \delta(t - t'). \tag{8.41}$$

The argument of the delta function only vanishes at the upper limit of this integral. What is the value of an integral like this? If we rewrite this integral, changing its upper limit to some instant *after* t and multiplying the integrand by the step function $\Theta(t - t')$, we see that the problem here is equivalent to that discussed in Appendix F: how to determine the value of the product $\Theta(u)\delta(u)$.

To prevent such ambiguities, the idea of a stochastic differential equation had to be put into a more rigorous basis than that provided by Langevin's pioneering paper. Here we discuss briefly this more rigorous formulation. We will see that there are an infinite number of ways in which one can construct a rigorous stochastic calculus where equations such as (8.41) are no longer ambiguous. But the two most popular alternative ways are those proposed by Ito [312] and Stratonovich [580]. In Itô calculus, (8.41) vanishes as Langevin had assumed originally. On the other hand, the ordinary calculus chain rule changes. In Stratonovich calculus, the ordinary rules of calculus are maintained, but (8.41) is taken to be $D/2$.

A completely rigorous mathematical presentation of stochastic calculus in the usual axioms–theorem–proof style would be completely out of touch with the spirit of this book and its intended readership. Instead, we have a more concrete discussion of stochastic calculus based on an approach developed by Gillespie [232]. This discussion differs from Gillespie's only because we deal explicitly with the inconsistency problem that leads to Stratonovich's and Ito's calculus, whereas [232] does not. The reader interested in a more mathematical approach can find some useful references in the recommended reading section at the end of the chapter, where even some books about applications of stochastic calculus in quantitative finance are mentioned.

Any differential equation can be rewritten as a system of first-order differential equations. Let us look at the simplest case, where there is just one equation in this system and see how we can generalize it to make it stochastic. The general form of a first-order differential equation is

$$\frac{d}{dt}x(t) = A\left(x(t), t\right), \tag{8.42}$$

where A is a smooth function of x and t.

What (8.42) is supposed to tell us is how x evolves in time. The form (8.42) is really shorthand for the update formula,

$$x(x + dt) = x(t) + A\left(x(t), t\right) dt, \tag{8.43}$$

which describes the information content of (8.42) explicitly. In the language of statistics, a variable that evolves in time, such as, x, is called a *process*. This sort of update formula describes a "memoryless" process in the sense that only the present value of x is required to calculate its update rather than requiring its past values. This process is also deterministic and continuous.[9] Then the natural stochastic generalization of a first-order differential equation is *a continuous memoryless stochastic process*, usually referred to as a continuous Markov process.

Let us look for the most general form the increment of a continuous Markov process $X(t)$ can take.[10] As the process $X(t)$ is "memoryless," this increment can only depend on the time step dt, the present time t, and $X(t)$ the *present* value of our process. So we can write it as a function of dt, t, and $X(t)$: say,

$$\Upsilon(dt, X(t), t) \equiv X(t + dt) - X(t). \tag{8.44}$$

Moreover, as $X(t)$ is a continuous process, its increment will also be continuous [i.e., $\Upsilon(dt, x, t) \to 0$ as $dt \to 0$], and it will be a smooth (differentiable) function of $X(t)$ and t.

Now let us divide the interval of time from t to $t + dt$ into s subintervals of equal length dt/s delimited by the times $t_k = t + (dt/s)k$, where $k = 0, \ldots, s$. Then we see that Υ must satisfy the following self-consistency condition:

$$\Upsilon(dt, X(t), t) = \sum_{k=1}^{s} \left\{ X(t_k) - X(t_{k-1}) \right\} = \sum_{k=1}^{s} \Upsilon(dt/s, X(t_{k-1}), t_{k-1}). \tag{8.45}$$

Here it is quite tempting to argue the following way. As $\Upsilon(dt, X(t), t)$ must be on the order of dt and the various $X(t_k)$ will differ from $X(t)$ at most by something on the order of dt, up to this order we can replace $\Upsilon(dt/s, X(t_{k-1}), t_{k-1})$ by $\Upsilon(dt/s, X(t), t)$. But this argument is wrong, because $\Upsilon(dt, X(t), t)$ can be of a *lower* order than dt. Why? Because although $X(t)$ is continuous, its randomness leads to kinks in its sampled paths $x(t)$, making $X(t)$ nondifferentiable (see the inset in Figure 8.2). Thus, if we approximate[11] $X(t_{k-1})$ in the right end of (8.45) by $\bar{X}(t)$, where $\bar{X}(t)$ is some interpolated value between $X(t)$ and $X(t + dt)$, we will be making a change in (8.45) of higher order than dt.

The important point is, however, that this part of $\Upsilon(dt, X(t), t)$, which is of lower order than dt, is equally likely to be negative or positive and is mostly canceled on

---

[9]x can be expressed as the integral of A, so it is continuous.

[10]To avoid confusing a stochastic variable with one of its particular instances (i.e., one of its sampled values), we will use capitals for the random variable and lowercase letter for its sampled values.

[11]Notice that there is no such problem for times t_{k-1} themselves, as they are not random variables. All the time instants t_{k-1} can be approximated by t.

average. Whatever comes out of an average will never have any trace of this order lower than dt because averaged paths have no kinks. This means that as long as we do it consistently throughout the calculations, we are free to approximate $X(t_{k-1})$ by any arbitrary interpolation of $X(t)$ and $X(t + dt)$, such as

$$\bar{X}(t) \equiv p_b X(t) + p_e X(t + dt), \tag{8.46}$$

where p_b and p_e are positive and $p_b + p_e = 1$. The error of order lower than dt does not alter the statistical information we can extract from the stochastic process. Therefore,

$$\Upsilon(dt, X(t), t) = \sum_{k=1}^{s} \Upsilon(dt/s, \bar{X}(t), t). \tag{8.47}$$

Taking s to be arbitrarily large, it follows from the central limit theorem [224] that $\Upsilon(dt, X(t), t)$ is a normal (Gaussian) random variable. Normal random variables are characterized completely by their means and their variances. Using (8.47), we find the following expressions for the mean and variance of $\Upsilon(dt, X(t), t)$:

$$\text{mean}\left\{\Upsilon(dt, X(t), t)\right\} = s \text{ mean}\left\{\Upsilon(dt/s, \bar{X}(t), t)\right\}, \tag{8.48}$$
$$\text{var}\left\{\Upsilon(dt, X(t), t)\right\} = s \text{ var}\left\{\Upsilon(dt/s, \bar{X}(t), t)\right\}. \tag{8.49}$$

Assuming that the variance and the mean are smooth functions of dt/s, we can expand them into a power series in dt/s, keeping only the lowest-order term [i.e., the term linear in dt as $\Upsilon(0, \bar{X}(t), t) = 0$], so that mean $\left\{\Upsilon(0, \bar{X}(t), t)\right\} = 0$ and var $\left\{\Upsilon(0, \bar{X}(t), t)\right\} = 0$,

$$\text{mean}\left\{\Upsilon(dt, X(t), t)\right\} \equiv A\left(\bar{X}(t), t\right) dt \tag{8.50}$$
$$\text{var}\left\{\Upsilon(dt, X(t), t)\right\} \equiv D\left(\bar{X}(t), t\right) dt. \tag{8.51}$$

Now, a normal variable $G(\bar{m}, v)$ of mean $\bar{m}$ and variance v can always be written as the sum of $\bar{m}$ and $\sqrt{v}\,W$, where W is the normal variable of zero mean and unit variance (Problem 8.4). Thus, the stochastic generalization (continuous Markov process) of the deterministic update formula (8.43) is given by[12]

$$X(t + dt) = X(t) + A\left(\bar{X}(t), t\right) dt + \sqrt{D\left(\bar{X}(t), t\right)}\, W(t) \sqrt{dt}, \tag{8.52}$$

where dt is understood never to be negative,[13] and $W(t)$ is a continuous Markov process which at any given time is a normal variable of mean zero and unity variance.

---

[12]You can now see clearly why mathematicians do not like the Langevin way (8.29) of writing down a stochastic differential equation, which is so popular among physicists. The Langevin force in (8.29) corresponds to the last term on the right-hand side of (8.52) divided by dt. As this term is proportional to $\sqrt{dt}$, such a force diverges (that is why its self-correlation function yields a delta function). As a mathematician would say, a Wiener process does not have a derivative.
[13]Mathematicians avoid such considerations by writing (8.52) in more abstract but less transparent form, where $W(t)\sqrt{dt}$ is replaced by $dW(t)$.

The process $W(t)$ is named after Norbert Wiener, who studied it in great detail, that is why we have chosen the letter W to represent it here.

The two most popular choices for $\bar{X}(t)$ are $X(t)$, which leads to Ito's stochastic calculus, and $[X(t + dt) + X(t)]/2$, which leads to Stratonovich's. Itô's seems to come first in this popularity list, perhaps because it makes calculations and proofs less cumbersome than in Stratonovich's stochastic calculus.

Itô's stochastic differential equations have the form

$$X(t + dt) = X(t) + A\left(X(t), t\right) dt + \sqrt{D\left(X(t), t\right)}\, W(t)\, \sqrt{dt}. \qquad (8.53)$$

To understand why this is such a great choice of interpolation style, notice that as $X(t)$ and $W(t)$ are independent (uncorrelated) processes, for any function $\mathcal{G}_1$ of $X(t)$ and $\mathcal{G}_2$ of $W(t)$, we have

$$\langle \mathcal{G}_1\left(X(t)\right) \mathcal{G}_2\left(W(t)\right) \rangle = \langle \mathcal{G}_1\left(X(t)\right) \rangle \langle \mathcal{G}_2\left(W(t)\right) \rangle. \qquad (8.54)$$

This "disentanglement," which is such a great help in calculations, would not happen, if instead of $X(t)$ we had $X(t + dt)$ above, because the latter is correlated to $W(t)$. Using (8.54), we see that Langevin's assumption of a vanishing $\langle x(t)F(t) \rangle$ holds true in Ito's calculus:

$$\left\langle X(t)W(t)\sqrt{dt} \right\rangle = \langle X(t) \rangle \underbrace{\langle W(t) \rangle}_{0} \sqrt{dt} = 0. \qquad (8.55)$$

Another useful tool for making calculations with (8.53) is this: Consider $W(t)\sqrt{dt}\, W(t')\sqrt{dt'}$. Due to the differentials, the variance and all higher moments of $W(t)\sqrt{dt}\, W(t')\sqrt{dt'}$ will be of higher order than dt and dt'. In other words, any fluctuation in $W(t)\sqrt{dt}\, W(t')\sqrt{dt'}$ will be negligible. Thus, $W(t)\sqrt{dt}\, W(t')\sqrt{dt'}$ can be regarded as a sure quantity (deterministic) equal to its mean. As $\langle W(t)W(t') \rangle$ vanishes for $t \neq t'$ and is unity for $t = t'$, we have

$$W(t)\sqrt{dt}\, W(t')\sqrt{dt'} = \begin{cases} 0 & \text{for } t \neq t' \\ dt & \text{for } t = t' \end{cases} \qquad (8.56)$$

Now let us show that Ito's calculus has a different rule for chain differentiation (change of variables) than ordinary calculus. Let f be an arbitrary function of $X(t)$: What is the expression for

$$df(X(t)) = f(X(t + dt)) - f(X(t)) \qquad (8.57)$$

according to Ito's calculus? Assuming that f is continuous and reasonably smooth,[14] we can expand $f(X(t + dt))$ in a Taylor series and neglect all terms of order higher than dt:

$$df(X(t)) = \left. \frac{df}{dx} \right|_{X(t)} dX(t) + \frac{1}{2} \left. \frac{d^2 f}{dt^2} \right|_{X(t)} \{dX(t)\}^2$$

[14]Small enough to have a first and a second derivative.

$$= \left\{ A\left(X(t), t\right) \left.\frac{df}{dx}\right|_{X(t)} + \frac{1}{2} D\left(X(t), t\right) \left.\frac{d^2 f}{dt^2}\right|_{X(t)} \right\} dt$$

$$+ \sqrt{D\left(X(t), t\right)} \left.\frac{d^2 f}{dt^2}\right|_{X(t)} W(t)\sqrt{dt}, \qquad (8.58)$$

where we have used (8.53) in $dX(t) \equiv X(t + dt) - X(t)$ and kept terms only up to the order of dt. Equation (8.58) is known as Ito's formula or Ito's rule. It differs from the usual rule for chain differentiation in ordinary calculus by the presence of the second term inside the braces on the second line of (8.58). Like the chain rule in ordinary calculus, this is a very important tool in Ito calculus. Next, we use it to derive a differential equation for the probability distribution associated with $X(t)$.

If we take the ensemble average of (8.58), the term on the last line vanishes, so that dividing by dt, we find that

$$\frac{d}{dt} \langle f\left(X(t)\right)\rangle = \left\langle A\left(X(t), t\right) \left.\frac{df}{dx}\right|_{X(t)} \right\rangle + \frac{1}{2} \left\langle D\left(X(t), t\right) \left.\frac{d^2 f}{dt^2}\right|_{X(t)} \right\rangle. \quad (8.59)$$

Now let us call $P(x, t)$ the probability distribution of $X(t)$. In terms of $P(x, t)$, we can rewrite (8.59) as

$$\frac{d}{dt} \int_{-\infty}^{\infty} dx\, P(x, t) f(x) = \int_{-\infty}^{\infty} dx\, P(x, t) \left\{ A(x, t) \frac{df(x)}{dx} + \frac{1}{2} D(x, t) \frac{d^2 f(x)}{dt^2} \right\}. \quad (8.60)$$

Integrating by parts, we obtain that

$$\int_{-\infty}^{\infty} dx \left\{ \frac{\partial}{\partial t} P(x, t) + \frac{\partial}{\partial x} [A(x, t) P(x, t)] - \frac{1}{2} \frac{\partial^2}{\partial x^2} [D(x, t) P(x, t)] \right\} f(x) = 0. \quad (8.61)$$

As this must vanish for any $f(x)$, the expression between the braces in the integrand must vanish at every point x. So we find the following partial differential equation for $P(x, t)$:

$$\frac{\partial}{\partial t} P(x, t) + \frac{\partial}{\partial x} [A(x, t) P(x, t)] - \frac{1}{2} \frac{\partial^2}{\partial x^2} [D(x, t) P(x, t)] = 0, \qquad (8.62)$$

which is known as the *Fokker–Planck equation*. The meaning of the coefficients A and D becomes clear here: $A(x, t)$ tells us how fast the probability distribution $P(x, t)$ drifts away from a given point x, and $D(x, t)$ tells us how fast it diffuses from x.

Let us look into Stratonovich's version now. His stochastic differential equations have the following form:

$$dX(t) = \mathcal{A}\left(\frac{X(t + dt) + X(t)}{2}, t\right) dt + \mathcal{B}\left(\frac{X(t + dt) + X(t)}{2}, t\right) W(t)\sqrt{dt}, \qquad (8.63)$$

where we have used the shorthand notation $dX(t) \equiv X(t + dt) - X(t)$. With Stratonovich's choice of $\bar{X}$ as the algebraic mean of $X(t)$ and $X(t+dt)$, Langevin's assumption of a vanishing $\langle x(t)F(t) \rangle$ does not hold; instead,

$$
\begin{aligned}
\left\langle XW\sqrt{dt} \right\rangle_S &= \left\langle \frac{X(t+dt) + X(t)}{2} W(t)\sqrt{dt} \right\rangle \\
&= \underbrace{\left\langle X(t)W(t)\sqrt{dt} \right\rangle}_{\langle X(t) \rangle \underbrace{\langle W(t) \rangle}_{0} \sqrt{dt}} + \frac{1}{2} \langle \mathcal{B}\left(X(t), t\right) \rangle \, dt \\
&= \frac{1}{2} \langle \mathcal{B}\left(X(t), t\right) \rangle \, dt,
\end{aligned}
\tag{8.64}
$$

where the subscript S is for Stratonovich, and we have used (8.63), keeping only terms up to the order of dt.

We can use the same expansion technique to find the Ito stochastic equation that is equivalent to (8.63). Expanding $\mathcal{A}$ and $\mathcal{B}$ and neglecting all terms of order higher than dt, we obtain

$$
dX(t) = \left\{ \mathcal{A}\left(X(t), t\right) + \frac{1}{2}\mathcal{B}\left(X(t), t\right) \frac{\partial}{\partial x}\mathcal{B}\left(X(t), t\right) \right\} dt + \mathcal{B}\left(X(t), t\right) W(t)\sqrt{dt}.
\tag{8.65}
$$

So Stratonovich's stochastic differential equation (8.63) is equivalent to an Ito stochastic differential equation where the drift coefficient is given by

$$
A\left(X(t), t\right) = \mathcal{A}\left(X(t), t\right) + \frac{1}{2}\mathcal{B}\left(X(t), t\right)\frac{\partial}{\partial x}\mathcal{B}\left(X(t), t\right)
\tag{8.66}
$$

and the diffusion coefficient is given by

$$
D\left(X(t), t\right) = \mathcal{B}^2\left(X(t), t\right).
\tag{8.67}
$$

We can use this equivalence to show that Strotonovich's chain rule, unlike Ito's, is the same as that of ordinary calculus [224]. From Ito's rule (8.58), (8.66), and (8.67), we find that

$$
\begin{aligned}
df(X(t)) = \Bigg\{ &\left[\mathcal{A}\left(X(t), t\right) + \frac{1}{2}\mathcal{B}\left(X(t), t\right)\frac{\partial}{\partial x}\mathcal{B}\left(X(t), t\right) \right]\frac{d}{dx}f\left(X(t)\right) \\
&+ \frac{1}{2}\mathcal{B}^2\left(X(t), t\right)\frac{d^2}{dx^2}f\left(X(t)\right) \Bigg\} dt + \mathcal{B}\left(X(t), t\right)\frac{d}{dx}f\left(X(t)\right) W(t)\sqrt{dt}.
\end{aligned}
\tag{8.68}
$$

Calling $f\left(X(t)\right)$ a new stochastic variable,

$$
Y(t) \equiv f\left(X(t)\right),
\tag{8.69}
$$

and assuming that there is an inverse transformation $g\left(Y\left(t\right)\right)$ that changes $Y\left(t\right)$ back into $X\left(t\right)$,

$$X\left(t\right) = g\left(Y\left(t\right)\right), \tag{8.70}$$

we see that

$$\frac{df}{dx} = \left(\frac{dg}{dy}\right)^{-1} \quad \text{and} \quad \frac{d^2f}{dx^2} = -\frac{d^2g}{dy^2}\left(\frac{dg}{dy}\right)^{-3}. \tag{8.71}$$

Thus, (8.68) becomes

$$
\begin{aligned}
dY\left(t\right) = & \left\{\left[\mathcal{A} + \frac{\mathcal{B}}{2}\frac{\partial\mathcal{B}}{\partial x}\left(\frac{dg}{dy}\right)^{-1}\right]\left(\frac{dg}{dy}\right)^{-1} - \frac{\mathcal{B}^2}{2}\frac{d^2g}{dy^2}\left(\frac{dg}{dy}\right)^{-3}\right\}dt \\
& + \mathcal{B}\left(\frac{dg}{dy}\right)^{-1}W\left(t\right)\sqrt{dt}.
\end{aligned}
\tag{8.72}
$$

Inverting (8.66) and (8.67), we find that the Itô's stochastic differential equation (8.72) corresponds to the following Stratonovich's form:

$$dY\left(t\right) = \left\{\mathcal{A}\,dt + \mathcal{B}W\left(t\right)dt\right\}\left(\frac{dg}{dy}\right)^{-1} \tag{8.73}$$

or, in other words, to

$$
\begin{aligned}
d_Sf\left(X\left(t\right)\right) = & \frac{d}{dx}f\left(\frac{X\left(t+dt\right)+X\left(t\right)}{2}, t\right)\left\{\mathcal{A}\left(\frac{X\left(t+dt\right)+X\left(t\right)}{2}, t\right)dt\right. \\
& \left. + \mathcal{B}\left(\frac{X\left(t+dt\right)+X\left(t\right)}{2}, t\right)W\left(t\right)\sqrt{dt}\right\},
\end{aligned}
\tag{8.74}
$$

which is analogous to the chain rule of ordinary calculus.

This is all we are going to mention about stochastic calculus in this brief introduction to the subject. See some of the references in the recommended reading for a more comprehensive discussion of this vast field of mathematics.

8.3 RATE EQUATIONS WITH NOISE: THE EFFECT OF SPONTANEOUS EMISSION

Now we can go back to our simple back-of-the-envelope rate equations for a laser and see the effect of noise. There are various sources of noise in a laser. There is noise from spontaneous emission, noise from the intrinsic quantum mechanical nature of light, pump noise, noise from the random decay processes that take the atoms or molecules from the lower energy level of the lasing transition to other levels that do not take part in the laser,[15] and so on. Here we consider only spontaneous emission noise. This is often a good approximation [611].

---

[15]When chance leaves an atom/molecule in this lower level, it can absorb a photon from the cavity mode.

To describe spontaneous emission noise, we add a Langevin force $F(t)$ to the rate equation for the photon number n. So (8.7) remains unchanged, but instead of (8.6), we now have

$$\dot{n} = -\kappa n + \beta \gamma_{\parallel} N n + R_{\mathrm{sp}} + F, \tag{8.75}$$

where

$$\langle F(t)F(t') \rangle = D\,\delta(t - t'). \tag{8.76}$$

What value should D have? As we have seen in Section 8.2, D is the diffusion coefficient associated with the Brownian motion due to F. Thus, the variance of n is Dt. We can calculate this variance using a simple heuristic argument.

As the only source of noise that we are considering here is spontaneous emission, if we separate from n the photons n_{sp} that have been generated spontaneously, we will be able to treat the remaining number of photons as an ordinary (deterministic) variable. Let us write the average of the square of the photon number in the following way:

$$\begin{aligned}
\langle n^2 \rangle &= \langle (n - n_{\mathrm{sp}} + n_{\mathrm{sp}})(n - n_{\mathrm{sp}} + n_{\mathrm{sp}}) \rangle \\
&= \langle n - n_{\mathrm{sp}} \rangle \langle n - n_{\mathrm{sp}} \rangle + 2 \langle n - n_{\mathrm{sp}} \rangle \langle n_{\mathrm{sp}} \rangle + \langle n_{\mathrm{sp}}^2 \rangle \tag{8.77}
\end{aligned}$$

Now, in a laser the dominant contribution for the photon number should be that of stimulated emission, so we can approximate $\langle n - n_{\mathrm{sp}} \rangle$ in (8.77) by $\langle n \rangle$ and neglect $\langle n_{\mathrm{sp}}^2 \rangle$. The variance of n will then be

$$\langle n^2 \rangle - \langle n \rangle^2 = 2\langle n \rangle \langle n_{\mathrm{sp}} \rangle, \tag{8.78}$$

and as by definition of spontaneous emission rate, the number of photons emitted spontaneously up to time t is $\langle n_{\mathrm{sp}} \rangle = R_{\mathrm{st}}t$, we have

$$\langle n^2 \rangle - \langle n \rangle^2 = 2\langle n \rangle R_{\mathrm{st}}t, \tag{8.79}$$

so we can conclude that

$$D = 2\langle n \rangle R_{\mathrm{st}}. \tag{8.80}$$

With spontaneous emission noise incorporated in our simple rate equations, let us see how this noise affects the steady state of the laser. The steady state can be determined by taking the ensemble average of the rate equations and setting $\langle \dot{n} \rangle$ and $\langle \dot{N} \rangle$ to zero. Then we get the following equations for the steady-state photon number n_s and the steady-state population of the upper lasing level N_s:

$$\kappa n_s = (n_s + 1)\gamma_{\parallel}\beta N_s, \tag{8.81}$$

$$\Lambda = \gamma_{\parallel}N_s + \gamma_{\parallel}\beta n_s N_s. \tag{8.82}$$

Solving (8.81) for n_s and (8.82) for N_s, we obtain

$$n_s = \frac{(\gamma_{\parallel}\beta/\kappa)N_s}{1 - (\gamma_{\parallel}\beta/\kappa)N_s} \quad \text{and} \quad N_s = \frac{\Lambda/\gamma_{\parallel}}{1 + \beta n_s}. \tag{8.83}$$

This is not a solution for n_s and N_s, as each of them is given in terms of the other, but it yields some insight into the workings of a laser.

From the second equation in (8.83) we see that when n_s becomes on the order of $n_0 \equiv 1/\beta$, the steady-state atomic population starts decreasing. This is the phenomenon of gain saturation, and we call n_0 the saturation photon number. When the photon number grows near n_0, the gain (proportional to N_s) starts decreasing, preventing the photon number from growing indefinitely. Thanks to saturation, the laser is not a bomb.

From the first equation in (8.83) we see that when N_s approaches the value $N_0 \equiv \kappa/(\gamma_\| \beta)$, the steady-state photon number grows indefinitely. Saturation, however, prevents N_s from ever approaching N_0 and n_s diverging, as we can see by substituting N_s given by the second equation in (8.83) into the first and solving for n_s,

$$ n_s = \frac{n_0}{N_0} \left\{ \frac{1}{2} \left(\frac{\Lambda}{\gamma_\|} - N_0 \right) + \sqrt{ \frac{1}{4} \left(\frac{\Lambda}{\gamma_\|} - N_0 \right)^2 + \frac{N_0}{n_0} \frac{\Lambda}{\gamma_\|} } \right\}. \tag{8.84} $$

If N_0 is much larger than both $\Lambda/\gamma_\|$ and $\Lambda/(\gamma_\| n_0)$, we can see by expanding the square root in (8.84) that the laser will not lase (i.e., $n_s \approx 0$). If, on the other hand, N_0 is much smaller than both $\Lambda/\gamma_\|$ and $\Lambda/\gamma_\| n_0$, expanding the square root in (8.84) appropriately, we see that the laser will lase (i.e., have $n_s > 1$) for pump rates Λ larger than $(N_0/n_0)\gamma_\|$. So the smaller N_0, the smaller the number of inverted atoms that we need to get the laser to work. For this reason, we call N_0 the critical atom number.

Assuming that the fluctuations around the steady state caused by the spontaneous emission noise term F are small, we write

$$ N = N_s + \delta N \quad \text{and} \quad n = n_s + \delta n, \tag{8.85} $$

where $\delta N \ll N_s$ and $\delta n \ll n_s$. Substituting (8.85) in (8.7) and (8.75), we find that

$$ \delta \dot{n} = (\gamma_\| \beta N_s - \kappa)\delta n + (n_s + 1)\gamma_\| \beta \, \delta N + F, \tag{8.86} $$
$$ \delta \dot{N} = -\gamma_\| N_s \beta \, \delta n - (\beta n_s + 1)\gamma_\| \, \delta N. \tag{8.87} $$

These equations can be solved using a Fourier transform (Problem 8.5). From this solution, we obtain the following expression for the variance of the photon number:

$$ \langle \delta n^2 \rangle = \left\{ 1 - \frac{(n_s + 1)\beta n_s}{(n_s + 2)\beta n_s + 1} \frac{(n_s + 1)(1 + \beta n_s)}{(n_s + 1)(\beta n_s + 1) + N_0/n_0} n_s \right\} n_s \tag{8.88} $$

Figure 8.3 shows how the steady-state photon number and its fluctuations behave as we vary the rate at which energy is fed into the laser (i.e., the pump parameter Λ). The plot on the left shows the typical behavior of a laser (class A). For $\Lambda < \kappa/\beta$, the laser does not oscillate. In this regime, spontaneous emission is dominant and the laser behaves like an ordinary thermal source of light. For $\Lambda > \kappa/\beta$, the laser oscillates and the steady-state photon number increases linearly with the pump. The region around $\Lambda = \kappa/\beta$, which separates the nonoscillatory from the oscillatory regime, is called the *laser threshold*. As the laser crosses its threshold, the fluctuations in the photon number peak. This is the signature of a phase transition, as when water boils. In fact, there is an analogy [147] between the one-photon laser phase threshold and a second-order phase transition.[16] The laser is in this sense a macroscopic device, operating with a large number (thermodynamic limit) of atoms (or molecules) and photons.

As β becomes larger, however, this analogy begins to break because the very notion of a laser threshold starts losing its meaning [518]. Large β can turn a laser into a microscopic device operating with only a few photons, for which the thermodynamic limit of a large number of particles no longer applies. The plot on the left in Figure 8.3 illustrates this point. You can see the kink at $M = 1$ in the photon number curve become smoother and start disappearing. The peak in the Fano parameter curve also becomes wider, and this curve starts becoming flat for large M. But this is all that this simple rate equation analysis can reveal. We cannot go to the limit of $\beta \rightarrow 1$ here, because the rate equations break down well before this limit. We will go back to this problem later using a fully quantum theory.

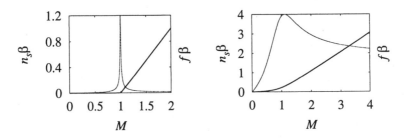

Figure 8.3 The solid line represents the steady-state photon number and the dashed line the Fano parameter $f \equiv \langle \delta n^2 \rangle / n_s$, a good measure of the fluctuations in the photon number. Both n_s and f are plot against the normalized pump parameter $M \equiv \beta \Lambda / \kappa$. On the left plot we used a typical spontaneous emission factor of $\beta \sim 10^{-5}$. On the right plot we used $\beta \sim 0.05$. On both plots, we assumed that $\kappa \ll \gamma_{\parallel}$, so that $N_0 / n_0 \ll 1$.

The cavity QED cousins of the laser and the micromaser are in this $\beta \rightarrow 1$ regime. For them, β, defined earlier as the fraction of spontaneous emission that goes into the lasing mode, even loses its meaning, because in these devices, *all* spontaneous emission photons go into the lasing (or masing) mode. It makes more sense to talk about the saturation photon number n_0 and the critical atom number N_0. The cavity

---

[16]A second-order phase transition is continuous. There are also discontinuous phase transitions, called *first-order transitions*. A laser based on a two-photon transition would have a threshold analogous to a first-order phase transition. This is reflected in the two-photon micromaser, whose nonoscillatory steady state is actually stable. The two-photon micromaser gets out of this stable nonoscillatory steady state by tunneling over the barrier that separates it from the stable oscillatory steady state. For this reason, a two-photon micromaser has a starting time that is quite long compared to other characteristic times in the system, reaching the order of seconds.

QED cousins of the laser and maser have both n_0 and N_0 smaller than 1. They are truly microscopic devices: able to lase (or mase) with less than one atom and less than one photon on average in the cavity!

8.4 IDEAL LASER LIGHT: WHAT QUANTUM STATE OF LIGHT WOULD BE GENERATED IN AN IDEAL LASER?

Even after including noise, so far our analysis of the laser has been classical. In this section we take a first look into the quantum mechanics of a laser by considering it as an ideal counterpart of a radio transmitter in the optical regime. This rough approximation will allow us to work out exactly what sort of quantum state of light an ideal laser would generate.

In Section 8.1 we saw how the laser can be viewed as a radio transmitter for optical frequencies. This optical transmitter is based on an analogy, where the LC resonant circuit of the radio transmitter is replaced by a cavity in the optical regime, and the amplifying element (vacuum tube) is replaced by excited atoms or molecules. Essential for the performance of the atoms or molecules as amplifying elements in the optical regime is stimulated emission. Although spontaneous emission is required to start the laser, later it becomes a sort of parasite that only contributes to noise. In an ideal optical transmitter, spontaneous emission should be negligible. This is roughly what happens in a laser operating far above threshold. To neglect spontaneous emission while keeping a quantum description of light, we will replace the amplifying atoms or molecules by classical dipoles.

One ingredient that we cannot fail to include, though, is cavity damping. Without it, the energy pumped into the laser just keeps piling up, making the intensity of the light inside the cavity grow indefinitely, quite unlike any real device. A perfect cavity is the analog of a lossless LC circuit: It does not need an amplifying element to become a self-sustained oscillator.

We deal with cavity damping in the following phenomenological way. Consider the closed system formed by the cavity and its environment. This environment is whatever continuum set of degrees of freedom the energy that escapes the cavity dissipates into (e.g., the external electromagnetic modes in the case of damping exclusively through transmissivity losses, or phonons in the cavity walls). The important thing is that this environment can be regarded as a set of harmonic oscillators. For concreteness, we assume that the losses are due only to the transmissivity of the cavity walls.

In the zeroth-order approximation (in the transmissivity), the cavity is perfect and the Hamiltonian for the cavity and the outside is just the sum of the single-mode cavity Hamiltonian,

$$\hat{H}_{\text{cav}} = \hbar\omega_c \hat{a}^\dagger \hat{a}, \tag{8.89}$$

and the Hamiltonian for the external continuum of modes,

$$\hat{H}_{\text{out}} = \hbar \int_{-\infty}^{\infty} d\eta\, \eta \hat{b}^\dagger(\eta)\hat{b}(\eta), \tag{8.90}$$

which we model as a one-dimensional continuum and for mathematical convenience extend to *negative* frequencies.[17]

Now as the damping increases, we would expect to have corrections to the cavity and external modes themselves, as both cavity and outside would no longer satisfy the perfect reflector's boundary conditions (see Appendix B). We will assume that at least up to the first correction to this zeroth-order approximation, the modes of the cavity and outside are still those that are obtained using the boundary conditions of a perfect reflector. This correction then accounts for cavity damping through a simple interaction term $\hat{H}_{\text{int}}$, where the annihilation of a photon in the cavity mode leads to the creation of a photon in one of the external modes, and vice versa (as it must be Hermitian),

$$\hat{H}_{\text{int}} = \hbar V \int_{-\infty}^{\infty} d\eta \left\{ \hat{a}\hat{b}^{\dagger}(\eta) + \hat{b}(\eta)\hat{a}^{\dagger} \right\}, \tag{8.91}$$

where $V \equiv \sqrt{\kappa/2\pi}$, with κ being the cavity decay rate introduced in the rate equations earlier in the chapter.

The zeroth-order approximation with this correction,

$$\hat{H}_{\text{GC}} = \hat{H}_{\text{cav}} + \hat{H}_{\text{out}} + \hat{H}_{\text{int}}, \tag{8.92}$$

is the *Gardiner–Collett Hamiltonian* [109, 225], which is widely used in quantum optics to model damping in high-Q cavities. In Chapter 9 we discuss it in more detail and even derive it from first principles.

Using the Gardiner–Collett Hamiltonian to describe the cavity damping, the Hamiltonian of our ideal laser can be written as

$$\hat{H} = \hat{H}_{\text{GC}} + \hbar\chi \left\{ e^{i\omega_c t}\hat{a} + \hat{a}^{\dagger}e^{-i\omega_c t} \right\}, \tag{8.93}$$

where we have made the rotating-wave approximation. The classical dipole is resonant with the cavity mode and χ is Rabi frequency obtained from the product of its dipole moment, the vacuum strength in the cavity, and the cavity mode function at the position of the dipole.

Using the technique of Fano diagonalization introduced in Chapter 7, we can rewrite the Gardiner–Collett Hamiltonian as (Problem 8.6)

$$\hat{H}_{\text{GC}} = \hbar \int_{-\infty}^{\infty} d\omega \, \hat{A}^{\dagger}(\omega)\hat{A}(\omega), \tag{8.94}$$

where the global annihilation operator $\hat{A}(\omega)$ is given in terms of the cavity and external annihilation operators by

$$\hat{A}(\omega) = \alpha(\omega)\hat{a} + \int d\eta \, \beta(\omega, \eta)\hat{b}(\eta), \tag{8.95}$$

---

[17]This is a reasonable approximation as long as ω_c is much larger than V^2 in (8.91). See, for example, the discussion in Sec. II.A of [225].

with

$$\alpha(\omega) = \frac{V}{\omega - \omega_c + i\pi V^2}, \tag{8.96}$$

$$\beta(\omega, \eta) = \left\{ V\mathcal{P}\frac{1}{\omega - \eta} + \frac{\omega - \omega_c}{V}\delta(\omega - \eta) \right\}\alpha(\omega). \tag{8.97}$$

As

$$\hat{a} = \int_{-\infty}^{\infty} d\omega\, \alpha^*(\omega)\hat{A}(\omega), \tag{8.98}$$

the ideal laser Hamiltonian becomes

$$\hat{H} = \int_{-\infty}^{\infty} d\omega\, \hat{A}^\dagger(\omega)\hat{A}(\omega) + \hbar\chi \int_{-\infty}^{\infty} d\omega \left\{ e^{i\omega_c t}\alpha^*(\omega)\hat{A}(\omega) + e^{-i\omega_c t}\alpha(\omega)\hat{A}^\dagger(\omega) \right\}. \tag{8.99}$$

In the interaction picture, the time evolution operator $\hat{U}_I$ derived from this Hamiltonian is given by

$$\hat{U}_I(t) = \exp\left(-\frac{i}{\hbar}\int_0^t dt'\, \hat{H}_I(t') \right), \tag{8.100}$$

where

$$\hat{H}_I(t) = \hbar\chi \int_{-\infty}^{\infty} d\omega\, e^{i(\omega_c - \omega)t}\alpha^*(\omega)\hat{A}(\omega) + e^{-i(\omega_c - \omega)t}\alpha(\omega)\hat{A}^\dagger(\omega). \tag{8.101}$$

Integrating (8.101) over time and using (8.95), we find that

$$-\frac{i}{\hbar}\int_0^t dt'\, \hat{H}_I(t') = -i\chi\left\{ C_a(t)\hat{a} + \int_{-\infty}^{\infty} d\eta\, C_b(\eta, t)\hat{b}(\eta) + \text{H.c.} \right\}, \tag{8.102}$$

where

$$C_a(t) = \int_{-\infty}^{\infty} d\omega\, \frac{e^{i(\omega_c\omega)t} - 1}{i(\omega_c - \omega)}|\alpha(\omega)|^2, \tag{8.103}$$

$$C_b(\eta, t) = \int_{-\infty}^{\infty} d\omega\, \frac{e^{i(\omega_c\omega)t} - 1}{i(\omega_c - \omega)}\beta(\omega, \eta)\alpha^*(\omega). \tag{8.104}$$

Now, neglecting spontaneous emission is only a reasonable approximation to a real laser in the steady state, and even then only if this steady state is far above threshold. Taking the limit $t \to \infty$ of (8.103) and (8.104) to obtain the steady state, we find that[18] (Problem 8.7)

$$\lim_{t\to\infty} C_a(t) = \frac{2}{\kappa} \quad \text{and} \quad \lim_{t\to\infty} C_b(\eta, t) = \frac{2V}{i(\omega_c\eta)\kappa}. \tag{8.105}$$

---

[18]Provided that $V \neq 0$.

Thus, in the steady state, the evolution operator is a product of Glauber displacement operators (the Glauber displacement operator was introduced in Chapter 4),

$$\lim_{t \to \infty} \hat{U}_I(t) = \hat{D}_{\hat{a}}\left(-i2\frac{\chi}{\kappa}\right) \prod_{\{\eta\}} \hat{D}_{\hat{b}(\eta)d\eta}\left(2\frac{\chi}{\kappa}\frac{V}{\omega_c - \eta}\right), \tag{8.106}$$

where the subscript in each displacement operator denotes the mode it corresponds to (e.g., $\hat{D}_{\hat{a}}$ corresponds to the cavity mode), and the productory is over the continuum of external modes.

The steady-state quantum state of our closed system is obtained by applying (8.106) to the initial quantum state. Assuming that we started with the quantum vacuum before switching on our ideal laser, both the cavity and outside are in a product of coherent states in the steady state.[19] In particular, the cavity mode is in the coherent state

$$\left|-i2\frac{\chi}{\kappa}\right\rangle = e^{-2(\chi/\kappa)^2} \sum_{j=0}^{\infty} \frac{1}{j!}\left(-i2\frac{\chi}{\kappa}\right)^j |j\rangle, \tag{8.107}$$

where $|j\rangle$ is a number state with j photons. The main effect of spontaneous emission is to make the phase of this coherent state perform a random walk which leads to a very slow phase diffusion. For times short enough compared with this diffusion time, a coherent state is a good approximation to the quantum state of laser light.

8.5   THE SINGLE-ATOM MASER

The first ASER was the ammonia-beam maser, whose idea came to Charles Townes in 1951 on a park bench [602] and was realized three years later by James P. Gordon, Herbert J. Zeiger, and Townes at Columbia University [242]. It consisted of a microwave cavity through which a beam of excited ammonia molecules was sent (Figure 8.4). At any given time there would be a large number of ammonia molecules in the cavity; otherwise, it would not mase. Incoherent processes such as cavity damping, thermal blackbody radiation, decay from the molecular energy levels of the masing transition, and the velocity spread of the molecular beam played a major role in the dynamics of the device.

It took 31 years to reduce most of these incoherent processes to a negligible level and to build a much simpler device[20]: a maser that mases in the quantum regime of $n_0, N_0 \ll 1$. The first such micromaser was realized by Dieter Meschede, Herbert Walther, and Günter Müller at the Max-Planck Institute for Quantum Optics in Garching, Germany [440].

As in the first maser, the gain medium was also a material beam that flowed across the cavity rather than something that remained inside it (Figure 8.5). But the

---

[19]Notice that, in general, a dissipative system and its environment would become *entangled* rather than remain in a product state through time evolution.
[20]From the point of view of theory.

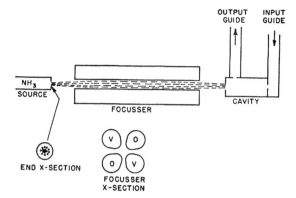

Figure 8.4 Schematic diagram of the first maser. A beam of ammonia emerges from an oven and goes into the focuser, which separates out the molecules in the lower state. As the beam enters the microwave cavity, it consists basically of molecules in the upper state of the masing transition. (Reproduced from [242], ©1955 The American Physical Society.)

similarities stop there, because a micromaser, as the name suggests, is a device where a field can be build inside a cavity due to interaction of the cavity mode with only a few atoms at a time—ideally, one atom at a time, as we assume in this chapter.

For atomic decay times much larger than the transit time through the cavity, there are three parameters that determine whether such a regime of operation can be achieved: the interaction time t_{int}, which is the time an atom spends in the cavity, the interval of time between successive atoms t_{at}, and the cavity decay time t_{cav}. In order to have at most one atom at a time in the cavity, we must require that

$$t_{at} > t_{int}. \tag{8.108}$$

Now, the net gain will depend on the average number of atoms crossing the cavity during its decay time t_{cav}. This number is given by

$$N_{ex} = \frac{t_{cav}}{t_{at}}, \tag{8.109}$$

and to build a field in the cavity, we must require at least that

$$N_{ex} > 1. \tag{8.110}$$

Inequalities (8.108) and (8.110), together with the definition of N_{ex} (8.109), yield the following condition, which must be satisfied to realize a micromaser:

$$t_{int} < t_{at} < t_{cav}. \tag{8.111}$$

The interaction time, t_{int}, can be controlled by selecting the velocities of the atoms in the atomic beam. The time interval between the atoms, t_{at}, can be controlled by

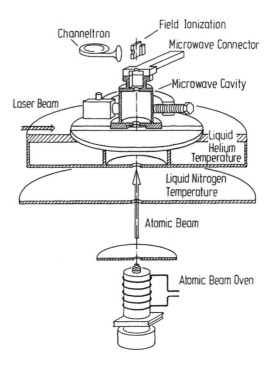

Figure 8.5 Schematic diagram of the first micromaser. A beam of ^{85}Rb emerges from an oven and goes into a niobium superconducting cavity. A laser beam that shines on the atoms just before they enter the cavity prepares them in the upper masing level $63p_{3/2}$ (the lower level of the masing transition is $61p_{3/2}$). When the atoms exit the cavity they go through two field ionization detectors. The first one has a field strength high enough to ionize atoms in the $63p$ state but not those in the $61p$ state. The second has a higher field strength, to ionize atoms in the $61p$ state. In this first experimental demonstration, there was no velocity selection and the thermal velocity spread of the atomic beam prevented direct observation of the Rabi nutation of the atoms. Later a Fizeau-type velocity selector was fitted between the oven and the cavity to allow reasonably fixed interaction times. (Reproduced from [440], ©1985 The American Physical Society.)

the excitation mechanism that determines how many atoms in the beam are prepared in the right state to interact with the cavity mode. Random fluctuations in t_{at} and t_{int} will be neglected in our model.

In addition to condition (8.111), a micromaser is also supposed to operate in an intrinsically nonlinear regime, where each atom performs many Rabi oscillations before exiting the cavity [274]. This means that the coupling constant g between the atoms and the cavity mode must be large enough.[21] Micromasers have only become

---

[21]The Rabi period must be much shorter than the lifetime of a photon in the cavity mode (i.e., $g \gg t_{cav}$).

possible basically because of two reasons: the ability to prepare Rydberg states with large electric dipole moments and small linewidths, and the development of high-Q superconducting microwave cavities.

With these operational conditions satisfied, the interaction between atom and cavity mode is given to a very good approximation by the Jaynes–Cummings model (discussed in Chapter 6). Field damping is nonnegligible only in between the passage of one atom and the next through the cavity (i.e., when the cavity is empty of atoms). In the first micromaser experiment [440], the cavity Q value was about 10^8, giving a photon lifetime of a few milliseconds, $t_{\text{int}} \sim 80\,\mu\text{s}$, $g \sim 43$ kHz, and the atomic lifetime was a few milliseconds. In 1990, the velocity selectivity improved substantially and the cavity Q value grew to about 3×10^{10}, corresponding to $t_{\text{cav}} \sim 0.2$ s [516].

Let us call $\hat{\rho}(t_i)$ the reduced density matrix for the field just before the passage of the ith atom through the cavity. The reduced density matrix after that atom goes through the cavity is given by

$$\hat{\rho}(t_i + t_{\text{int}}) = G(t_{\text{int}})\hat{\rho}(t_i), \tag{8.112}$$

where $G(t_{\text{int}})$ is the superoperator[22] that produces the Jaynes–Cummings time evolution. Now, with the atoms injected in the upper level, the reduced-density matrix for the field remains diagonal throughout the interaction between the atom and the field.[23] So calling $\langle n|\hat{\rho}(t)|n\rangle \equiv p_n(t)$, from our discussion on the Jaynes–Cummings model in Chapter 6, we can see that the $(\langle n|, |n\rangle)$-matrix element of (8.112) is given by

$$p_n(t_i + t_{\text{int}}) = \{1 - C_{n+1}(\tau)\}\, p_n(t_i) + C_n(\tau)p_{n-1}(t_i), \tag{8.113}$$

where

$$C_n(\tau) = \sin^2\left(g\tau\sqrt{n}\right). \tag{8.114}$$

For the time t_p between the exit of the ith atom from the cavity and the arrival of the $(i+1)$th atom at time t_{i+1}, the field dynamics is governed by the Lindblad master equation for cavity damping introduced in Chapter 6:

$$\frac{d}{dt}\hat{\rho} = L\hat{\rho}, \tag{8.115}$$

where L is the Lindblad superoperator, which acts on $\hat{\rho}$ in the following way[24]:

$$L\hat{\rho} \equiv \frac{1}{2t_{\text{cav}}}\left\{2\hat{a}\hat{\rho}\hat{a}^\dagger - \hat{a}^\dagger\hat{a}\hat{\rho} - \hat{\rho}\hat{a}^\dagger\hat{a}\right\}. \tag{8.116}$$

We can integrate (8.115) formally and write the reduced-density matrix for the field at the time of the arrival of the $(i+1)$th atom at the cavity as

$$\hat{\rho}(t_{i+1}) = e^{t_p L}\hat{\rho}(t_i + t_{\text{int}}). \tag{8.117}$$

---

[22] A superoperator is an operator that operates on ordinary operators.

[23] It does not remain diagonal if the atoms are injected in a *quantum superposition* of upper and lower levels.

[24] For simplicity, we assume that the temperature is low enough for the average number of thermal photons to be negligible.

It is very difficult to control experimentally the time between successive atoms in an atomic beam. Here we follow Filipowicz et al. [209] and assume that the time interval t_p has an exponential probability distribution $(1/\bar{t}_p)\exp(-t/\bar{t}_p)$, with mean $\bar{t}_p$. Assuming that the fluctuations of each t_p are statistically independent, $\exp(t_p L)$ and $\hat{\rho}(t_i + t_{\text{int}})$ in (8.117) will not be correlated and the average yields the product of two separate averages: one over the present t_p, the other over previous t_p's on which $\hat{\rho}(t_i + t_{\text{int}})$ alone depends (see the appendix of [209] for a formal proof of this). Calling $\bar{\rho}$ such an averaged $\hat{\rho}$, we can write the evolution of the field in the micromaser as

$$\bar{\rho}(t_{i+1}) = \frac{1}{\bar{t}_p}\int_0^\infty dt_p\ \exp\left(-\left[\frac{1}{\bar{t}_p} - L\right]t_p\right)F(t_{\text{int}})\bar{\rho}(t_i)$$

$$= \frac{1/\bar{t}_p}{(1/\bar{t}_p) - L}F(t_{\text{int}})\bar{\rho}(t_i). \tag{8.118}$$

The steady state of the mapping (8.118) is given by[25]

$$\frac{1}{\bar{t}_p}\left\{1 - F(t_{\text{int}})\right\}\bar{\rho} = L\bar{\rho}. \tag{8.119}$$

Using that the $(\langle n|, |n\rangle)$-matrix element of (8.116) is

$$\langle n|L\bar{\rho}|n\rangle = \frac{1}{t_{\text{cav}}}\left\{np_n - (n+1)p_{n+1}\right\}, \tag{8.120}$$

we find that (Problem 8.8)

$$\bar{p}_n = \bar{p}_0\prod_{k=1}^n \frac{N_{\text{ex}}}{k}\sin^2\left(gt_{\text{int}}\sqrt{k}\right), \tag{8.121}$$

where we have also approximated $\bar{t}_p = \bar{t}_{i+1} - \bar{t}_i - t_{\text{int}} \approx \bar{t}_{i+1} - \bar{t}_i = t_{\text{at}}$.

Notice that N_{ex} is the counterpart in a micromaser of the number of excited atoms of the gain medium in a laser. In a micromaser, however, the interaction time can be varied by selecting a different speed for the atomic beam, whereas in a laser it is always fixed by the short lifetime of the atomic energy levels. So, in the micromaser, the quantity that plays the role of dimensionless pumping parameter is

$$\theta \equiv gt_{\text{int}}\sqrt{N_{\text{ex}}}. \tag{8.122}$$

Figure 8.6 shows the average photon number and its fluctuations as functions of θ obtained numerically using (8.121). The narrow dips in the curve of the average photon number are the signature of trapping states. A *trapping state* is a unique

---

[25]This steady-state equation holds even under more general assumptions than we have made here, see [417].

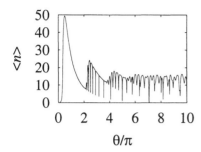 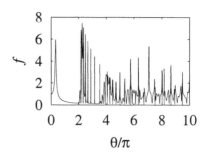

Figure 8.6 The plot on the left shows the average photon number as a function of the dimensionless pumping parameter θ (8.122) in a micromaser with N_{ex} fixed at 50. On the right-hand side you see the corresponding Fano parameter $f \equiv \left(\langle n^2 \rangle - \langle n \rangle^2\right) / \langle n \rangle$. The several dips on the average photon number curve, reflected also in the behavior of the Fano parameter, are the tail signatures of trapping states.

consequence of the coherent quantum dynamics of the micromaser that is not found in ordinary masers and lasers. It is formed when N_{ex} and θ are such that each atom undergoes an integer number of Rabi oscillations in the cavity and exits in the excited state, without changing the number of photons in the cavity. All subsequent atoms will also exit in the excited state (unless N_{ex} and θ are varied), stopping the growth of the photon population at a given maximum photon number. For this reason, the micromaser fields corresponding to trapping states are highly nonclassical. They are also very sensitive to thermal photons, as any fluctuation in the photon number increasing it beyond the maximum photon number will take the micromaser out of a trapping state. Fluctuations in the interaction time, t_{int}, can also take the micromaser out of a trapping state. This need for very low temperatures and for precise control of the interaction times has prevented the realization of trapping states until very recently. With recent technological advances, however, trapping states are even being exploited in schemes for the generation of Fock states.

Trapping states can be characterized by two numbers: the maximum number of photons and the number q of complete (2π) Rabi flips each atom undergoes when the micromaser reaches that trapping state. The trapping-state condition

$$gt_{\text{int}} \sqrt{n_q + 1} = 2\pi q \tag{8.123}$$

then gives us the interaction time that would yield that particular trapping state.

Now, to fix N_{ex} and vary θ, as in Figure 8.6, amounts to *fixing the number of excited atoms in the gain medium* and *varying the interaction time*. In a laser, however, it is the (effective) interaction time that is fixed by the short lifetime of the atomic energy levels involved in optical transitions, as we mentioned above. Only the number of excited atoms in the gain medium (pump rate) can be varied in a laser. So to make a comparison with the typical laser features discussed in Section 8.3, we will break with micromaser tradition and plot the average photon number and its fluctuations for a *fixed interaction time, varying* N_{ex}.

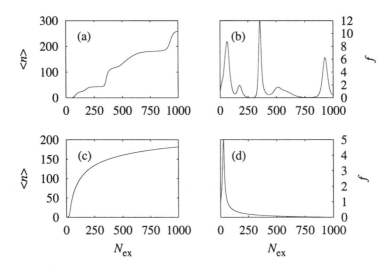

Figure 8.7 To compare with the laser, the average photon number and the Fano parameter $f \equiv \left(\langle n^2 \rangle - \langle n \rangle^2 \right) / \langle n \rangle$ in a micromaser are plotted here as a function of N_{ex} for a fixed interaction time. In (a) and (b), $gt_{\text{int}} = 22$, corresponding to about 3.5 Rabi cycles. In (c) and (d), $gt_{\text{int}} = 0.2$, which corresponds to a Rabi oscillation of only about 11.5 degrees.

Figure 8.7(a) shows the average photon number as a function of N_{ex} for $gt_{\text{int}} = 22$. Instead of the nice straight line above threshold seen in Figure 8.3 for a typical laser, here we have a highly nonlinear curve. The curve of the fluctuations in the photon number, shown in Figure 8.7(b), also has a much richer structure than its laser counterpart in Figure 8.3. Unlike the case of a laser, $gt_{\text{int}} = 22$ corresponds to an impressive coherent quantum evolution where each atom undergoes about three and a half Rabi cycles before exiting the cavity. Figure 8.7(c) and (d) show the average photon number and its fluctuations for the case of a more modest[26] $gt_{\text{int}} = 0.2$. As you can see, the photon number curve is still nonlinear. Moreover, although the curve of the fluctuations in the photon number has a single peak in this region, indicating the threshold as its laser counterpart, unlike Figure 8.3, it becomes sub-Poissonian far above threshold (large N_{ex}).

---

[26]In a laser the effective gt_{int} is typically still much smaller than the modest value used in Figure 8.7(c) and (d). Using the experimental data published in [611] for a 0.2-mm-thin Nd:YVO$_4$ laser crystal, and taking t_{int} to be the lifetime of the lower level of the lasing transition, we find that the effective gt_{int} is about 10^{-4}.

8.6   THE THRESHOLDLESS LASER: A STAGEPOST ON THE ROAD FROM MICROMASER TO LASER

Imagine that technology were so advanced that we could build a micromaser in the optical regime.[27] How could we get it to behave more like a typical laser? The answer is that we would have to spoil the nice coherent quantum dynamics by introducing the dissipative processes that are so common in lasers. Among the many dissipative processes, often the most significant is the extremely fast decay of the polarization of the gain medium. This is why we were able to start our discussion on lasers in this chapter with two rate equations (one for the field intensity, the other for the atomic population) rather than three Maxwell–Bloch equations [533] (where the atomic polarization is also accounted for). What would be the dynamics of an optical micromaser if it had only the incoherent decay of the atomic polarization added to its usual features?[28]

There is a simple way to describe the polarization decay of a two-level atom. First, we just write down the Schrödinger equation in the form of a master equation. Then we add a Lindblad term, which is identical to that introduced in Chapter 6 to account for cavity decay except that $\hat{a}$ must be replaced by the Pauli matrix $\hat{\sigma}_z$ (see the recommended reading).

Thus, the master equation for a two-level atom interacting with the field in a resonant single-mode cavity, under some dephasing mechanism that damps the atomic polarization, is given by

$$\frac{d}{dt}\hat{\rho} = -\frac{i}{\hbar}[\hat{H}, \hat{\rho}] + \mathcal{L}\hat{\rho}, \tag{8.124}$$

where $\hat{H}$ is the Jaynes–Cummings Hamiltonian,

$$\hat{H} = \hbar\hat{a}^\dagger\hat{a} + \hbar\frac{\omega}{2}\hat{\sigma}_z + \hbar g\left(\hat{a}^\dagger\hat{\sigma} + \hat{\sigma}^\dagger\hat{a}\right), \tag{8.125}$$

and

$$\mathcal{L}\hat{\rho} = -\frac{\gamma_\perp}{2}\hat{\rho} + \frac{\gamma_\perp}{2}\hat{\sigma}_z\hat{\rho}\hat{\rho}_z \tag{8.126}$$

is the Lindblad term describing polarization decay.

Before tackling our optical micromaser, it is instructive to look at the effect of polarization decay on spontaneous emission. Without polarization decay, an initially excited two-level atom in a single-mode (perfect) cavity undergoes vacuum Rabi oscillations; that is, the expectation value of its inversion is given by

$$\langle\hat{\sigma}_z(t)\rangle = \cos 2gt. \tag{8.127}$$

---

[27]There was an experiment in 1994 [16] which ultimately aimed at approaching this goal, but it was not possible to reach the strong-coupling regime at the time.
[28]This idea was first considered by the author and collaborators in [174].

For a polarization decay rate $\gamma_\perp$ much larger than the Rabi frequency, you can easily show (Problem 8.9) that spontaneous emission becomes irreversible with

$$\langle \hat{\sigma}_z(t) \rangle \approx e^{-4g^2 t/\gamma_\perp}. \tag{8.128}$$

The time scale $\gamma_\perp/(4g^2)$ of this irreversible spontaneous emission is, however, much longer than that of the coherent Rabi oscillations $1/g$. This long-time scale is a consequence of the sudden switching off of coupling between atom and field due to the destruction of the atomic dipole (polarization). As the atom starts decaying, its dipole is destroyed and the decay has to stop. The time at which the dipole is destroyed is random, leading to an expectation value of $\langle \hat{\sigma}_z \rangle$ with an exponential time decay.

Now, what is the consequence of polarization decay in a micromaser? In other words, what will it do to *stimulated* rather than *spontaneous* emission? The most general way to tackle this problem is that of *damping bases* or the *dissipation picture method* [36, 73]. Instead, as the only decay we want to introduce is that of the polarization, we can afford to use a much simpler and more direct method, which is sort of complementary to that of damping bases.

Here we can use as a basis the eigenstates of coherent part of the dynamics, that is, of the Jaynes–Cummings Hamiltonian,

$$|n\pm\rangle \equiv \frac{1}{\sqrt{2}} \left(|\uparrow\rangle|n\rangle \pm |\downarrow\rangle|n+1\rangle \right), \tag{8.129}$$

where $|\uparrow\rangle$ is the eigenstate of $\hat{\sigma}_z$ with eigenvalue $+1$ (upper atomic state), $|\downarrow\rangle$ is the eigenstate of $\hat{\sigma}_z$ with eigenvalue -1 (the lower atomic state), and $|n\rangle$ is the n-photon state. In this basis, the matrix elements of the two terms on the right-hand side of (8.124) are given by

$$\frac{1}{\hbar} \langle n_s | \hat{H} | n'_{s'} \rangle = \left\{ \left(n + \frac{1}{2} \right) \omega + sg\sqrt{n+1} \right\} \delta_{n,n'} \delta_{s,s'} \tag{8.130}$$

and

$$\frac{2}{\gamma_\perp} \langle n_s | \mathcal{L}\hat{\rho} | n'_{s'} \rangle = -\rho_{n_s,n'_{s'}} + \rho_{n_{-s},n'_{-s'}}, \tag{8.131}$$

where $\rho_{n_p,m_q} \equiv \langle n_p | \hat{\rho} | m_q \rangle$. Then we can rewrite (8.124) as

$$\dot{\rho}_{n_s,n'_{s'}} = -i \left\{ (n-n')\omega + \left(s\sqrt{n+1} - s'\sqrt{n'+1} \right) g \right\} \rho_{n_s,n'_{s'}}$$
$$- \frac{\gamma_\perp}{2} \rho_{n_s,n'_{s'}} + \frac{\gamma_\perp}{2} \rho_{n_{-s},n'_{-s'}}. \tag{8.132}$$

Notice that the "equation of motion" for $\rho_{n_s,n'_{s'}}$, (8.132), is coupled to that for $\rho_{n_{-s},n'_{-s'}}$. To solve these coupled equations, we introduce the slowly varying quadrature-like variables,

$$u_{n,n',s,s'} \equiv \left(\rho_{n_s,n'_{s'}} + \rho_{n_{-s},n'_{-s'}} \right) e^{i(n-n')\omega t}, \tag{8.133}$$

$$v_{n,n',s,s'} \equiv \left(\rho_{n_s,n'_{s'}} - \rho_{n_{-s},n'_{-s'}} \right) e^{i(n-n')\omega t}. \tag{8.134}$$

In terms of these variables, the system of coupled equations (8.132) can be written as

$$\dot{u}_{n,n',s,s'} = -i\left(s\sqrt{n+1} - s'\sqrt{n'+1}\right)gv_{n,n',s,s'}, \tag{8.135}$$

$$\dot{v}_{n,n',s,s'} = -i\left(s\sqrt{n+1} - s'\sqrt{n'+1}\right)gu_{n,n',s,s'} - \gamma_\perp v_{n,n',s,s'}. \tag{8.136}$$

Differentiating (8.136) and using (8.135) to eliminate $\dot{u}_{n,n',s,s'}$, we find a second-order differential equation for $v_{n,n',s,s'}$ alone,

$$\ddot{v}_{n,n',s,s'} + \gamma_\perp \dot{v}_{n,n',s,s'} + \left(s\sqrt{n+1} - s'\sqrt{n'+1}\right)^2 g^2 v_{n,n',s,s'} = 0. \tag{8.137}$$

This is the equation of motion of a damped harmonic oscillator, and the general solution is given by

$$v_{n,n',s,s'}(t) = A_{n,n',s,s'}e^{\lambda_{-,n,n',s,s'}t} + B_{n,n',s,s'}e^{\lambda_{+,n,n',s,s'}t}, \tag{8.138}$$

where

$$\lambda_{\pm,n,n',s,s'} = -\frac{\gamma_\perp}{2} \pm \sqrt{\left(\frac{\gamma_\perp}{2}\right)^2 - \left(s\sqrt{n+1} - s'\sqrt{n'+1}\right)^2 g^2} \tag{8.139}$$

and $A_{n,n',s,s'}$ and $B_{n,n',s,s'}$ are two constants to be determined by the initial conditions.

Assuming that each atom always enters the cavity in the excited state and that the field started with a diagonal density matrix $\rho_{n,n'}(0)$ in the photon-number basis, we find that

$$\rho_{n_s,n'_{s'}} = \frac{1}{2}\exp\left(-\left[\frac{\gamma_\perp}{2} + i(n-n')\omega\right]t_{\text{int}}\right)\left\{\cosh\left(\Omega_{n,n',ss'}t_{\text{int}}\right)\right.$$

$$\left. + \frac{\gamma_\perp/2 - i\left(s\sqrt{n+1} - s'\sqrt{n'+1}\right)g}{\Omega_{n,n',ss'}}\sinh\left(\Omega_{n,n',ss'}t_{\text{int}}\right)\right\}\rho_{n,n'}(0), \tag{8.140}$$

where

$$\Omega_{n,n',ss'} \equiv \sqrt{\left(\frac{\gamma_\perp}{2}\right)^2 - \left(\sqrt{n+1} - s's\sqrt{n'+1}\right)^2 g^2}. \tag{8.141}$$

To convert this density matrix in the dressed basis to the reduced density matrix for the field in the photon number basis, we notice that

$$\rho_{n,m} = \frac{1}{2}\left\{\rho_{n_+,m_+} + \rho_{n_+,m_-} + \rho_{n_-,m_+} + \rho_{n_-,m_-} + \rho_{(n-1)_+,(m-1)_+}\right.$$

$$\left. - \rho_{(n-1)_+,(m-1)_-} - \rho_{(n-1)_-,(m-1)_+} + \rho_{(n-1)_-,(m-1)_-}\right\}. \tag{8.142}$$

Thus, an atom entering the cavity at time t_i and interacting with the field for a time t_{int} before leaving the cavity changes the diagonal reduced density matrix from $\rho(t_i)$

to $\rho(t_i + t_{\text{int}})$, given by

$$
\begin{aligned}
p_{n,m}(t_i + t_{\text{int}}) = \frac{1}{2} \exp\left(-\left[\frac{\gamma_\perp}{2} + i(n - n')\omega\right] t_{\text{int}}\right) \\
\times \Big\{ [K_{n,m,+}(t_{\text{int}}) + K_{n,m,-}(t_{\text{int}})]\, p_{n,m}(t_i) \\
+ [K_{n-1,m-1,+}(t_{\text{int}}) - K_{n-1,m-1,-}(t_{\text{int}})]\, p_{n-1,m-1}(t_i) \Big\},
\end{aligned}
$$
(8.143)

where

$$
K_{n,m,s}(t_{\text{int}}) \equiv \cosh\left(\Omega_{n,m,s} t_{\text{int}}\right) + \frac{\gamma_\perp/2}{\Omega_{n,m,s}} \sinh\left(\Omega_{n,m,s} t_{\text{int}}\right).
$$
(8.144)

Repeating the steps that lead to the steady-state photon distribution (8.121) in Section 8.5, with (8.113) now replaced by (8.143), we find the simple expression

$$
p_n = p_0 \prod_{k=1}^{n} \frac{N_{\text{ex}}}{2k} \left(1 - G_k\right),
$$
(8.145)

where

$$
G_k \equiv e^{-\Gamma w} \left\{ \cosh\left(w\sqrt{\Gamma^2 - n}\right) + \frac{\Gamma}{\sqrt{\Gamma^2 - n}} \sinh\left(w\sqrt{\Gamma^2 - n}\right) \right\},
$$
(8.146)

with the dimensionless parameters Γ and w given by

$$
\Gamma \equiv \frac{\gamma_\perp}{4g} \quad \text{and} \quad w \equiv 2g t_{\text{int}}.
$$
(8.147)

Now suppose that Γ is large enough for Γ^2 always to be much larger than n. In other words, we assume that whatever steady-state field is established inside the cavity, only numbers of photons $n \ll \Gamma^2$ have a nonnegligible probability. As the field in the cavity is only being generated by the incoming excited atoms, this criterion is satisfied when $N_{\text{ex}}/2 \ll \Gamma^2$. With this assumption, we can approximate (8.146) by

$$
G_k \approx \exp\left(-n\frac{w}{2\Gamma}\right),
$$
(8.148)

where we are neglecting terms on the order of n/Γ^2 and higher, but not terms of order w/Γ, as w can be comparable to Γ. If, however, $N_{\text{ex}} \ll 4\Gamma/w$, we can safely take $nw/2\Gamma \ll 1$, and approximate (8.148) further by

$$
G_k \approx 1 - n\frac{w}{2\Gamma}.
$$
(8.149)

In that case, substituting (8.149) in (8.145), we get

$$
p_n = \left(\frac{N_{\text{ex}} w}{4\Gamma}\right)^n p_0.
$$
(8.150)

This means that for $N_{\text{ex}} \ll 4\Gamma/w$, the photon statistics is thermal. From the requirement that the probabilities have to add up to 1, we can determine p_0 and then find the following expression for the average number of photons in the cavity,

$$\langle n \rangle = \frac{N_{\text{ex}}w/(4\Gamma)}{1 - N_{\text{ex}}w/(4\Gamma)}. \tag{8.151}$$

This is the same average photon number that a thermal field would have if its temperature were

$$T = -\frac{\hbar\omega}{k_B} \frac{1}{\ln\left(N_{\text{ex}}w/[4\Gamma]\right)}. \tag{8.152}$$

On the other hand, for $N_{\text{ex}} \gg 4\Gamma/w$, we can approximate[29] (8.148) by $G_k \approx 0$, and then

$$p_n = \left(\frac{N_{\text{ex}}}{2}\right)^n \frac{1}{n!} p_0. \tag{8.153}$$

This photon probability distribution is Poissonian. So $N_{\text{ex}} = 4\Gamma/w$ is the threshold of this laser.

So far, our hypothetical device works like a normal laser with a well-defined threshold. But what if $w/2\Gamma$ is so large that even for $n = 1$, $\exp(-nw/2\Gamma)$ is already almost zero? Then the photon statistics will be Poissonian for any pump power (i.e., even for arbitrarily small values of N_{ex}). In other words, this device will become thresholdless.

What does $w/2\Gamma \gg 1$ mean? Rewriting it in terms of the interaction time, the vacuum Rabi frequency, and the decay rate of the polarization, we find that this condition means that

$$t_{\text{int}} \gg \left(\frac{4g^2}{\gamma_\perp}\right)^{-1}. \tag{8.154}$$

As $4g^2/\gamma_\perp$ is the spontaneous emission rate into the cavity mode derived earlier, this condition says that the interaction time must be longer than the average time required for an atom to emit a photon spontaneously into the cavity. So now we can get a physical picture of what is going on. When an atom enters the cavity, the strong coupling conditions $g \gg t_{\text{cav}}^{-1}, \gamma_\parallel$ will ensure that all emission will be in the cavity mode. But as $\gamma_\perp > g$, the atom can go through the cavity without emitting. Only when the transit time t_{int} is much larger than the spontaneous emission time will the atom emit before leaving the cavity. Notice that this gives us a way of controlling β, the fraction of spontaneous emission into the lasing mode, just by varying the transit time t_{int}. Condition (8.154) takes us to the $\beta \to 1$ regime, which, together with $4g^2/\gamma_\perp \gg t_{\text{cav}}^{-1}$, makes the laser become an ideal thresholdless laser. This transition from the small β regime to the $\beta \to 1$ regime is illustrated in Figure 8.8.

---

[29]The key assumption here is that the steady-state photon distribution will be peaked around $N_{\text{ex}}/2$, so that even though the approximation $G_k \approx 0$ breaks down for small k, it does not matter because these low photon numbers have a very low probability anyway. The only photon numbers that have a nonnegligible probability are those near $N_{\text{ex}}/2$, and for those the approximation $G_k \approx 0$ holds well.

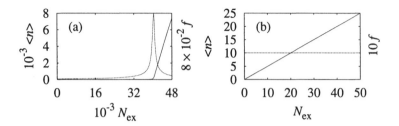

Figure 8.8 Here we repeat the plots of Figure 8.7, adding polarization decay according to the theory described in Section 8.6. The solid line is the curve of the average number of photons, and the dotted line that of the Fano parameter. In (a), $\Gamma = 10^3$ and $w = 10^{-1}$, resulting in the typical ordinary laser behavior with a well-defined threshold. In (b), $\Gamma = 1.8$ and $w = 20$, leading to complete disappearance of the laser threshold [174].

8.7 THE ONE-AND-THE-SAME ATOM LASER

Recent advances in cooling and in optical mirror coating technologies have allowed the realization of a long-dreamed goal in cavity QED: to build a laser with a gain medium consisting of a single atom trapped inside a high-Q cavity, where it interacts with the single-mode radiation field in the strong-coupling regime [436].

In ordinary lasers, masers, and even in the micromaser, the field is built up in the cavity by the contributions of many individual atoms. In lasers, these atoms are usually all inside the cavity at the same time. But the distances between atoms are large enough for them to interact with the field independently rather than cooperatively as in superradiance [255]. This lack of cooperativity leads to the same sort of field buildup that is achieved in the micromaser, where atoms go through the cavity one at a time [214, 417]. The differences between the highly quantum fields produced in the micromaser and the quasiclassical fields generated in ordinary lasers are due primarily to the strong coupling achieved in the micromaser rather than to there being one atom at a time in the cavity.

The one-and-the-same atom laser had been studied extensively [234, 299, 328, 355, 412, 443, 444, 459, 487, 488] before its realization in Jeff Kimble's quantum optics group at Caltech [436]. There was even a detailed experimental proposal of an ion-trap laser in 1997 by Meyer, Briegel, and Walther from the Max-Planck Institute for Quantum Optics [444]. In the long ran, however, neutral atom trapping using a FORT (far-off resonance trap) [442] proved to be a more workable technology for intracavity trapping.

As the atom is held inside the cavity for a long time compared to the lifetime of its energy levels and of a photon in the cavity mode, the incoherent decay dynamics cannot be neglected during its interaction with the field. So unlike that of the micromaser, the theory of the one-and-the-same atom laser must be solved numerically. The usual approach is to write down the master equation and then solve it in a truncated Fock space. In the theoretical analysis of their experiment [65, 437], McKeever, Boca, Boozer, Buck, and Kimble used the MATLAB®[30] quantum-optics toolbox developed by Sze M. Tan [589], and in some cases the truncation of Fock space was done at two photons. For more details on the numerics, see [534, 535, 589].

The experiment [436] showed that such a cavity QED laser has indeed no ordinary laser threshold. There is, however, a *second threshold*, due to the Stark shifting of the upper lasing level out of resonance with the cavity as the pump strength increases beyond a certain value. Both the photon number and the Fano parameter, as functions of the pump intensity, exhibit a second threshold. It is interesting to compare this with the result of the ion-trap laser simulations, as the latter showed both the ordinary laser threshold and the second threshold [444]. We can understand why this happens by looking at how the time scales compare in the two cases.

In the ion-trap laser proposal [444], the best realistic values given the constraint of finding a suitable ion for trapping, were $g \approx 14.8$ MHz, $t_{cav}^{-1} \approx 1$ MHz, and $\gamma_\parallel \approx 10.6$ MHz. Notice how the atomic decay rate $\gamma_\parallel$ is dangerously close to the Rabi frequency. In the one-and-the-same atom laser experiment [436], however, $g \approx 16$ MHz, $t_{cav}^{-1} \approx 4.2$ MHz, and $\gamma_\parallel \approx 2.6$ MHz. So whereas in the ion-trap laser proposal the saturation photon number was $n_0 \approx 0.2$ and the critical atom number $N_0 \approx 0.09$, in the Caltech experiment $n_0 \approx 0.01$ and $N_0 \approx 0.08$.

Another important experimental result is the demonstration of sub-Poissonian photon statistics. The physics here is the same as for a multiatom laser [275, 354, 459, 504, 505, 521, 522]. This is a consequence of the atomic and field dynamics having similar time scales that do not allow the adiabatic elimination of the nonlasing atomic levels involved in the pumping scheme[31] [74]. Without strong coupling, however, it is not possible to get sub-Poissonian photon statistics with a single atom in this way [459]. So this result provides firm evidence that the Caltech experiment has indeed achieved the strong-coupling regime.

RECOMMENDED READING

- For an entertaining and instructive discussion on radio oscillators and their history, see [460]. The van der Pol oscillator is reviewed in [609]. Also worth checking out is [272], where a simple model is developed for the self-sustained oscillator circuit in Figure 8.1, and it is shown that the steady-state oscillations can built up from a nonoscillatory state by noise.

- The laser bible is Siegman's book [560]. Svelto's [586] and Milonni and Eberly's [454] books provide excellent introductions to the subject that complement each other (Svelto's is more semiclassical and Milonni and Eberly's

---

[30]MATLAB is a registered trademark of The MathWorks, Inc.

[31]If these levels are formally eliminated, the resulting master equation will describe a non-Markovian pumping. It is this non-Markovian character that leads to noise reduction and the sub-Poissonian photon statistics.

more quantum mechanical). For a book that puts a greater emphasis on techno-logical applications of low-noise lasers, see [469]. Specifically on the quantum theory of the laser, I would recommend the following books. Haken's two-volume work [260] is an excellent introduction. A more specialized treatment is given in [261], [341], and [533]. For more exotic lasers such as the correlated emission laser (CEL), see Scully and Zubairy's book [552].

- The history of the laser is nicely told by one of its creators, Charles Townes, in [602]. See also the review paper [375].

- In 1900 a student of Jules Henri Poincaré, Louis Bachelier, anticipated many of Einstein's ideas on Brownian motion by studying the apparently completely different problem of the changes in the price of shares in the Parisian stock market [25]. Bachelier assumed that share prices obeyed a Gaussian distri-bution. This means that in his model, unlike in reality, share prices could become *negative*.[32] It was Sprenkle [574] who, in the 1960s, first corrected Bachelier's assumption, eliminating negative prices with the adoption of a *log-normal* distribution instead of a Gaussian. This corresponds to a geometric Brownian motion, where the step size is proportional to the distance position rather than the ordinary arithmetic Brownian motion seen in this chapter. For more on these ideas, see Osborne's excellent book [473]. These developments eventually lead to the Black and Scholes theory of option pricing [59], which practically began the field of quantitative finance. For more on quantitative finance, see the following excellent books [304, 476, 653].

- To know more about stochastic calculus from a physicist's perspective, see [224, 232, 580, 612–614]. If you prefer a rigorous approach in the axioms–theorem–proof style, have a look at [340, 371]. Many of the important historical references can be found in the collection [623]. The bible of numerical methods to solve stochastic differential equations is Kloeden and Platen's book [358].

- An important milestone before the realization of the one-and-the-same atom laser was Kyangwon An and Michael S. Feld's single-atom laser [16]. This device is similar to the micromaser except that it operates in the optical regime and the Rabi frequency is comparable to the decay rates. As in the micromaser, the cavity field is built by a succession of atoms rather than the same single atom as in the Caltech experiment [436]. For more on An and Feld's single-atom laser, see [16–19, 192, 405, 423].

- The combination of strong coupling in the optical regime with advanced trap-ping technology also allows the generation of single-photon states on demand (i.e., deterministically). See [356, 372, 373] and the recent realization of a deterministic source of single photons [438].

---

[32]Japanese interest rates, however, seem to fit a Gaussian distribution better than a lognormal one.

Problems

8.1   When a child on a swing lowers and raises her center of gravity, the effective length of the pendulum string corresponding to the swing (see the model presented in Chapter 2) changes. Considering that this change in the effective length is small in comparison with the length itself, assuming that the swing only swings by small angles and that the child lowers and raises her center of gravity sinusoidally, derive the following equation for the pumped-up swing:

$$\frac{d^2\theta}{dt^2} + \gamma\frac{d\theta}{dt} + \{1 + \epsilon\cos(\omega t)\}\,\omega_0\theta = 0,$$

where θ is the angle of the swing with the vertical, γ is the coefficient of viscous friction, $\epsilon \ll 1$, ω_0 is the ordinary resonance frequency of the swing, and ω is the pumping frequency due to the lowering and raising of the center of gravity. Assuming that $\gamma << \omega_0$, $\theta(0) = 0$, and $\dot{\theta}(0) = v$, solve this equation by Laplace transform. Show that the oscillations are amplified when $\omega = 2\omega_0$.

8.2   Solve (8.20) assuming that the particle starts the diffusion at $x = 0$ at $t = 0$ [i.e., $p(x,0) = \delta(x)$].
Hint: Apply a Fourier spatial transform to get an ordinary first-order differential equation from (8.20). You can easily solve this equation for the transform of $p(x)$. To invert the transform, complete the squares in the exponential to obtain a Gaussian.

8.3   Using (8.31) and (8.35) in (8.34), obtain (8.36).

8.4   For a function $Z \equiv f(Y_1,\ldots,Y_n)$ of the n random variables $Y_1,\ldots,Y_n$, the probability distribution $Q(z)$ of Z is given in terms of that of $Y_1,\ldots,Y_n$, $P(y_1,\ldots,y_n)$, by [232]

$$Q(z) = \int_{-\infty}^{\infty} dy_1 \cdots \int_{-\infty}^{\infty} dy_n\, P(y_1,\ldots,y_n)\,\delta\left(z - f(y_1,\ldots,y_n)\right).$$

Using this, show that if $Z = \bar{m} + \sqrt{v}\,Y$, where $\bar{m}$, v are deterministic variables and Y is a normal variable of mean zero and variance 1 whose probability distribution is given by

$$P(y) = \frac{e^{-y^2/2}}{\sqrt{2\pi}},$$

the probability distribution of Z is

$$Q(z) = \frac{1}{\sqrt{2\pi v}}\exp\left(-\frac{(z-\bar{m})^2}{2v}\right),$$

(i.e., Z is a normal variable of mean $\bar{m}$ and variance v).

8.5   Defining the Fourier transform $\tilde{x}(\omega)$ of a variable $x(t)$ by

$$\tilde{x}(\omega) \equiv \int_{-\infty}^{\infty}\frac{dt}{2\pi}\,e^{-i\omega t}x(t),$$

show that for $F(t)$ in (8.76), $\left\langle \tilde{F}(\omega)\tilde{F}^*(\omega') \right\rangle = (D/[2\pi])\delta(\omega - \omega')$. Then take the Fourier transforms of (8.86) and (8.87) and solve them. Show that $\langle \delta n^2 \rangle$ is given by (8.88).

8.6 Using the technique of Fano, diagonalize (8.92) and obtain (8.96) and (8.97).

8.7 Do the integrals in (8.103) and (8.104). Then, assuming that $V \neq 0$, take the limit of $t \rightarrow \infty$ and obtain (8.105).

8.8 Write down the $(\langle n|, |n\rangle)$-matrix element of (8.119) as a recursion relation and show that the solution is given by (8.121).

8.9 Consider a two-level atom under polarization damping interacting with the radiation field in a single-mode cavity according to (8.124). To focus only on spontaneous emission, take the initial field as the vacuum. Then the total density matrix will be of the form

$$\hat{\rho} = |0\rangle\langle 0| \otimes \hat{\rho}_{00} + |1\rangle\langle 1| \otimes \hat{\rho}_{11} + |0\rangle\langle 1| \otimes \hat{\rho}_{01} + |1\rangle\langle 0| \otimes \hat{\rho}_{01}^{\dagger},$$

where $\hat{\rho}_{jk}$ is an atomic density matrix, $\otimes$ stands for the tensor product, $|0\rangle$ is the vacuum, and $|1\rangle$ is the one-photon state. Assuming that the atom is initially in the excited state, solve (8.124) and show that

$$\langle \hat{\sigma}_z(t) \rangle = \frac{(Y_- + \gamma_\perp)\, e^{Y_- t} - (Y_+ + \gamma_\perp)\, e^{Y_+ t}}{Y_- - Y_+},$$

where

$$Y_\pm \equiv -\frac{\gamma_\perp}{2} \pm \sqrt{\left(\frac{\gamma_\perp}{2}\right)^2 - 4g^2}.$$

9

What is a mode of an open cavity?

The narrower the channel of communication between the interior of a vessel and the external medium, the greater does the independence become. Such cavities constitute resonators; in the presence of an external source of sound, the contained air vibrates in unison, and with an amplitude dependent upon the relative magnitudes of the natural and forced periods, rising to great intensity in the case of approximate isochronism. When the original cause of sound ceases, the resonator yields back the vibrations stored up as it were within it, thus becoming itself for a short time a secondary source. The theory of resonators constitutes an important branch of our subject.

—Lord Rayleigh [509]

Dissipation is ever present in real devices but is not a fundamental phenomenon. It arises because real physical systems are never completely isolated and thus interact with their surroundings. Real physical systems are open. This view of dissipation is reflected in the usual way it is dealt with in quantum mechanics,[1] which we described in Chapter 4. Instead of treating the open system directly, we treat the enlarged system composed by it and its environment. This large system is closed, so we can apply quantum mechanics in the usual way. The effect of dissipation then comes to light when we trace over the inaccessible degrees of freedom of the environment.

---

[1]For a discussion as to why this approach became dominant and a review of other approaches, see the recommended reading section at the end of the chapter.

Cavities are no exception. A real cavity is always dissipative. In this chapter we consider this dissipation from a fundamental perspective and address some key conceptual questions related to it. For concreteness and simplicity, we neglect absorption at the cavity walls and consider the leakage of electromagnetic radiation out of the cavity as the only source of dissipation. In that case, a first approximation to the problem is to quantize the field inside the cavity and outside as if the cavity is perfect (i.e., completely closed by perfectly reflecting walls) and then deal with the transmissivity losses by including an ad hoc coupling term in the Hamiltonian of the total system.

This coupling term annihilates a photon in one of the discrete cavity modes and creates another in one of the outside continuum of modes. It should also do the reverse[2] (otherwise, it will not be Hermitian). To make it even simpler, we can assume that the coupling strength V with which a cavity mode couples to the outside is a constant (i.e., the same for all cavity mode–external mode pairs). The Hamiltonian is then given by

$$\hat{H} = \underbrace{\sum_n \hbar\omega_n \hat{a}_n^\dagger \hat{a}_n}_{\text{cavity}} + \underbrace{\int dk\ \hbar ck\ \hat{a}^\dagger(k)\hat{a}(k)}_{\text{outside}} + \underbrace{\sum_n \int dk\ \left\{ V\hat{a}^\dagger(k)\hat{a}_n + V^*\hat{a}_n^\dagger\hat{a}(k) \right\}}_{\text{coupling}}.$$

(9.1)

In the quantum optics literature this simple Hamiltonian is called the *Gardiner–Collett Hamiltonian* owing to its appearance in the two famous papers [109, 225], where their input–output approach to damped quantum systems was developed.[3] From (9.1), one can deduce (see Problem 9.1) the Lindblad master equation for high-Q cavity damping (6.68) introduced in Chapter 6.

This is fine as a first approximation that holds well for high-Q cavities, where as we have seen in Section 6.2, the photon round trip-time in the cavity is much shorter than the average time it takes to escape to the outside (lifetime). But as soon as the cavity Q decreases a little, this approximation starts to break down [37, 43, 228, 390, 391, 408, 603]. An exact solution can be obtained by expanding the fields in terms of the normal modes of the entire closed system formed by the cavity and the outside, as we did in Section 2.5.2. This is called the *modes-of-the-universe approach*. Unfortunately, these modes can only be calculated for the simplest of structures. Moreover, as they are modes of the entire system, a photon in one of these modes of the continuum is completely delocalized. What one would like is an exact treatment where cavity modes were singled out explicitly. Then a deeper conceptual question also arises: What is a mode of an open cavity? This is the main theme of this chapter.

---

[2]That is, annihilating a photon in one of the outside continuum of modes and creating a photon in one of the discrete cavity modes.

[3]To be more precise, the Gardiner–Collett Hamiltonian has a generic coupling constant that can be different for each pair of coupled modes.

In quantum electronics, this question came to the forefront with the invention of the laser. The first masers used closed[4] microwave cavities whose modes are calculated as in Appendix A. Any absorption at the cavity walls or transmissivity was modeled by assuming that there was an absorptive medium distributed homogeneously inside the cavity. The delocalization of absorption inside the cavity allows use of the Gardiner–Collett approach.[5] These closed cavities had to have dimensions comparable to the wavelength of the maser (centimeter) for the various cavity modes to be much farther apart from each other in frequency than the bandwidth of the molecular beam resonance. Otherwise, more than one frequency (cavity resonance) would be amplified and the maser would not be monochromatic. As the wavelength was reduced from microwaves (centimeter, millimeter) to the optical regime (nanometer), it was no longer possible at the time (late 1950s) to make a closed cavity of dimensions of such minute wavelength. The great idea that was put forward independently by Schawlow and Townes, Gordon, and Prokhorov was to effectively reduce the number of modes by selectively increasing the losses of all but one or just a few of them. This selective increase of losses would be done by removing the sidewalls of the closed cavity, transforming it into a Fabry–Pérot. So the Fabry–Pérot, which started life in 1899 [187] as an interferometer (see Figure 9.1), was now used as a resonator for the first time.[6]

At that time, as Bennett [47] wrote: "It was not entirely obvious that normal modes would even exist in these open-walled structures. ..." But as the citation in the prologue to this chapter suggests, open cavities have modes, too. Anyone who has experienced echo in a valley knows that intuitively. In quantum electronics this was clarified by the work of Fox and Li [220, 221], Vaĭnshteĭn [605–607], and others.

This, however, is yet another example of an important early result being forgotten by the scientific community, only to be rediscovered much later. Long before Fox, Li, and Vaĭnshteĭn tackled the problem of modes of open cavities, Joseph John Thomson, discoverer of the first subatomic particle, the electron, had already shown how to define and calculate such modes [594, 595]. We discuss Thomson's ideas in Section 9.2.

In quantum optics, the problem of open cavities is deeper than in the usual semiclassical approach of laser physics. Any modes used to expand the quantized electromagnetic field must form a complete set. Unfortunately, the usual Fox–Li or Vaĭnshteĭn modes *do not* form a complete set (see, e.g., Sec. 63 p. 330, and Probs. 10.4 and 10.5 of [607]), due to adoption of the Sommerfeld radiation condition (discussed in Section 9.2), which excludes incoming radiation from infinity. In a semiclassical treatment, this is a problem only when excitation of the resonator by an external field is considered (see, e.g., Chap. 10 of [607]). With the quantized radiation field,

---

[4]They had a small hole for the molecular beam to enter. But as Sommerfeld showed, as long as the dimensions of the hole are small in comparison with the wavelength of the maser, leakage of radiation through the hole is negligible. It is not so for optical frequencies.

[5]As mentioned in Section 6.2, when the photon round-trip time is much shorter than its lifetime in the cavity, the losses appear to be distributed evenly throughout the cavity rather than being concentrated at the semitransparent walls.

[6]As far as the author knows.

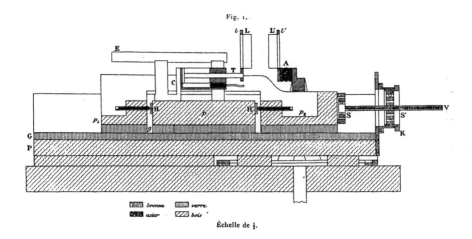

Figure 9.1 The Fabry–Perot was originally designed as an interferometer. This is a schematic diagram of the first Fabry–Perot. (Reproduced from [187].)

however, the cavity is always being fed external vacuum fluctuations, and those must be describable in a mode expansion.[7] In Section 9.4 we discuss how completeness can be achieved by supplementing Thomson's natural modes with a different sort of mode.

Before embarking on our discussion of the mode of an open cavity, we show how the Gardiner–Collett Hamiltonian emerges from a first-principles theory as a high-Q approximation to the general problem.

9.1 THE GARDINER–COLLETT HAMILTONIAN

Even though the Gardiner–Collett Hamiltonian (9.1) was introduced as a phenomenological Hamiltonian, it is possible to deduce it from first principles [171]. Then it emerges as a high-Q approximation to the first-principles description. We now review this derivation in greater detail than in [171], as this will help us understand the need for using natural modes rather than perfect cavity modes when the cavity is open.

To focus only on the essential features of the problem, let us consider the simple one-dimensional model of a leaky cavity depicted in Figure 9.2. Adding the multiple reflections and transmissions, as in Section 2.5.2, we find that the quantized electric

---

[7]Otherwise, the commutation relations will not be the correct ones required by quantum mechanics.

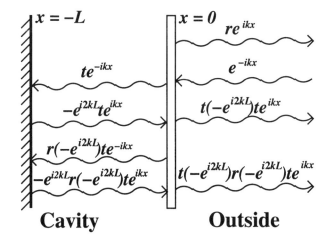

Figure 9.2 The cavity is formed by the perfect mirror at $x = -L$ and the semitransparent mirror at $x = 0$. The perfect mirror has an amplitude reflectivity coefficient of -1 and vanishing transmissivity, of course. The semitransparent mirror has an amplitude reflectivity r and transmissivity t, which can be complex in general. We assume that there is no absorption, so $|r|^2 + |t|^2 = 1$ and $r^*t + t^*r = 0$ (see [317]). The modes of the universe can be determined using a scattering approach, as in Section 2.5.2. In this schematic drawing, we depict the incident plane wave e^{-ikx} coming from infinity, its first reflection on the semitransparent mirror re^{ikx}, and some of its multiple reflections inside the cavity as well as some of its partial transmissions to the outside.

and magnetic fields are given by

$$\hat{E}_{\text{in}}(x) = -i2 \int_0^\infty dk\, \mathcal{E}_{\text{vac}}(k) e^{ikL} \mathcal{L}(k)\, \sin\left([x + L]k\right) \hat{a}(k) + \text{H.c.} \quad (9.2)$$

$$\hat{B}_{\text{in}}(x) = -2 \int_0^\infty dk\, \mathcal{E}_{\text{vac}}(k) e^{ikL} \mathcal{L}(k)\, \cos\left([x + L]k\right) \hat{a}(k) + \text{H.c.} \quad (9.3)$$

inside the cavity, and by

$$\hat{E}_{\text{out}}(x) = \int_0^\infty dk\, \mathcal{E}_{\text{vac}}(k) \left\{ e^{-ikx} + e^{ikx} \left[r - te^{i2kL} \mathcal{L}(k) \right] \right\} \hat{a}(k) + \text{H.c.} \quad (9.4)$$

$$\hat{B}_{\text{out}}(x) = \int_0^\infty dk\, \mathcal{E}_{\text{vac}}(k) \left\{ -e^{-ikx} + e^{ikx} \left[r - te^{i2kL} \mathcal{L}(k) \right] \right\} \hat{a}(k) + \text{H.c.} \quad (9.5)$$

outside the cavity, where

$$\mathcal{E}_{\text{vac}}(k) = \sqrt{2\pi \hbar c k} \quad (9.6)$$

is the vacuum field strength, and

$$\mathcal{L}(k) = t \sum_{l=0}^\infty \left(-r\, e^{i2kL} \right)^l \quad (9.7)$$

is the modulation factor introduced by the multiple reflections on the cavity mirrors. The annihilation and creation operators obey the continuum commutation relations

$$[\hat{a}(k), \hat{a}^\dagger(k')] = \delta(k - k') \quad \text{and} \quad [\hat{a}(k), \hat{a}(k')] = 0. \tag{9.8}$$

The perfect mirror at $x = 0$ forces us to restrict k to the positive half space (see [448, 495].

The Hamiltonian is the total energy and is given by

$$\hat{H} = \int_{-L}^{0} dx\, \hat{U}_{\text{in}}(x) + \int_{0}^{\infty} dx\, \hat{U}_{\text{out}}(x) + \lim_{\delta \to 0+} \int_{-\delta}^{\delta} dx\, \hat{U}_{\text{s-t}}(x), \tag{9.9}$$

where $\hat{U}_{\text{in}}(x)$ is the energy density inside the cavity, $\hat{U}_{\text{out}}(x)$ outside, and $\hat{U}_{\text{s-t}}(x)$ inside the semitransparent mirror. There is no contribution from inside the perfect mirror because the radiation fields vanish there. We know that the energy density inside the cavity is given by

$$\hat{U}_{\text{in}}(x) = \frac{1}{8\pi} \left\{ \hat{E}_{\text{in}}^2(x) + \hat{B}_{\text{in}}^2(x) \right\}, \tag{9.10}$$

and outside by

$$\hat{U}_{\text{out}}(x) = \frac{1}{8\pi} \left\{ \hat{E}_{\text{out}}^2(x) + \hat{B}_{\text{out}}^2(x) \right\}. \tag{9.11}$$

But we do not know what the energy density is inside the semitransparent mirror, as we have not specified any microscopic model for its internal structure. We prefer to regard this mirror as a nonabsorptive "black box" scatterer with reflection and transmission coefficients (in general, functions of k) restricted only by the general conditions derived from energy conservation [317]:

$$|r|^2 + |t|^2 = 1 \quad \text{and} \quad r^*t + t^*r = 0. \tag{9.12}$$

Fortunately, we can find an expression for the integral of $\hat{U}_{\text{s-t}}(x)$ over the semitransparent mirror (i.e., for the energy stored in the mirror) using Poynting's theorem:

$$\frac{\partial}{\partial x}\hat{S}(x, t) + \frac{\partial}{\partial t}\hat{U}(x, t) = 0, \tag{9.13}$$

where the Poynting "vector" $\hat{S}$ is known both inside and outside the cavity [532]:

$$\hat{S}_{\substack{\text{in} \\ \text{out}}} = \frac{1}{8\pi} \left\{ \hat{E}_{\substack{\text{in} \\ \text{out}}}(x)\hat{B}_{\substack{\text{in} \\ \text{out}}}(x) + \hat{B}_{\substack{\text{in} \\ \text{out}}}(x)\hat{E}_{\substack{\text{in} \\ \text{out}}}(x) \right\}. \tag{9.14}$$

From (9.13) we find that

$$\frac{d}{dt}\left\{ \lim_{\delta \to 0+} \int_{-\delta}^{\delta} dx\, \hat{U}_{\text{s-t}}(x) \right\} = \hat{S}_{\text{in}}(0) - \hat{S}_{\text{out}}(0). \tag{9.15}$$

Integrating in time, we obtain

$$
\lim_{\delta \to 0^+} \int_{-\delta}^{\delta} dx \, \hat{U}_{s-t}(x) = -\frac{1}{4\pi} \int_0^\infty dk \int_0^\infty dk' \, \mathcal{E}_{\text{vac}}(k)\mathcal{E}_{\text{vac}}(k')
$$
$$
\times \Bigg(\hat{a}(k)\hat{a}(k')\frac{1}{k+k'}\Big\{ \mathcal{L}(k)\mathcal{L}(k')e^{i(k+k')L} \sin\left([k+k']L\right)
$$
$$
+ \frac{i}{2}\Big[-1 + \{r - te^{i2kL}\mathcal{L}(k)\}\{r - te^{i2k'L}\mathcal{L}(k')\}\Big] \Big\}
$$
$$
+ \hat{a}(k)\hat{a}^\dagger(k')\mathcal{P}\frac{1}{k-k'}\Big\{ \mathcal{L}(k)\mathcal{L}^*(k')e^{i(k-k')L} \sin\left([k-k']L\right)
$$
$$
+ \frac{i}{2}\Big[-1 + \{r - te^{i2kL}\mathcal{L}(k)\}\{r^* - t^*e^{-i2kL}\mathcal{L}^*(k')\}\Big] \Big\} \Bigg) + \text{H.c.} \qquad (9.16)
$$

Using (9.10), (9.11), and (9.16) in (9.9), we find the expected[8] Hamiltonian for the modes of the universe creation and annihilation operators (see Problem 9.2),

$$
\hat{H} = \frac{\hbar c}{2} \int_0^\infty dk \, k \left\{ \hat{a}^\dagger(k)\hat{a}(k) + \hat{a}(k)\hat{a}^\dagger(k) \right\}. \qquad (9.17)
$$

Before tackling our main goal, it is useful to examine $\mathcal{L}(k)$ a bit more closely. As $|r| < 1$, the function

$$
f(\lambda) \equiv \left(-r\, e^{i2kL}\right)^\lambda \qquad (9.18)
$$

is bounded in the interval $0 \leq \lambda < \infty$ and $\lim_{\lambda \to \infty} f(\lambda) = 0$. So we can use Poisson's sum formula, [599] which states that

$$
\sum_{l=0}^\infty f(l) = \frac{1}{2}f(0) + \sum_{n=-\infty}^\infty \int_0^\infty d\lambda \, f(\lambda)e^{-i2\pi n\lambda}. \qquad (9.19)
$$

Then we find that (see Problem 9.3)

$$
\mathcal{L}(k) = \frac{t}{2} + \sqrt{\frac{\pi}{L}} \sum_{n=-\infty}^\infty \mathcal{L}_n(k), \qquad (9.20)
$$

where

$$
\mathcal{L}_n(k) \equiv \frac{it/2\sqrt{\pi L}}{k - n\pi/L + [\arg(r) - \pi]/2L - i\ln|r|/2L}. \qquad (9.21)
$$

Moreover, it is possible to show (Problem 9.4) that

$$
|\mathcal{L}(k)|^2 = \frac{1}{L} \sum_{n=-\infty}^\infty \frac{-\ln|r|/2L}{(k - n\pi/L + [\arg(r) - \pi]/2L)^2 + (\ln|r|/2L)^2}. \qquad (9.22)
$$

---

[8]Each mode of the universe is equivalent to an uncoupled harmonic oscillator.

In other words, $|\mathcal{L}(k)|^2$ is a sum of Lorentzians of half width at half maximum $-\ln|r|/2L$ centered at the cavity resonances $n\pi/L + [\pi - \arg(r)]/2L$.

Now let us go back to our main goal. To describe the effect of cavity damping correctly (e.g., to give the correct energy loss rate), the coupling strength V in the Gardiner–Collett Hamiltonian (9.1) must be of first order in the cavity mirror transmissivity t [109]. So, in principle, for this scheme to work, the field in the cavity should be describable, up to first order in the transmissivity, as a linear superposition of perfect cavity modes. Unfortunately, this will not happen for every cavity. In fact, one such example is already found in the literature.

Barnett and Radmore [37] used the Fano diagonalization method to compare the time evolution of the electric field inside a one-dimensional cavity given by the Gardiner–Collett Hamiltonian (9.1) with that derived from a rigorous modes-of-the-universe approach. Their one-dimensional cavity model consisted of a perfect mirror and a dielectric-delta-function mirror.[9] They found that for this cavity model, the cavity resonances already differ from the perfect cavity resonances in first order of the transmissivity. In their model, the complex reflectivity has a phase that depends on the transmissivity already in first order, making it impossible to recover the reflectivity of a perfect cavity mirror in that order.[10]

There is, however, a simple choice of semitransparent mirror for which the reflectivity reduces to that of a perfect cavity up to first order in the transmissivity. This reflectivity leads to cavity resonances that agree up to first order with those of a perfect cavity. Let the reflectivity be real and the transmissivity be pure imaginary; specifically,

$$r = -\sqrt{1 - \epsilon^2} \quad \text{and} \quad t = i\epsilon, \tag{9.23}$$

where ϵ, as the name suggests, is a small (less than 1) positive number. Then $r = -1$ up to first order in $|t|$, yielding the perfect cavity modes in that approximation. We will now use this choice of r and t to derive (9.1) as a first-order approximation to an exact modes-of-the-universe description. To get further insight into this problem, however, the expansion in powers of ϵ will be delayed until the final stages of the calculation.

The perfect cavity modes for the electric field form a complete set for the space inside the cavity. Any physical configuration of the field located in $-L < x < 0$, that is, everywhere inside apart from the *boundaries* (this is a very important point!), can be expanded as a Fourier sine series. So we can always write the internal field as

$$\hat{E}_{\text{in}}(x) = -i2 \sum_{n'=1}^{\infty} \mathcal{E}_{\text{vac}}(k_{n'}) \left(\hat{a}_{n'} - \hat{a}_{n'}^\dagger \right) \sin\left([x + L]k_{n'} \right), \tag{9.24}$$

where $k_{n'} \equiv n'\pi/L$.

---

[9]This dielectric-delta-function mirror was introduced in [573] and subsequently adopted in other works, including [43, 391].

[10]As this is only a shift of the resonances, one could in principle still use (9.1) if, instead of the usual perfect cavity modes, one uses perfect cavity modes with the resonances shifted by the appropriate amount first order in the transmissivity.

The operators $\hat{a}_{n'}$ are not necessarily annihilation operators. In fact, as we know $\hat{E}_{\text{in}}(x)$ in terms of the global operators $\hat{a}(k)$ [i.e., (9.2)], (9.24) is really a definition of $\hat{a}_{n'}$. However, as a definition of $\hat{a}_{n'}$, it is incomplete. It defines only the "imaginary" quadrature of $\hat{a}_{n'}$. As we are free, then, to define the "real" quadrature of $\hat{a}_{n'}$, let us define it in such a way that the internal magnetic field can also be described by an expression formally similar to its perfect cavity expression. We are allowed to do so because the perfect cavity modes for the magnetic field also form a complete set for the space inside the cavity. So we write the magnetic field as

$$\hat{B}_{\text{in}}(x) = -2 \sum_{n'=1}^{\infty} \mathcal{E}_{\text{vac}}(k_{n'}) \left(\hat{a}_{n'} + \hat{a}_{n'}^{\dagger} \right) \cos\left([x + L]k_{n'} \right), \qquad (9.25)$$

providing the required definition of the "real" quadrature of $\hat{a}_{n'}$.

Together, (9.24) and (9.25) define the non-Hermitian operators $\hat{a}_{n'}$ uniquely. We can now obtain an expression for $\hat{a}_{n'}$ in terms of the global operators $\hat{a}(k)$ in the following way. First, multiply (9.24) by $\sin([x + L]k_n)$ and integrate over x from $-L$ to 0 using (9.2). We find the "imaginary" quadrature of $\hat{a}_{n'}$ this way,

$$\hat{a}_{n'} - \hat{a}_{n'}^{\dagger} = \frac{1}{\sqrt{\pi L}} \int_0^{\infty} dk \sqrt{\frac{k}{k_n}} \left\{ \frac{\sin\left([k - k_n] L \right)}{k - k_n} - \frac{\sin\left([k + k_n] L \right)}{k + k_n} \right\}$$
$$\times \left\{ e^{ikL} \mathcal{L}(k)\hat{a}(k) - e^{-ikL} \mathcal{L}^*(k)\hat{a}^{\dagger}(k) \right\}. \quad (9.26)$$

To find the "real" quadrature, we multiply (9.25) by $\cos([x + L]k_n)$ and integrate over x from $-L$ to 0 using (9.3). We obtain

$$\hat{a}_{n'} + \hat{a}_{n'}^{\dagger} = \frac{1}{\sqrt{\pi L}} \int_0^{\infty} dk \sqrt{\frac{k}{k_n}} \left\{ \frac{\sin\left([k - k_n] L \right)}{k - k_n} + \frac{\sin\left([k + k_n] L \right)}{k + k_n} \right\}$$
$$\times \left\{ e^{ikL} \mathcal{L}(k)\hat{a}(k) + e^{-ikL} \mathcal{L}^*(k)\hat{a}^{\dagger}(k) \right\}. \quad (9.27)$$

From (9.26) and (9.27), we find the following expression for $\hat{a}_n$:

$$\hat{a}_n = \frac{1}{\sqrt{\pi L}} \int_0^{\infty} dk \sqrt{\frac{k}{k_n}} \left\{ \frac{\sin\left([k - k_n] L \right)}{k - k_n} e^{ikL} \mathcal{L}(k)\hat{a}(k) \right.$$
$$\left. + \frac{\sin\left([k + k_n] L \right)}{k + k_n} e^{-ikL} \mathcal{L}^*(k)\hat{a}^{\dagger}(k) \right\}. \quad (9.28)$$

The phenomenological Gardiner–Collett approach not only assumes that the radiation field inside the cavity is given by the same expressions as for a perfect cavity, but also assumes that the quantization formalism is the same. In other words, it assumes that the $\hat{a}_n$ obey the usual commutation rules,

$$[\hat{a}_n, \hat{a}_{n'}^{\dagger}] = \delta_{nn'} \quad \text{and} \quad [\hat{a}_n, \hat{a}_{n'}] = 0. \qquad (9.29)$$

It is possible to show (Problem 9.5) that these commutation rules follow from (9.8) and (9.28). From a more fundamental stand, they can also be seen to be a consequence of the perfect cavity modes orthonormality properties and the canonical commutation relations for the radiation fields (Problem 9.6).

So far so good, but to carry on and find the coupling, we must describe the modes of the outside. We proceed in the same way as for the intracavity fields, expanding the outside fields in terms of free-space modes with a perfect mirror at $x = 0$. Then

$$\hat{E}_{\text{out}}(x) = -i2 \int_0^\infty dk \, \mathcal{E}_{\text{vac}}(k) \left\{ \hat{b}(k) - \hat{b}^\dagger(k) \right\} \sin kx, \tag{9.30}$$

and

$$\hat{B}_{\text{out}}(x) = -2 \int_0^\infty dk \, \mathcal{E}_{\text{vac}}(k) \left\{ \hat{b}(k) + \hat{b}^\dagger(k) \right\} \cos kx. \tag{9.31}$$

These expressions define the operators $\hat{b}(k)$ and $\hat{b}^\dagger(k)$. With this definition, $\hat{b}(k)$ and $\hat{b}^\dagger(k)$ automatically satisfy the usual commutation relations of annihilation and creation operators of the continuum (Problem 9.7):

$$[\hat{b}(k), \hat{b}^\dagger(k')] = \delta(k - k') \quad \text{and} \quad [\hat{b}(k), \hat{b}(k')] = 0. \tag{9.32}$$

Substituting (9.4) and (9.5) in (9.30) and (9.31), multiplying by the respective mode function and integrating over x outside the cavity, we find that

$$\hat{b}(k) - \hat{b}^\dagger(k) = \frac{i}{\pi \mathcal{E}_{\text{vac}}(k)} \int_0^\infty dx \, \sin kx$$
$$\times \left\{ \int_0^\infty dk' \, \mathcal{E}_{\text{vac}}(k') \left[e^{-ik'x} + e^{ik'x} R(k') \right] \hat{a}(k') + \text{H.c.} \right\}, \tag{9.33}$$

$$\hat{b}(k) + \hat{b}^\dagger(k) = -\frac{1}{\pi \mathcal{E}_{\text{vac}}(k)} \int_0^\infty dx \, \cos kx$$
$$\times \left\{ \int_0^\infty dk' \, \mathcal{E}_{\text{vac}}(k') \left[-e^{-ik'x} + e^{ik'x} R(k') \right] \hat{a}(k') + \text{H.c.} \right\}, \tag{9.34}$$

where

$$R(k) = \frac{r}{r^*} e^{i2kL} \frac{\left(1 + r e^{i2kL} \right)^*}{1 + r e^{i2kL}}. \tag{9.35}$$

Adding (9.33) and (9.34), we obtain

$$\hat{b}(k) = \frac{-1}{2\pi \mathcal{E}_{\text{vac}}(k)} \int_0^\infty dx \int_0^\infty dk' \mathcal{E}_{\text{vac}}(k')$$
$$\times \left\{ \left[e^{-i(k-k')x} R(k') - e^{i(k-k')x} \right] \hat{a}(k') \right.$$
$$\left. + \left[e^{-i(k+k')x} R^*(k') - e^{i(k+k')x} \right] \hat{a}^\dagger(k') \right\}. \tag{9.36}$$

For our choice of real r and imaginary t [i.e., (9.23)], we find that

$$R(k) = -1 - i2 \left| \frac{1+r}{t} \right| e^{ikL} \mathcal{L}(k) \cos kL. \tag{9.37}$$

Using (9.37) in (9.36) and doing the integral over x, we obtain

$$\hat{b}(k) = \hat{a}(k) + \frac{1}{\pi} \left| \frac{1+r}{t} \right| \int_0^\infty dk' \sqrt{\frac{k'}{k}} e^{ik'L} \cos k'L\, \mathcal{L}(k') \lim_{\delta \to 0+} \frac{1}{k - k' - i\delta} \hat{a}(k')$$
$$- \frac{1}{\pi} \left| \frac{1+r}{t} \right| \int_0^\infty dk' \sqrt{\frac{k'}{k}} e^{-ik'L} \cos k'L\, \mathcal{L}^*(k') \lim_{\delta \to 0+} \frac{1}{k + k' - i\delta} \hat{a}^\dagger(k'). \tag{9.38}$$

Now, if the perfect[11] intracavity and free-space modes that we have used in expansions (9.24), (9.25), (9.30), and (9.31) were together able to describe any physically attainable configuration of the fields, $\hat{a}_n$, $\hat{b}(k)$, and their Hermitian conjugates would span the space of the global operators $\hat{a}(k)$ and $\hat{a}^\dagger(k)$ completely. In other words, we should have

$$\hat{a}(k) = \sum_{n=1}^\infty \left\{ \alpha_{n1}(k)\hat{a}_n + \alpha_{n2}(k)\hat{a}_n^\dagger \right\}$$
$$+ \int_0^\infty dk' \left\{ \beta_1(k, k')\hat{b}(k') + \beta_2(k, k')\hat{b}^\dagger(k') \right\}, \tag{9.39}$$

where, as in the dressed operator version of Fano diagonalization [37, 38, 306, 307],

$$\alpha_{n1}(k) = [\hat{a}(k), \hat{a}_n^\dagger]$$
$$= \frac{1}{\sqrt{\pi L}} \sqrt{\frac{k}{k_n}} \frac{\sin([k - k_n]L)}{k - k_n} e^{-ikL} \mathcal{L}^*(k), \tag{9.40}$$
$$\alpha_{n2}(k) = [\hat{a}_n, \hat{a}(k)]$$
$$= -\frac{1}{\sqrt{\pi L}} \sqrt{\frac{k}{k_n}} \frac{\sin([k + k_n]L)}{k + k_n} e^{-ikL} \mathcal{L}^*(k), \tag{9.41}$$

using (9.28), and

$$\beta_1(k, k') = [\hat{a}(k), \hat{b}^\dagger(k')]$$
$$= \delta(k - k') + \frac{1}{\pi} \left| \frac{1+r}{t} \right| \sqrt{\frac{k}{k'}} e^{-ikL} \mathcal{L}^*(k) \cos kL \lim_{\delta \to 0+} \frac{1}{k' - k + i\delta}, \tag{9.42}$$

$$\beta_2(k, k') = [\hat{b}(k'), \hat{a}(k)]$$
$$= \frac{1}{\pi} \left| \frac{1+r}{t} \right| \sqrt{\frac{k}{k'}} e^{-ikL} \mathcal{L}^*(k) \cos kL \lim_{\delta \to 0+} \frac{1}{k' + k - i\delta}, \tag{9.43}$$

---

[11]*Perfect* here is used just to indicate that these are the modes obtained when the semitransparent mirror is regarded as if it were a *perfectly* reflecting mirror.

using (9.38). However, we must not forget that the perfect intracavity and free-space modes *do not* describe correctly, beyond first order in ϵ, the field configuration *on the boundary* formed by the semitransparent mirror. Because of *this single point,* the perfect mode projections that define $\hat{a}_n$, $\hat{a}_n^\dagger$, $\hat{b}(k)$, and $\hat{b}^\dagger(k)$ will miss some of the global $\hat{a}(k)$ and $\hat{a}^\dagger(k)$. In other words, the $\hat{a}_n$, $\hat{a}_n^\dagger$, $\hat{b}(k)$, and $\hat{b}^\dagger(k)$ are not enough to cover the entire continuum of Fock spaces spanned by the global operators $\hat{a}(k)$ and $\hat{a}^\dagger(k)$. Something is missing. Thus, (9.39) *does not* hold in general, only up to first order in ϵ.

As the breakdown of (9.39) beyond first order in ϵ is a key point in our first-principles derivation of the Gardiner–Collett Hamiltonian (9.1), let us examine it a bit more closely. Let us call

$$\hat{a}_C(k) \equiv \int_0^\infty dk' \left\{ \beta_1(k,k')\hat{b}(k') + \beta_2(k,k')\hat{b}^\dagger(k') \right\} \tag{9.44}$$

the continuum part of $\hat{a}(k)$ according to (9.39), and

$$\hat{a}_D(k) \equiv \sum_{n=1}^\infty \left\{ \alpha_{n1}(k)\hat{a}_n + \alpha_{n2}(k)\hat{a}_n^\dagger \right\} \tag{9.45}$$

the discrete part. Then if (9.39) holds,

$$[\hat{a}(k), \hat{a}^\dagger(k')] = [\hat{a}_D(k), \hat{a}_D^\dagger(k')] + [\hat{a}_C(k), \hat{a}_C^\dagger(k')]. \tag{9.46}$$

Using (9.42) and (9.43), we find that

$$\begin{aligned}
[\hat{a}_C(k), \hat{a}_C^\dagger(k')] &= \delta(k-k') \\
&+ \frac{1}{\pi}\left|\frac{1+r}{t}\right|\sqrt{\frac{k}{k'}}e^{-ikL}\mathcal{L}^*(k)\cos kL \lim_{\delta\to0+}\frac{1}{k'-k+i\delta} \\
&+ \frac{1}{\pi}\left|\frac{1+r}{t}\right|\sqrt{\frac{k'}{k}}e^{ik'L}\mathcal{L}(k')\cos k'L \lim_{\delta\to0+}\frac{1}{k-k'-i\delta} \\
&+ \frac{i}{\pi}\left|\frac{1+r}{t}\right|^2\sqrt{kk'}e^{i(k'-k)L}\mathcal{L}^*(k)\mathcal{L}(k')\cos kL\cos k'L \\
&\times \left\{ \mathcal{P}\frac{1}{k}\mathcal{P}\frac{1}{k'} + 2\mathcal{P}\frac{1}{k'}\lim_{\delta\to0+}\frac{1}{k'-k+i\delta} \right\}.
\end{aligned} \tag{9.47}$$

Using (9.40) and (9.41) yields

$$\begin{aligned}
[\hat{a}_D(k), \hat{a}_D^\dagger(k')] &= \frac{e^{i(k'-k)L}}{\pi}\mathcal{L}^*(k)\mathcal{L}(k')\mathcal{P}\frac{1}{k-k'} \\
&\times \left\{ \sqrt{\frac{k}{k'}}\sin kL\cos k'L - \sqrt{\frac{k'}{k}}\sin k'L\cos kL \right\} \\
&- \frac{e^{i(k'-k)L}}{\pi L}\mathcal{L}^*(k)\mathcal{L}(k')\frac{k+k'}{kk'\sqrt{kk'}}\sin kL\sin k'L. \tag{9.48}
\end{aligned}$$

Now then, in general, (9.46) is not verified. But for $\epsilon \ll 1$, up to first order in ϵ, we find that

$$\mathcal{L}(k) \approx \frac{1}{\sqrt{L}} \sum_{n=0}^{\infty} \frac{-\epsilon/2\sqrt{L}}{k - k_n + i\epsilon^2/4L} \tag{9.49}$$

and

$$\alpha_{n1}(k) \approx \frac{(-1)^n}{\sqrt{\pi}} \frac{-\epsilon/2\sqrt{L}}{k - k_n - i\epsilon^2/4L}, \qquad \alpha_{n2}(k) \approx 0. \tag{9.50}$$

Under this approximation,

$$[\hat{a}_C(k), \hat{a}_C^\dagger(k')] \approx \delta(k - k') - \frac{1}{\pi} \sum_{n=1}^{\infty} \frac{-\epsilon/2\sqrt{L}}{k - k_n - i\epsilon^2/4L} \frac{-\epsilon/2\sqrt{L}}{k' - k_n + i\epsilon^2/4L} \tag{9.51}$$

and

$$[\hat{a}_D(k), \hat{a}_D^\dagger(k')] \approx \frac{1}{\pi} \sum_{n=1}^{\infty} \frac{-\epsilon/2\sqrt{L}}{k - k_n - i\epsilon^2/4L} \frac{-\epsilon/2\sqrt{L}}{k' - k_n + i\epsilon^2/4L}. \tag{9.52}$$

From (9.51) and (9.52), we find that (9.46) does hold for $\epsilon \ll 1$ up to first order. In an analogous way, we can show that

$$[\hat{a}(k), \hat{a}(k')] = [\hat{a}_D(k), \hat{a}_D(k')] + [\hat{a}_C(k), \hat{a}_C(k')] \tag{9.53}$$

also holds up to first order. Then (9.39) is valid up to first order and our expansion into perfect modes makes sense, but only up to first order (see also Problem 9.8). To go beyond the high-Q regime and be able to deal with larger losses, we have to use modes that satisfy the exact boundary conditions at the semitransparent mirror. These are the natural modes of the cavity and, unlike perfect modes, they are nonorthogonal. We discuss natural modes in the rest of this chapter. But now let us determine the Hamiltonian and the coupling within this high-Q approximation.

As mentioned before, (9.39) does suggest a parallel with the dressed operator version of Fano diagonalization [38]. In the Fano diagonalization method, however, the global operators $\hat{a}(k)$ and $\hat{a}^\dagger(k)$ that diagonalize the Hamiltonian are the unknowns that we wish to determine, whereas the $\hat{a}_n$, $\hat{a}_n^\dagger$, $\hat{b}(k)$, $\hat{b}^\dagger(k)$ together with the non-diagonal Hamiltonian that couples then are known. In contrast, here we have the opposite, $\hat{a}(k)$ and $\hat{a}^\dagger(k)$ and the diagonal Hamiltonian are known, while the $\hat{a}_n$, $\hat{a}_n^\dagger$, $\hat{b}(k)$, $\hat{b}^\dagger(k)$ and their nondiagonal Hamiltonian are not. So here we wish to do the reverse of the Fano diagonalization that was done in [37].

Using (9.39) in (9.17), we can write the Hamiltonian in terms of the operators $\hat{a}_n$, $\hat{a}_n^\dagger$, $\hat{b}(k)$, and $\hat{b}^\dagger(k)$ instead of $\hat{a}(k)$ and $\hat{a}^\dagger(k)$. As we have seen above, this will be correct only up to first order in ϵ. So we should use the approximate expressions (9.50) for $\alpha_{n1}(k)$ and $\alpha_{n2}(k)$, as well as

$$\beta_1(k, k') \approx \delta(k - k') + \frac{1}{\sqrt{\pi}} \lim_{\delta \to 0+} \frac{1}{k' - k + i\delta} \sum_{n=1}^{\infty} \frac{(-1)^n \epsilon}{2\sqrt{L}} \alpha_{n1}(k) \tag{9.54}$$

and

$$\beta_2(k, k') \approx 0. \tag{9.55}$$

Doing so, we find that

$$\hat{H} = \sum_{n=1}^{\infty} \frac{\hbar c k_n}{2} \left(\hat{a}_n^\dagger \hat{a}_n + \hat{a}_n \hat{a}_n^\dagger\right) + \int_0^\infty dk \frac{\hbar c k}{2} \left\{\hat{b}^\dagger(k)\hat{b}(k) + \hat{b}(k)\hat{b}^\dagger(k)\right\}$$
$$- \sum_{n=1}^{\infty} \int_0^\infty dk \left\{\epsilon \frac{\hbar c}{2\sqrt{\pi L}} e^{-ikL} \frac{\sin([k-k_n]L)}{(k-k_n)L} \hat{a}_n \hat{b}^\dagger(k) + \text{H.c.}\right\}. \tag{9.56}$$

Equations (9.50) and (9.54)–(9.56) agree with Barnett and Radmore's expressions [37] for coupling strength[12]:

$$V_n(k) = -\epsilon \frac{\hbar c}{2\sqrt{\pi L}} e^{-ikL} \frac{\sin([k-k_n]L)}{(k-k_n)L}. \tag{9.57}$$

So we have derived from first principles the phenomenological Gardiner–Collett Hamiltonian (9.1), have demonstrated that it holds only for high-Q cavities, and have found the reason why it breaks down away from the high-Q regime. Moreover, we have derived an explicit expression for the coupling strength within our simple one-dimensional model. Let us look into what this expression tells us.

As the cavity photons can only escape to the external continuum of modes, there is really just a single reservoir (this external continuum) for all cavity modes. In quantum optics, however, it is often tacitly assumed [261] that each mode is coupled to an independent reservoir. The explicit expression (9.57) for the coupling strength that we have found shows how this effective independent reservoir behavior emerges from a single reservoir in the high-Q regime. The sinc function in (9.57) splits the continuum of external modes into a collection of smaller reservoirs, each having a finite frequency width and coupled to a single-cavity mode. Their independency arises from the smallness of the overlap of a given cavity mode with neighboring cavity resonances in the high-Q regime (see Figure 9.3).

Grangier and Poizat [247, 248] have shown that the breakdown of the independent reservoir assumption is essential for the appearance of excess quantum noise in lasers. This is a curious phenomenon first found in gain-guided semiconductor lasers by Petermann [491], where a laser has a linewidth larger than the usual Schawlow–Townes linewidth, as if there were more than one spontaneous emission noise photon per mode. Later, Siegman developed a semiclassical theory [561, 562], where excess noise emerges as a consequence of mode nonorthogonality. Siegman showed that for those lasers the perfect cavity mode expansion of the radiation field breaks down and one must use instead the actual modes of the open laser cavity, which are nonorthogonal. Thus, the breakdown of the Gardiner–Collett Hamiltonian, the need to adopt the

---

[12]Notice that unlike the constant coupling strength advocated by Barnett and Radmore [37] and Gardiner and Collett [225], among others, the frequency-dependent $V_n(k)$ given by (9.57) does not make the phase shift integral in the Fano diagonalization method diverge. Moreover, (9.57) is practically flat (constant) within the cavity resonance peaks but falls off between peaks.

natural cavity modes rather than perfect cavity modes, and the emergence of excess noise in lasers are all connected.

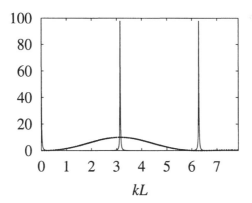

Figure 9.3 Here we show the coupling strength $V_n(k)$ in relation to the cavity resonances. The thin line is a plot of $|\mathcal{L}(k)|^2$; the peaks show the cavity resonances corresponding to k_0, k_1, and k_2. The thick dotted line is a plot of $\sin^2([k - k_1]L)/(k - k_1)^2$ properly scaled to appear in the figure. Notice how $V_1(k)$ is practically flat inside the $n = 1$-mode linewidth and falls off as it approaches the next modes. We have used an energy reflectivity $|r|^2$ of 96% (i.e., $\epsilon = 0.2$).

Another interesting feature of this derivation is that even though we have not adopted the rotating-wave approximation, there are no counterrotating terms in (9.56). This absence of counterrotating terms is a consequence of $\alpha_{n2}(k)$ and $\beta_2(k, k')$ vanishing in the high-Q approximation. It guarantees that the master equation derived from (9.56) is of the Lindblad form [411], as (6.68), derived in Chapter 6.

9.2 SOMMERFELD'S RADIATION CONDITION

Sommerfeld pointed out that for wave motion in an infinite medium, the usual boundary conditions at the finite surfaces involved in the problem are not enough to determine the solution uniquely. He was then led to propose restrictions on the behavior of the wave at infinity to guarantee a unique solution. These restrictions, known as *Sommerfeld's radiation condition,* amount to an asymptotic boundary condition at infinity.

Sommerfeld's argument (see Sec. 28 of [571]) is this: Consider a spherical cavity of radius R and a scalar field ψ whose boundary condition is that it vanishes on the surface of the sphere. Then for spherically symmetric radial oscillations, the eigenfunctions ψ_n of the Helmholtz equation

$$\nabla^2\psi + k^2\psi = 0 \tag{9.58}$$

are given by

$$\psi_n(r) = \frac{\sin k_n r}{r}, \quad \text{where} \quad k_n = \frac{\pi}{R}n, \quad \text{with} \quad n = \pm 1, \pm 2, \ldots \qquad (9.59)$$

Now, as $R \to \infty$, the various eigenvalues k_n become closer and closer, resulting in a continuous spectrum, and the cavity becomes free space. Thus, we can consider

$$\psi(r, k) = \frac{\sin kr}{r}, \qquad (9.60)$$

which is everywhere regular and vanishes at infinity as an eigenfunction of free space. This means that in any problem where a given source $f(\mathbf{r})$ in a finite domain drives oscillations in free space, obeying the driven Helmholtz equation

$$\nabla^2 \psi + k^2 \psi = f(\mathbf{r}), \qquad (9.61)$$

we can always add the solution (9.60) of the homogeneous Helmholtz equation (9.58) to the particular solution of (9.61) and then obtain a different solution of (9.61). So, unlike potential problems, oscillation problems *are not* determined *uniquely* by their prescribed sources in a finite domain.

The solution that Sommerfeld proposed to this "paradoxical result" is to demand not only that the field vanishes at infinity, but also that the sources really be sources rather than sinks of energy. Rewriting (9.60) as

$$\psi(r, k) = \frac{1}{i2} \left(\frac{e^{ikr}}{r} - \frac{e^{-ikr}}{r} \right), \qquad (9.62)$$

he argues that if we add back the time dependence $\exp(-i\omega t)$ of the wave motion, we see that (9.62) contains both a radiated spherical wave $(1/r)\exp(ikr)$, which moves away to infinity, and an absorbed spherical wave $(1/r)\exp(-ikr)$, which comes from infinity back to the source. He then proposes to exclude the latter component so that only the waves radiated by the source be present. This is done by imposing the following asymptotic condition at infinity:

$$\lim_{r \to \infty} r \left(\frac{\partial \psi}{\partial r} - ik\,\psi \right) = 0, \qquad (9.63)$$

which Sommerfeld named the *general condition of radiation*.

9.3 THOMSON'S NATURAL MODES

In the later part of the nineteenth century, the hottest new theory in physics was Maxwell's classical electrodynamics. As we know, one of the outstanding features of Maxwell's theory is the prediction that light is an electromagnetic wave. Lord Kelvin's discovery in 1853 of the oscillatory discharge of a Leyden jar [597] gave a way of producing electrical waves. Still, before Hertz's successful experimental confirmation of Maxwell's prediction, physicists were eager to study any system producing electrical oscillations. So in 1883, 27-year-old J. J. Thomson[13] set to work on the problem of finding electrical oscillations that can occur on the surface of a conducting sphere.[14] Even though he is widely respected now for his achievements in experimental physics, Thomson always had a knack for theory [576]. His solution appeared in a paper published in the *Proceedings of the London Mathematical Society* in 1884, the same year he was offered the chair of Cavendish Professor at Cambridge and elected a Fellow of the Royal Society [594].

Thomson considered a thin spherical shell and calculated the resulting electromagnetic oscillations when the cause that first produced an irregular charge distribution on the surface of the shell is suddenly withdrawn. These oscillations are due to the interchange of energy between the currents that start off to cancel the charge distribution and the potential energy of this distribution. He looked into two cases: one where the initial irregular charge distribution is created by some electrical system inside the sphere, and another where it is created by a system outside. In the latter case, Thomson realized that "since the currents fade away, owing to the resistance of the conductor," the field radiated outside the sphere will be damped. In his analysis, this meant that the only solution for the field outside the sphere had to be an outgoing spherical wave. So from a physical consideration, Thomson was led to Sommerfeld's radiation condition before Sommerfeld formulated it in all its generality.

What Thomson calculated were the natural modes of oscillation of his spherical shell "cavity." These are the modes in which a cavity naturally resonates after the source that drove it initially is suddenly withdrawn. They result from a direct application to the case of an open cavity of the physical idea of mode discussed in Section 2.3. To calculate these modes, one must not only use the boundary conditions on the cavity walls, but also impose Sommerfeld's radiation condition at infinity.

9.4 INCOMING VACUUM MODES AND COMPLETENESS

As they obey Sommerfeld's radiation condition, natural modes on their own do not form a complete set. So we cannot expand the quantized electromagnetic radiation field in terms of them alone. In this section we formulate this problem more precisely and suggest how they can be supplemented by a different sort of mode in a way that will make the total set complete. But before starting this discussion, it is convenient to introduce first a fascinating result that is not widely taught in electrodynamics lecture courses today: Although we normally use one scalar potential and a vector potential to describe the electromagnetic field, the same job can be done using only two real scalar potentials or, equivalently, a single complex scalar potential.

---

[13] He was later to become the discoverer of the first elementary particle, the electron.
[14] See page 95, "Researches, 1880–84" in Chapter III of [596].

9.4.1 Whittaker's scalar potentials: From a vector and a scalar potential to just two scalar potentials

We know from elementary electrodynamics that a description in terms of electromagnetic potentials introduces redundant degrees of freedom, whose main consequence is giving us the possibility of choosing the gauge that suits us best. Can we use this redundancy to reduce the number of electromagnetic potentials needed to describe the fields? Yes, we can. We will now show that this redundancy can be exploited to reduce the usual description in terms of a scalar potential and a vector (with three scalar components) potential to one involving only two scalar potentials. This is not only useful to simplify calculations, but also seems to have some fundamental significance, as we know that light has only two independent polarizations (see Chapters 2 and 3).

One can derive this representation in a general way, in the presence of matter [466] and without making specific assumptions about the cavity shape and so on. As we just wish to develop the basic ideas of such representation, however, we will assume vacuum and take a specific cavity geometry that will lead to a simple presentation of these ideas.

Consider a cylindrical cavity geometry, where the surfaces of the cavity walls are everywhere parallel to the z-axis. Assuming that the cavity is just sitting in the vacuum, the frequency $\omega = ck$ components of the fields satisfy the following Maxwell equations:

$$\nabla \wedge \mathbf{B} + ik\,\mathbf{E} = \mathbf{0}, \tag{9.64}$$

$$\nabla \cdot \mathbf{E} = 0, \tag{9.65}$$

$$\nabla \wedge \mathbf{E} - ik\,\mathbf{B} = \mathbf{0}, \tag{9.66}$$

$$\nabla \cdot \mathbf{B} = 0, \tag{9.67}$$

with the boundary conditions (see Appendix B)

$$\hat{\mathbf{n}} \wedge \mathbf{E}|_S = \mathbf{0} \quad \text{and} \quad \hat{\mathbf{n}} \cdot \mathbf{B}|_S = 0, \tag{9.68}$$

where $\hat{\mathbf{n}}$ is the unit normal to a cavity wall and $|_S$ denotes that these quantities are to be evaluated on the cavity walls.

To satisfy the vanishing divergence equations (9.65) and (9.67) automatically, we can take $\mathbf{E}$ and $\mathbf{B}$ as the curl of some vector fields. It is convenient to write this as

$$\mathbf{E} = ik\nabla \wedge (v\hat{\mathbf{z}}) + \nabla \wedge \boldsymbol{\mathcal{E}} \quad \text{and} \quad \mathbf{B} = -ik\nabla \wedge (u\hat{\mathbf{z}}) + \nabla \wedge \boldsymbol{\mathcal{B}}, \tag{9.69}$$

where u and v are two scalar fields and $\boldsymbol{\mathcal{E}}$ and $\boldsymbol{\mathcal{B}}$ are two vector fields. When $\boldsymbol{\mathcal{E}}$ vanishes, $\mathbf{E}$ is entirely in the xy-plane. When $\boldsymbol{\mathcal{B}}$ vanishes, $\mathbf{B}$ is also entirely in the xy-plane. Moreover, notice that there is, of course, a gauge freedom: We can add a gradient of an arbitrary function of position to $ikv\hat{\mathbf{z}} + \boldsymbol{\mathcal{E}}$ and another to $-iku\hat{\mathbf{z}} + \boldsymbol{\mathcal{B}}$, without changing $\mathbf{E}$ and $\mathbf{B}$.

Substituting the first equation of (9.69) in (9.64), we find that

$$\nabla \wedge \left\{ \mathbf{B} - k^2\, v\,\hat{\mathbf{z}} + ik\,\boldsymbol{\mathcal{E}} \right\} = \mathbf{0}. \tag{9.70}$$

This means that $\mathbf{B} - k^2 \, v \, \hat{\mathbf{z}} + ik \, \boldsymbol{\mathcal{E}}$ is the gradient of some function of position. The gauge freedom in $ikv\hat{\mathbf{z}} + \boldsymbol{\mathcal{E}}$ amounts to this function being arbitrary. Then, substituting the right-hand side of the second equation in (9.69) for $\mathbf{B}$, we obtain

$$ik\boldsymbol{\mathcal{E}} - k^2 v\hat{\mathbf{z}} + \nabla\phi - ik\nabla \wedge (u\hat{\mathbf{z}}) + \nabla \wedge \boldsymbol{\mathcal{B}} = \mathbf{0}, \qquad (9.71)$$

where ϕ is an arbitrary function of position. Applying the same reasoning to (9.66), we find that

$$-ik\boldsymbol{\mathcal{B}} - k^2 u\hat{\mathbf{z}} + \nabla\phi' + \nabla \wedge \boldsymbol{\mathcal{E}} + ik\nabla \wedge (v\hat{\mathbf{z}}) = \mathbf{0}, \qquad (9.72)$$

where ϕ' is another arbitrary scalar function. Thus, we can choose ϕ and ϕ' such that

$$\nabla \wedge \boldsymbol{\mathcal{E}} + \nabla\phi' - k^2 u\hat{\mathbf{z}} = \mathbf{0}, \qquad (9.73)$$
$$\nabla \wedge \boldsymbol{\mathcal{B}} + \nabla\phi - k^2 v\hat{\mathbf{z}} = \mathbf{0}. \qquad (9.74)$$

With this choice in (9.71) and (9.72), we find that

$$\boldsymbol{\mathcal{B}} = \nabla \wedge (v \, \hat{\mathbf{z}}), \qquad (9.75)$$
$$\boldsymbol{\mathcal{E}} = \nabla \wedge (u \, \hat{\mathbf{z}}). \qquad (9.76)$$

More specifically, we can choose ϕ' to satisfy (9.73) in the following way. Substituting (9.76) in (9.73) and expanding $\nabla \wedge \{\nabla \wedge (u\hat{\mathbf{z}})\}$, we obtain

$$\nabla\left[(\hat{\mathbf{z}} \cdot \nabla) u\right] - \hat{\mathbf{z}}\nabla^2 u + \nabla\phi' - k^2 \, u \, \hat{\mathbf{z}} = \mathbf{0}. \qquad (9.77)$$

So if we choose

$$\nabla\phi' = -\nabla\left[(\hat{\mathbf{z}} \cdot \nabla) u\right] \qquad (9.78)$$

and require that

$$\nabla^2 u + k^2 \, u = 0, \qquad (9.79)$$

(9.73) will be satisfied. Notice that (9.78) does not fix the gauge completely, because it chooses only the gradient of ϕ'. Later we fix the gauge by specifying ϕ' itself.

Repeating this argument with (9.74), we find that we can satisfy (9.74) by requiring that

$$\nabla\phi = -\nabla\left[(\hat{\mathbf{z}} \cdot \nabla) v\right], \qquad (9.80)$$
$$\nabla^2 v + k^2 \, v = 0, \qquad (9.81)$$

where just as in the case of ϕ', (9.80) does not fix the gauge completely, as it determines $\nabla\phi$ rather than ϕ itself.

We now fix the gauge by choosing $\phi = \phi' = 0$. Then, from (9.78) and (9.80), we see that our gauge choice implies the following conditions on the two scalar potentials

$$\nabla\left[(\hat{\mathbf{z}} \cdot \nabla) u\right] = \mathbf{0}, \qquad (9.82)$$
$$\nabla\left[(\hat{\mathbf{z}} \cdot \nabla) v\right] = \mathbf{0}. \qquad (9.83)$$

To summarize this description of the fields in terms of only two scalar potentials, besides satisfying (9.82) and (9.83), they also satisfy the Helmholtz equation,

$$\nabla^2 u + k^2\, u = 0, \tag{9.84}$$
$$\nabla^2 v + k^2\, v = 0, \tag{9.85}$$

and the fields are given in terms of them by

$$\mathbf{E} = ik\nabla \wedge (v\, \hat{\mathbf{z}}) + \nabla \wedge \{\nabla \wedge (u\, \hat{\mathbf{z}})\}, \tag{9.86}$$
$$\mathbf{B} = -ik\nabla \wedge (u\, \hat{\mathbf{z}}) + \nabla \wedge \{\nabla \wedge (v\, \hat{\mathbf{z}})\}. \tag{9.87}$$

We will see now that as the normal to the cavity walls is always orthogonal to the z-axis, the boundary conditions assume a rather simple form in terms of u and v. Using $\hat{\mathbf{n}} \cdot \hat{\mathbf{z}} = 0$, we notice that

$$\hat{\mathbf{n}} \wedge \{\nabla \wedge (v\, \hat{\mathbf{z}})\} = (\hat{\mathbf{n}} \cdot \hat{\mathbf{z}})\, \nabla v - (\hat{\mathbf{n}} \cdot \nabla v)\, \hat{\mathbf{z}} = -(\hat{\mathbf{n}} \cdot \nabla v)\, \hat{\mathbf{z}} = -\frac{\partial v}{\partial n}\hat{\mathbf{z}}. \tag{9.88}$$

Using (9.82), we notice that

$$\hat{\mathbf{n}} \wedge \{\nabla \wedge [\hat{\mathbf{z}} \wedge (\nabla u)]\} = \hat{\mathbf{n}} \wedge \{(\nabla^2 u)\, \hat{\mathbf{z}} - \nabla\, [(\hat{\mathbf{z}} \cdot \nabla)\, u]\} = (\hat{\mathbf{n}} \wedge \hat{\mathbf{z}})\, (\nabla^2 u)$$
$$= -(\hat{\mathbf{n}} \wedge \hat{\mathbf{z}})\, k^2 u. \tag{9.89}$$

So substituting (9.86), (9.88), and (9.89) in the first boundary condition in (9.68) yields

$$ik\, \frac{\partial v}{\partial n}\bigg|_S + k^2\, (\hat{\mathbf{n}} \wedge \hat{\mathbf{z}})\, u|_S = 0. \tag{9.90}$$

As $\hat{\mathbf{z}}$ and $\hat{\mathbf{n}} \wedge \hat{\mathbf{z}}$ are two orthogonal and thus independent directions, it follows that the first boundary condition in (9.68) is equivalent to

$$u|_S = 0 \quad \text{and} \quad \frac{\partial v}{\partial n}\bigg|_S = 0. \tag{9.91}$$

The boundary condition (9.68) for the magnetic field yields $\hat{\mathbf{n}} \cdot \{\hat{\mathbf{z}} \wedge (\nabla u)\}|_S = 0$, because $\hat{\mathbf{n}} \cdot \{\nabla \wedge [\nabla \wedge (v\hat{\mathbf{z}})]\}|_S = 0$. Now as u is constant over the cavity walls (i.e., $u|_S = 0$), the gradient of u over the walls can only point in the direction of $\hat{\mathbf{n}}$, so this boundary condition is trivially satisfied.

9.4.2 General formulation of the problem

For a given frequency $\omega = ck$, describing the electromagnetic radiation field by two scalar potentials, we can reduce the general *three-dimensional* problem of finding the natural modes of an open cavity to two *scalar* problems [84]:

1. Solving the Helmholtz equation

$$\nabla^2 u + k^2\, u = 0, \tag{9.92}$$

with the boundary condition

$$u|_S = 0 \tag{9.93}$$

over the cavity walls and Sommerfeld's radiation condition (9.63) at infinity, which amounts to[15]

$$\lim_{r \to \infty} u(\mathbf{r}) = g(\theta, \phi) \frac{e^{ikr}}{r}. \tag{9.94}$$

2. Solving the Helmholtz equation

$$\nabla^2 v + k^2 \, v = 0, \tag{9.95}$$

with the boundary condition

$$\left. \frac{\partial v}{\partial n} \right|_S = 0 \tag{9.96}$$

over the cavity walls and the asymptotic condition

$$\lim_{r \to \infty} v(\mathbf{r}) = f(\theta, \phi) \frac{e^{ikr}}{r}. \tag{9.97}$$

The radiation condition implies that a linear combination of natural modes can only reproduce outgoing fields. Thus, it does not form a complete set capable of representing every conceivable quantum vacuum fluctuation which can also have incoming fields from infinity.

To get a complete set of modes for the quantized radiation field, we must change the asymptotic boundary condition at infinity to allow for incoming and outgoing waves together, as in a scattering problem. The resulting modes will form a continuum. They are the *modes of the universe*, which were determined in Section 2.5.2 using precisely such a scattering formulation. There it was convenient to work with plane waves. Here, however, it is more convenient to formulate the asymptotic boundary condition in terms of spherical waves. So in place of Sommerfeld's radiation condition, we demand that [607]

$$\lim_{r \to \infty} \Phi(\mathbf{r}) = \chi_I(\theta, \varphi) \frac{e^{-ikr}}{r} + \chi_R(\theta, \varphi) \frac{e^{ikr}}{r}, \tag{9.98}$$

where Φ is u in case 1 and v in case 2. The first term on the right-hand side of (9.98) represents an incoming wave from infinity, and the second term, the scattered outgoing wave. So χ_I and χ_R are related by the scattering matrix S (i.e., $\chi_R = S\chi_I$).

So there lies the problem of using natural modes in a quantum theory. The Sommerfeld radiation condition satisfied by natural modes is only part of the asymptotic boundary condition at infinity (9.98) that defines the set of all possible solutions for the radiation field. To make a complete set out of natural modes, we must supplement them with other modes that, to put it crudely, satisfy the remaining required asymptotic boundary conditions (9.98). This is, as far as I know, still an open problem in

---

[15]See Section 9.2 or Chapter 10 of [607].

the general three-dimensional case. There is a one-dimensional solution [172, 173], though. In a nutshell, this solution consists of dividing the modes-of-the-universe boundary problem into two separate boundary problems, one of which yields the natural modes and the other, their required complement to make a complete set. For more on this one-dimensional case, see the recommended reading.

RECOMMENDED READING

- The description of dissipative systems directly by a Lagrangian or even a Hamiltonian is a polemic topic in the literature. In the Newtonian formulation of classical mechanics, friction forces are dealt with just like any other force. The variational formulation of mechanics, however, requires that the forces be derivable from a work function (they must be "monogenic" rather than "polygenic" forces; see, e.g., [386]). As quantum mechanics is derived from the latter formulation rather than from Newton's, it inherited this requirement. In the case of the Hamiltonian formulation of quantum mechanics, the requirement is even more stringent: The Hamiltonian must be the total energy of the system (see Secs. 27 and 28 of [159]). This is the main reason for the usual treatment of dissipation in quantum mechanics where the environment is included to make the total system closed and nondissipative. Feynman's 1948 "proof" of Maxwell equations[16] has been reinterpreted as a derivation that the most general kind of force the Hamiltonian–Lagrangian formalism can deal with is one whose form is as that of the Lorentz force [295, 302] (see also Sec. 1-5, pp. 21–24 of [240]). For a simple pedagogic discussion of soluble models of quantum dumping in cavity QED, see [410]. For more on the polemic attempts to describe damping directly by a Lagrangian or Hamiltonian, see [45, 76, 150, 251, 295, 339, 365, 368, 507, 508, 512, 513, 520, 582].

- The "Gardiner–Collett" Hamiltonian was known in the literature before Gardiner and Collett's two papers [109, 225] on their input–output theory. An analogous Hamiltonian was introduced in 1962 by Cohen et al. to explain experiments involving tunneling of electrons between superconducting films covered by a thin oxide layer [107]. This Hamiltonian was ultimately based [29] on some early work by Bardeen [28] on a many-body theory of tunneling. For this reason, in solid-state physics this Hamiltonian is known as the *Bardeen Hamiltonian*. The Bardeen Hamiltonian was deduced from first principles (for the solid-state case of tunneling of particles) by Prange [496], who pointed out that in general it gives rise to mode nonorthogonality. Gardiner and Collett's input–output formalism has been extended to bosonic matter fields to model the output coupling of atoms (as in an atom laser) from a Bose–Einstein conden-

---

[16]In 1948, Feynman derived Maxwell equations as a mathematical consequence from two apparently unconnected ingredients: Newton's second law and the quantum commutation relations for position and momentum! See Dyson's account of Feynman's derivation [176].

sate in a single-mode trap [296]. Recently, the input–output theory has been generalized to fermions [553], where the Pauli exclusion principle forces us to take into account all "cavity modes."

- In the quantum optics literature, the modes of open cavities are often referred to as *quasimodes* to distinguish them from the normal modes of closed (perfect) cavities. There is an impressive lack of consensus on how such modes should be defined. Most authors miss the essential point of the problem of using natural cavity modes to describe the quantized radiation field, which we discussed in Section 9.4, and the ideas of classical resonator theory (particularly Thomson's [594, 595]). For a review of some of these approaches, see [173]. Here I will just list the main papers: [30, 31, 43, 78–80, 102, 103, 132–140, 169, 193, 194, 226–228, 247, 248, 258, 259, 294, 362, 374, 378–382, 390, 391, 400–404, 408, 603, 604, 608], [618, Chapter 8],

- For a pedagogical account of the radiation condition from Sommerfeld himself, see Sec. 28 of [571]. For a treatment of the underlying mathematical issues, see [24, 651].

- The possibility of representing the electromagnetic field by two scalar potentials alone was discovered by Sir Edmund Whittaker in 1903 [640]. Two other ways of representing the field by only two scalar potentials were developed later by Green and Wolf [250] and Bouwkamp and Casimir [70]. For a review, see Nisbet's paper [466] or Section 10 of Bremmer's article in the *Handbuch der Physik*[17] [72].

Problems

9.1 Use the Gardiner–Collett Hamiltonian, (9.1), to derive the Lindblad master equation for cavity damping, (6.68).
Hint: Write down a general expression for the total density matrix in terms of number states of the cavity mode and the external modes. Then use (9.1) to obtain the master equation for this total density matrix and trace over the external modes to get a master equation for the reduced density matrix for the cavity mode.

9.2 Calculate the first two integrals on the right-hand side of (9.9) by substituting (9.10) and (9.11) and using

$$\lim_{\epsilon \to 0} \int_0^\infty dx\, e^{-i(k \pm k' - i\epsilon)x} = -i\mathcal{P}\frac{1}{k \pm k'} + \pi\,\delta(k \pm k'),$$

$$\lim_{\epsilon \to 0} \int_0^\infty dx\, e^{i(k \pm k' + i\epsilon)x} = i\mathcal{P}\frac{1}{k \pm k'} + \pi\,\delta(k \pm k'),$$

$$\int_{-L}^0 dx\, \cos\left([k \pm k'][x + L]\right) = \frac{\sin\left([k \pm k']L\right)}{k \pm k'}.$$

---

[17]Bremmer's article is in English.

Then notice from (9.12) that $t^2 = r^2 - r/r^*$, so that $|r - t\exp(i2kL)\,\mathcal{L}(k)|^2 = 1$. Now use (9.16) to show that (9.9) yields (9.17).

9.3   In section 9.1 (page 261) we have used Poisson's sum formula to show that $\mathcal{L}(k)$ can be expressed as a constant term, $t/2$, plus a series of complex functions $\mathcal{L}_n(k)$. This series is very suggestive, as the modulus square of each $\mathcal{L}_n(k)$ is a Lorentzian. Using the series summation result,

$$2x \sum_{n=1}^{\infty} \frac{1}{x^2 - n^2} = \pi\cot(\pi x) - \frac{1}{x},$$

derived in Chapter 2 by considering the Fourier expansion[18] of $\cos\pi x$, sum the series in (9.20) and show that (9.20) is indeed equal to (9.7).

Hint: Notice that as $|r| < 1$, (9.7) can be summed, its closed expression being given by $\mathcal{L}(k) = t/\left[1 + r\exp\left(i2Lk\right)\right]$.

9.4   Writing r as $|r|\exp[i\arg(r)]$ in (9.7), show that

$$|\mathcal{L}(k)|^2 = |t|^2 \sum_{m=0}^{\infty} |r|^{2m} \sum_{l=-\infty}^{\infty} |r|^{|l|} \exp\left\{i\left[2kL + \arg(r) - \pi\right]l\right\}.$$

Do the first sum and then use (9.12) to show that

$$|\mathcal{L}(k)|^2 = \sum_{l=0}^{\infty} \left\{|r|\,e^{i[2kL+\arg(r)-\pi]}\right\}^l + \sum_{l=1}^{\infty} \left\{|r|\,e^{-i[2kL+\arg(r)-\pi]}\right\}^l.$$

Now use Poisson's sum formula again in each of the two series on the right-hand side of the preceding equation and deduce (9.22).

9.5   From (9.8) and (9.28) show that

$$[\hat{a}_n, \hat{a}_{n'}^\dagger] = \frac{1}{\pi L\sqrt{k_n k_{n'}}} \underbrace{\int_{-\infty}^{\infty} dk\, k\, \frac{\sin\left(\left[k - k_n\right]L\right)}{k - k_n} \frac{\sin\left(\left[k - k_{n'}\right]L\right)}{k - k_{n'}} |\mathcal{L}(k)|^2}_{c_{nn'}}$$

and

$$[\hat{a}_n, \hat{a}_{n'}] = \frac{1}{\pi L\sqrt{k_n k_{n'}}} \underbrace{\int_{-\infty}^{\infty} dk\, k\, \frac{\sin\left(\left[k - k_n\right]L\right)}{k - k_n} \frac{\sin\left(\left[k + k_{n'}\right]L\right)}{k + k_{n'}} |\mathcal{L}(k)|^2}_{s_{nn'}}$$

Solve the integrals $c_{nn'}$ and $s_{nn'}$ by contour integration and show that the solutions are $c_{nn'} = (1/2)\operatorname{Re}\mathcal{J}_{nn'-}$ and $s_{nn'} = (1/2)\operatorname{Re}\mathcal{J}_{nn'+}$, where

$$\mathcal{J}_{nn'-} = 2\pi L|\mathcal{L}(k_n)|^2 k_n \delta_{nn'}$$

[18]See also [433].

$$+ \frac{\pi}{L} \sum_{m=-\infty}^{\infty} (\bar{k}_m + i\gamma) \underbrace{\frac{e^{i(k_n - k_{n'})L} - e^{i2\bar{k}_m L} e^{-i(k_n + k_{n'})L} e^{-2\gamma L}}{(\bar{k}_m - k_n + i\gamma)(\bar{k}_m - k_{n'} + i\gamma)}}_{S}$$

and

$$\mathcal{J}_{nn'+} = \frac{\pi}{L} \sum_{m=-\infty}^{\infty} (\bar{k}_m + i\gamma) \underbrace{\frac{e^{i(k_n + k_{n'})L} - e^{i2\bar{k}_m L} e^{-i(k_n - k_{n'})L} e^{-2\gamma L}}{(\bar{k}_m - k_n + i\gamma)(\bar{k}_m + k_{n'} + i\gamma)}}_{S'},$$

where $k_n \equiv n\pi/L$, $\gamma \equiv -\ln|r|/2L$, $\bar{k}_n \equiv k_n - \Delta$, and $\Delta \equiv [\arg(r) - \pi]/2L$. To sum the series S and S', use the series summation result (see Chap. XII, Sec. 125, p. 370 of [77])

$$\sum_{m=-\infty}^{\infty} \frac{e^{ik_m x}}{\bar{k}_m - k_n + i\gamma} = -i2L \frac{e^{i(k_n + \Delta - i\gamma)x}}{e^{i2L(k_n + \Delta - i\gamma)} - 1}.$$

But be careful that it is valid only on the window $0 < x < 2L$, as the series above is discontinuous at the boundaries of this window. To avoid the discontinuities and remain within the window, you can adopt a sort of regularization procedure writing S as

$$S = \lim_{\xi \to 0^+} \left(e^{i(k_n - k_{n'})L} \left\{ \sum_{m=-\infty}^{\infty} \frac{e^{ik_m \xi}}{\bar{k}_m - k_n + i\gamma} \right. \right.$$

$$+ \frac{k_{n'}}{k_n - k_{n'}} \left[\sum_{m=-\infty}^{\infty} \frac{e^{ik_m \xi}}{\bar{k}_m - k_n + i\gamma} - \sum_{m=-\infty}^{\infty} \frac{e^{ik_m \xi}}{\bar{k}_m - k_{n'} + i\gamma} \right] \right\}$$

$$- e^{-i2L\Delta} e^{-i(k_n + k_{n'})L} e^{-2\gamma L} \left\{ \sum_{m=-\infty}^{\infty} \frac{e^{i(2L-\xi)k_m}}{\bar{k}_m - k_n + i\gamma} \right.$$

$$+ \left. \left. \frac{k_{n'}}{k_n - k_{n'}} \left[\sum_{m=-\infty}^{\infty} \frac{e^{i(2L-\xi)k_m}}{\bar{k}_m - k_n + i\gamma} - \sum_{m=-\infty}^{\infty} \frac{e^{i(2L-\xi)k_m}}{\bar{k}_m - k_{n'} + i\gamma} \right] \right\} \right)$$

and S' as

$$S' = \lim_{\xi \to 0^+} \left(e^{i(k_n + k_{n'})L} \left\{ \sum_{m=-\infty}^{\infty} \frac{e^{ik_m \xi}}{\bar{k}_m - k_n + i\gamma} \right. \right.$$

$$- \frac{k_{n'}}{k_n + k_{n'}} \left[\sum_{m=-\infty}^{\infty} \frac{e^{ik_m \xi}}{\bar{k}_m - k_n + i\gamma} - \sum_{m=-\infty}^{\infty} \frac{e^{ik_m \xi}}{\bar{k}_m - k_{n'} + i\gamma} \right] \right\}$$

$$- e^{-i2L\Delta} e^{-i(k_n - k_{n'})L} e^{-2\gamma L} \left\{ \sum_{m=-\infty}^{\infty} \frac{e^{i(2L-\xi)k_m}}{\bar{k}_m - k_n + i\gamma} \right.$$

$$- \left. \left. \frac{k_{n'}}{k_n + k_{n'}} \left[\sum_{m=-\infty}^{\infty} \frac{e^{i(2L-\xi)k_m}}{\bar{k}_m - k_n + i\gamma} - \sum_{m=-\infty}^{\infty} \frac{e^{i(2L-\xi)k_m}}{\bar{k}_m - k_{n'} + i\gamma} \right] \right\} \right).$$

Remembering that $\sin([k_n - k_{n'}]L)/(k_n - k_{n'}) = L\delta_{nn'}$, show that

$$S = 2k_{n'}L^2\delta_{nn'}\left\{1 + i \cot\left([k_{n'} + \Delta - i\gamma]L\right)\right\}$$

and that

$$|\mathcal{L}(k_n)|^2 = -\mathrm{Re}\left\{i \cot\left([k_n + \Delta - i\gamma]L\right)\right\}.$$

Conclude then that $[\hat{a}_n, \hat{a}_{n'}^\dagger] = \delta_{nn'}$. Proceed in an analogous way with S' to conclude that $[\hat{a}_n, \hat{a}_{n'}] = 0$.

9.6   Using the canonical commutation relations for the radiation field inside the cavity,[19]

$$\left[\hat{B}_{\mathrm{in}}(x), \hat{E}_{\mathrm{in}}(x')\right] = -i2h\frac{\partial}{\partial x}\left\{\delta(x - x') - \delta(x + x' + 2L)\right\},$$
$$\left[\hat{B}_{\mathrm{in}}(x), \hat{B}_{\mathrm{in}}(x')\right] = \left[\hat{E}_{\mathrm{in}}(x), \hat{E}_{\mathrm{in}}(x')\right] = 0,$$

show that $[\hat{a}_n, \hat{a}_{n'}^\dagger] = \delta_{nn'}$ and $[\hat{a}_n, \hat{a}_{n'}] = 0$.

9.7   Using the canonical commutation relations for the radiation field outside the cavity,

$$\left[\hat{B}_{\mathrm{out}}(x), \hat{E}_{\mathrm{out}}(x')\right] = -i2h\frac{\partial}{\partial x}\delta(x - x'),$$
$$\left[\hat{B}_{\mathrm{out}}(x), \hat{B}_{\mathrm{out}}(x')\right] = \left[\hat{E}_{\mathrm{out}}(x), \hat{E}_{\mathrm{out}}(x')\right] = 0,$$

show that $[\hat{b}(k), \hat{b}(k')^\dagger] = \delta(k - k')$ and $[\hat{b}(k), \hat{b}(k')] = 0$.

9.8   Another way to test (9.39) is to substitute it in (9.2) and see if the resulting intracavity electric field agrees with (9.24). Substituting (9.39) in (9.2), we find that $\hat{E}_{\mathrm{in}}(x) = \hat{E}_{\mathrm{in,D}}(x) + \hat{E}_{\mathrm{in,C}}(x)$, where

$$\hat{E}_{\mathrm{in,D}}(x) \equiv -i2\sum_{n=1}^{\infty}\hat{a}_n$$

$$\times \underbrace{\int_0^\infty dk\,\mathcal{E}_{\mathrm{vac}}(k)\sin\left([x + L]\,k\right)\left\{\mathcal{L}(k)e^{ikL}\alpha_{n1}(k) - \mathcal{L}^*(k)e^{-ikL}\alpha_{n2}(k)\right\}}_{D_n(x)} + \mathrm{H.c.}$$

and

$$\hat{E}_{\mathrm{in,C}}(x) \equiv -i2\int_0^\infty dk'\,\hat{b}(k')$$

---

[19]The extra delta function is there because of the perfect mirror; see [448, 495].

$$\times \int_0^\infty dk \, \mathcal{E}_{vac}(k) \sin\left(\left[x + L\right]k\right) \left\{ \underbrace{\mathcal{L}(k)e^{ikL}\beta_1(k, k') - \mathcal{L}^*(k)e^{-ikL}\beta_2^*(k, k')}_{I(k', k)} \right\}$$

$$+ \text{ H.c.}$$

1. Using (9.40) and (9.41), show that $D_n(x)$ is given by

$$D_n(x) = \sqrt{\frac{\hbar c}{\pi L k_n}} \int_{-\infty}^\infty dk \, k|\mathcal{L}(k)|^2 \sin\left(\left[x + L\right]k\right) \frac{\sin\left(\left[k - k_n\right]L\right)}{k - k_n}$$

and that the integral above can be rewritten as

$$\frac{(-1)^n}{2} \text{Re} \underbrace{\int_{-\infty}^\infty dk \, k \, |\mathcal{L}(k)|^2 \mathcal{P}\frac{1}{k - k_n} \left\{ e^{ikx} - e^{i(x+2L)k} \right\}}_{I_n(x)}.$$

Now noticing that inside the cavity x varies from $-L$ to 0 only, do the integral $I_n(x)$ by contour integration and verify that it is given by[20]

$$I_n(x) = -i2\pi e^{ik_n x} k_n |\mathcal{L}(k_n)|^2$$

$$+ \frac{\pi}{L} \underbrace{\sum_{m=-\infty}^\infty \left\{ \frac{k_m - i\gamma}{k_m - k_n - i\gamma}e^{i(k_m - i\gamma)x} - \frac{k_m + i\gamma}{k_m - k_n + i\gamma}e^{i(x+2L)(k_m + i\gamma)} \right\}}_{S_n(x)}.$$

Show that $S_n(x)$ can be rewritten as $S_n(x) = S_{1n}(x) + k_n S_{2n}(x)$, where

$$S_{1n}(x) = \sum_{m=-\infty}^\infty \left\{ e^{i(k_m - i\gamma)x} - e^{i(k_m + i\gamma)(x+2L)} \right\},$$

$$S_{2n}(x) = \sum_{m=-\infty}^\infty \left\{ \frac{e^{ik_m x}}{k_m - k_n - i\gamma}e^{\gamma x} - \frac{e^{i(x+2L)k_m}}{k_m - k_n + i\gamma}e^{-(x+2L)\gamma} \right\}.$$

Now, using

$$\sum_{m=-\infty}^\infty e^{ik_m u} = 2L \sum_{m'=-\infty}^\infty \delta(u - 2Lm'),$$

and noticing that $-L < x < 0$, show that

$$S_{1n}(x) = 2L \left\{ 1 - e^{-2L\gamma} \right\} \delta(x).$$

To sum $S_{2n}(x)$, use the Fourier series for $\cos([x + \pi]a)$ and its derivative with respect to x. Verify that

$$\sum_{m=-\infty}^\infty \frac{e^{ik_m x}}{k_m - k_n + i\gamma} = -i2L\frac{e^{-L\gamma}}{e^{L\gamma} - e^{-L\gamma}}e^{\gamma u}e^{ik_n u},$$

---

[20]The notation here is the same as in Problem 9.5.

where $0 < u < 2L$. Now using that[21] $|\mathcal{L}(k_n)|^2 = -\text{Re}\{i\cotan(-i\gamma L)\} = [\exp(L\gamma) + \exp(-L\gamma)]/[\exp(L\gamma) - \exp(-L\gamma)]$, finally show that

$$\int_{-\infty}^{\infty} dk\, k|\mathcal{L}(k)|^2 \sin\left([x + L]\,k\right) \frac{\sin\left([k - k_n]\,L\right)}{k - k_n}$$

$$= (-1)^n \pi \left\{1 - e^{-2L\gamma}\right\} \delta(x) + \pi k_n \sin\left([x + L]\,k_n\right).$$

2. Analogously, show that

$$I(k', x) = \frac{1}{2}\sqrt{\frac{\hbar c}{\pi k'}} \delta(x) \left\{1 - e^{-2L\gamma}\right\} e^{ik'L} \mathcal{L}(k') \left\{\cos k'L - \frac{1}{k'L}\sin k'L\right\}$$

$$+ \frac{1}{4\pi}\sqrt{\frac{\hbar c}{\pi k'}} \left|\frac{1 + r}{t}\right| \delta(x) \left\{1 - e^{-2L\gamma}\right\} \int_0^{\infty} dk'' |\mathcal{L}(k'')|^2 \cos k''L$$

$$\times \lim_{\delta \to 0+} \left[\frac{1}{k'' + k' + i\delta}\left\{\cos k''L - \frac{1}{k''L}\sin k''L\right\}\right.$$

$$+ \frac{1}{k'' - k' - i\delta}\left\{\cos k''L - \frac{1}{k''L}\sin k''L\right\}\bigg].$$

[21] See Problem 9.5.

Appendix A
Modes of a perfectly conducting closed cavity: A quick review

There are many different modes, each of which will have a different resonant frequency corresponding to some particular complicated arrangement of the electric and magnetic fields. Each of these arrangements is called a resonant *mode*. The resonance frequency of each mode can be calculated by solving Maxwell's equations for the electric and magnetic fields in the cavity.

—Richard Phillips Feynman [206]

In Section 2.2 we showed that Maxwell equations involve only fields at the same point in Fourier space. Electrodynamics then becomes extremely simple. But that was all done in free space. For a cavity, the fields have to satisfy the constraints imposed by the boundary conditions on the walls. Here we show how the Fourier-space procedure that simplified electrodynamics in free space can be generalized for a perfectly conducting closed cavity (a perfect cavity for short). We will see that the plane waves appearing in the Fourier transforms in free space can just be replaced by a different sort of wave, whose shape is determined by the boundary conditions of the cavity. These waves are the cavity modes. They are extremely useful, especially for quantizing the electromagnetic field, because Maxwell equations for the electromagnetic field in the cavity involve only fields at the same point in mode space. From this point of view, the plane waves that appeared in the Fourier transforms in Section 2.2 are just modes of free space.

Let us consider a closed cavity of any shape without any matter inside it, and made up of perfectly conducting walls. Any electromagnetic radiation inside it can be described by Maxwell equations without sources:

$$\nabla \cdot \mathbf{E} = 0, \tag{A.1}$$

$$\nabla \cdot \mathbf{B} = 0, \tag{A.2}$$

$$\nabla \wedge \mathbf{E} = -\frac{1}{c}\frac{\partial}{\partial t}\mathbf{B}, \tag{A.3}$$

$$\nabla \wedge \mathbf{B} = \frac{1}{c}\frac{\partial}{\partial t}\mathbf{E}, \tag{A.4}$$

with the boundary conditions[1]

$$\mathbf{n} \wedge \mathbf{E}|_S = 0 \tag{A.5}$$

and

$$\mathbf{n} \cdot \mathbf{B}|_S = 0, \tag{A.6}$$

where $\mathbf{n}$ is the normal to the surface of the cavity. Taking the curl of (A.3), using (A.4) and (A.1), we find that $\mathbf{E}$ obeys the wave equation

$$\nabla^2 \mathbf{E} = \frac{1}{c^2}\frac{\partial^2}{\partial t^2}\mathbf{E}. \tag{A.7}$$

It is possible to show [458] that any solution of (A.7) can be written as a linear combination of special solutions that factor out as a product of a function of $\mathbf{r}$ alone by a function of t alone, that is,

$$\mathbf{E}(\mathbf{r}, t) = \chi(\mathbf{r})T(t). \tag{A.8}$$

So if we can find these special solutions, we can obtain any solution of (A.7).

Substituting (A.8) in (A.7), we find that

$$\frac{\nabla^2 \chi_i(\mathbf{r})}{\chi_i(\mathbf{r})} = \frac{\ddot{T}(t)}{c^2 T(t)}, \tag{A.9}$$

where the subscript $i = 1, 2, 3$ stands for the independent components of χ in any arbitrary coordinate system. Because the left-hand side of (A.9) depends only $\mathbf{r}$ and

---

[1]If you are not familiar with these boundary conditions, see Appendix B.

the right-hand side only on t, they can only agree for all values of $\mathbf{r}$ and t if they are both equal to a constant. Moreover, as the right-hand side of (A.9) is independent of i, this constant should be the same for every value of i. Calling this constant η, we obtain

$$\nabla^2\chi(\mathbf{r}) = \eta\chi(\mathbf{r}), \tag{A.10}$$
$$\ddot{T}(t) = c^2\eta T(t). \tag{A.11}$$

In principle, η can be any complex number. The boundary conditions on the cavity walls, however, determine what kind of complex number η can be. First, we show that η must be a negative real number in the case of a perfect cavity. For that, we notice that if we take the scalar product of (A.10) with χ and integrate, the right-hand side of (A.10) will be a positive number times η. Now look at the left-hand side of (A.10). As $\nabla \cdot \chi = 0$, we can replace $\nabla^2\chi$ by $-\nabla \wedge (\nabla \wedge \chi)$, so that

$$\eta \int dV(\chi)^2 = -\int dV\, \chi \cdot [\nabla \wedge (\nabla \wedge \chi)]$$
$$= \underbrace{\int dV\, \nabla \cdot [\chi \wedge (\nabla \wedge \chi)]}_{\substack{\displaystyle \oint d\mathbf{S} \cdot [\chi \wedge (\nabla \wedge \chi)] \\ \displaystyle \oint (\nabla_\wedge \chi) \cdot \underbrace{(d\mathbf{S} \wedge \chi)}_{\text{vanishes}}}} - \int dV\, (\nabla \wedge \chi)^2$$
$$= -\int dV\, (\nabla \wedge \chi)^2 \leq 0. \tag{A.12}$$

We have used the Gauss integral theorem to transform the first volume integral on the second line of (A.12) into an integral over the surface of the cavity which can be rewritten, with the help of a vector identity, in a form that brings out the boundary condition. Then we let

$$\eta = -k^2, \tag{A.13}$$

where k is real, so that k^2 is a positive real number. In Chapter 9 we see that in the case of leaky or open cavities, η can have an imaginary part.

Another consequence of having a cavity is that k must be discrete. To see this, let us introduce the subscript k^2 to indicate that χ_{k^2} is the eigenfunction corresponding to the eigenvalue k^2. Then we can rewrite (A.10) as

$$\nabla \wedge (\nabla \wedge \chi_{k^2}) = k^2\chi_{k^2}. \tag{A.14}$$

Subtracting this equation from the corresponding equation for a different eigenvalue k'^2, we find that

$$(k^2 - k'^2)\int dV\chi_{k'^2}^* \cdot \chi_{k^2}$$

$$= \int dV \left\{ \chi^*_{k'^2} \cdot [\nabla \wedge (\nabla \wedge \chi_{k^2})] - \chi_{k^2} \cdot [\nabla \wedge (\nabla \wedge \chi^*_{k'^2})] \right\}$$

$$= \int dV \nabla \cdot \left\{ \chi_{k^2} \wedge (\nabla \wedge \chi^*_{k'^2}) - \chi^*_{k'^2} \wedge (\nabla \wedge \chi_{k^2}) \right\}$$

$$= \oint dS \cdot \left\{ \chi_{k^2} \wedge (\nabla \wedge \chi^*_{k'^2}) - \chi^*_{k'^2} \wedge (\nabla \wedge \chi_{k^2}) \right\}$$

$$= \underbrace{\oint (\nabla \wedge \chi^*_{k'^2}) \cdot (dS \wedge \chi_{k^2})}_{\text{vanishes}} - \underbrace{\oint (\nabla \wedge \chi_{k^2}) \cdot (dS \wedge \chi^*_{k'^2})}_{\text{vanishes}}. \qquad \text{(A.15)}$$

Then

$$(k^2 - k'^2) \int dV \, \chi^*_{k'^2}(\mathbf{r}') \cdot \chi_{k^2}(\mathbf{r}) = 0, \qquad (A.16)$$

where the integral is over the cavity volume. The integral in (A.16) must vanish whenever $k \neq k'$. Now suppose that k and k' are continuous. Then this integral must equal a linear combination of $\delta(k^2 - k'^2)$ and its derivatives. But as (A.16) also states explicitly that the product of this integral by $k^2 - k'^2$ must always vanish, the integral cannot yield any derivative of the delta function. So if k and k' are continuous, we must have

$$\int dV \, \chi^*_{k'^2}(\mathbf{r}') \cdot \chi_{k^2}(\mathbf{r}) = C \, \delta(k^2 - k'^2), \qquad (A.17)$$

where C is a constant. As we know that this integral does not diverge when $k' = k$ because both χ and the cavity volume are finite, k and k' must be discrete[2].

So far, without even specifying the shape of our closed cavity, we were able to deduce that k must be real and discrete. Unfortunately, analytical expressions for these mode functions can only be obtained when (A.10) is separable [458]. We consider only such cavity geometries now. In this case, k^2 depends on three discrete eigenvalues, say u, v, and w, associated with each of the three separate equations obtained from (A.10) in the separable coordinate system. An orthogonality condition similar to (A.16) is obtained from each of these three equations. So we can label k and the mode functions with a subscript n that stands for three integers corresponding to each of these three discrete eigenvalues. Moreover, as the divergence of the mode functions vanishes because of (A.1), there are only two independent directions in three-dimensional space for the mode functions. We use the subscript $s = 1, 2$ to label any two such directions that are orthogonal to each other. Then

$$\int dV \, \chi^*_{n',s'}(\mathbf{r}') \cdot \chi_{n,s}(\mathbf{r}) = \delta_{n',n}\delta_{s',s}, \qquad (A.18)$$

where $\delta_{n',n}$ stands for $\delta_{n_u',n_u}\delta_{n_v',n_v}\delta_{n_w',n_w}$, with $\delta_{i,j}$ being the Kronecker delta, which vanishes for $i \neq j$ and equals 1 for $i = j$.

---

[2]For a more rigorous proof of discreteness, try Problems 5.12 and 5.13 in [40]. If you can read German and like going to original sources whenever possible, have a look at the two papers [634, 635] in which Weyl demonstrates that for these perfect cavity boundary conditions, the eigenvalues are discrete and infinite with a lower bound.

The χ are the modes of the electric field, but we also need the modes of the magnetic field. They are connected to each other by Maxwell equations; thus once the χ are determined, the magnetic field modes follow. To see this, we notice that substitution of the separable expression (A.8) for $\mathbf{E}$ in (A.3) and (A.4) suggests that $\mathbf{B}$ is also separable. Taking $\mathbf{B}$ to be given by

$$\mathbf{B}(\mathbf{r}, t) = \phi(\mathbf{r})M(t), \tag{A.19}$$

and using it and (A.8) in (A.3) and (A.4), we find that

$$\dot{M}(t)\phi(\mathbf{r}) = -cT(t)\nabla \wedge \chi(\mathbf{r}) \tag{A.20}$$

$$M(t)\nabla \wedge \phi(\mathbf{r}) = \frac{1}{c}\dot{T}(t)\chi(\mathbf{r}). \tag{A.21}$$

Differentiating (A.21) in time, taking the curl, and using (A.11) and (A.2), we find that

$$\nabla^2\phi(\mathbf{r}) + k^2\phi(\mathbf{r}) = 0. \tag{A.22}$$

Taking the time derivative and the curl of (A.20), and using (A.10) and (A.21), we obtain

$$\ddot{M}(t) = -c^2 k^2 M(t). \tag{A.23}$$

So ϕ is a solution of Helmholtz equation just as χ is, and M is a solution of the harmonic oscillator equation just as T is. Moreover, they have the same eigenvalue, $-k^2$. They will be different solutions, however, because ϕ obeys a different boundary condition [i.e., (A.6) rather than (A.5)].

But can we obtain ϕ and M from χ and T? Sure, we just have to apply the technique of separation of variables to (A.20). Then we can write

$$\phi(\mathbf{r}) = \zeta\nabla \wedge \chi(\mathbf{r}) \tag{A.24}$$

and

$$T(t) = -\frac{\zeta}{c}\dot{M}(t), \tag{A.25}$$

We can put (A.25) in a more explicit form by differentiating in time and using (A.23) to eliminate $\ddot{M}$:

$$M(t) = \frac{1}{c\zeta k^2}\dot{T}(t). \tag{A.26}$$

The separation constant ζ can be chosen at will; it is completely arbitrary. Each particular choice will yield a different set of modes. But if we demand that both χ and ϕ be normalized, our freedom to choose ζ at will is restricted to a phase. In other words, to have

$$\int dV\ \phi^*(\mathbf{r}) \cdot \phi(\mathbf{r}) = 1, \tag{A.27}$$

we must choose $1/|\zeta|^2$ to be

$$\frac{1}{|\zeta|^2} = \int dV\ \{\nabla \wedge \chi^*(\mathbf{r})\} \cdot \{\nabla \wedge \chi(\mathbf{r})\}, \tag{A.28}$$

as can be seen by using (A.24) in (A.27). Now using the vector identity

$$(\nabla \wedge \chi^*) \cdot (\nabla \wedge \chi) = \nabla \cdot \{\chi^* \wedge (\nabla \wedge \chi)\} + \chi^* \cdot \{\nabla \wedge (\nabla \wedge \chi)\} \quad \text{(A.29)}$$

in (A.28) we can transform the volume integral of the first term on the right-hand side of (A.29) into a surface integral using Gauss's theorem:

$$\frac{1}{|\zeta|^2} = \oint dS \cdot \underbrace{\{\chi^* \wedge (\nabla \wedge \chi)\}}_{(\nabla \wedge \chi) \cdot \underbrace{\{dS \wedge \chi^*\}}_{\text{vanishes}}} + \int dV \, \chi^* \cdot \underbrace{\{\nabla \wedge (\nabla \wedge \chi)\}}_{-\nabla^2 \chi = k^2 \chi}. \quad \text{(A.30)}$$

Thus,

$$\frac{1}{|\zeta|^2} = k^2 \int dV \, \chi^*(\mathbf{r}) \cdot \chi(\mathbf{r}), \quad \text{(A.31)}$$

and if we have normalized χ as well, the square modulus of ζ must be given by

$$|\zeta|^2 = \frac{1}{k^2}. \quad \text{(A.32)}$$

For convenience, we choose the arbitrary phase so that ζ is real and coincides with the inverse of k, that is,

$$\zeta = \frac{1}{k}. \quad \text{(A.33)}$$

As a consistency check that we can really obtain ϕ from χ, we must show that $\phi = (\nabla \wedge \chi)/k$ obeys the proper boundary condition (A.6) for the magnetic field[3] (i.e., $\mathbf{n} \cdot \phi|_S = 0$). To show this, we notice that

$$\mathbf{n} \cdot (\nabla \wedge \chi)|_S = \nabla \cdot \underbrace{(\chi \wedge \mathbf{n})|_S}_{\text{vanishes}} + \chi \cdot (\nabla \wedge \mathbf{n})|_S$$

$$= \chi \cdot (\nabla \wedge \mathbf{n})|_S, \quad \text{(A.34)}$$

where we have used (A.5). Now, by applying Stokes's integral theorem to $\mathbf{n}$ over an infinitesimal area of the surface,

$$\int dS \, (\nabla \wedge \mathbf{n}) \cdot \mathbf{n} = \oint \mathbf{n} \cdot d\mathbf{l}, \quad \text{(A.35)}$$

and noticing that $\mathbf{n}$ is perpendicular to $d\mathbf{l}$ (because $d\mathbf{l}$ lies on the surface), we conclude that $\nabla \wedge \mathbf{n}|_S$ must be tangent to the surface.[4] Then, as χ has no component tangential

---

[3]Naively, we could think that ϕ obeys the boundary condition $\mathbf{n} \wedge (\nabla \wedge \phi)|_S = 0$, which can be derived by wrongly inverting (A.24). This inversion is achieved by taking the curl of (A.24) and using (A.10) to yield $\chi = (\nabla \wedge \phi)/k$. Then as $\mathbf{n} \wedge \chi|_S = 0$, we would deduce that $\mathbf{n} \wedge (\nabla \wedge \phi)|_S = 0$. The flaw is that $(\nabla \wedge \phi)/k$ gives only the transverse component of χ, and on the surface of the cavity (only there!) χ has a nonvanishing longitudinal component because of the surface charges.

[4]The same conclusion can be drawn simply by rewriting $\mathbf{n} \cdot (\nabla \wedge \mathbf{n})$ as $(\mathbf{n} \wedge \nabla) \cdot \mathbf{n}$ and noticing that $(\mathbf{n} \wedge \nabla)$ is perpendicular to $\mathbf{n}$.

to the surface [i.e., from (A.5)], it follows that $\chi \cdot (\nabla \wedge \mathbf{n})|_S = 0$. Using this result in (A.34), we conclude that $\mathbf{n} \cdot \phi|_S = 0$, so that ϕ satisfies the correct boundary condition for the magnetic field.

We can now get back to the comparison with Fourier space mentioned at the beginning of this appendix. Instead of (2.46), we have for the cavity fields

$$e_{n,s} = \int dV \, \chi^*_{n,s} \cdot \mathbf{E}, \tag{A.36}$$

$$b_{n,s} = \int dV \, \frac{1}{k_n} \left(\nabla \wedge \chi^*_{n,s} \right) \cdot \mathbf{B}, \tag{A.37}$$

with the inverse transformations given by

$$\mathbf{E} = \sum_{n,s} \chi_{n,s} e_{n,s}, \tag{A.38}$$

$$\mathbf{B} = \sum_{n,s} \frac{1}{k_n} \nabla \wedge \chi_{n,s} b_{n,s}. \tag{A.39}$$

In this *mode space*, Maxwell's equations then become much simpler. In fact, they reduce to just two coupled ordinary differential equations:

$$\dot{b}_{n,s} = -ck_n e_{n,s}, \tag{A.40}$$

$$\dot{e}_{n,s} = ck_n b_{n,s}, \tag{A.41}$$

as can be seen by substituting (A.38) and (A.39) in (A.1) to (A.4) and using the orthogonality properties of the mode functions. Together, these equations yield harmonic oscillator equations for the electric and magnetic fields, like (2.51) and (2.52):

$$\ddot{e}_{n,s} + c^2 k_n^2 e_{n,s} = 0, \tag{A.42}$$

$$\ddot{b}_{n,s} + c^2 k_n^2 b_{n,s} = 0. \tag{A.43}$$

Appendix B
Perfect cavity boundary conditions

Electrical conductors are substances with freely movable electrons. Electric force components inside a conductor produce charge displacements—a current flows. The conditions for no current flow ... are (1) no field may be present inside the conductor: $\mathbf{E} = 0$; (2) at the surface of the conductor the tangential component of the field vanishes. ...

—Wolfgang Ernst Pauli [481]

In Appendix A we used the boundary conditions at a perfect conductor, (A.5) and (A.6), to find the modes of a closed cavity made of perfect mirrors. These boundary conditions can be found in any good book on electrodynamics. However, to make our derivation of cavity modes self-contained and save those readers who might be unfamiliar with it the trouble of looking it up somewhere else, we summarize briefly here how these boundary conditions arise.

The standard procedure to derive these boundary conditions is to consider an infinitesimally small cylinder positioned so that one of its facets lies on the surface of the cavity wall and the other inside the wall (see Figure B.1). Integrating (2.44)

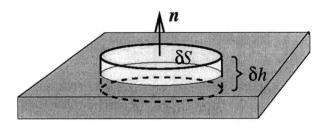

Figure B.1 Infinitesimal cylinder with one of its facets on the surface of a cavity wall and the other inside the wall. The height of the cylinder is δh and δS is the area of each of its facets.

over the volume of the cylinder and using

$$\int dV\, \boldsymbol{\nabla} \wedge \mathbf{F} = \oint d\mathbf{S} \wedge \mathbf{F}, \tag{B.1}$$

we find that

$$\mathbf{n} \wedge \mathbf{E}|_S - \mathbf{n} \wedge \mathbf{E}|_{\text{inside}} = -\frac{1}{c}\frac{\partial}{\partial t}\mathbf{B}\,\delta h. \tag{B.2}$$

As $\delta h \to 0$, we find that

$$\mathbf{n} \wedge \mathbf{E}|_S - \mathbf{n} \wedge \mathbf{E}|_{\text{inside}} = 0. \tag{B.3}$$

Now inside a perfect conductor there can be no electric field, then (B.3) reduces to (A.5).

To obtain the boundary condition for the magnetic field (A.6), we integrate (2.43) over the volume of the cylinder. Using the Gauss integral theorem, taking $\delta h \to 0$ we obtain

$$\mathbf{n} \cdot \mathbf{B}|_S - \mathbf{n} \cdot \mathbf{B}|_{\text{inside}} = 0. \tag{B.4}$$

Now we are only interested in dynamic fields here, and from (2.44) we see that as the electric field vanishes inside the wall, the dynamic magnetic field must vanish as well. Then (B.4) implies that the boundary condition for the magnetic field is that its component normal to the wall must vanish [i.e., (A.6)].

Appendix C
Quaternions and special relativity

THE TETRACTYS,
Or high Matheis, with her charm severe
Of line and number, was our theme; and we
Sought to behold her unborn progeny,
And thrones reserved in Truth's celestial sphere:
While views, before attained, became more clear;
And how the *One of Time*, of *Space the Three*,
Might, in a chain Symbol, girdled be:
And when my eager and reverted ear
Caught some faint echoes of an ancient strain,
Some shadowy outlines of old thoughts sublime,
Gently he smiled to see, revived again,
In a later age, and occidental clime,
A dimly traced Pythagorean lore,
A westward floating, mystic dream of FOUR.

—William Rowan Hamilton [327]

Back in the first half of the nineteenth century, equations in physics were still all written out in terms of their Cartesian components. This is how Maxwell first wrote his equations of electrodynamics.[1] Quaternions were one of the first attempts to develop a coordinate-free system of mathematics. Unfortunately, as the Latin root of their name[2] tells us, quaternions have *four* components rather than three. So they are a bit awkward to use to represent ordinary three-dimensional physical quantities.[3] This prompted Heaviside and Gibbs to develop the vector system that is widely used in physics and engineering today.[4] Heaviside and Gibbs basically removed "by force" the fourth component of the quaternions. This then led them to dismember the original quaternionic product into two different kinds of products, known nowadays as scalar and vector products. In terms of modern vectors, quaternions can be regarded as a vector plus a scalar. Heaviside hated quaternions. In a review of a book on vector analysis [284], he referred in the following way to the problem of adding a scalar to a vector:

> It is really quite legitimate to add together all sorts of different things. Everybody does it. My washerwoman is always doing it. She adds and subtracts all sorts of things, and performs various operations upon them (including linear operations), and at the end of the week this poor ignorant woman does an equation in multiplex algebra by equating the sum of a number of different things in the basket at the beginning of the week to a number of things she puts in the basket at the end of the week. Sometimes she makes mistakes in her operations. So do mathematicians.

Why talk about this mathematical dinosaur in a book about modern physics? For the simple reason that it is not a dinosaur at all. The dismissal of quaternions in the later part of the nineteenth century in favor of vectors is due primarily to one thing: Relativity had not been discovered yet. Quaternions, or rather biquaternions, to be more precise, are ideally suited for dealing with relativistic quantities [386, 506, 566], which have *three spatial* dimensions and *one time* dimension.[5] As the photon is perhaps the most relativistic of all particles in the sense that it has no nonrelativistic limit, biquaternions are very useful for developing some basic equations of quantum electrodynamics. In this book, however, they are used only in Chapter 3.

---

[1]The vectorial form of Maxwell's equations that we use today was developed by Oliver Heaviside.

[2]The name *quaternion* means "a set of four" and originates from a passage in the bible where the apostle Peter is described as having been delivered by Herod to the charge of four quaternions of soldiers [641]. According to Tait [588], Hamilton was probably influenced by the recollection of its Greek equivalent, the Pythagorean Tetractys, the mystic source of all things.

[3]Still, they are quite useful for computing three-dimensional rotations, and for that they are actively applied nowadays in robotics [104, 105, 622, 659], computer graphics [554], and spacecraft orientation [633] (e.g., the space shuttle's flight software does many important calculations for guidance navigation and flight control using quaternions).

[4]See, for instance, the chapter on the "great quaternionic war" in Paul Nahin's biography of Oliver Heaviside [461].

[5]Quaternions and Clifford algebras are used in field theory (e.g., see [244, 256, 257]). Even a version of quantum mechanics where complex numbers are replaced by quaternions has been proposed [6, 210–212, 338, 489, 658]. Note, please, that this quaternion quantum mechanics is very different from what we do in Chapter 3. In that chapter we only use biquaternions to replace Dirac matrices and spinors.

In the next section we give a heuristic introduction to quaternions. In Section C.2 we discuss some useful results in quaternion calculus. Finally, in Section C.3 we apply biquaternions to special relativity.

C.1   WHAT ARE QUATERNIONS?

Quaternions are a generalization of complex numbers to four dimensions. Complex numbers can be seen as a two-dimensional space in which a line of imaginary numbers (real multiples of $i = \sqrt{-1}$) represents a second dimension perpendicular to the one-dimensional space formed by the line of real numbers. This is none other than our old friend, the complex plane (see Figure C.1).

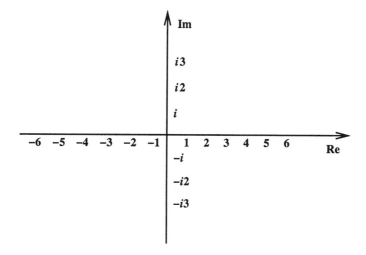

Figure C.1   We can represent any complex number $z = x + iy$ as a point on a plane using a pair of Cartesian axes: one axis for the real components and another axis, orthogonal to the first, for the imaginary components.

This geometrical interpretation is motivated by the following reasoning.[6] If we imagine a point along the real line as a vector, multiplying it by -1 will correspond to a rotation of 180°, which keeps it on the same one-dimensional space (the line of real numbers) but makes it point in the opposite direction. Assuming that this connection between multiplication and rotation is general, there must be a number x that makes a vector along the line of real numbers rotate by 90°. This rotation by 90° will take our

---

[6]This geometrical interpretation seems to have first been put forward in 1685 by Wallis in his *Treatise of Algebra*. It was later developed further (apparently in an independent way) by Abbé Buée in 1804. The modern form of these ideas is almost identical to that published by Argand in 1806 [23]. For more details on the historical development, see [588].

vector out of its original one-dimensional space in the line of real numbers into a new second dimension. Moreover, as $90° + 90° = 180°$, two consecutive multiplications by x must make this vector rotate by $180°$. Thus, $x^2 = -1$, so that x cannot be a real number but must be the imaginary unit $i = \sqrt{-1}$.

To rotate our vector by an arbitrary angle, the straightforward approach would be just to calculate the appropriate root of -1. But a much more powerful method which will also strengthen this analogy between algebra and geometry emerges from the following very simple argument. Consider a complex number z forming a vector of length ρ, making an arbitrary angle θ with the line of real numbers as in Figure C.2. From basic trigonometry, it follows that

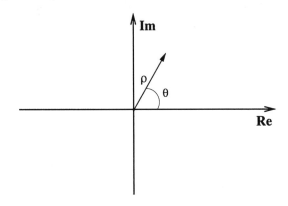

Figure C.2 We can think of any point, $z = x + iy$, on the complex plane as a vector from the origin to that point, whose length is $\rho \equiv \sqrt{x^2 + y^2}$ and which forms an angle $\theta \equiv \arctan(y/x)$ with the x-axis.

$$z = (\cos\theta + i\sin\theta)\,\rho. \tag{C.1}$$

Now consider the Taylor series of $\cos\theta$ and of $\sin\theta$ around $\theta = 0$,

$$\cos\theta = 1 - \frac{\theta^2}{2!} + \frac{\theta^4}{4!} - \frac{\theta^6}{6!} + \cdots \tag{C.2}$$

$$\sin\theta = \theta - \frac{\theta^3}{3!} + \frac{\theta^5}{5!} - \frac{\theta^7}{7!} + \cdots \tag{C.3}$$

Then as $i^2 = -1$, we find that

$$\begin{aligned}
z &= (\cos\theta + i\sin\theta)\,\rho \\
&= \rho\left\{1 + \frac{(i\theta)^1}{1!} + \frac{(i\theta)^2}{2!} + \frac{(i\theta)^3}{3!} + \frac{(i\theta)^4}{4!} + \cdots\right\} \\
&= \rho\,e^{i\theta}. \tag{C.4}
\end{aligned}$$

From the right-hand side of the first line of (C.4) and the last line, we obtain, after canceling out ρ,

$$e^{i\theta} = \cos\theta + i\sin\theta. \tag{C.5}$$

This is Euler's famous equation that unites algebra and geometry. It is such a remarkable formula that Feynman even called it a mathematical jewel (see Chap. 22 of [205]).

Hamilton knew this beautiful geometrical interpretation of complex numbers and wanted to generalize it to our three-dimensional space. He long tried to find the number that would produce a rotation out of the two-dimensional complex plane into a third dimension. When he finally cracked this problem, he discovered that in order to carry on with the assumed connection between rotation and multiplication, his generalization of complex numbers had to have *four* rather than just *three* dimensions. This is one of the best documented discoveries in mathematics. The very moment of discovery is almost something out of a Hollywood film, as can be seen from this extract from a letter by Hamilton to his son Archibald telling of the quaternion discovery [267]:

> ... (you) used to ask me, "Well, Papa, can you multiply triplets?" Whereto I was always obliged to reply, with a sad shake of the head: "No, I can only add and subtract them." But on the 16th day of the same month—which happened to be a Monday, and a Council day of the Royal Irish Academy—I was walking in to attend and preside, and your mother was walking with me, along the Royal Canal ... An electric circuit seemed to close; and a spark flashed forth, the herald (as I foresaw, immediately) of many long years to come of definitely directed thought and work, by myself if spared, and at all events on the part of others, if I should even be allowed to live long enough distinctly to communicate the discovery. Nor could I resist the impulse—unphilosophical as it may have been—to cut with a knife on a stone of Brougham Bridge,[7] as we passed it, the fundamental formula with the symbols, i, j, k; namely, $i^2 = j^2 = k^2 = ijk = -1$ which contains the solution of the Problem. ...

So what is this solution of the problem? Hamilton's line of thought is clearly described in a letter he wrote to his friend John T. Graves, which was later published in the *Philosophical Magazine* [264]. Here we will argue differently.

To extend the connection between multiplication by i and rotation in two dimensions to three dimensions, we must generalize the two-dimensional complex plane to include a third axis, orthogonal to both the real line and the ordinary i-imaginary line, as shown in Figure C.3. To rotate a vector on the real line by 90 degrees to take it to this new axis, we multiply the vector by a new number which we will call j. So any vector on the third axis is a real number multiplied by j. Notice that this third axis must be an imaginary rather than real axis; that is,

$$j^2 = -1, \qquad (C.6)$$

as multiplying by j twice, just like multiplying by i twice, must correspond to rotation by 180 degrees because the new axis is also at 90 degrees from the real line.

So far, all we have done has been to apply the same old idea of the ordinary imaginary line to a third dimension. But we cannot just apply that idea in a straightforward way, simply because with three dimensions we end up with two imaginary

---

[7]"Brougham Bridge" is properly referred to as Broome Bridge, named after a local family (see Graves's biography of Hamilton [267]).

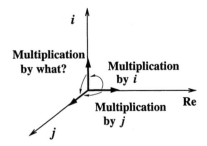

Figure C.3   The complex plane is generalized to three dimensions by introducing a second imaginary axis j. Just as multiplication by i rotates a vector along the real line out of its one-dimensional world into the second dimension, multiplication by j rotates such a vector out of the real line into the third dimension. Just as multiplication by i twice means rotating 180 degrees (i.e., inverting the direction of the vector), multiplication by j twice must also mean rotating 180 degrees. So just as $i^2 = -1$, j^2 must also be -1. But by which number can we multiply a vector on the ordinary imaginary line to rotate it out of this line into the new imaginary line?

lines rather than just one. Now we can go not only from the real line to any one of the other two imaginary lines, but we can also go from one of the imaginary lines to the other imaginary line. The former is solved by our straightforward extension of the old idea of multiplying by i: To get from the real line to the new imaginary line, we just multiply by the new imaginary number j. The latter, however, is still unresolved. There are two key questions that we must answer in order to complete our three-dimensional generalization of the connection in two dimensions between multiplication of complex numbers and rotation:

1. By what number can we multiply a vector that lies on the i-imaginary line to rotate it to the j-imaginary line?

2. Analogously, by what number can we multiply a vector that lies on the j-imaginary line to rotate it to the i-imaginary line?

Let us concentrate on question 1 first. Clearly, this number cannot be real, for it would only change the length or invert the direction of the vector without making it leave the i-imaginary line. Neither can it be i nor j. It cannot be i, because $ii = -1$, which takes the vector to the real axis. It cannot be j because $ji = j$ would imply that $i = 1$.

So we are forced to assume the existence of a *fourth* axis with a new unit number k, whose multiplication takes a vector on the i-imaginary line onto the j-imaginary line; that is,

$$ki = j. \tag{C.7}$$

Being a fourth dimension, this new axis is at 90 degrees to the other axes. Then multiplication by k twice must mean a rotation of 180 degrees, which implies that this forth axis is also an imaginary axis; that is,

$$k^2 = -1. \tag{C.8}$$

Notice that in order for this scheme to make any sense, multiplication must be non-commutative[8]; otherwise, squaring (C.7), we would find $k^2 = 1$ instead of (C.8).

At this point, the reader might be wondering whether we will find out next that in order to have a consistent scheme, we require yet another fifth, sixth, seventh dimension, and so on, never stopping. Well, part of the beauty of this is that four dimensions is just enough to make the entire scheme self-consistent, as we will show now.

Multiplying (C.7) by k from the left and using (C.8), we find that

$$kj = -i. \tag{C.9}$$

So we have answered question (2) as well: Multiplication by minus this third imaginary number k takes a vector on the j-imaginary line to the i-imaginary line. But what about the new problem introduced by this fourth axis of rotating a vector that lies on one of the old imaginary lines to this new imaginary line? By what number can we multiply such vector to achieve this rotation? To answer these questions we first multiply (C.9) from the right by j; we find that

$$ij = k. \tag{C.10}$$

In other words, multiplication by i takes a vector on the j-imaginary line to the new k-imaginary line. Next, we multiply (C.7) by i from the right and find that

$$ji = -k. \tag{C.11}$$

This means that multiplication by $-j$ takes a vector on the i-imaginary line to the k-imaginary line.

Thus, with a real line and three orthogonal imaginary lines, we have a closed algebraic system whose multiplication rules can be summarized by the simple expression that Hamilton carved on the stone of Brougham Bridge[9]:

$$i^2 = j^2 = k^2 = ijk = -1. \tag{C.12}$$

Looking back now, we can see that both the need for three imaginary units and the noncommutability can be understood as a consequence of the various properties of rotation in three dimensions as compared to two dimensions. In two dimensions we only need to specify one number, the rotation angle. Moreover, rotation is commutative in two dimensions (i.e., to rotate by an angle α and then by an angle β achieves

---

[8]This was apparently the first time that noncommutative multiplication appeared in mathematics. Matrix algebra, which is also noncommutative, was developed only after Hamilton's discovery.

[9]Notice that even though quaternion multiplication is not commutative, it is still associative [i.e., given the quaternions $\mathcal{A}$, $\mathcal{B}$, and $\mathcal{C}$, $\mathcal{A}(\mathcal{B}\mathcal{C}) = (\mathcal{A}\mathcal{B})\mathcal{C} = \mathcal{A}\mathcal{B}\mathcal{C}$].

the same result as rotating first by β and then by α). Now, in three dimensions, we need to specify three numbers: the direction (two numbers) of the axis of rotation and the angle (one number) of rotation about this axis. Three-dimensional rotations are also, in general, noncommutative (i.e., to rotate by an angle α about an axis $\hat{a}$ and then by an angle β about an axis $\hat{b}$ is, in general, different from rotating first by β about $\hat{b}$ followed by α about $\hat{a}$).

An important property of quaternions is that they obey the law of the norms; that is,

$$|QQ'| = |Q||Q'|, \tag{C.13}$$

where

$$|\mathcal{Z}| = \sqrt{z_0^2 + z_1^2 + z_2^2 + z_3^2} \tag{C.14}$$

is the norm of a quaternion $\mathcal{Z} \equiv z_0 + z_1 i + z_2 j + z_3 k$. Equation (C.13) can easily be verified if we define the conjugate of $\mathcal{Z}$ as

$$\bar{\mathcal{Z}} \equiv z_0 - z_1 i - z_2 j - z_3 k. \tag{C.15}$$

Notice that given two quaternions, $\mathcal{Z}_1$ and $\mathcal{Z}_2$,

$$\overline{\mathcal{Z}_1 \mathcal{Z}_2} = \bar{\mathcal{Z}}_2 \bar{\mathcal{Z}}_1. \tag{C.16}$$

The square of the norm of a quaternion can simply be written as $|\mathcal{Z}|^2 = \mathcal{Z}\bar{\mathcal{Z}}$. Then you can see that

$$|QQ'|^2 = QQ'\overline{QQ'} = Q\underbrace{Q'\bar{Q}'}_{|Q'|^2}\bar{Q} = |Q|^2|Q'|^2. \tag{C.17}$$

The law of norms implies that if for two quaternions $\mathcal{A}$ and $\mathcal{B}$, $\mathcal{A}\mathcal{B} = 0$, then $\mathcal{A} = 0$ or $\mathcal{B} = 0$. When the norm is always positive (i.e., when the coefficients of the quaternion are all real), this property allows one to introduce the operation of division.

C.2 QUATERNION CALCULUS

As quaternions do not commute, not all the niceties of calculus with complex numbers transfer to quaternions. In particular, we cannot derive a (right) quaternion derivative of a quaternion function $\mathcal{F}(Q)$ because

$$\lim_{\Delta Q \to 0} \frac{\mathcal{F}(Q + \Delta Q) - \mathcal{F}(Q)}{\Delta Q} \tag{C.18}$$

depends on the particular path through which ΔQ approaches zero [265]. No such path dependence arises for complex numbers, allowing us to define the complex derivative of a complex function $f(z) \equiv u(z) + iv(z)$ of the complex number $z \equiv x + iy$ as

$$\frac{df}{dz} \equiv \lim_{\Delta z \to 0} \frac{f(z + \Delta z) - f(z)}{\Delta z}. \tag{C.19}$$

An important consequence of this path independence are the Cauchy–Riemann equations. If we make Δz approach zero along the x-axis, $df/dz = \partial u/\partial x + i\partial v/\partial x$. If we make Δz approach zero along the y-axis, $df/dz = \partial v/\partial y - i\partial u/\partial y$. Thus,

$$\frac{\partial u}{\partial x} = \frac{\partial v}{\partial y} \quad \text{and} \quad \frac{\partial v}{\partial x} = -\frac{\partial u}{\partial y}. \tag{C.20}$$

The Cauchy–Riemann equations (C.20) have a powerful application in physics for finding solutions to the two-dimensional Laplace equation. Differentiating the first Cauchy–Riemann equation in (C.20) by x, the second by y and adding, we find that

$$\frac{\partial^2 u}{\partial x^2} + \frac{\partial^2 u}{\partial y^2} = 0. \tag{C.21}$$

Analogously, differentiating the first by y and the second by x, we get

$$\frac{\partial^2 v}{\partial x^2} + \frac{\partial^2 v}{\partial y^2} = 0. \tag{C.22}$$

So if $f(z)$ is analytic in some domain D, its components u and v are harmonic functions (i.e., solutions of Laplace's equation) in D.

But do not despair. Despite the path dependence of limits such as (C.18), it is still possible to define a quaternion derivative, to have a quaternion analog of the Cauchy–Riemann equations, and quaternion functions satisfying this analog will be solutions of Laplace's equation in four dimensions [146].

Given a unit[10] quaternion $n = n_0 + in_1 + jn_2 + kn_3$, we can define the directional quaternion derivative of $\mathcal{F}$ in the direction of n, $\partial\mathcal{F}/\partial n$, as we define a directional derivative in vectorial calculus,

$$\frac{\partial}{\partial n}\mathcal{F}(\mathcal{Q}) \equiv \lim_{\epsilon \to 0} \frac{\mathcal{F}(\mathcal{Q} + \epsilon\, n) - \mathcal{F}(\mathcal{Q})}{\epsilon}. \tag{C.23}$$

In vectorial calculus, such a directional derivative is given by the scalar product of the gradient with the unit vector that defines the given direction. Here, too, a quaternion gradient can be introduced in an analogous way. Viewing $\mathcal{F}(\mathcal{Q})$ as $\mathcal{F}(q_0, q_1, q_2, q_3)$, where $\mathcal{Q} \equiv q_0 + q_1 i + q_2 j + q_3 k$, we can expand $\mathcal{F}(\mathcal{Q} + \epsilon\, n) = \mathcal{F}(q_0 + \epsilon n_0, q_1\epsilon n_1, q_2 + \epsilon n_2, q_3 + \epsilon n_3)$ in a Taylor series around $\mathcal{F}(\mathcal{Q}) = \mathcal{F}(q_0, q_1, q_2, q_3)$ and rewrite (C.23) as

$$\begin{aligned}
\frac{\partial}{\partial n}\mathcal{F}(\mathcal{Q}) &= \lim_{\epsilon \to 0} \frac{\mathcal{F}(\mathcal{Q} + \epsilon\, n) - \mathcal{F}(\mathcal{Q})}{\epsilon} \\
&= \lim_{\epsilon \to 0} \left\{ n_0 \frac{\partial \mathcal{F}}{\partial x_0} + n_1 \frac{\partial \mathcal{F}}{\partial x_1} + n_2 \frac{\partial \mathcal{F}}{\partial x_2} + n_3 \frac{\partial \mathcal{F}}{\partial x_3} + O(\epsilon^2) \right\} \\
&= S\left(\bar{n}\Diamond\right)\mathcal{F}(\mathcal{Q}), \tag{C.24}
\end{aligned}$$

---

[10]That is, n is such that $\bar{n}n = 1$.

where $S(\mathcal{AB})$ denotes the scalar part of the quaternion product $\mathcal{AB}$ and

$$\Diamond \equiv \frac{\partial}{\partial x_0} + i\frac{\partial}{\partial x_1} + j\frac{\partial}{\partial x_2} + k\frac{\partial}{\partial x_3} \tag{C.25}$$

is the quaternion gradient.

To obtain the quaternion analog of the Cauchy–Riemann equations, we first derive two quaternion integral theorems that will be very useful in Chapter 3. These are derived in a very similar way to our derivation of Gauss's theorem in Appendix E. Consider an infinitesimal hypercube (a four-dimensional cube) of sides dx_0, dx_1, dx_2, and dx_3, whose lower corner is located at x_0, x_1, x_2, x_3. To simplify the notation, from now on we drop Hamilton's historical symbols i, j, k and instead adopt the symbols i_1, i_2, and i_3 for the quaternion imaginary units. This not only prevents people confusing the quaternion i with the complex imaginary unit i, or the quaternion imaginary units i, j, k with the Cartesian unit vectors **i**, **j**, **k**, but also allows for much compacter expressions. The multiplication rules for the quaternion imaginary units, for example, become simply

$$i_n i_m = \sum_{l=1}^{3} \epsilon_{lnm} i_l - \delta_{nm}. \tag{C.26}$$

Now as in Appendix E, consider the flux of a quaternion function $\mathcal{F}$ through the surface of this hypercube:

$$\begin{aligned}
d\mathcal{Q}\,\mathcal{F} = &-dx_1 dx_2 dx_3\,\mathcal{F}(x_0, x_1, x_2, x_3) + dx_1 dx_2 dx_3\,\mathcal{F}(x_0 + dx_0, x_1, x_2, x_3)\\
&-dx_0 dx_2 dx_3\,i_1\mathcal{F}(x_0, x_1, x_2, x_3) + dx_0 dx_2 dx_3\,i_1\mathcal{F}(x_0, x_1 + dx_1, x_2, x_3)\\
&-dx_0 dx_1 dx_3\,i_2\mathcal{F}(x_0, x_1, x_2, x_3) + dx_0 dx_1 dx_3\,i_2\mathcal{F}(x_0, x_1, x_2 + dx_2, x_3)\\
&-dx_0 dx_1 dx_2\,i_3\mathcal{F}(x_0, x_1, x_2, x_3) + dx_0 dx_1 dx_2\,i_3\mathcal{F}(x_0, x_1, x_2, x_3 + dx_3).
\end{aligned} \tag{C.27}$$

Expanding terms such as $\mathcal{F}(x_0 + dx_0, x_1, x_2, x_3)$ in a Taylor series about $\mathcal{F}(x_0, x_1, x_2, x_3)$ and discarding terms higher than second order in the infinitesimal displacements, we find that

$$d\mathcal{Q}\,\mathcal{F} = dx_0\,dx_1\,dx_2\,dx_3\left\{\frac{\partial}{\partial x_0}\mathcal{F} + \sum_{n=1}^{3} i_n\,\frac{\partial}{\partial x_n}\mathcal{F}\right\}$$

$$= d^4 r\,\Diamond\mathcal{F}, \tag{C.28}$$

where $d^4 r \equiv dx_0\,dx_1\,dx_2\,dx_3$. Notice that given a closed four-dimensional volume σ bounded by a hypersurface $\delta\sigma$, we can obtain the flux through $\delta\sigma$ adding infinitesimal hypercubes. The flux through adjacent hypercube walls will cancel out, leaving only the flux through the outer hypercube walls that form the hypersurface $\delta\sigma$. Thus, we have obtained the following quaternion integral theorem [146]:

$$\int_{\sigma} d^4 r\,\Diamond\mathcal{F} = \oint_{\delta\sigma} d\mathcal{Q}\,\mathcal{F}. \tag{C.29}$$

Repeating this derivation for the flux $\mathcal{F}\,dQ$ through an infinitesimal hypercube, we obtain another integral theorem, [146]

$$\int_\sigma d^4r\ \mathcal{F}\Diamond = \oint_{\delta\sigma} \mathcal{F}\,dQ. \tag{C.30}$$

In the complex plane, a function $f(z)$ is analytic in a domain D when

$$\oint_C dz\ f(z) = 0 \tag{C.31}$$

for every closed contour C in D. Analogously, we define for quaternion functions the ideas of left regular and right regular (now the order is important). We call $cal F$ a left regular function in a domain D if for every closed hypersurface $\delta\sigma$ in D,

$$\oint_{\delta\sigma} dQ\ \mathcal{F} = 0. \tag{C.32}$$

Similarly, we call $cal F$ a right regular function in a domain D if for every closed hypersurface $\delta\sigma$ in D,

$$\oint_{\delta\sigma} \mathcal{F}\,dQ = 0. \tag{C.33}$$

In complex analysis, (C.31) leads to the Cauchy–Riemann equations. Here, left regularity leads to $\Diamond\mathcal{F} = 0$, which can be written in terms of components as

$$\frac{\partial f_0}{\partial x_0} = \nabla \cdot \mathbf{f}, \tag{C.34}$$

$$\nabla f_0 = -\frac{\partial \mathbf{f}}{\partial x_0} - \nabla \wedge \mathbf{f}, \tag{C.35}$$

where $\mathcal{F} \equiv f_0 + \sum_n i_n f_n$ and $\mathbf{f}$ is the three-dimensional vector formed with the imaginary components of the quaternion $\mathcal{F}$. Right regularity means that $\mathcal{F}\Diamond = 0$, which in terms of components reads

$$\frac{\partial f_0}{\partial x_0} = \nabla \cdot \mathbf{f}, \tag{C.36}$$

$$\nabla f_0 = -\frac{\partial \mathbf{f}}{\partial x_0} + \nabla \wedge \mathbf{f}. \tag{C.37}$$

Notice that the conditions for left regularity (C.35) and (C.35) only differ from those for right regularity (C.37) and (C.37) through the sign of the curl[11] of $\mathbf{f}$. So if we differentiate (C.35) or (C.37) with respect to x_0 and take the divergency of the

---

[11]If a quaternion function $\mathcal{F}$ is both left and right regular, the curl of $\mathbf{f}$ must vanish (i.e., $\mathbf{f}$ must be the gradient of a scalar function).

corresponding second equation [i.e., either (C.35) or (C.37)], and add the result, we find that

$$\sum_{n=0}^{3} \frac{\partial^2 f_0}{\partial x_n^2} = 0. \tag{C.38}$$

In other words, if $\mathcal{F}$ is left or right regular, the scalar component of $\mathcal{F}$ satisfies the four-dimensional Laplace equation. Similarly, you can easily show [146] that the other components of $\mathcal{F}$ also satisfy a four-dimensional Laplace equation if $\mathcal{F}$ is left or right regular.

There is much more that can be done in quaternion calculus. There is, for instance, an analog of the Cauchy integral formula (the basis of the technique of contour integration); however, limitations of space and time do not allow us to delve too deeply into the delights of quaternion calculus here. We refer the interested reader to [146] and [584].

C.3 BIQUATERNIONS AND LORENTZ TRANSFORMATIONS

The use of quaternions in special relativity was pioneered by Conway in 1911 [118] and, independently, by Silberstein in 1912 [566]. For an extended list of references, see for instance, the review paper by Rastall [506]. Here we adopt a similar approach to that of Lanczos in Chapter 9 of [386].

Let us regard the quaternion

$$\mathcal{R} \equiv x_0 + \sum_{n=1}^{3} x_n i_n \tag{C.39}$$

as describing a point in four-dimensional space. Now consider another coordinate system obtained from the present one by a general four-dimensional rotation. Let $\mathcal{R}'$ be the quaternion that represents the same point in the rotated coordinate system. Then $\mathcal{R}$ and $\mathcal{R}'$ are related by a linear transformation with six degrees of freedom, which in quaternion notation can be written as [96]

$$\mathcal{R}' = \mathcal{A}\mathcal{R}\mathcal{B}, \tag{C.40}$$

where $\mathcal{A}$ and $\mathcal{B}$ are two quaternions of unit norm, that is,

$$\mathcal{A}\bar{\mathcal{A}} = \mathcal{B}\bar{\mathcal{B}} = 1. \tag{C.41}$$

Lorentz transformations are particular four-dimensional rotations that preserve the Minkowskian length. The Minkowskian length can be written in a simple way using the quaternion norm if we let the quaternion coefficients be complex rather than real. Quaternions with complex coefficients are called *biquaternions*. We represent a point in space time by (C.39) with $x_0 = ct$, $x_1 = ix$, $x_2 = iy$, and $x_3 = iz$, where c is the speed of light, t is the time coordinate, and x, y, z are the three Cartesian spatial coordinates. The square of the invariant Minkowskian length is given then by

$$|\mathcal{R}|^2 = c^2 t^2 - x^2 - y^2 - z^2, \tag{C.42}$$

where $\mathcal{R}$ has the special property

$$\mathcal{R}^\dagger = \mathcal{R}. \tag{C.43}$$

The dagger in (C.43) denotes the operation that Hamilton called *biconjugation*, which is nowadays called *Hermitian conjugation*. This operation is defined as

$$\mathcal{R}^\dagger \equiv \bar{\mathcal{R}}^*. \tag{C.44}$$

So (C.43) simply says that $\mathcal{R}$ is Hermitian.

For a four-dimensional rotation to keep the Minkowskian length invariant, it is both necessary and sufficient that it preserve the hermicity of $\mathcal{R}$. Suppose that the Minkowskian length remains invariant, then so does its square and we can write

$$c^2 t'^2 - x'^2 - y'^2 - z'^2 = c^2 t^2 - x^2 - y^2 - z^2 = |\mathcal{R}|^2 \tag{C.45}$$

as $\bar{\mathcal{R}} = \mathcal{R}^*$. But as the rotation preserves the norm of $\mathcal{R}$, $|\mathcal{R}|^2 = |\mathcal{R}'|^2$, so that (C.45) implies that

$$c^2 t'^2 - x'^2 - y'^2 - z'^2 = |\mathcal{R}'|^2. \tag{C.46}$$

Therefore, $\bar{\mathcal{R}}'$ must be equal to $\mathcal{R}'^*$ (i.e., if the rotation keeps the Minkowskian length invariant, it must also preserve the hermicity of $\mathcal{R}$). To see that preserving the hermicity of $\mathcal{R}$ is sufficient to keep the Minkowskian length invariant, notice that when the hermicity is preserved,

$$c^2 t'^2 - x'^2 - y'^2 - z'^2 = |\mathcal{R}'|^2. \tag{C.47}$$

As the rotation preserves the norm, $|\mathcal{R}'|^2 = |\mathcal{R}|^2$, which together with (C.47) then implies that

$$c^2 t'^2 - x'^2 - y'^2 - z'^2 = c^2 t^2 - x^2 - y^2 - z^2 \tag{C.48}$$

(i.e., the Minkowskian length remains invariant).

Now if a four-dimensional rotation given by (C.40) preserves the hermicity of $\mathcal{R}$, we must have

$$\underbrace{\mathcal{A}^* \mathcal{R}^* \mathcal{B}^*}_{\mathcal{R}'^*} = \underbrace{\bar{\mathcal{B}} \bar{\mathcal{R}} \bar{\mathcal{A}}}_{\bar{\mathcal{R}}'} = \bar{\mathcal{B}} \mathcal{R}^* \bar{\mathcal{A}}. \tag{C.49}$$

Then $\mathcal{A}^* = \bar{\mathcal{B}}$, so that a general Lorentz transformation can be represented by

$$\mathcal{R}' = \mathcal{U} \mathcal{R} \mathcal{U}^\dagger, \tag{C.50}$$

where $\mathcal{U}$ is a unitary biquaternion, that is,

$$\mathcal{U} \bar{\mathcal{U}} = \bar{\mathcal{U}} \mathcal{U} = 1. \tag{C.51}$$

An interesting special case of (C.50) is when $\mathcal{U}^\dagger = \bar{\mathcal{U}}$ (i.e., when $\mathcal{U}$ is a pure quaternion rather than a biquaternion). Then the time component of $\mathcal{R}$ remains unchanged under the transformation $\mathcal{R}' = \mathcal{U} \mathcal{R} \mathcal{U}^\dagger$, because

$$\mathcal{R}' = \mathcal{U} \mathcal{R} \mathcal{U}^\dagger$$
$$= \mathcal{U} \mathcal{R} \bar{\mathcal{U}}$$
$$= ct + \sum_{n=1}^{3} \mathcal{U} i_n \bar{\mathcal{U}} \, x_n, \tag{C.52}$$

and $\mathcal{U}i_n\bar{\mathcal{U}}$ is a pure imaginary unit number, as

$$|\mathcal{U}i_n\bar{\mathcal{U}}|^2 = \mathcal{U}i_n\underbrace{\bar{\mathcal{U}}\mathcal{U}}_{1}(-i_n)\bar{\mathcal{U}} = 1 \tag{C.53}$$

and

$$(\mathcal{U}i_n\bar{\mathcal{U}})^2 = \mathcal{U}i_n\underbrace{\bar{\mathcal{U}}\mathcal{U}}_{1}i_n\bar{\mathcal{U}} = -1. \tag{C.54}$$

Thus, when $\mathcal{U}^\dagger = \bar{\mathcal{U}}$, $\mathcal{R}' = \mathcal{U}\mathcal{R}\bar{\mathcal{U}}$ must be an ordinary three-dimensional rotation. In fact, Cayley showed [95, 96] that for a rotation through an angle θ around an axis passing through the origin whose direction is given by the unit vector $\hat{\Omega}$, $\mathcal{U}$ has the following simple form:

$$\mathcal{U} = \cos\frac{\theta}{2} + \sum_{n=1}^{3} i_n\hat{\Omega}_n \sin\frac{\theta}{2}, \tag{C.55}$$

where $\hat{\Omega}_n$ is the nth Cartesian component of Ω. This expression[12] is quite easy to derive using quaternion calculus, and we shall go through the derivation just as a demonstration of the power of quaternion notation.

If the vector $\mathbf{r}$ rotates round $\hat{\Omega}$ by an infinitesimal angle $\delta\theta$, the displacement $\delta\mathbf{r} \equiv \mathbf{r}' - \mathbf{r}$ will be perpendicular to both $\mathbf{r}$ and $\hat{\Omega}$. The length of $\delta\mathbf{r}$ will be given by $\delta\theta$ times the projection $r\sin\alpha$ of $\mathbf{r}$ on the plane perpendicular to $\hat{\Omega}$ (see Figure C.4). These two sentences are summarized in the equation

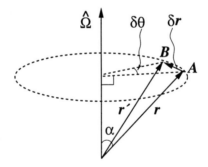

Figure C.4 For an infinitesimal rotation angle $\delta\theta$, the length of $\delta\mathbf{r}$ approximates that of the arch AB, which is given by the product of $\delta\theta$ and the projection of $\mathbf{r}$ on the plane orthogonal to $\hat{\Omega}$, $r\sin\alpha$. As $\delta\mathbf{r}$ is perpendicular to both $\mathbf{r}$ and to $\hat{\Omega}$, it can be written as $\delta\mathbf{r} = \hat{\Omega} \wedge \mathbf{r}\, \delta\theta$ for the direction of rotation indicated in the figure, which obeys the corkscrew rule (i.e., if you place a corkscrew along $\hat{\Omega}$ with its tip in the same direction as the arrow in the figure, the direction of rotation is that which you would rotate the corkscrew to perforate a cork).

$$\delta\mathbf{r} = \hat{\Omega} \wedge \mathbf{r}\, \delta\theta, \tag{C.56}$$

[12]It is also very important in computer graphics and animation, as numerical implementations of this are faster than those for the usual matrix multiplications with Euler angles.

which leads to the well-known differential equation

$$\frac{d\mathbf{r}}{d\theta} = \hat{\Omega} \wedge \mathbf{r}. \tag{C.57}$$

In quaternion notation, (C.57) becomes

$$\frac{dR}{d\theta} = -\frac{1}{2}\left(R\bar{\Omega} + \Omega R\right), \tag{C.58}$$

where R and Ω are the quaternion counterparts of $\mathbf{r}$ and $\hat{\Omega}$, that is,

$$R = \sum_{n=1}^{3} i_n r_n \quad \text{and} \quad \Omega = \sum_{n=1}^{3} i_n \hat{\Omega}_n. \tag{C.59}$$

Multiplying (C.58) on the right by $\exp(\bar{\Omega}\,\theta/2)$ and on the left by $\exp(\Omega\,\theta/2)$, we find that

$$e^{\Omega\theta/2}\left(\frac{dR}{d\theta}\right)e^{\bar{\Omega}\theta/2} = -e^{\Omega\theta/2}Re^{\bar{\Omega}\theta/2}\frac{\bar{\Omega}}{2} - \frac{\Omega}{2}e^{\Omega\theta/2}Re^{\bar{\Omega}\theta/2}. \tag{C.60}$$

Noticing that $\exp(\bar{\Omega}\theta/2)\bar{\Omega}/2$ and $(\Omega/2)\exp(\Omega\theta/2)$ are the derivatives with respect to θ of $\exp(\bar{\Omega}\theta/2)$ and $\exp(\Omega\theta/2)$, respectively, we see that we can rewrite (C.60) as

$$\frac{d}{d\theta}\left(e^{\Omega\theta/2}Re^{\bar{\Omega}\theta/2}\right) = 0. \tag{C.61}$$

Integrating both sides of (C.61) over θ from 0 to θ, we obtain

$$e^{\Omega\theta/2}R(\theta)e^{\bar{\Omega}\theta/2} - R(0) = 0, \tag{C.62}$$

which can also be written as

$$R(\theta) = e^{-\Omega\theta/2}R(0)e^{-\bar{\Omega}\theta/2}. \tag{C.63}$$

As $\hat{\Omega}$ is a unit vector, we could choose a particular rotated set of Cartesian coordinates such that Ω would coincide with one of the imaginary units. Let us call this imaginary unit $\imath$. In this rotated set of coordinates,

$$e^{-\Omega\theta/2} = e^{-\imath\theta/2} = \cos\frac{\theta}{2} - \imath\sin\frac{\theta}{2}, \tag{C.64}$$

as $-\imath^2 = -1$ just like the ordinary imaginary number i. Going back now to the old unrotated coordinates, we can write

$$e^{-\Omega\theta/2} = \cos\frac{\theta}{2} - \Omega\sin\frac{\theta}{2}. \tag{C.65}$$

Using this equation in (C.63), we have

$$R(\theta) = \left\{\cos\frac{\theta}{2} - \Omega\sin\frac{\theta}{2}\right\}R(0)\left\{\cos\frac{\theta}{2} - \bar{\Omega}\sin\frac{\theta}{2}\right\}. \tag{C.66}$$

Now notice that to rotate the coordinate axes by θ round $\hat{\Omega}$ keeping $\mathbf{r}$ fixed is the same as to rotate $\mathbf{r}$ by $-\theta$ round $\hat{\Omega}$ keeping the coordinate axes fixed. Moreover, as $(\cos\theta/2 + \Omega\sin\theta/2)(\cos\theta/2 + \bar{\Omega}\sin\theta/2) = 1$, the time component does not change and we can write

$$\mathcal{R}' = \left\{\cos\frac{\theta}{2} + \Omega\sin\frac{\theta}{2}\right\}\mathcal{R}\left\{\cos\frac{\theta}{2} + \bar{\Omega}\sin\frac{\theta}{2}\right\}, \tag{C.67}$$

which is the formula discovered by Cayley.

Last but not least, let us obtain an explicit formula for a simple Lorentz transformation, say a Lorentz boost. An arbitrary Lorentz transformation (excluding reflections) can be obtained by applying a boost and then an ordinary three-dimensional rotation.

In vector notation, a Lorentz boost can be written as [48, 566]

$$\mathbf{r}' = \mathbf{r} + (\gamma - 1)(\mathbf{r}\cdot\hat{\mathbf{v}})\hat{\mathbf{v}} - \beta\gamma ct\hat{\mathbf{v}}, \tag{C.68}$$

$$ct' = \gamma\left\{ct - \beta\mathbf{r}\cdot\hat{\mathbf{v}}\right\}, \tag{C.69}$$

where $\mathbf{r}$ and t are the spacial position and time, respectively, relative to the old reference frame, their primed counterparts refer to the new reference frame, which moves at constant velocity $v\hat{\mathbf{v}}$ relative to the old frame, v being the speed and $\hat{\mathbf{v}}$ the direction of the velocity, $\beta = v/c$, and $\gamma = 1/\sqrt{1 - \beta^2}$. A straightforward way to obtain $\mathcal{U}$ for this Lorentz boost is to argue that (C.68) and (C.69) imply that $\mathcal{U}^\dagger = \mathcal{U}$ and then write down the individual components of $\mathcal{R}' = \mathcal{U}\mathcal{R}\mathcal{U}$ and use (C.68) and (C.69) to identify each component of $\mathcal{U}$. This is the approach adopted by Silberstein in [566]. Here, we will proceed in a different way to illustrate yet again the power of quaternion calculus.

Consider an infinitesimal Lorentz boost; that is, let the speed be $\delta v \ll c$ and call $\delta\beta \equiv \delta v/c$. Then $\gamma \approx 1$ and

$$\mathbf{r}' = \mathbf{r} - ct\hat{\mathbf{v}}\,\delta\beta, \tag{C.70}$$

$$ct' = ct - \mathbf{r}\cdot\hat{\mathbf{v}}\,\delta\beta. \tag{C.71}$$

Defining the unit velocity biquaternion

$$\mathcal{V} = \sum_{n=1}^{3} i_n\,i\hat{v}_n, \tag{C.72}$$

where $\hat{v}_n$ is the nth Cartesian component of $\hat{\mathbf{v}}$, we can rewrite (C.70) and (C.71) in quaternion notation as

$$\mathcal{R}' - \bar{\mathcal{R}}' = \mathcal{R} - \bar{\mathcal{R}} - (\mathcal{R} + \bar{\mathcal{R}})\mathcal{V}\,\delta\beta, \tag{C.73}$$

$$\mathcal{R}' + \bar{\mathcal{R}}' = \mathcal{R} + \bar{\mathcal{R}} - (\mathcal{R}\mathcal{V} + \mathcal{V}\mathcal{R})\,\delta\beta. \tag{C.74}$$

Equation (C.73) is a straightforward rewriting of (C.70). To obtain (C.74) from (C.71), we used

$$\mathbf{r}\cdot\hat{\mathbf{v}} = -\frac{1}{2}\left(\mathcal{R}\bar{\mathcal{V}} + \mathcal{V}\bar{\mathcal{R}}\right) = \frac{1}{2}\left(\mathcal{R}\mathcal{V} + \mathcal{V}\mathcal{R}\right), \tag{C.75}$$

as $\bar{\mathcal{V}} = -\mathcal{V}$.

Adding (C.73) and (C.74), and noticing that

$$\bar{\mathcal{R}}\mathcal{V} = \mathcal{V}\mathcal{R} - 2\mathbf{r} \cdot \hat{\mathbf{v}} = \mathcal{V}\mathcal{R} - (\mathcal{R}\mathcal{V} + \mathcal{V}\mathcal{R}), \tag{C.76}$$

we find that

$$\mathcal{R}' = \mathcal{R} - \frac{1}{2}(\mathcal{R}\mathcal{V} + \mathcal{V}\mathcal{R})\,\delta\beta. \tag{C.77}$$

Now, for an infinitesimal boost, $\mathcal{U}$ will differ from 1 by something on the order of $\delta\beta$ and higher. To first order in $\delta\beta$,

$$\mathcal{U} = 1 + \mathcal{G}\,\delta\beta \tag{C.78}$$

and

$$\mathcal{R}' = \mathcal{U}\mathcal{R}\mathcal{U}^\dagger \approx \mathcal{R} + \left(\mathcal{G}\mathcal{R} + \mathcal{R}\mathcal{G}^\dagger\right)\delta\beta. \tag{C.79}$$

Comparing (C.77) and (C.79), we see that

$$\mathcal{G} = -\frac{1}{2}\mathcal{V}. \tag{C.80}$$

A boost with a finite speed $c\beta$ in the direction $\mathcal{V}$ can be approximated by a series of N infinitesimal boosts applied in succession. This approximation will be exact when $N \to \infty$ and $\beta \to 0$, with the sum of all the N infinitesimal velocities yielding the desired finite velocity, that is, in the limit

$$\mathcal{U}(\beta) = \lim_{\substack{N\to\infty \\ \delta\beta\to 0}} \left\{1 - \frac{1}{2}\mathcal{V}\,\delta\beta\right\}^N. \tag{C.81}$$

To calculate this limit, we must find the $\delta\beta$ such that the sum of N of them add up to β. Notice that we cannot simply write $N\delta\beta = \beta$ as in classical Newtonian mechanics because relativistic velocities add in a different way. Given two velocities, $v_1\hat{\mathbf{v}}$ and $v_2\hat{\mathbf{v}}$, along the same direction $\hat{\mathbf{v}}$, their sum is $v\hat{\mathbf{v}}$ with the speed v given by (see pp. 42–44 in Chapter 4 of [48])

$$v = \frac{v_1 + v_2}{1 + v_1 v_2/c^2}. \tag{C.82}$$

Then for our succession of infinitesimal boosts, we can write the following relation between β_n, the resulting speed in units of the speed of light for the nth application of the infinitesimal boost, and β_{n-1}, for the $(n-1)$th application:

$$\beta_n = \frac{\beta_{n-1} + \delta\beta}{1 + \beta_{n-1}\delta\beta}. \tag{C.83}$$

Up to first order in $\delta\beta$, (C.83) can be rewritten as

$$\delta\beta \approx \frac{\beta_n - \beta_{n-1}}{1 - \beta_{n-1}^2}. \tag{C.84}$$

The sum of all the N steps yields

$$\sum_{n=1}^{N} \frac{\beta_n - \beta_{n-1}}{1 - \beta_{n-1}^2} \approx N\, \delta\beta. \tag{C.85}$$

Notice that $\beta_n - \beta_{n-1}$ in (C.85) is the step increment in β_{n-1} to get to the next term in the series (i.e., β_n). So, in the limit of an infinitesimally small step, the left-hand side of (C.85) becomes an integral. We can anticipate that and approximate (C.85) by

$$\int_0^{\beta} \frac{d\xi}{1 - \xi^2} \approx N\, \delta\beta. \tag{C.86}$$

To do the integral on the left-hand side of (C.86), we can change the integration variable from ξ to ζ with ζ defined by $\xi \equiv \tanh\zeta$. Then as $1 - \tanh^2\zeta = \operatorname{sech}^2\zeta$ and $d\xi = d\zeta\operatorname{sech}^2\zeta$, we find that

$$\delta\beta \approx \frac{1}{N}\operatorname{arctanh}\beta. \tag{C.87}$$

Substituting (C.87) in (C.81), we obtain

$$\mathcal{U}(\beta) = \lim_{\substack{N \to \infty \\ \delta \to 0}} \left\{ 1 - V\frac{1}{2N}\operatorname{arctanh}\beta \right\}^N = \exp\left(-V\frac{1}{2}\operatorname{arctanh}\beta \right). \tag{C.88}$$

Appendix D
The Baker–Hausdorff
formula

The algebra of operators is noncommutative, so that all of the ordinary algebra, calculus, and analysis with ordinary numbers becomes of small utility for operators. Thus, for a single operator, α, ordinary functions of this operator, such as $A = \exp\alpha$, can be defined, for example, by power series. These functions obey the rules of ordinary analysis even though α is an operator. But if another operator β is introduced with which α does not commute, the question of functions of the two variables α, β is beset with commutation difficulties and the simplest theorems of analysis are lost. For example, if $B = \exp\beta$, it is not true that BA, that is $\exp\beta\exp\alpha$, is equal to $\exp(\beta + \alpha)$.

—Richard Phillips Feynman [200]

This is a very useful tool to disentangle exponentials of certain operators often appear in quantum optics.[1] Suppose that $\hat{A}$ and $\hat{B}$ are two operators such that $[\hat{A}, \hat{B}] =$

---

[1] For more general disentanglement techniques, look at [142, 200, 369]

$\hat{C}$, where $\hat{C}$ commutes with both $\hat{A}$ and $\hat{B}$. Then the Baker–Hausdorff formula holds:

$$\exp\left(\hat{A}\alpha + \hat{B}\beta\right) = \exp\left(\hat{B}\beta\right)\exp\left(\hat{A}\alpha\right)\exp\left(\frac{\beta\alpha}{2}\hat{C}\right). \tag{D.1}$$

Notice that, in general, a function of $\hat{A}\alpha + \hat{B}\beta$ can always be written as a *sum* of products of a function of $\hat{A}\alpha$, a function of $\hat{B}\beta$, and a function of $\hat{C}$. What the Baker–Hausdorff formula states that is not immediately obvious is that when the function of $\hat{A}\alpha + \hat{B}\beta$ is an exponential, all the other functions are exponentials, too, and moreover, there is just one term in the sum. To see that it is really so, let us take β, $\hat{A}$, and $\hat{B}$ as given parameters for the moment and consider $\hat{F}(\alpha) = \exp\left(\hat{A}\alpha + \hat{B}\beta\right)$ as a function of α only. The derivative of $\hat{F}$ with respect to α can then be calculated using the definition of $\hat{F}$ in terms of its Taylor series expansion:

$$\frac{d}{d\alpha}\hat{F} = \frac{d}{d\alpha}\sum_{n=0}^{\infty}\frac{1}{n!}\left(\hat{A}\alpha + \hat{B}\beta\right)^n$$

$$= \sum_{n=0}^{\infty}\frac{1}{n!}\left\{\hat{A}\left(\hat{A}\alpha + \hat{B}\beta\right)^{n-1} + \left(\hat{A}\alpha + \hat{B}\beta\right)\hat{A}\left(\hat{A}\alpha + \hat{B}\beta\right)^{n-2}\right.$$

$$\left. + \cdots + \left(\hat{A}\alpha + \hat{B}\beta\right)^{n-1}\hat{A}\right\}$$

$$= \sum_{n=0}^{\infty}\frac{1}{n!}\left\{n\left(\hat{A}\alpha + \hat{B}\beta\right)^{n-1}\hat{A} + (n-1)\beta\hat{C}\left(\hat{A}\alpha + \hat{B}\beta\right)^{n-2}\right.$$

$$\left. + (n-2)\beta\hat{C}\left(\hat{A}\alpha + \hat{B}\beta\right)^{n-2} + \cdots + \beta\hat{C}\left(\hat{A}\alpha + \hat{B}\beta\right)^{n-2}\right\}$$

$$= \hat{F}\hat{A} + \frac{\beta}{2}\hat{F}\hat{C}. \tag{D.2}$$

Integrating (D.2) over α from 0 to α, we find that

$$\hat{F}(\alpha) = \hat{F}(0)\exp\left(\hat{A}\alpha + \frac{\beta\alpha}{2}\hat{C}\right). \tag{D.3}$$

As $\hat{F}(0) = \exp(\hat{B}\beta)$ and $\hat{C}$ commutes with $\hat{A}$, we obtain (D.1).

Appendix E
Trade secrets: Tools for dealing with vectors and vector identities

If anything has magnitude and direction, its magnitude and direction taken together constitute what is called a vector.

—Josiah Willard Gibbs [230]

Del treats them all the same

—Steve Clark [106].

Before the last decades of the nineteenth century, vectors were not used in physics. Back then, physicists would write clumsy expressions, all in terms of components. That is how Maxwell wrote most of the equations in his famous treatise (the now popular vector form of Maxwell's equations is due to Oliver Heaviside). Then Hamilton

*

invented quaternions (see Appendix C), which offered a very compact way of writing four-dimensional quantities. The vector analysis that we use in physics today was introduced by Gibbs and Heaviside, who based it in Hamilton's quaternions and in Clifford's and Grassmann's work.

In this appendix we review briefly some useful techniques for working with vectorial expressions and for deriving vector identities. If you are familiar with the vector identities used throughout this book, you should probably skip this appendix. However, you might still want to give it a quick scan first, for there is some material here that you would not normally find collected together in other books and which most lecture courses do not usually cover.

E.1 RELATION BETWEEN VECTOR PRODUCTS AND DETERMINANTS

A useful tool for working out vector products is to write them down as determinants. Here is how the relation between vector products and determinants comes about. Let x_1, x_2, and x_3 be the unit vectors defining a right-handed Cartesian coordinate system, as in Figure E.1. Then the vector products between these vectors is given by

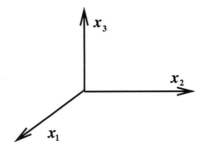

Figure E.1 Right-handed Cartesian coordinate axes.

$$x_1 \wedge x_2 = x_3, \tag{E.1}$$

$$x_2 \wedge x_3 = x_1, \tag{E.2}$$

$$x_3 \wedge x_1 = x_2. \tag{E.3}$$

Also, for any vectors $\mathbf{u}$, $\mathbf{v}$, $\mathbf{w}$, the vector product is distributive,

$$\mathbf{u} \wedge (\mathbf{v} + \mathbf{w}) = \mathbf{u} \wedge \mathbf{v} + \mathbf{u} \wedge \mathbf{w}, \tag{E.4}$$

but nonassociative, with

$$\mathbf{u} \wedge (\mathbf{v} \wedge \mathbf{w}) \neq (\mathbf{u} \wedge \mathbf{v}) \wedge \mathbf{w}, \tag{E.5}$$

in general, and noncommutative, with

$$\mathbf{u} \wedge \mathbf{v} = -\mathbf{v} \wedge \mathbf{u}. \tag{E.6}$$

As a consequence of (E.6), the vector product of two parallel vectors vanishes.

Now for $\mathbf{a} = \sum_i a_i \mathbf{x}_i$ and $\mathbf{b} = \sum_i b_i \mathbf{x}_i$, we find from (E.2), (E.3), and (E.6),

$$\mathbf{a} \wedge \mathbf{b} = (a_2 b_3 - a_3 b_2)\mathbf{x}_1 + (a_3 b_1 - a_1 b_3)\mathbf{x}_2 + (a_1 b_2 - a_2 b_1)\mathbf{x}_3. \qquad \text{(E.7)}$$

The right-hand side of (E.7) can easily be recognized as the determinant

$$\begin{vmatrix} \mathbf{x}_1 & \mathbf{x}_2 & \mathbf{x}_3 \\ a_1 & a_2 & a_3 \\ b_1 & b_2 & b_3 \end{vmatrix}, \qquad \text{(E.8)}$$

so that we can write

$$\mathbf{a} \wedge \mathbf{b} = \begin{vmatrix} \mathbf{x}_1 & \mathbf{x}_2 & \mathbf{x}_3 \\ a_1 & a_2 & a_3 \\ b_1 & b_2 & b_3 \end{vmatrix}. \qquad \text{(E.9)}$$

From (E.9) it follows that for $\mathbf{c} = \sum_i c_i \mathbf{x}_i$,

$$\mathbf{c} \cdot (\mathbf{a} \wedge \mathbf{b}) = \begin{vmatrix} c_1 & c_2 & c_3 \\ a_1 & a_2 & a_3 \\ b_1 & b_2 & b_3 \end{vmatrix}. \qquad \text{(E.10)}$$

E.2 VECTOR PRODUCTS AND THE LEVY–CIVITA TENSOR

It is often convenient to write vector products in terms of Cartesian components with the help of the Levy–Civita tensor ε_{ijk}. Let $\mathbf{x}_i$, $\mathbf{x}_j$, and $\mathbf{x}_k$ be some rearrangement of the unit vectors $\mathbf{x}_1$, $\mathbf{x}_2$, $\mathbf{x}_3$ that define a right-handed Cartesian coordinate system as in Figure E.1. This rearrangement can change the order (e.g., $i = 3$, $j = 1$, $k = 2$, or even repeat some of the vectors as in $i = 1$, $j = 1$, $k = 3$). The Levy–Civita tensor is defined as

$$\begin{aligned} \varepsilon_{ijk} &= \mathbf{x}_i \cdot (\mathbf{x}_j \wedge \mathbf{x}_k) \\ &= (\delta_{2j}\delta_{3k} - \delta_{2k}\delta_{3j})\delta_{1i} + (\delta_{1k}\delta_{3j} - \delta_{1j}\delta_{3k})\delta_{2i} + (\delta_{1j}\delta_{2k} - \delta_{1k}\delta_{2j})\delta_{3i}. \end{aligned}$$
$$\text{(E.11)}$$

This tensor vanishes whenever two of the indices are the same, is 1 when they are $i = 1$, $j = 2$, $k = 3$, or any cyclic permutation of that, and is -1 when the indices are $i = 2$, $j = 1$, $k = 3$, or any cyclic permutation of that. With the help of the Levy–Civita tensor, we can write the vector product of $\mathbf{a} = \sum_i a_i \mathbf{x}_i$ and $bfb = \sum_i b_i \mathbf{x}_i$ as

$$\mathbf{a} \wedge \mathbf{b} = \sum_{ijk} \mathbf{x}_i \varepsilon_{ijk} a_j b_k. \qquad \text{(E.12)}$$

E.3 THE PRODUCT OF TWO LEVY–CIVITA TENSORS AS A DETERMINANT

A very powerful rule of thumb for calculations involving the Levy–Civita tensor relates the product of two such tensors to a determinant of Kronecker deltas. We can derive this rule of thumb in the following way. Let x_i, x_j, x_k and x_l, x_m, x_n be two rearrangements of x_1, x_2, x_3 as defined in Section E.2. Now suppose that we try to expand each of the vectors x_i, x_j, x_k in terms of x_l, x_m, x_n:

$$\mathbf{x}'_i \equiv (\mathbf{x}_i \cdot \mathbf{x}_l)\mathbf{x}_l + (\mathbf{x}_i \cdot \mathbf{x}_m)\mathbf{x}_m + (\mathbf{x}_i \cdot \mathbf{x}_n)\mathbf{x}_n, \tag{E.13}$$

$$\mathbf{x}'_j \equiv (\mathbf{x}_j \cdot \mathbf{x}_l)\mathbf{x}_l + (\mathbf{x}_j \cdot \mathbf{x}_m)\mathbf{x}_m + (\mathbf{x}_j \cdot \mathbf{x}_n)\mathbf{x}_n, \tag{E.14}$$

$$\mathbf{x}'_k \equiv (\mathbf{x}_k \cdot \mathbf{x}_l)\mathbf{x}_l + (\mathbf{x}_k \cdot \mathbf{x}_m)\mathbf{x}_m + (\mathbf{x}_k \cdot \mathbf{x}_n)\mathbf{x}_n. \tag{E.15}$$

These expansions will coincide with x_i, x_j, x_k only if the set x_l, x_m, x_n contains all the unit vectors x_1, x_2, x_3. If any index is repeated in l, m, n, one of the unit vectors x_1, x_2, x_3 will be missing, so that one of x'_i, x'_j, x'_k will vanish. This means that $x'_i \cdot (x'_j \wedge x'_k)$ will be equal to ε_{ijk}, as in (E.11), if l, m, n contain all three indices $1, 2, 3$; otherwise, $x'_i \cdot (x'_j \wedge x'_k)$ will vanish. We can write this statement as

$$\mathbf{x}'_i \cdot (\mathbf{x}'_j \wedge \mathbf{x}'_k) = \varepsilon_{ijk}\varepsilon^2_{lmn}, \tag{E.16}$$

where the square of the Levy–Civita is 1 if l, m, n is any permutation of $1, 2, 3$, and it vanishes if any index is repeated. Now, using (E.13)–(E.15), we find that

$$
\begin{aligned}
\mathbf{x}'_i \cdot (\mathbf{x}'_j \wedge \mathbf{x}'_k) = {} & (\mathbf{x}_i \cdot \mathbf{x}_l)\{[\mathbf{x}_l \cdot (\mathbf{x}_m \wedge \mathbf{x}_n)](\mathbf{x}_j \cdot \mathbf{x}_m)(\mathbf{x}_k \cdot \mathbf{x}_n) \\
& + [\mathbf{x}_l \cdot (\mathbf{x}_n \wedge \mathbf{x}_m)](\mathbf{x}_j \cdot \mathbf{x}_n)(\mathbf{x}_k \cdot \mathbf{x}_m)\} \\
& + (\mathbf{x}_i \cdot \mathbf{x}_m)\{[\mathbf{x}_m \cdot (\mathbf{x}_l \wedge \mathbf{x}_n)](\mathbf{x}_j \cdot \mathbf{x}_l)(\mathbf{x}_k \cdot \mathbf{x}_n) \\
& + [\mathbf{x}_m \cdot (\mathbf{x}_n \wedge \mathbf{x}_l)](\mathbf{x}_j \cdot \mathbf{x}_n)(\mathbf{x}_k \cdot \mathbf{x}_l)\} \\
& + (\mathbf{x}_i \cdot \mathbf{x}_n)\{[\mathbf{x}_n \cdot (\mathbf{x}_l \wedge \mathbf{x}_m)](\mathbf{x}_j \cdot \mathbf{x}_l)(\mathbf{x}_k \cdot \mathbf{x}_m) \\
& + [\mathbf{x}_n \cdot (\mathbf{x}_m \wedge \mathbf{x}_l)](\mathbf{x}_j \cdot \mathbf{x}_m)(\mathbf{x}_k \cdot \mathbf{x}_l)\} \\
= {} & \{(\delta_{jm}\delta_{kn} - \delta_{jn}\delta_{km})\delta_{il} + (\delta_{jn}\delta_{kl} - \delta_{jl}\delta_{kn})\delta_{im} \\
& + (\delta_{jl}\delta_{km} - \delta_{jm}\delta_{kl})\delta_{in}\}\varepsilon_{lmn}. \tag{E.17}
\end{aligned}
$$

Substituting (E.17) in (E.16), we finally obtain the promised result: an expression for the product of two Levy–Civita tensors as a determinant of Kronecker deltas,

$$\varepsilon_{ijk}\varepsilon_{lmn} = \begin{vmatrix} \delta_{il} & \delta_{im} & \delta_{in} \\ \delta_{jl} & \delta_{jm} & \delta_{jn} \\ \delta_{kl} & \delta_{km} & \delta_{kn} \end{vmatrix}. \tag{E.18}$$

E.4 THE VECTOR PRODUCT OF THREE VECTORS

We can demonstrate the usefulness of (E.18) by deriving a very popular vector identity (used many times in this book) for the vector product of three vectors. Let $\mathbf{a} =$

$\sum_i a_i \mathbf{x}_i$, $\mathbf{b} = \sum_i b_i \mathbf{x}_i$, and $\mathbf{c} = \sum_i c_i \mathbf{x}_i$. Then

$$
\mathbf{a} \wedge (\mathbf{b} \wedge \mathbf{c}) = \sum_{ijk} \sum_{lm} \mathbf{x}_i \varepsilon_{ijk} a_j \varepsilon_{klm} b_l c_m
$$

$$
= \sum_{ijlm} \mathbf{x}_i a_j b_l c_m \sum_k \varepsilon_{ijk} \varepsilon_{klm}. \tag{E.19}
$$

Now using (E.18), we find that

$$
\sum_k \varepsilon_{ijk} \varepsilon_{klm} = \sum_k \Big\{ (\delta_{jl}\delta_{km} - \delta_{jm}\delta_{kl})\delta_{ik} + (\delta_{kk}\delta_{jm} - \delta_{jk}\delta_{km})\delta_{il}
$$

$$
+ (\delta_{jk}\delta_{kl} - \delta_{jl}\delta_{kk})\delta_{im} \Big\}
$$

$$
= \delta_{jm}\delta_{il} - \delta_{jl}\delta_{im}, \tag{E.20}
$$

so that

$$
\mathbf{a} \wedge (\mathbf{b} \wedge \mathbf{c}) = \sum_{ij} \{ a_j c_j b_i - a_j b_j c_i \} \mathbf{x}_i
$$

$$
= (\mathbf{a} \cdot \mathbf{c})\mathbf{b} - (\mathbf{a} \cdot \mathbf{b})\mathbf{c}. \tag{E.21}
$$

E.5 VECTORIAL EXPRESSIONS INVOLVING DEL

Feynman [206] invented a nice trick to manipulate vectorial expressions involving the del operator $\boldsymbol{\nabla} \equiv \sum_i \mathbf{x}_i \partial/\partial x_i$. Given a vectorial expression containing a del operator that acts on only some of the vector fields in the expression, we can define a new del operator that acts on those functions no matter how it is placed in reaction to them within the vectorial expression. This new del operator has exactly the same effect as the ordinary one, so that we can substitute it for the ordinary one. Then we can treat it as an ordinary vector and move it around using ordinary vector identities. After manipulating the expression, we can position this new operator just before the vector fields it acts on, so that it can be replaced by the ordinary del operator.

This technique is best explained by working out a particular example. Let us derive the vector identity (A.29) by defining $\boldsymbol{\nabla}_F$, and $\boldsymbol{\nabla}_G$ del operators that only act on $\mathbf{F}$ and $\mathbf{G}$, respectively:

$$
(\boldsymbol{\nabla} \wedge \mathbf{F}) \cdot (\boldsymbol{\nabla} \wedge \mathbf{G}) = (\boldsymbol{\nabla}_F \wedge \mathbf{F}) \cdot (\boldsymbol{\nabla} \wedge \mathbf{G})
$$

$$
= \boldsymbol{\nabla}_F \cdot \{ \mathbf{F} \wedge (\boldsymbol{\nabla} \wedge \mathbf{G}) \}
$$

$$
= (\boldsymbol{\nabla} - \boldsymbol{\nabla}_G) \cdot \{ \mathbf{F} \wedge (\boldsymbol{\nabla} \wedge \mathbf{G}) \}
$$

$$
= \boldsymbol{\nabla} \cdot \{ \mathbf{F} \wedge (\boldsymbol{\nabla} \wedge \mathbf{G}) \} - \underbrace{\boldsymbol{\nabla}_G \cdot \{ \mathbf{F} \wedge (\boldsymbol{\nabla} \wedge \mathbf{G}) \}}_{-\mathbf{F} \cdot \{ \boldsymbol{\nabla}_G \wedge (\boldsymbol{\nabla} \wedge \mathbf{G}) \}}
$$

$$
= \boldsymbol{\nabla} \cdot \{ \mathbf{F} \wedge (\boldsymbol{\nabla} \wedge \mathbf{G}) \} + \mathbf{F} \cdot \{ \boldsymbol{\nabla} \wedge (\boldsymbol{\nabla} \wedge \mathbf{G}) \}. \tag{E.22}
$$

E.6 SOME USEFUL INTEGRAL THEOREMS

Throughout this book we make extensive use of Gauss's divergence theorem and Stokes's theorem. Here is how these integral theorems can be derived, the physicists' way.

For the divergence theorem, consider the flux of a vector field $\mathbf{F}$ through an infinitesimal cube. Taking the Cartesian coordinates appropriately oriented, as in Figure E.2, the flux through the whole cube is $\oint_{\delta S} \mathbf{F} \cdot d\mathbf{S}$, where $d\mathbf{S} = \mathbf{n}\, dS$, with dS being

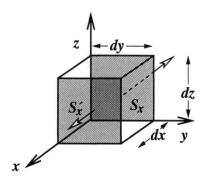

Figure E.2 Infinitesimal cube.

an infinitesimal surface element and $\mathbf{n}$ the outward unit vector normal to the surface. This flux is the sum of fluxes through each of the six sides of the cube. For the S_x and S'_x sides, the flux is

$$-F_x(x, y, z)\, dy\, dz + F_x(x + dx, y, z)\, dy\, dz.$$

Assuming that $\mathbf{F}$ is differentiable, we can expand $F_x(x + dx, y, z)$ in a Taylor series up to first order:

$$F_x(x + dx, y, z) = F_x(x, y, z) + \frac{\partial}{\partial x} F_x(x, y, z)\, dx + O(dx^2).$$

Then the flux through S_x and S'_x is given by

$$\frac{\partial}{\partial x} F_x(x, y, z)\, dx\, dy\, dz.$$

Doing the same for the other two pairs of sides, we find that

$$\oint_{\delta S} \mathbf{F} \cdot d\mathbf{S} = \nabla \cdot \mathbf{F}\, dV. \tag{E.23}$$

Notice that even though we have started with the adoption of a specific coordinate system, (E.23) is independent of the coordinate system.

Any arbitrary volume enclosed by a surface S can be decomposed into several infinitesimal volumes for which (E.23) will hold. Adding up all these infinitesimal volumes, and noticing that the fluxes through adjacent surfaces cancel each other, we find that

$$\int_V \nabla \cdot \mathbf{F}\, dV = \oint_S \mathbf{F} \cdot d\mathbf{S}. \tag{E.24}$$

Other integral theorems can be derived from (E.24) in a very simple way. For instance, if we let $\mathbf{F} = \mathbf{c}f$, where $\mathbf{c}$ is a constant vector, (E.24) becomes

$$\mathbf{c} \cdot \int_V \nabla f\, dV = \mathbf{c} \cdot \oint_S f\, d\mathbf{S}.$$

As this equation must hold for any $\mathbf{c}$, the following *gradient integral theorem* must hold

$$\int_V \nabla f\, dV = \oint_S f\, d\mathbf{S}. \tag{E.25}$$

Similarly, by taking $\mathbf{F} = \mathbf{c} \wedge \mathbf{G}$, we obtain the following *curl integral theorem*

$$\int_V \nabla \wedge \mathbf{G}\, dV = \oint d\mathbf{S} \wedge \mathbf{G}. \tag{E.26}$$

To derive Stokes's theorem, consider the line integral of a vector field $\mathbf{F}$ through an infinitesimal square. Taking Cartesian coordinates oriented as in Figure E.3, we

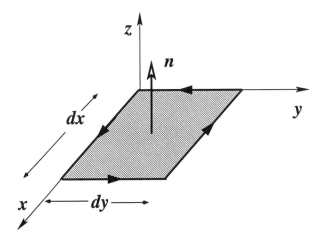

Figure E.3 Infinitesimal area element on the xy-plane.

find that

$$\oint \mathbf{F} \cdot d\mathbf{l} = F_x(x, y, z)\, dx + F_y(x + dx, y, z)\, dy$$
$$- F_x(x, y + dy, z)\, dx - F_y(x, y, z)\, dy. \tag{E.27}$$

Assuming that $\mathbf{F}$ is differentiable, we can expand $F_x(x, y+dy, z)$ and $F_y(x+dx, y, z)$ in a Taylor series about x, y, z up to first order in dy and dx, respectively. Then

$$\oint \mathbf{F} \cdot d\mathbf{l} = \frac{\partial}{\partial x} F_y(x, y, z) \, dx \, dy - \frac{\partial}{\partial y} F_x(x, y, z) \, dx \, dy$$
$$= (\nabla \wedge \mathbf{F}) \cdot d\mathbf{S}. \tag{E.28}$$

As with Gauss's theorem, Stokes's theorem is obtained by adding up all the infinitesimal area elements that make up a finite surface S. Contributions from adjacent line segments will cancel, leaving only the contribution from the line integral along the edge of the surface:

$$\int_S (\nabla \wedge \mathbf{F}) \cdot d\mathbf{S} = \oint \mathbf{F} \cdot d\mathbf{l}. \tag{E.29}$$

As for the divergence theorem, we can easily derive other integral theorems from (E.29). Letting $\mathbf{F} = \mathbf{c}f$, where $\mathbf{c}$ is an arbitrary constant vector, we find a gradient version of Stokes's theorem,

$$\int_S d\mathbf{S} \wedge \nabla f = \oint f \, d\mathbf{l}. \tag{E.30}$$

Letting $\mathbf{F} = \mathbf{c} \wedge \mathbf{G}$, we find that

$$\int_S (d\mathbf{S} \wedge \nabla) \wedge \mathbf{G} = \oint d\mathbf{l} \wedge \mathbf{G}. \tag{E.31}$$

Appendix F
The Good, the Bad, and the Ugly: Principal Parts, Step and Delta Functions

Is this nonsense? ... Not at all. ... We have to note that if Q is any function of the time, then $(\partial/\partial t)Q$ is its rate of increase. If, then, as in the present case, Q is zero before and constant after $t = 0$, $(\partial/\partial t)Q$ is zero except when $t = 0$. It is then infinite. But its total amount is Q. That is to say, $(\partial/\partial t)\theta(t)$ means a function of t which is wholly concentrated at the moment $t = 0$, of total amount 1. It is impulsive, so to speak. The idea of an impulse is well known in mechanics, and it is essentially the same here. ... the function $(\partial/\partial t)\theta(t)$ ... involves only the ordinary ideas of differentiation and integration pushed to their limit.

—Oliver Heaviside [281]

Most physics books have either a superficial nonrigorous introduction to this subject, or a very rigorous mathematical treatment. The former often leaves readers

*

pray to many pitfalls. The latter is often difficult to follow, leaving readers when doing practical calculations to work out for themselves how to connect to their real needs the abstract theorems proved. This appendix intends to be different. Here we present a number of nonrigorous physicist-type derivations that connect together many key results regarding delta functions, step functions, principal parts, and their various representations. But rather than presenting only a superficial treatment, our aim is to guide the reader through a number of subtleties and pitfalls that often arise in real calculations involving these generalized "functions." This appendix is a "bag of tools" that, to the best of our knowledge, could so far only be found scattered in the literature.

F.1   CONNECTION BETWEEN VARIOUS REPRESENTATIONS

Let us start with the ugly. Even though it was Dirac [159] who introduced the now famous notation $\delta(x)$ and who stated some of the most important properties of the *delta function,* the idea of a delta function is much older than Dirac. At the end of the nineteenth century, Oliver Heaviside was using delta functions in his operational calculus [282, 283, 419], which later became a standard technique in electrical engineering. It might be no coincidence that Dirac, who popularized the delta function in physics, was also an electrical engineer by training. But long before, Heaviside, Cauchy, Poisson, Kirchhoff, Helmholtz, and Hermite had already used a delta function.[1] According to Lützen [420], the idea behind the delta function can even be traced back to Ancient Greece, but a mathematical expression describing it probably appeared for the first time in Fourier's 1822 book *Théorie analytique de la chaleur.*[2] The main feature of the delta function is the *sampling* or *sifting property,* where for an arbitrary continuous[3] function $f(x)$,

$$\int_a^b dx \, f(x)\delta(x) = \begin{cases} 0 & \text{if } (a,b) \text{ does not contain the origin.} \\ f(0) & \text{if } (a,b) \text{ contains the origin.} \end{cases} \tag{F.1}$$

To be "mathematically correct," the delta function is not really a function because it does not map numbers into numbers but rather is a distribution that maps functions into numbers. It can, however, be seen as a limit of certain ordinary functions. Physics textbooks frequently mention the rectangular barrier, the sinc function, the Lorentzian, and the Gaussian as functions that become a delta function when we keep their area constant but decrease their widths to zero. In practical calculations, it is important to know which functions become delta functions this way. In this section we present simple physicist-type derivations that connect the most common representations of

---

[1]For more on history, see [419, 420, 610].
[2]"L'expression $(1/2) + \sum_{n=1}^{\infty} \cos([x - \alpha]n)$ représente une fonction de x et de α telle que, si on la multiplie par un fonction quelconque $F(\alpha)$ et si, après avoir écrit $d\alpha$, on intègre entre les limites $\alpha = -\pi$ et $\alpha = \pi$, on aura changé la fonction proposée $F(\alpha)$ en une pareille fonction de x, multipliée par la demi-circonférence π."
[3]If $f(x)$ is not continuous, the situation is more complex (see Section F.3).

the delta function appearing in physics. At the end of the section we describe briefly a general method of identifying and constructing representations of delta functions [8].

F.1.1 The rectangular barrier

The rectangular barrier $y_\varepsilon(x)$ is perhaps the simplest function one can think of whose fixed-area vanishing-width limit yields a delta function. It is defined (see Figure F.1) by

$$y_\varepsilon(x) = \begin{cases} 0, & |x| > \varepsilon/2. \\ 1/\varepsilon, & -\varepsilon/2 < x < \varepsilon/2. \end{cases} \tag{F.2}$$

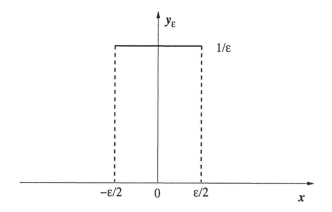

Figure F.1 A plot of the rectangular barrier (F.2).

We can use the intuitive relation $\lim_{\varepsilon \to 0} y_\varepsilon(x) = \delta(x)$ to deduce a very popular representation of the delta function in physics. Consider the Fourier transform of $y_\varepsilon(x)$,

$$\mathcal{F}\left[y_\varepsilon(x)\right](k) = \int_{-\varepsilon/2}^{\varepsilon/2} dx \, \frac{e^{-ikx}}{\varepsilon}$$

$$= \frac{2}{\varepsilon k} \sin\left(\frac{\varepsilon k}{2}\right). \tag{F.3}$$

Now if we invert (F.3) and take the limit $\varepsilon \to 0$, we find that

$$\delta(x) = \lim_{\varepsilon \to 0} y_\varepsilon(x)$$

$$= \frac{1}{\pi} \int_{-\infty}^{\infty} dk \, \frac{e^{ikx}}{k} \underbrace{\lim_{\varepsilon \to 0} \frac{1}{\varepsilon} \sin\left(\frac{\varepsilon k}{2}\right)}_{k/2}$$

$$= \frac{1}{2\pi} \int_{-\infty}^{\infty} dk\, e^{ikx}. \tag{F.4}$$

F.1.2   The sinc function

The sinc function is another common representation of the delta function. We can obtain it very simply by realizing that the integral on the last line of (F.4) is actually ill defined, as $\exp(ikx)$, rather than assuming a well-defined value at $k = -\infty$ and $k = \infty$, just oscillates rapidly. A more mathematically correct statement than (F.4) is

$$\delta(x) = \lim_{\eta \to \infty} \int_{-\eta}^{\eta} dk\, \frac{e^{ikx}}{2\pi}$$

$$= \lim_{\eta \to \infty} \frac{1}{\pi} \frac{\sin(\eta x)}{x}. \tag{F.5}$$

F.1.3   The Lorentzian representation and the principal part

The Lorentzian representation and Cauchy's principal part can be obtained from another attempt of making (F.4) more mathematically correct. Instead of making the endpoints of the ill-defined integral finite, we can damp the oscillations of the exponential at infinity. Consider the integral

$$\xi(x) = \lim_{\varepsilon \to 0+} \int_0^{\infty} dk\, \frac{e^{ik(x+i\varepsilon)}}{2\pi}$$

$$= \frac{i}{2\pi} \lim_{\varepsilon \to 0+} \frac{1}{x + i\varepsilon}. \tag{F.6}$$

The alternative more mathematically correct way of writing (F.4) is to rewrite the ill-defined integral in the last line of (F.4) as twice the real part of (F.6):

$$\delta(x) = 2\Re\{\xi(x)\}$$

$$= \frac{i}{2\pi} \lim_{\varepsilon \to 0+} \left\{ \frac{1}{x + i\varepsilon} - \frac{1}{x - i\varepsilon} \right\}$$

$$= \frac{1}{\pi} \lim_{\varepsilon \to 0+} \frac{\varepsilon}{x^2 + \varepsilon^2}. \tag{F.7}$$

This Lorentzian representation of the delta function was first used by Cauchy and Poisson, independently, in 1815 [610]. The imaginary part of $\xi(x)$ is proportional to

Cauchy's principal part (a way of getting rid of the infinity at the origin in $1/x$)[4]:

$$2i\Im\{\xi(x)\} = \frac{i}{2\pi}\lim_{\varepsilon\to 0+}\left\{\frac{1}{x+i\varepsilon}+\frac{1}{x-i\varepsilon}\right\}$$

$$= \frac{1}{\pi}\lim_{\varepsilon\to 0+}\frac{x}{x^2+\varepsilon^2}$$

$$= \frac{1}{\pi}\mathcal{P}\frac{1}{x}. \tag{F.8}$$

Joining (F.7) and (F.8), we obtain another very useful expression (especially in scattering theory):

$$\lim_{\varepsilon\to 0+}\frac{1}{x+i\varepsilon} = \mathcal{P}\frac{1}{x} - i\pi\delta(x), \tag{F.9}$$

which can also be written in a less rigorous way as

$$\int_0^\infty dk\, e^{ikx} = \pi\delta(x) + i\mathcal{P}\frac{1}{x}. \tag{F.10}$$

Those readers who like contour integration would be glad to know that the apparently enigmatic relation (F.9) can also be obtained from contour integration. Let $f(x)$ have no poles in the upper half of the complex plane and be such that $\lim_{R\to\infty} f\left(Re^{i\theta}\right) = 0$ for $0 \le \theta \le \pi$. Then

$$\lim_{\varepsilon\to 0+}\int_{-\infty}^\infty dx\,\frac{f(x)}{x+i\varepsilon} = \lim_{\varepsilon\to 0+}\int_{C_\varepsilon} dz\,\frac{f(z)}{z}$$

$$= \lim_{\varepsilon\to 0+}\left\{\int_\varepsilon^\infty dx\,\frac{f(x)}{x}+\int_{-\infty}^{-\varepsilon} dx\,\frac{f(x)}{x}\right\}$$

$$+ \lim_{\varepsilon\to 0+}\int_{\delta_\varepsilon} dz\,\frac{f(z)}{z}, \tag{F.11}$$

where the contours C_ε and δ_ε are shown in the Figure F.2. The two integrals in the middle line of (F.11) yield the principal part. The integral in last line yields

$$\lim_{\varepsilon\to 0+}\int_{\delta_\varepsilon} dz\,\frac{f(z)}{z} = \lim_{\varepsilon\to 0+}\int_\pi^0 d\left(\varepsilon e^{i\theta}\right)\frac{f\left(\varepsilon e^{i\theta}\right)}{\varepsilon e^{i\theta}}$$

$$= \lim_{\varepsilon\to 0+}\int_\pi^0 i\,d\theta\, f\left(\varepsilon e^{i\theta}\right)$$

$$= -i\pi f(0). \tag{F.12}$$

---

[4] A more transparent but equivalent definition of Cauchy's principal part is

$$\int_{-\infty}^\infty dx\,\mathcal{P}\frac{1}{x}\,f(x) = \lim_{\varepsilon\to 0+}\left\{\int_\varepsilon^\infty dx\,\frac{1}{x}\,f(x)+\int_{-\infty}^{-\varepsilon} dx\,\frac{1}{x}\,f(x)\right\}.$$

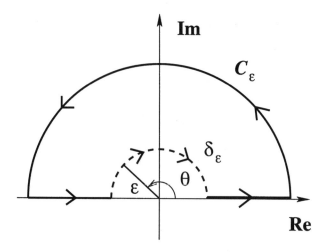

Figure F.2 C_ε is the whole closed contour and δ_ε is the small doted arch.

F.1.4 The Gaussian

The Gaussian representation of the delta function, introduced by Kirchhoff in his formulation of Huygens's principle [610], can also be obtained as another attempt of making (F.4) more mathematically correct. Instead of separating the integral in the last line of (F.4) into a positive and a negative k part in order to introduce the positive imaginary part $i\varepsilon$ to x for positive k and the negative imaginary part $-i\varepsilon$ to x for negative k, we can introduce the imaginary part $i\varepsilon k$ to x such that when multiplied by ik, it always gives rise to a converging negative real term in the exponential regardless of the sign of k.

$$
\begin{aligned}
\delta(x) &= \lim_{\varepsilon \to 0+} \int_{-\infty}^{\infty} dk \; \frac{e^{ik(x+i\varepsilon k)}}{2\pi} \\
&= \lim_{\varepsilon \to 0+} \int_{-\infty}^{\infty} dk \; \frac{e^{-x^2/(4\varepsilon)}}{2\pi} \; \exp\left(-\varepsilon k^2 + ixk + \frac{x^2}{4\varepsilon}\right) \\
&= \lim_{\varepsilon \to 0+} \frac{e^{-x^2/(4\varepsilon)}}{2\pi} \int_{\infty}^{\infty} dk \; \exp\left(-\left[\sqrt{\varepsilon}k - i\frac{x}{2\sqrt{\varepsilon}}\right]^2\right) \\
&= \lim_{\varepsilon \to 0+} \frac{e^{-x^2/(4\varepsilon)}}{2\sqrt{\pi\varepsilon}} \\
&= \lim_{n \to \infty} \sqrt{\frac{n}{\pi}} \; e^{-nx^2}.
\end{aligned}
\tag{F.13}
$$

F.1.5 The Laplacian of $1/r$

This is another common representation of the delta function in physics. To derive it, consider first the Fourier transform of $1/r$:

$$\mathcal{F}\left[\frac{1}{r}\right](\mathbf{k}) = \int d^3r \, \frac{e^{i\mathbf{k}\cdot\mathbf{r}}}{r}$$

$$= -\int_0^\infty r \, dr \int_0^{2\pi} d\phi \int_0^\pi d(\cos\theta) e^{ikr\cos\theta}$$

$$= 2\pi \int_0^\infty r \, dr \int_{-1}^1 d\zeta \, e^{ikr\zeta}$$

$$= 2\pi \int_0^\infty r \, dr \, \frac{e^{ikr} - e^{-ikr}}{ikr}$$

$$= 2\pi \int_0^\infty dr \, \frac{e^{ikr} - e^{-ikr}}{ik}$$

$$= \frac{4\pi}{k^2}. \tag{F.14}$$

Using (F.14), we can write the Laplacian of $1/r$ as

$$\nabla^2\left(\frac{1}{r}\right) = \nabla^2 \frac{1}{(2\pi)^3} \int d^3k \, e^{-i\mathbf{k}\cdot\mathbf{r}} \mathcal{F}\left[\frac{1}{r}\right](\mathbf{k})$$

$$= \frac{1}{(2\pi)^3} \int d^3k \, (-k^2) e^{-i\mathbf{k}\cdot\mathbf{r}} \frac{4\pi}{k^2}$$

$$= -4\pi\delta(\mathbf{r}). \tag{F.15}$$

Related to this representation of the delta function, Frahm [222] has presented two lesser known identities,

$$\frac{\partial}{\partial x_i} \frac{\partial}{\partial x_j} \frac{1}{r} = -\frac{4\pi}{3} \delta_{ij}\delta(\mathbf{r}) + \frac{3x_i x_j - r^2\delta_{ij}}{r^5}, \tag{F.16}$$

$$\frac{\partial}{\partial x_i} \frac{\partial}{\partial x_j} \frac{\partial}{\partial x_k} \frac{1}{r} = -\frac{4\pi}{5}\left\{\delta_{ij}\frac{\partial}{\partial x_k}\delta(\mathbf{r}) + \delta_{jk}\frac{\partial}{\partial x_i}\delta(\mathbf{r}) + \delta_{ki}\frac{\partial}{\partial x_j}\delta(\mathbf{r})\right\}$$

$$+ \frac{3(\delta_{ij}x_k + \delta_{jk}x_i + \delta_{ki}x_j)}{r^5} - \frac{15x_i x_j x_k}{r^7}. \tag{F.17}$$

F.1.6 The comb function

This is yet another very common representation of the delta function in physics. It is actually not a representation of a single delta function but of a periodic train of delta functions. It can be derived by constructing a Fourier series for $y_\varepsilon(x)$ in a one-dimensional box defined by $-L < x < L$,

$$y_\varepsilon^p(x) = \frac{1}{2L} \sum_{n=-\infty}^{\infty} e^{ix n\pi/L} \frac{2L}{\varepsilon\pi n} \sin\left(\frac{\varepsilon}{2}\frac{n\pi}{L}\right). \tag{F.18}$$

As the Fourier series $y_\varepsilon^p(x)$ produces a copy of the function $y_\varepsilon(x)$ in the interval

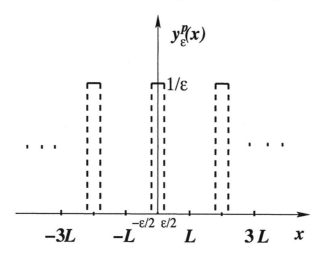

Figure F.3   A plot of the comb function (F.18).

$-L < x < L$ which is reproduced periodically throughout the entire real line (see Figure F.3), if we now take the limit of vanishing ε, we obtain a periodic train of delta functions,

$$
\begin{aligned}
\lim_{\varepsilon \to 0} y_\varepsilon^p(x) &= \sum_{n=-\infty}^{\infty} e^{ixn\pi/L} \lim_{\varepsilon \to 0} \frac{1}{\varepsilon \pi n} \sin\left(\frac{\varepsilon}{2}\frac{n\pi}{L}\right) \\
&= \frac{1}{2L} \sum_{n=-\infty}^{\infty} e^{ixn\pi/L} \\
&= \sum_{m=-\infty}^{\infty} \delta\left(x - 2Lm\right).
\end{aligned}
\tag{F.19}
$$

F.1.7   A general rule to find representations

Aguirregabiria and collaborators [8] have proposed a general method to find new representations of the delta function. Their idea is based on a simplified version of a convergence theorem well known to mathematicians working with generalized functions [229, 329, 370]. This theorem says that if the function g is such that its integral

$$
\mathcal{G} = \int_{-\infty}^{\infty} dx\, g(x)
\tag{F.20}
$$

is finite, we can obtain a delta function from it by taking the limit

$$
\lim_{\gamma \to \infty} \gamma\, g(\gamma x) = \mathcal{G}\delta(x).
\tag{F.21}
$$

To show that (F.21) really holds, consider the integral

$$\int_{-\infty}^{x} du \lim_{\gamma \to \infty} \gamma g(\gamma u) = \lim_{\gamma \to \infty} \int_{-\infty}^{\gamma x} d\xi \, g(\xi). \tag{F.22}$$

If x is positive, the upper limit of the integral on the right-hand side of (F.22) is $+\infty$, and according to (F.20), the integral will be equal to $\mathcal{G}$. If, however, x is negative, the upper limit of the integral on the right-hand side of (F.22) is $-\infty$, making the integral vanish. Thus,

$$\int_{-\infty}^{x} du \lim_{\gamma \to \infty} \gamma g(\gamma u) = \mathcal{G} \, \theta(x). \tag{F.23}$$

As the delta function is the derivative of the step function, differentiation of (F.23) with respect to x yields (F.21).

We can also verify (F.21) in the following way. For arbitrary a and b so that $a \leq b$ and a continuous function $f(x)$, using the same change of variables used in (F.22), we can write

$$\int_{a}^{b} dx \, f(x) \lim_{\gamma \to \infty} \gamma \, g(\gamma x) = \lim_{\gamma \to \infty} \int_{a\gamma}^{b\gamma} d\xi \, f\left(\frac{\xi}{\gamma}\right) g(\xi). \tag{F.24}$$

Now if $ab > 0$, the two limits of the integral on the right-hand side of (F.24) will be either both ∞ or both $-\infty$ in the limit of $\gamma \to \infty$, making the integral vanish. But if $ab < 0$, the lower limit will become $-\infty$ and the upper ∞ when $\gamma \to \infty$. Moreover, $f(\xi/\gamma)$ will become $f(0)$ as $\gamma \to \infty$, so that we can write

$$\int_{a}^{b} dx \, f(x) \lim_{\gamma \to \infty} \gamma \, g(\gamma x) = \begin{cases} 0 & \text{for } ab > 0, \\ \mathcal{G} \, f(0) & \text{for } ab < 0. \end{cases} \tag{F.25}$$

This is just another way of writing the basic property (F.1) of the delta function. So it justifies (F.21). As we mentioned before, we delay to Section F.3 any considerations about what happens when $ab = 0$.

Aguirregabiria and collaborators [8] show how (F.21) can be used to obtain a "quick and dirty" derivation of the identities (F.16) and (F.17), which is much simpler and shorter than Frahm's original derivation [222]. But rather than presenting this derivation, we will now reproduce their derivation of a much lesser known identity involving delta functions which was first obtained by Aichelburg and Sexl [11] in the context of general relativity. First note that if $a = \lim_{\gamma \to \infty} h(\gamma)$, then

$$\lim_{\gamma \to \infty} \gamma g\left(\{x - h(\gamma)\}\gamma\right) = \mathcal{G} \, \delta(x - a). \tag{F.26}$$

Now take

$$g(x) = \frac{1}{\sqrt{x^2 + \rho^2}} - \frac{1}{\sqrt{x^2 + 1}} \tag{F.27}$$

and

$$h(\gamma) \equiv v(\gamma)$$
$$= c\sqrt{1 - \frac{1}{\gamma^2}}, \tag{F.28}$$

where ρ is a positive real number, v is the speed, c is the speed of light, and $\gamma = 1/\sqrt{1-(v/c)^2}$. Then, as $\lim_{\gamma \to \infty} h(\gamma) = c$, we find that

$$\lim_{v \to c} \left\{ \frac{\gamma}{\sqrt{(x-vt)\gamma^2 + \rho^2}} - \frac{\gamma}{\sqrt{(x-vt)\gamma^2 + 1}} \right\} = \mathcal{G}\,\delta(x - ct), \qquad (F.29)$$

where in this case,

$$\mathcal{G} = \int_{-\infty}^{\infty} dx \left\{ \frac{1}{\sqrt{x^2 + \rho^2}} - \frac{1}{\sqrt{x^2 + 1}} \right\}. \qquad (F.30)$$

At first sight the integral in (F.30) seems to vanish, as a change of variables $x' = x\rho$ for the integral of the first term would make it identical to the integral of the second term. However, we must take extra care here because the separate integration of each term diverges. So the right procedure is first to consider

$$\int_{-\eta}^{\eta} \frac{dx}{\sqrt{x^2 + b^2}} = \int_{-\eta/b}^{\eta/b} \frac{dx'}{\sqrt{x'^2 + 1}}$$

$$= \ln \left| \frac{\sqrt{1 + (\eta/b)^2} + \eta/b}{\sqrt{1 + (\eta/b)^2} - \eta/b} \right|, \qquad (F.31)$$

and take the limit of $\eta \to \infty$ only at the end. Using (F.31) in (F.30) with $b = \rho$ for the first term in the integrand of (F.30) and $b = 1$ for the second, we obtain

$$\int_{-\infty}^{\infty} dx \left\{ \frac{1}{\sqrt{x^2 + \rho^2}} - \frac{1}{\sqrt{x^2 + 1}} \right\}$$

$$= \lim_{\eta \to \infty} \int_{-\eta}^{\eta} \left\{ \frac{1}{\sqrt{x^2 + \rho^2}} - \frac{1}{\sqrt{x^2 + 1}} \right\}$$

$$= \lim_{\eta \to \infty} \ln \left| \frac{\sqrt{1 + (\eta/\rho)^2} + \eta/\rho \sqrt{1 + \eta^2} + \eta}{\sqrt{1 + (\eta/\rho)^2} - \eta/\rho \sqrt{1 + \eta^2} - \eta} \right|$$

$$= \lim_{\eta \to \infty} \ln \left| \frac{2\eta/\rho}{(1/2)(\rho/\eta)} \frac{(1/2)(1/\eta)}{2\eta} \right|$$

$$= -2 \ln \rho. \qquad (F.32)$$

So, finally, we obtain the following exotic identity:

$$\lim_{v \to c} \left\{ \frac{\gamma}{\sqrt{(x-vt)\gamma^2 + \rho^2}} - \frac{\gamma}{\sqrt{(x-vt)\gamma^2 + 1}} \right\} = -2 \ln \rho \,\, \delta(x - ct). \qquad (F.33)$$

F.2 A USEFUL IDENTITY IN FANO DIAGONALIZATION: THE PRODUCT OF TWO PRINCIPAL PARTS

The product of two principal parts is defined as

$$
\int_{-\infty}^{\infty} dx \, f(x) \, \mathcal{P}\frac{1}{x-a}\mathcal{P}\frac{1}{x-b}
$$

$$
= \lim_{\varepsilon \to 0^+} \left\{ \int_{-\infty}^{a-\varepsilon} dx \, f(x) \, \frac{1}{x-a}\frac{1}{x-b} + \int_{a+\varepsilon}^{b-\varepsilon} dx \, f(x) \, \frac{1}{x-a}\frac{1}{x-b} \right.
$$

$$
\left. + \int_{b+\varepsilon}^{\infty} dx \, f(x) \, \frac{1}{x-a}\frac{1}{x-b} \right\}, \tag{F.34}
$$

where we have assumed that $a < b$. Now, if we rewrite this as in (F.8), we find that

$$
\mathcal{P}\frac{1}{x-a}\mathcal{P}\frac{1}{x-b} = \lim_{\varepsilon \to 0^+} \frac{x-a}{(x-a)^2+\varepsilon^2}\frac{x-b}{(x-b)^2+\varepsilon^2}
$$

$$
= \lim_{\varepsilon \to 0^+} \frac{1}{a-b}\left\{ \frac{x-a}{(x-a)^2+\varepsilon^2} - \frac{x-b}{(x-b)^2+\varepsilon^2} \right\}
$$

$$
+ \lim_{\varepsilon \to 0^+} \frac{\varepsilon^2}{\{(x-a)^2+\varepsilon^2\}\{(x-b)^2+\varepsilon^2\}}
$$

$$
= \frac{1}{a-b}\left\{ \mathcal{P}\frac{1}{x-a} - \mathcal{P}\frac{1}{x-b} \right\} + \pi^2 \delta(x-a)\delta(x-b), \tag{F.35}
$$

which was deduced in another way by Fano in his paper on Fano diagonalization [38, 190].

F.3 WHEN THE SAMPLING HAPPENS AT A POINT OF DISCONTINUITY

As we mentioned before, the sampling property of the delta "function" (F.1) is usually defined only for functions $f(x)$ that are continuous at the point where the delta function explodes.[5] If the function is discontinuous at that point but the discontinuity is removable, we can just redefine the sampling property in terms of the limit of the function at that point [300]:

$$
\int_a^b dx \, f(x) \, \delta(x) = \begin{cases} 0 & \text{if } (a,b) \text{ does not contain the origin.} \\ \lim_{\zeta \to 0} f(\zeta) & \text{if } (a,b) \text{ contains the origin.} \end{cases}
$$

$$\tag{F.36}$$

---

[5]A similar problem happens in Langevin equations with white noise if the white noise term appears in the equation multiplied by a function of the position (see Section 8.2.4 and [224, 613]). There are two popular solutions to the ambiguity that results. One leads to the Langevin equation being treated as an Itô stochastic differential equation [312]. The other leads to it being treated as a Stratonovich stochastic differential equation [580].

The real difficulty happens when $f(x)$ has a jump discontinuity at the point where the delta function explodes. Then what should we do? Should we take the limit from the left or from the right? Should we take the average of the two limits? As the delta function is even, one is tempted to write

$$\int_{-\infty}^{\infty} dx \ f(x)\delta(x) = \int_{-\infty}^{\infty} dx \ f(-x) \ \delta(x), \tag{F.37}$$

so that

$$\int_{-\infty}^{\infty} dx \ f(x)\delta(x) = \int_{-\infty}^{\infty} dx \ \frac{f(x) + f(-x)}{2} \ \delta(x). \tag{F.38}$$

Then, as $f(x) + f(-x)$ is continuous at $x = 0$, we can write

$$\int_{-\infty}^{\infty} dx \ f(x)\delta(x) = \frac{f(0^-) + f(0^+)}{2}. \tag{F.39}$$

Another way to justify (F.39) is to remember that the delta function is the derivative of the Heaviside step function $\theta(x)$. So if we write

$$\theta^2(x) = \theta(x) \tag{F.40}$$

and differentiate both sides, we find that

$$\theta(x)\delta(x) = \frac{1}{2}\delta(x), \tag{F.41}$$

which is just another way of writing (F.39). But blind application of (F.39) or (F.41) can lead to trouble, as was pointed out by Griffiths and Walborn [252]. Consider the differential equation

$$\frac{df}{dx} = f(x)\delta(x). \tag{F.42}$$

Except for $x = 0$, this equation reads

$$\frac{df}{dx} = 0. \tag{F.43}$$

So the solution should be

$$f(x) = \begin{cases} c_1 & \text{for } x < 0 \\ c_2 & \text{for } x > 0, \end{cases} \tag{F.44}$$

where c_1 and c_2 are two constants. The question is: How do you match these two constants? This is a first-order differential equation and there must be only one integration constant, so there must be a relation between c_1 and c_2. If we use (F.39) or (F.41), we conclude that this relation should be

$$c_2 = 3c_1. \tag{F.45}$$

If, on the other hand, we regard the delta function as the limit of a very thin and large barrier, we easily find that [252]

$$c_2 = e\, c_1. \tag{F.46}$$

Which one is correct, (F.45) or (F.46)? We can easily see that (F.46) is the correct relation because as Griffiths and Walborn [252] pointed out, (F.42) can be solved directly

$$\frac{df}{f} = \delta(x)\, dx \;\Rightarrow\; f(x) = c\, e^{\theta(x)}, \tag{F.47}$$

where c is the single integration constant.

Can we trace down the mistake in the steps that lead to (F.39) and (F.41)? We notice that if we apply the steps that lead us to (F.41) to the product of four step functions instead of just two as in (F.40), we find that

$$\theta^3(x)\delta(x) = \theta(x)\delta(x) = \frac{1}{4}\delta(x), \tag{F.48}$$

which together with (F.41) implies the absurd conclusion that $1/4 = 1/2$. In (F.39), the mistake is the assumption that as $\delta(x)$ is even, the total integrand would also be even. Depending on the microscopic behavior of both $\delta(x)$ and $f(x)$ at $x = 0$, however, the delta function can sample the function more at one side of the discontinuity than at the other side, as we showed in the example above.

Another way of writing (F.39) or (F.41) that often appears in the literature [610, 620] is

$$\int_a^b dx\, f(x)\, \delta(x) = \begin{cases} 0 & \text{if } ab > 0, \\ (1/2)f(0) & \text{if } ab = 0, \\ f(0) & \text{if } ab < 0. \end{cases} \tag{F.49}$$

This was even suggested in a paper [664] as an alternative, more precise definition of the delta function. It is embarrassing that even though the problem with the blind use of (F.39) or (F.41) or (F.49) has been known for a long time, papers misusing these relations keep reappearing in the literature.[6] Griffiths and Walborn [252] mention that in the physics literature, the problem was spotted at least as far back as 1981 [86, 126, 439, 530, 585]. But well before that, in 1954, Schwartz [542] had already pointed out the problems with the multiplication of a distribution by another distribution. In that paper he shows that the associative property of multiplication and the concept of derivative are incompatible, within a mathematical framework where multiplication of distributions by each other is allowed. This incompatibility can be seen easily from our short demonstration about the flaw in our deduction of (F.41). There, the only logical steps are the use of the associative property of multiplication to infer that $\theta^3(x) = \theta(x)$, and the rule for the derivative of a product. Together, these two simple steps lead to the absurd conclusion that $1/4 = 1/2$. So they are incompatible within the context of multiplication of distributions.

---

[6]For some papers where this mistake was made, as well as their subsequent discussion and correction, see [41, 86, 126, 127, 253, 393, 425, 426, 570].

Schwartz's conclusion in that paper [542] was that the multiplication of distributions is impossible. But there is another solution to this problem. This solution, as we have already suggested, is to abandon the associative law for multiplication of distributions. There are several step functions and delta functions differing in their microscopic behavior. Then, even though the product of any number of step functions is also a step function, it is yet a step function of a different sort than the ones that were multiplied to originate it. Its microscopic behavior at its discontinuity point is different from that of its originators. This is basically the idea proposed by Colombeau in his theory of multiplication of distributions [110–112, 114]. We do not discuss Colombeau's theory in detail here, but we recommend to the interested reader his paper [112], his book [113], and another book by Biagioni [52], all of which are quite accessible to physicists and engineers.

References

1. B. Abbott et al. Search for heavy pointlike Dirac monopoles. *Physical Review Letters*, **81**(3):524–529, July 20, 1998.

2. I. Abonyi, J. F. Bito, and J. K. Tar. A quaternion representation of the Lorentz group for classical physical applications. *Journal of Physics A*, **24**:3245–3254, 1991.

3. J. R. Ackerhalt, P. L. Knight, and J. H. Eberly. Radiation reaction and radiative frequency shifts. *Physical Review Letters*, **30**(10):456–460, March 5, 1973.

4. C. Adlard, E. R. Pike, and S. Sarkar. Localization of one-photon states. *Physical Review Letters*, **79**(9):1585–1587, September 1, 1997.

5. R. J. Adler, B. Casey, and O. C. Jacob. Vacuum catastrophe: An elementary exposition of the cosmological constant problem. *American Journal of Physics*, **63**(7):620–626, July 1995.

6. S. L. Adler. Quaternionic quantum field theory. *Physical Review Letters*, **55**(8): 783–786, August 19, 1985. See also the erratum for this letter: S. L. Adler, *Physical Review Letters* **55**(13):1430, September 23, 1985.

7. G. S. Agarwal. *Quantum Statistical Theories of Spontaneous Emission and Their Relation to Other Approaches*, volume 70 of *Springer Tracts in Modern Physics*. Springer-Verlag, Heidelberg, 1974.

8. J. M. Aguirregabira, A. Hernandez, and M. Rivas. δ-Function converging sequences. *American Journal of Physics*, **70**(2):180–185, February 2002.

9. Y. Aharonov and D. Bohm. Significance of electromagnetic potentials in the quantum theory. *Physical Review*, **115**(3):485–491, August 1, 1959.

10. Y. Aharonov and D. Bohm. Time in the quantum theory and the uncertainty relation for time and energy. *Physical Review*, **122**(5):1649–1658, June 1, 1961.

11. P. C. Aichelburg and R. U. Sexl. On the gravitational field of a massless particle. *General Relativity and Gravitation*, **4**:303–312, 1971.

12. A. I. Akhiezer and V. B. Berestetsky. *Kvantovaya Elektrodinamika*. Gosudarstvennoe Izdatelstvo Tekhniko-Teoreticheskoi Literatury, Moscow, 1953. There is an English translation in two volumes: A. I. Akhiezer and V. B. Berestetsky, *Quantum Electrodynamics*, U.S. Atomic Energy Commission (originally available from the Office of Technical Services, U.S. Department of Commerce, Washington, DC, for the incredible price of $2.65).

13. L. Allen and J. H. Eberly. *Optical Resonance and Two-Level Atoms*. Dover, New York, 1975.

14. E. Altewischer, M. P. van Exter, and J. P. Woerdman. Plasmon-assisted transmission of entangled photons. *Nature*, **418**:304–306, July 18, 2002.

15. W. O. Amrein. Localizability for particles of mass zero. *Helvetica Physica Acta*, **42**(1):149–190, February 15, 1969.

16. K. An, J. J. Childs, R. R. Dasari, and M. S. Feld. Microlaser: A laser with one atom in an optical resonator. *Physical Review Letters*, **73**(25):3375–3378, December 19, 1994.

17. K. An and M. S. Feld. Role of standing-wave mode structure in microlaser emission. *Physical Review A*, **52**(2):1691–1697, August 1995.

18. K. An and M. S. Feld. Quantum trajectory analysis of a thresholdlike transition in the microlaser. *Physical Review A*, **55**(6):4492–4500, June 1997.

19. K. An and M. S. Feld. Semiclassical four-level single-atom laser. *Physical Review A*, **56**(2):1662–1665, August 1997.

20. J. Anandan. Classical and quantum interaction of the dipole. *Physical Review Letters*, **85**(7):1354–1357, August 14, 2000.

21. J. D. Anderson, E. L. Lau, T. P. Krisher, D. A. Dicus, D. C. Rosenbaum, and V. L. Teplitz. Improved bounds on nonluminous matter in solar orbit. *Astrophysical Journal*, **448**(2):885–892, August 1, 1995. See also the erratum J. D. Anderson et al. Astrophysical Journal **464**(2):1054, June 20, 1996.

22. R. Apfel. Sonoluminescence: And there was light! *Nature*, **398**:378–379, April 1, 1999.

23. J. R. Argand. *Essai sur une manière de représenter les quantités imaginaires dans les constructions géométriques.* Hoüel, Paris, 2nd edition, 1874.

24. F. V. Atkinson. On Sommerfeld's radiation condition. *Philosophical Magazine,* **1**(1):645–651, 1949.

25. L. Bachelier. Théorie de la spéculation. PhD dissertation. Faculté des Sciences de Paris, Paris, 1900. Translated to English in pp. 17–78 of [122].

26. H.-A. Bachor. *A Guide to Experiments in Quantum Optics.* Wiley-VCH, Weinheim, Germany, 1998.

27. T. Banks. The cosmological constant problem. *Physics Today,* **57**(3):46–51, March 2004.

28. J. Bardeen. Tunnelling from a many-particle point of view. *Physical Review Letters,* **6**(2):57–59, January 15, 1961.

29. J. Bardeen. Tunneling into superconductors. *Physical Review Letters,* **9**(4):147–149, August 15, 1962.

30. P. J. Bardroff and S. Stenholm. Quantum theory of excess noise. *Physical Review A,* **60**(3):2529–2533, September 1999.

31. P. J. Bardroff and S. Stenholm. Quantum Langevin theory of excess noise in lasers. *Physical Review A,* **61**(2):023806, February 2000.

32. V. Bargmann and E. P. Wigner. Group theoretical discussion of relativistic wave equations. *Proceedings of the National Academy of Sciences of the United States of America,* **34**:211–223, 1948.

33. S. M. Barnett. The quantum optics of dielectrics. In S. Reynaud, E. Giacobino, and J. Zinn-Justin, editors, *Quantum Fluctuations,* Les Houches, Session LXIII, 1995, pp. 137–179. Elsevier Science, Amsterdam, 1997.

34. S. M. Barnett, P. Filipowicz, J. Javanainen, P. L. Knight, and P. Meystre. The Jaynes-Cummings model and beyond. In E. R. Pike and S. Sarkar, editors, *Frontiers in Quantum Optics,* pp. 485–520, Adam Hilger, Gloucestershire, England, 1986.

35. S. M. Barnett, C. R. Gilson, B. Huttner, and N. Imoto. Field commutation relations in optical cavities. *Physical Review Letters,* **77**(9):1739–1742, August 26, 1996.

36. S. M. Barnett and P. L. Knight. Dissipation in a fundamental model of quantum optical resonance. *Physical Review A,* **33**(4):2444–2448, April 1986.

37. S. M. Barnett and P. M. Radmore. Quantum theory of cavity quasimodes. *Optics Communications,* **68**(5):364–368, November 1, 1988.

38. S. M. Barnett and P. M. Radmore. *Methods in Theoretical Quantum Optics*, volume 15 of *Oxford Series in Optical and Imaging Sciences*. Clarendon Press, Oxford, 1997.

39. G. Barton. Quantum electrodynamics of spinless particles between conducting plates. *Proceedings of the Royal Society of London, Series A*, **320**:251–275, 1970.

40. G. Barton. *Elements of Green's Functions and Propagation*. Oxford University Press, Oxford, 1989.

41. G. Barton and D. Waxman. Wave equations with point-support potentials having dimensionless strength parameters. Sussex Report, 1994.

42. A. O. Barut. *Electrodynamics and Classical Theory of Fields and Particles*. Dover, New York, 1964.

43. B. Baseia and H. M. Nussenzveig. Semiclassical theory of laser transmission loss. *Optica Acta*, **31**(1):39–62, January 1984.

44. L. Bass and E. Schrödinger. Must the photon mass be zero? *Proceedings of the Royal Society of London, series A*, **232**:1–6, October 11, 1955.

45. H. Bateman. On dissipative systems and related variational principles. *Physical Review*, **38**(4):815–819, August 15, 1931.

46. W. E. Baylis, J. Bonenfant, J. Derbyshire, and J. Huschilt. Light polarization: A geometric-algebra approach. *American Journal of Physics*, **61**(6):534–545, June 1993.

47. W. R. Bennett Jr. Gaseous optical masers. *Applied Optics, Supplement*, **1**:24–62, 1962. Reprinted in *Lasers: A Collection of Reprints with Commentary*, edited by J. Weber, Gordon & Breach, New York, 1969.

48. P. G. Bergmann. *Introduction to the Theory of Relativity*. Dover, New York, 1976.

49. P. R. Berman, editor. *Cavity Quantum Electrodynamics*, volume 2 of *Advances in Atomic, Molecular, and Optical Physics. Supplement*. Academic Press, Boston, 1994.

50. M. V. Berry. Evanescent and real waves in quantum billiards and Gaussian beams. *Journal of Physics A*, **27**:L391–L398, 1994.

51. H. A. Bethe. Theory of diffraction by small holes. *Physical Review*, **66**(7–8):163–182, October 1944.

52. H. A. Biagioni. *A Nonlinear Theory of Generalized Functions*, volume 1421 of *Lecture Notes in Mathematics*. Springer-Verlag, Berlin, 1990.

53. I. Bialynicki-Birula. On the wave function of the photon. *Acta Physica Polonica*, **86**:97–116, 1994. Online at *http://www.cft.edu.pl/~birula/publ.html*.

54. I. Bialynicki-Birula. Photon wave function. In E. Wolf, editor, *Progress in Optics*, volume XXXVI, Chap. 5, pp. 245–294. Elsevier Science, Amsterdam, 1996. This is a nice review of the literature on the wavefunction of the photon as well as an expanded version of the theory proposed in [53]. Online at *http://www.cft.edu.pl/~birula/publ.html*.

55. I. Bialynicki-Birula. The photon wave function. In L. Mandel and E. Wolf, editors, *Coherence and Quantum Optics VII*, pp. 313–323. Plenum, New York, 1996.

56. I. Bialynicki-Birula. Exponential localization of photons. *Physical Review Letters*, **80**(24):5247–5250, June 15, 1998.

57. N. D. Birrell and P. C. W. Davies. *Quantum Fields in Curved Space*. Cambridge University Press, Cambridge, 1982.

58. J. D. Bjorken and S. D. Drell. *Relativistic Quantum Mechanics. International Series in Pure and Applied Physics*. McGraw-Hill, New York, 1964.

59. F. Black and M. Scholes. The pricing of options and corporate liabilities. *Journal of Political Economy*, **81**(3):637–654, May–June 1973.

60. F. Bloch and A. Siegert. Magnetic resonance for nonrotating fields. *Physical Review*, **57**(6):522–527, March 15, 1940.

61. N. Bloembergen and R. V. Pound. Radiation damping in magnetic resonance experiments. *Physical Review*, **95**(1):8–12, July 1, 1954.

62. F. J. Bloore. Identity of commutator and Poisson bracket. *Journal of Physics A*, **6**:L7–L8, February 1973.

63. S. L. Boersma. A maritime analogy of the Casimir effect. *American Journal of Physics*, **64**(5):539–541, May 1996.

64. C. F. Bohren. Broken telephone games in the history of science. *American Journal of Physics*, **69**(12):1221–1222, December 2001.

65. A. D. Boozer, A. Boca, J. R. Buck, J. McKeever, and H. J. Kimble. Comparison of theory and experiment for a one-atom laser in a regime of strong coupling. *Physical Review A*, **70**(2):023814, August 25, 2004. Online at the Los Alamos e-print archive site (*http://www.arxiv.org*) under *quant-ph/0309133*.

66. M. Bordag, U. Mohideen, and V. M. Mostepanenko. New developments in the Casimir effect. *Physics Reports*, **353**:1–205, 2001.

67. A. M. Bork. Vectors versus quaternions: The letters in *Nature*. *American Journal of Physics*, **34**:202–211, 1966.

68. M. Born and P. Jordan. Zur Quantenmechanik. *Zeitschrift für Physik*, **34**:858–889, 1925.

69. M. Born and E. Wolf. *Principles of Optics*. Cambridge University Press, Cambridge, 7th (expanded) edition, 1999.

70. C. J. Bouwkamp and H. B. G. Casimir. On multipole expansions in the theory of electromagnetic radiation. *Physica*, **20**:539–554, 1954.

71. D. Bouwmeester, A. Ekert, and A. Zeilinger, editors. *The Physics of Quantum Information: Quantum Cryptography, Quantum Teleportation, Quantum Computation*. Springer-Verlag, Berlin, 2000.

72. H. Bremmer. Propagation of electromagnetic waves. In S. Flügge, editor, *Electric Fields and Waves*, volume XVI of *Handbuch der Physik*, pp. 423–639. Springer-Verlag, Berlin, 1958.

73. H.-J. Briegel and B.-G. Englert. Quantum optical master equations: The use of damping bases. *Physical Review A*, **47**(4):3311–3329, April 1993.

74. H.-J. Briegel, G. M. Meyer, and B.-G. Englert. Pump operator for lasers with multi-level excitation. *Europhysics Letters*, **33**(7):515–520, March 1, 1996.

75. O. L. Brill and B. Goodman. Causality in the Coulomb gauge. *American Journal of Physics*, **35**(9):832, 1967.

76. W. E. Brittin. A note on the quantization of dissipative systems. *Physical Review*, **77**(3):396–397, February 1, 1950.

77. T. J. I'A. Bromwich. *An Introduction to the Theory of Infinite Series*. Macmillan, London, 2nd edition, 1947.

78. S. A. Brown and B. J. Dalton. Generalized quasi mode theory of macroscopic canonical quantization in cavity quantum electrodynamics and quantum optics: I. Theory. *Journal of Modern Optics*, **48**(4):597–618, March 20, 2001.

79. S. A. Brown and B. J. Dalton. Generalized quasi mode theory of macroscopic canonical quantization in cavity quantum electrodynamics and quantum optics: II. Application to reflection and refraction at a dielectric interface. *Journal of Modern Optics*, **48**(4):639–669, March 20, 2001.

80. S. A. Brown and B. J. Dalton. Field quantization, photons and non-Hermitean modes. *Journal of Modern Optics*, **49**(7):1009–1041, June 15, 2002.

81. M. Brune, J. M. Raimond, P. Goy, L. Davidovich, and S. Haroche. Realization of a two-photon maser oscillator. *Physical Review Letters*, **59**(17):1899–1902, October 26, 1987.

82. M. Brune, F. Schmidt-Kaler, A. Maali, J. Dreyer, E. Hagley, J. M. Raimond, and S. Haroche. Quantum Rabi oscillation: A direct test of field quantization in a cavity. *Physical Review Letters*, **76**(11):1800–1803, March 11, 1996.

83. E. Buks and M. L. Roukes. Casimir force changes sign. *Nature*, **419**:119–120, September 12, 2002.

84. V. S. Buldyrev and E. E. Fradkin. Integral equations for open resonators. *Optics and Spectroscopy USSR*, **17**(4):314–320, 1964. This is an English translation of the Russian original: V. S. Buldyrev and E. E. Fradkin, *Optika i Spectroskopiya* **17**(4):583–596, 1964.

85. P. Bush. The time energy uncertainty relation. In J. G. Muga, R. Sala Mayato, and I. L. Egusquiza, editors, *Time in Quantum Mechanics*, pp. 69–98. Springer-Verlag, Berlin, 2002. Also available at the LANL site (*http://xxx.lanl.gov/*) under the eprint reference *quant-ph/0105049*.

86. M. G. Calkin, D. Kiang, and Y. Nogami. Proper treatment of the delta function potential in the one-dimensional Dirac equation. *American Journal of Physics*, **55**(8):737–739, August 1987.

87. L. Campbell. *The Life of James Clerk Maxwell with a Selection from His Correspondence and Occasional Writings and a Sketch of His Contributions to Science*. Macmillan, London, August 1882. Chapter VIII contains a reprint of Maxwell's essay on analogies in nature (Analogies: Are there real analogies in nature?) read on February 1856 to the "Apostles" club at Cambridge. Maxwell saw William Thomson's (Lord Kelvin) work on the heat flow analogy to electrostatics as the invention of an altogether new method for discovery in physics, which he (Maxwell) called the method of "physical analogy." Campbell's book is also available online at *http://www.hrshowcase.com/maxwell/preface.html*.

88. H. Carmichael. *An Open Systems Approach to Quantum Optics*, volume 18 of *Lecture Notes in Physics*. Springer-Verlag, Berlin, 1993.

89. P. Carruthers and M. M. Nieto. Phase and angle variables in quantum mechanics. *Reviews of Modern Physics*, **40**(2):411–440, April 1968.

90. W. B. Case. The field oscillator approach to classical electrodynamics. *American Journal of Physics*, **68**(9):800–811, September 2000.

91. H. B. G. Casimir. On the attraction between two perfectly conducting plates. *Proceedings of the Koninklijke Nederlandse Akademie van Wetenschappen*, **51**(7): 793–795, 1948.

92. H. B. G. Casimir. *Haphazard Reality*. Harper & Row, New York, 1983.

93. P. C. Causseé. *L'Album du marin*. Charpentier, Nantes, 1836.

94. C. M. Caves. Amplitude and phase in quantum optics. In F. Haake, L. M. Narducci, and D. Walls, editors, *Coherence, Cooperation and Fluctuations*, pp. 192–205. Cambridge University Press, Cambridge, 1986.

95. A. Cayley. On the application of quaternions to the theory of rotation. *Philosophical Magazine*, **33**:196–200, February 1848. Online at *http://home.pipeline.com/~hbaker1/quaternion/*.

96. A. Cayley. On certain results relating to quaternions. *Philosophical Magazine*, **26**(3):141–145, February 1945.

97. H. B. Chan, V. A. Aksyuk, R. N. Kleiman, D. J. Bishop, and F. Capasso. Quantum mechanical actuation of microelectromechanical systems by the casimir force. *Science*, **291**:1941–1944, March 9, 2001.

98. K. W. Chan, C. K. Law, and J. H. Eberly. Localized single-photon wave functions in free space. *Physical Review Letters*, **88**(10):100402, March 11, 2002.

99. K. W. Chan, C. K. Law, and J. H. Eberly. Quantum entanglement in photon–atom scattering. *Physical Review A*, **68**(2):022110, August 2003.

100. R. R. Chance, A. Prock, and R. Silbey. Molecular fluorescence and energy transfer near interfaces. *Advances in Chemical Physics*, 37:1–65, 1978.

101. K. Chandrasekharan, editor. *Hermann Weyl Gesammelte Abhandlungen*, volumes I–IV. Springer-Verlag, Berlin, 1968.

102. Y.-J. Cheng and A. E. Siegman. Generalized radiation-field quantization method and the Petermann excess-noise factor. *Physical Review A*, **68**(4):043808, October 2003.

103. E. S. C. Ching, P. T. Leung, A. Maassen van den Brink, W. M. Suen, S. S. Tong, and K. Young. Quasinormal-mode expansion for waves in open systems. *Reviews of Modern Physics*, **70**(4):1545–1554, October 1998.

104. J. C. K. Chou. Quaternion kinematic and dynamic differential equations. *IEEE Transactions on Robotics and Automation*, **8**:53–64, 1992.

105. J. C. K. Chou and M. Kamel. Finding the position and orientation of a sensor on a robot manipulator using quaternions. *International Journal of Robotics Research*, **10**:240–253, 1991.

106. S. Clark. *The Only Fools and Horses Story*. BBC, London, 1998. For more on this other del, see *http://www.comedyzone.beeb.com/ofah*.

107. M. H. Cohen, L. M. Falicov, and J. C. Phillips. Superconductive tunneling. *Physical Review Letters*, **8**(8):316–318, April 15, 1962.

108. C. Cohen-Tannoudji, J. Dupont-Roc, and G. Grynberg. *Photons and Atoms: Introduction to Quantum Electrodynamics*. Wiley-Interscience, New York, 1989.

109. M. J. Collett and C. W. Gardiner. Squeezing of intracavity and traveling-wave light fields produced in parametric amplification. *Physical Review A*, **30**(3):1386–1391, September 1984.

110. J. F. Colombeau. A multiplication of distributions. *Journal of Mathematical Analysis and Applications*, **94**:96–115, 1983.

111. J. F. Colombeau. The elastoplastic shock problem as an example of the resolution of ambiguities in the multiplication of distributions. *Journal of Mathematical Physics*, **30**(10):2273–2279, October 1989.

112. J. F. Colombeau. Multiplication of distributions. *Bulletin (New Series) of the American Mathematical Society*, **23**(2):251–268, October 1990.

113. J. F. Colombeau. *Multiplication of Distributions: A Tool in Mathematics, Numerical Engineering and Theoretical Physics*, volume 1532 of *Lecture Notes in Mathematics*. Springer-Verlag, Berlin, 1992.

114. J. F. Colombeau and A. Y. Le Roux. Multiplications of distributions in elasticity and hydrodynamics. *Journal of Mathematical Physics*, **29**(2):315–319, February 1988.

115. J. Combrisson, A. Honig, and C. H. Townes. Utilisation de la résonance de spins électroniques pour réaliser un oscillateur ou un amplificateur en hyperfréquences. *Comptes Rendus Hebdomadaires des Seances de l'Academie des Sciences*, **242**(20):2451–2453, 1956.

116. G. Compagno, R. Passante, and F. Persico. *Atom-Field Interactions and Dressed Atoms*, volume 13 of *Cambridge Studies in Modern Optics*. Cambridge University Press, Cambridge, 1995.

117. A. H. Compton. A quantum theory of the scattering of x-rays by light elements. *Physical Review*, **21**(5):483–502, May 1923. Online at http://www.aip.org/history/gap/PDF/compton.pdf.

118. A. W. Conway. On the application of quaternions to some recent developments of electrical theory. *Proceedings of the Royal Irish Academy A*, **29**:1, 1911.

119. R. J. Cook. Photon dynamics. *Physical Review A*, **25**(4):2164–2167, April 1982. In this paper, Cook rediscovered the Landau–Peierls wavefunction [387].

120. R. J. Cook. Lorentz covariance of photon dynamics. *Physical Review A*, **26**(5):2754–2760, November 1982. Continuation of [119], where Cook shows that his photon wavefunction formalism (basically, the Landau–Peierls wavefunction [387]) does not depend on the choice of Lorentz frame.

121. R. J. Cook and P. W. Milonni. Quantum theory of an atom near partially reflecting walls. *Physical Review A*, **35**(12):5081–5087, June 15, 1987.

122. P. H. Cootner, editor. *The Random Character of the Stock Market*. MIT Press, Cambridge, MA, 1964.

123. E. M. Corson. Second quantization and representation theory. *Physical Review*, **70**(9–10):728–748, November 1 and 15, 1946.

124. E. M. Corson. Reply to "a note on the paper 'Second quantization and representation theory'." *Physical Review*, **72**(8):737, October 15, 1947.

125. J. P. Costella, B. H. J. McKellar, and A. A. Rawlinson. Classical antiparticles. *American Journal of Physics*, **65**(9):835–841, September 1997.

126. F. A. B. Coutinho, Y. Nogami, and J. F. Perez. Generalized point interactions in one-dimensional quantum mechanics. *Journal of Physics A*, **30**:3937–3945, 1997.

127. F. A. B. Coutinho, Y. Nogami, and F. M. Toyama. Logarithmic perturbation expansion for the Dirac equation in one dimension: Application to the polarizability calculation. *American Journal of Physics*, **65**(8):788–794, August 1997.

128. R. E. Crandall. Photon mass experiment. *American Journal of Physics*, **51**(8): 698–702, August 1983.

129. M. Cray, M.-L. Shih, and P. W. Milonni. Stimulated emission, absorption, and interference. *American Journal of Physics*, **50**(11):1016–1021, November 1982.

130. F. W. Cummings. Stimulated emission of radiation in a single mode. *Physical Review*, **140**(4A):A1051–A1056, November 15, 1965.

131. J. Dalibard, J. Dupont-Roc, and C. Cohen-Tannoudji. Vacuum fluctuations and radiation reaction: Identification of their respective contributions. *Journal de Physique*, **43**(11):1617–1638, November 1982.

132. B. J. Dalton, S. M. Barnett, and B. M. Garraway. Theory of pseudomodes in quantum optical processes. *Physical Review A*, **64**(5):053813, November 2001.

133. B. J. Dalton, S. M. Barnett, and P. L. Knight. A quantum scattering theory approach to quantum-optical measurements. *Journal of Modern Optics*, **46**(7):1107–1121, June 15, 1999.

134. B. J. Dalton, S. M. Barnett, and P. L. Knight. Quasi mode theory of macroscopic canonical quantization in quantum optics and cavity quantum electrodynamics. *Journal of Modern Optics*, **46**(9):1315–1341, July 20, 1999.

135. B. J. Dalton, S. M. Barnett, and P. L. Knight. Macroscopic canonical quantization in quantum optics: Properties of quasi mode annihilation and creation operators. *Journal of Modern Optics*, **46**(10):1495–1502, August 15, 1999.

136. B. J. Dalton, S. M. Barnett, and P. L. Knight. Quasi mode theory of the beam splitter: A quantum scattering theory approach. *Journal of Modern Optics*, **46**(10):1559–1577, August 15, 1999.

137. B. J. Dalton and B. M. Garraway. Quasimode theory of quantum optical processes in photonic band gap materials. *Journal of Modern Optics*, **49**(5–6):947–958, April 15, 2002.

138. B. J. Dalton and B. M. Garraway. Non-Markovian decay of a three-level cascade atom in a structured reservoir. *Physical Review A*, **68**(3):033809, September 2003.

139. B. J. Dalton and P. L. Knight. The standard model in cavity quantum electrodynamics: I. General features of mode functions for a Fabry-Pérot cavity. *Journal of Modern Optics*, **46**(13):1817–1837, November 10, 1999.

140. B. J. Dalton and P. L. Knight. The standard model in cavity quantum electrodynamics: II. Coupling constants and atom field interaction. *Journal of Modern Optics*, **46**(13):1839–1868, November 10, 1999.

141. D. A. R. Dalvit and P. A. Maia Neto. Decoherence via the dynamical Casimir effect. *Physical Review Letters*, **84**(5):798–801, January 31, 2000.

142. A. Das Gupta. Disentanglement formulas: An alternative derivation and some applications to squeezed coherent states. *American Journal of Physics*, **64**(11): 1422–1427, November 1996.

143. K. Datta. The quantum Poisson bracket and transformation theory in quantum mechanics: Dirac's early work in quantum theory. *Resonance*, **8**(8):75–85, August 2003. Online at *http://www.ias.ac.in/resonance/August2003/August2003Contents.html*.

144. L. Davidovich. On the Weisskopf–Wigner approximation in atomic physics. Ph.D. dissertation. University of Rochester, Rochester, NY, 1975.

145. W. R. Davies, M. T. Chu, P. Dolan, J. R. McConnell, L. K. Norris, E. Ortiz, R. J. Plemmons, D. Ridgeway, B. K. P. Scaife, W. J. Stewart, J. W. York Jr., W. O. Doggett, B. M. Gellai, A. A. Gsponer, and C. A. Prioli, editors. *Cornelius Lanczos: Collected Published Papers with Commentaries*. North Carolina State University, Raleigh, NC, 1998.

146. C. A. Deavours. The quaternion calculus. *American Mathematical Monthly*, **80**(9):995–1008, November 1973.

147. V. DeGiorgio and M. O. Scully. Analogy between the laser threshold region and a second-order phase transition. *Physical Review A*, **2**(4):1170–1177, October 1970.

148. S. R. DeGroot and J. Vlieger. Derivation of Maxwell's equations: The statistical theory of the macroscopic equations. *Physica*, **31**:254–268, 1965.

149. R. P. Dellavalle, E. J. Hester, L. F. Heilig, A. L. Drake, J. W. Kuntzman, M. Graber, and L. M. Schilling. Going, going, gone: Lost Internet references. *Science*, **302**:787–788, October 31, 2003.

150. H. H. Denman. On linear friction in Lagrange's equation. *American Journal of Physics*, **34**(12):1147–1149, 1966.

151. I. H. Deutsch and J. C. Garrison. Paraxial quantum propagation. *Physical Review A*, **43**(5):2498–2513, March 1991. Here it is shown that the photon wavefunction and the photon position operator exist in the paraxial approximation.

152. R. H. Dicke. Molecular amplification and generation systems and methods. Patent 2,851,652, U.S. Patent Office, September 9, 1958. The application dates from May 21, 1956. The full text is given in J. Weber, *Lasers: A Collection of Reprints with Commentary*, Gordon & Breach, New York, 1968, pp. 645–664.

153. F. Dietrich, J. Krause, G. Rempe, M. O. Scully, and H. Walther. Laser experiments with single atoms as a test of basic physics. *IEEE Journal of Quantum Electronics*, **24**(7):1314–1319, July 1988.

154. B. Dietz and U. Smilansky. A scattering approach to the quantization of billiards. *Chaos*, **3**(4):581–589, 1993.

155. L. Diósi. Comment on "Nonclassical states: An observable criterion." *Physical Review Letters*, **85**(13):2841, September 25, 2000. See also [617].

156. P. A. M. Dirac. The quantum theory of emission and absorption of radiation. *Proceedings of the Royal Society of London, Series A*, **114**:243–265, 1927. Reprinted in [544], pp. 1–23.

157. P. A. M. Dirac. Quantized sigularities in the electromagnetic field. *Proceedings of the Royal Society of London, Series A*, **133**:60, 1931.

158. P. A. M. Dirac. *The Development of Quantum Mechanics*, Chap. 1, pp. 1–21. Wiley, New York, 1978. A public lecture given by Dirac at the School of Physics, University of New South Wales, Kensington, Sydney, Australia, August 25, 1975.

159. P. A. M. Dirac. *Principles of Quantum Mechanics*. Oxford University Press, Oxford, 4th edition, 1982.

160. P. A. M. Dirac. *Lectures on Quantum Mechanics*. Dover, New York, 2001.

161. V. V. Dodonov. Nonstationary Casimir effect and analytical solutions for quantum fields in cavities with moving boundaries. In M. W. Evans, editor, *Modern Nonlinear Optics*, volume 119, part 1 of *Advances in Chemical Physics*, pp. 309–394. Wiley, New York, 2001.

162. E. Doron and U. Smilansky. Chaotic spectroscopy. *Chaos*, **2**(1):117–124, 1992.

163. E. Doron and U. Smilansky. Semiclassical quantization of chaotic billiards: A scattering theory approach. *Nonlinearity*, **5**:1055–1084, 1992.

164. K. H. Drexhage. Interaction of light with monomolecular dye layers. *Progress in Optics*, 12:163–232, 1974.

165. S. M. Dutra. Generation and Detection of Fields in Cavity Quantum Electrodynamics. Ph.D. dissertation. Imperial College of Science, Technology and Medicine (University of London), London, 1995.

166. S. M. Dutra and K. Furuya. Macroscopic averages in QED in material media. *Physical Review A*, **55**(5):3832–3841, May 1997.

167. S. M. Dutra and K. Furuya. Relation between Huttner–Barnett QED in dielectrics and classical electrodynamics: Determining the dielectric permittivity. *Physical Review A*, **57**(4):3050–3058, April 1998.

168. S. M. Dutra and P. L. Knight. Atomic probe for quantum states of the electromagnetic field. *Physical Review A*, **49**(2):1506–1508, February 1994.

169. S. M. Dutra and P. L. Knight. Spontaneous emission in a planar Fabry–Pérot microcavity. *Physical Review A*, **53**(5):3587–3605, May 1996.

170. S. M. Dutra, P. L. Knight, and H. Moya-Cessa. Discriminating field mixtures from macroscopic superpositions. *Physical Review A*, **48**(4):3168–3173, October 1993.

171. S. M. Dutra and G. Nienhuis. Derivation of a Hamiltonian for photon decay in a cavity. *Journal of Optics B*, **2**(5):584–588, October 2000.

172. S. M. Dutra and G. Nienhuis. Quantized mode of a leaky cavity. *Physical Review A*, **62**(6):063805, November 2000.

173. S. M. Dutra and G. Nienhuis. What is a quantized mode of a leaky cavity? In M. Orszag and J. C. Retamal, editors, *Modern Challenges in Quantum Optics*, volume 575 of *Lecture Notes in Physics*, pp. 338–354. Springer-Verlag, Berlin, 2001.

174. S. M. Dutra, J. P. Woerdman, J. Visser, and G. Nienhuis. Route toward the ideal thresholdless laser. *Physical Review A*, **65**(3):033824, March 2002.

175. F. J. Dyson. Ground-state energy of a finite system of charged particles. *Journal of Mathematical Physics*, **8**:1538–1545, 1967.

176. F. J. Dyson. Feynman's proof of the Maxwell equations. *American Journal of Physics*, **58**(3):209–211, March 1990.

177. F. J. Dyson and A. Lenard. Stability of matter: I. *Journal of Mathematical Physics*, **8**:423–434, 1967.

178. J. H. Eberly, N. B. Narozhny, and J. J. Sanchez-Mondragon. Periodic spontaneous collapse and revival in a simple quantum model. *Physical Review Letters*, **44**(20):1323–1326, May 19, 1980.

179. J. D. Edmonds Jr. Quaternion quantum theory: New physics or number mysticism? *American Journal of Physics*, **42**(3):220–223, March 1974.

180. J. D. Edmonds Jr. Maxwell's eight equations as one quaternion equation. *American Journal of Physics*, **46**(4):430–431, April 1978.

181. A. Einstein. Über die von der molekular-kinetischen Theorie der Wärme geforderte Bewegung von in ruhenden Flüssigkeiten suspendieren Teilchen. *Annalen der Physik*, **17**:549, 1905. An English translation can be found in [183] pp. 1–18. German original online at *http://www.wiley-vch.de/berlin/journals/adp/549_560.pdf*.

182. A. Einstein. Quantentheorie der Strahlung. *Physikalische Zeitschrift*, **18**:121–128, 1917. Online at *http://www.rpi.edu/~schubert/More reprints/*.

183. A. Einstein. *Investigations on the Theory of the Brownian Movement*. Dover, New York, 1956. Edited by R. Fürth and translated by A. D. Cowper.

184. A. Einstein. *Albert Einstein: Autobiographical Notes*. Open Court Publishing, Peru, IL, 1996.

185. A. Ekert and P. L. Knight. Entangled quantum systems and the Schmidt decomposition. *American Journal of Physics*, **63**(5):415–423, May 1995.

186. L. Eyges. *The Classical Electromagnetic Field*. Dover, New York, 1972.

187. Ch. Fabry and A. Pérot. Théorie et applications d'une nouvelle méthode de spectroscopie interférentielle. *Annales de Chimie et de Physique*, **7**:115–144, 1899.

188. Z. Fang, N. Nagaosa, K. S. Takahashi, A. Asamitsu, R. Mathieu, T. Ogasawara, H. Yamada, M. Kawasaki, Y. Tokura, and K. Terakura. The anomalous Hall effect and magnetic monopoles in momentum space. *Science*, **302**:92–95, October 3, 2003.

189. U. Fano. Atomic theory of electromagnetic interactions in dense materials. *Physical Review*, **103**(5):1202–1218, September 1, 1956.

190. U. Fano. Effects of configuration interaction on intensities and phase shifts. *Physical Review*, **124**(6):1866–1878, December 15, 1961.

191. G. Feher, J. P. Gordon, E. Buehler, E. A. Gere, and C. D. Thurmond. Spontaneous emission of radiation from an electron spin system. *Physical Review*, **109**(1):221–222, January 1, 1958.

192. M. S. Feld and K. An. The single-atom laser. *Scientific American*, **279**(1):40–45, July 1998.

193. X.-P. Feng and K. Ujihara. Quantum theory of spontaneous emission from a two-level atom in a one-dimensional cavity with output coupling. *IEEE Journal of Quantum Electronics*, **25**(11):2332–2343, November 1989.

194. X.-P. Feng and K. Ujihara. Quantum theory of spontaneous emission in a one-dimensional optical cavity with two-side output coupling. *Physical Review A*, **41**(5):2668–2676, March 1, 1990.

195. E. Fermi. Quantum theory of radiation. *Reviews of Modern Physics*, **4**:87–132, January 1932.

196. A. L. Fetter and J. D. Waleska. *Quantum Theory of Many-Particles Systems.* McGraw-Hill, New York, 1971.

197. R. P. Feynman. A relativistic cut-off for classical electrodynamics. *Physical Review*, **74**(8):939–946, October 15, 1948.

198. R. P. Feynman. The theory of positrons. *Physical Review*, **76**(6):749–759, September 15, 1949.

199. R. P. Feynman. Space-time approach to quantum electrodynamics. *Physical Review*, **76**(6):769–789, September 15, 1949.

200. R. P. Feynman. An operator calculus having applications in quantum electrodynamics. *Physical Review*, **84**(1):108–128, 1951. See also [369].

201. R. P. Feynman. *QED: The Strange Theory of Light and Matter.* Princeton University Press, Princeton, NJ, 1985. These lectures have also been filmed and the streamed video is available online at *http://www.vega.org.uk/series/lectures/feynman/*.

202. R. P. Feynman. The development of the space-time view of quantum electrodynamics. In *Nobel Lectures: Physics 1963–1970*, pp. 155–178. World Scientific, Singapore, 1998. Online at *http://www.nobel.se*.

203. R. P. Feynman. *Quantum Electrodynamics. Advanced Book Classics.* Perseus Publishing, Boulder, CO, January 1998.

204. R. P. Feynman and A. R. Hibbs. *Quantum Mechanics and Path Integrals. International Series in Pure and Applied Physics.* McGraw-Hill, New York, 1965.

205. R. P. Feynman, R. B. Leighton, and M. Sands. *The Feynman Lectures on Physics*, volume 1. Addison-Wesley, Reading, MA, 1963.

206. R. P. Feynman, R. B. Leighton, and M. Sands. *The Feynman Lectures on Physics*, volume 2. Addison-Wesley, Reading, MA, 1963.

207. R. P. Feynman and S. Weinberg. *Elementary Particles and the Laws of Physics: The 1986 Dirac Memorial Lectures.* Cambridge University Press, Cambridge, 1987.

208. H. Fielding. *The History of Tom Jones. Penguin Popular Classics.* Penguin Books, London, 1994. This classic of eighteenth-century English comic novels is also available online at the Project Gutenberg Web site, *http://gutenberg.net*.

209. P. Filipowicz, J. Javanainen, and P. Meystre. Theory of a microscopic maser. *Physical Review A*, **34**(4):3077–3087, October 1986.

210. D. Finkelstein, J. M. Jauch, S. Schiminovich, and D. Speiser. Foundations of quaternion quantum mechanics. *Journal of Mathematical Physics*, **2**:207–220, 1962.

211. D. Finkelstein, J. M. Jauch, S. Schiminovich, and D. Speiser. Principle of general *Q* covariance. *Journal of Mathematical Physics*, **4**:788–796, 1963.

212. D. Finkelstein, J. M. Jauch, and D. Speiser. Quaternionic representations of compact groups. *Journal of Mathematical Physics*, **4**:136–140, 1963.

213. G. F. FitzGerald. On the possibility of originating wave disturbances in the ether by means of electric forces. *Scientific Transactions of the Dublin Royal Society*, 1883. Reprinted in *The Scientific Writings of the Late George Francis FitzGerald* edited by J. Larmor, Longmans, Green, London 1902, p. 92.

214. M. Fleischhauer. Relation between the n-atom laser and the one-atom laser. *Physical Review A*, **50**(3):2773–2776, September 1994.

215. V. Fock. Über die invariante Form der Wellen- und der Bewegungsgleichungen für einen geladenen Massenpunkt. *Zeitschrift für Physik*, **39**:226–232, October 2, 1926.

216. V. Fock. Konfigurationsraum und zweite Quantelung. *Zeitschrift für Physik*, **75**(9–10):622–647, 1932. See also the erratum [217].

217. V. Fock. Erratum: Konfigurationsraum und Zweite Quantelung. *Zeitschrift für Physik*, **76**(11–12):852, 1932.

218. V. Fock. A note on the paper "Second quantization and representation theory." *Physical Review*, **72**(8):737, October 15, 1947. See also Corson's reply [124].

219. E. J. S. Fonseca, C. H. Monken, and S. Pádua. Measurement of the de Broglie wavelength of a multiphoton wave packet. *Physical Review Letters*, **82**(14):2868–2871, April 5, 1999.

220. A. G. Fox and T. Li. Resonant modes in an optical maser. *Proceedings of the IRE*, **48**:1904–1905, November 1960.

221. A. G. Fox and T. Li. Resonant modes in a maser interferometer. *Bell System Technical Journal*, **40**:453–488, March 1961. Reprinted in *Lasers: A Collection of Reprints with Commentary,* edited by J. Weber, Gordon & Breach, New York, 1969.

222. C. P. Frahm. Some novel delta-function identities. *American Journal of Physics*, **51**(9):826–829, September 1983.

223. C. W. Gardiner. *Quantum Noise*, volume 56 of *Springer Series in Synergetics*. Springer-Verlag, Berlin, 1991.

224. C. W. Gardiner. *Handbook of Stochastic Methods for Physics, Chemistry, and the Natural Sciences. Springer Series in Synergetics.* Springer-Verlag, Berlin, 2nd edition, 2002.

225. C. W. Gardiner and M. J. Collett. Input and output in damped quantum systems: Quantum stochastic differential equations and the master equation. *Physical Review A*, **31**(6):3761–3774, June 1985.

226. B. M. Garraway. Nonperturbative decay of an atomic system in a cavity. *Physical Review A*, **55**(3):2290–2303, March 1997.

227. B. M. Garraway and P. L. Knight. Cavity modified quantum beats. *Physical Review A*, **54**(4):3592–3602, October 1996.

228. J. Gea-Banacloche, N. Lu, L. M. Pedrotti, S. Prasad, M. O. Scully, and K. Wódkiewicz. Treatment of the spectrum of squeezing based on the modes of the universe: I. Theory and a physical picture. *Physical Review A*, **41**(1):369–380, January 1990.

229. I. M. Gel'fand and G. E. Shilov. *Generalized Functions*, volume 1. Academic Press, New York, 1964. pp. 34–38.

230. J. W. Gibbs. *Elements of Vector Analysis.* Privately printed, New Haven, 1884. Online at *http://home.pipeline.com/~hbaker1/quaternion/*.

231. F. Gieres. Mathematical surprises and Dirac's formalism in quantum mechanics. *Reports on Progress in Physics*, **63**(12):1893–1931, 2000.

232. D. T. Gillespie. The mathematics of Brownian motion and Johnson noise. *American Journal of Physics*, **64**(3):225–240, March 1996.

233. V. L. Ginzburg. *Theoretical Physics and Astrophysics.* Gordon & Breach, New York, 1989. A translation of V. L. Ginzburg, *Teoreticheskaya Fizika i Astrofizika,* Nauka, Moscow, 1987.

234. C. Ginzel, H.-J. Briegel, U. Martini, B.-G. Englert, and A. Schenzle. Quantum optical master equations: The one-atom laser. *Physical Review A*, **48**(1):732–738, July 1993.

235. R. Glauber. Optical coherence and photon statistics. In C. de Witt, A. Bandin, and C. Cohen-Tannoudji, editors, *Quantum Optics and Electronics: Les Houches Lectures, 1964.* Gordon & Breach, New York, 1965.

236. R. J. Glauber. Photon correlations. *Physical Review Letters*, **10**(3):84–86, February 1, 1963.

237. R. J. Glauber and M. Lewenstein. Quantum optics in dielectric media. *Physical Review A*, **43**(1):467–491, January 1, 1991.

238. A. S. Goldhaber and M. M. Nieto. Terrestrial and extraterrestrial limits on the photon mass. *Reviews of Modern Physics*, **43**(3):277–296, July 1971.

239. A. S. Goldhaber and W. P. Trower. Resource letter mm-1: Magnetic monopoles. *American Journal of Physics*, **58**(5):429–439, May 1990.

240. H. Goldstein. *Classical Mechanics*. Addison-Wesley, Reading, MA, 2nd edition, 1980.

241. R. H. Good Jr. Particle aspect of the electromagnetic field equations. *Physical Review*, **105**(6):1914–1919, March 15, 1957.

242. J. P. Gordon, H. J. Zeiger, and C. H. Townes. Molecular microwave oscillator and new hyperfine structure in the microwave spectrum of NH_3. *Physical Review*, **95**(1):282–284, July 1, 1954.

243. K. Gottfried. *Quantum Mechanics*. Addison-Wesley, Redwood City, CA, 1989.

244. W. Gough. A quaternion expression for the quantum mechanical probability and current densities. *European Journal of Physics*, **10**:188–193, 1989.

245. P. Goy, J. M. Raimond, M. Gross, and S. Haroche. Observation of cavity-enhanced single-atom spontaneous emission. *Physical Review Letters*, **50**(24): 1903–1906, June 13, 1983.

246. R. Graham. Squeezing and frequency changes in harmonic oscillations. *Journal of Modern Optics*, **34**(6–7):873–879, 1987.

247. Ph. Grangier and J.-Ph. Poizat. A simple quantum picture for the Petermann excess noise factor. *European Physical Journal D*, **1**:97–104, January 1998.

248. Ph. Grangier and J.-Ph. Poizat. Quantum derivation of the excess noise factor in lasers with non-orthogonal eigenmodes. *European Physical Journal D*, **7**:99–105, August 1999.

249. R. D. Gray and D. H. Kobe. Gauge-invariant canonical quantisation of the electromagnetic field and duality transformations. *Journal of Physics A*, **15**:3145–3155, 1982.

250. H. S. Green and E. Wolf. A scalar representation of electromagnetic fields. *Proceedings of the Royal Society of London, Series A*, **66**:1129–1137, 1953.

251. D. M. Greenberger. A critique of the major approaches to damping in quantum theory. *Journal of Mathematical Physics*, **20**(5):762–770, May 1979.

252. D. Griffiths and S. Walborn. Dirac deltas and discontinuous functions. *American Journal of Physics*, **67**(5):446–447, May 1999.

253. D. J. Griffiths. Boundary conditions at the derivative of a delta function. *Journal of Physics A*, **26**:2265–2267, 1993.

254. R. B. Griffiths. Consistent resolution of some relativistic quantum paradoxes. *Physical Review A*, **66**(6):062101, December 2002.

255. M. Gross and S. Haroche. Superradiance: An essay on the theory of collective spontaneous emission. *Physics Report*, **93**(5):301–396, 1982.

256. A. Gsponer and J.-P. Hurni. The physical heritage of Sir W. R. Hamilton. In *Proceedings of the Conference: The Mathematical Heritage of Sir William Rowan Hamilton, Commemorating the Sesquicentennial of the Invention of Quaternions*. Trinity College, Dublin, August 17–20, 1993. Online at *arXiv:math-ph/0201058 v2 28 Nov 2002*.

257. A. Gsponer and J.-P. Hurni. Comment on formulating and generalizing Dirac's, Proca's, and Maxwell's equations with biquaternions or Clifford numbers. *Foundations of Physics Letters*, **14**(1):77–85, 2001.

258. G. Hackenbroich, C. Viviescas, and F. Haake. Field quantization for chaotic resonators with overlapping modes. *Physical Review Letters*, **89**(8):083902, August 19, 2002.

259. G. Hackenbroich, C. Viviescas, and F. Haake. Quantum statistics of overlapping modes in open resonators. *Physical Review A*, **68**(6):063805, December 2003.

260. H. Haken. *Light*, volumes 1 and 2. North-Holland, Amsterdam, 1981.

261. H. Haken. *Laser Theory*. Springer-Verlag, Berlin, 1983.

262. H. Haken and H. C. Wolf. *Molecular Physics and Elements of Quantum Chemistry: Introduction to Experiments and Theory*. Springer-Verlag, Berlin, 1995.

263. W. R. Hamilton. On a new species of imaginary quantities connected with a theory of quaternions. *Proceedings of the Royal Irish Academy*, **2**:424–434, 1844. Online at *http://www.maths.tcd.ie/pub/HistMath/People/Hamilton/*.

264. W. R. Hamilton. On quaternions; or a new system of imaginaries in algebra. *Philosofical Magazine, 3rd Series*, **25**:489–495, 1844. Copy of a letter from Sir William R. Hamilton to John T. Graves, Esq. describing how Hamilton got the idea of quanternions. Online at *http://www.maths.tcd.ie/pub/HistMath/People/Hamilton/*.

265. W. R. Hamilton. *Lectures on Quaternions*. Hodges and Smith, Dublin, 1853. Online at *http://historical.library.cornell.edu/*.

266. W. R. Hamilton. On the geometrical interpretation of some results obtained by calculation with biquaternions. *Proceedings of the Royal Irish Academy*, **5**:388–390, 1853. Online at *http://www.maths.tcd.ie/pub/HistMath/People/Hamilton/*.

267. W. R. Hamilton. Letter to his son Rev. Archibald H. Hamilton (August 5, 1865), reproduced in *Life of Sir William Rowan Hamilton*, by Robert P. Graves, volume

II, Chap. XXVII, Arno Press, New York, 1975. Online at *http://www.maths.tcd.ie/pub/HistMath/People/Hamilton/*.

268. C. L. Hammer and R. H. Good Jr. Wave equation for a massless particle with arbitrary spin. *Physical Review*, **108**(3):882–886, November 1, 1957.

269. C. L. Hammer and R. H. Good Jr. Quantization process for massless particles. *Physical Review*, **111**(1):342–345, July 1, 1958.

270. D. Han and Y. S. Kim. Little group for photons and gauge transformation. *American Journal of Physics*, **49**(4):348–351, April 1981.

271. T. Hård af Segerstad. Spoon files, chapter 54 "Go in and Tell It." Online at *http://www.konstvet.uu.se/onlinerebus/*.

272. G. P. Harnwell. *Principles of Electricity and Magnetism*. McGraw-Hill, New York, 1938.

273. S. Haroche. Rydberg atoms and radiation in a resonant cavity (a simple system to test basic quantum optics effects). In G. Grynberg and R. Stora, editors, *Les Houches 1982, Session XXXVIII: New Trends in Atomic Physics*, pp. 193–309. Elsevier Science, Amsterdam, 1984.

274. S. Haroche. Cavity quantum electrodynamics. In J. Dalibard, J. M. Raimond, and J. Zinn-Justin, editors, *Les Houches 1990, Session LIII: Fundamental Systems in Quantum Optics*, Chap. 13, pp. 767–940. North-Holland, Amsterdam, 1992. These are the lecture notes of the course on cavity QED given by Serge Haroche at the LIII Les Houches' summer school in 1990.

275. D. L. Hart and T. A. B. Kennedy. Quantum noise properties of the laser: Depleted pump regime. *Physical Review A*, **44**(7):4572–4577, October 1, 1991.

276. H. A. Haus. *Electromagnetic Noise and Quantum Optical Measurements*. Springer-Verlag, Berlin, 2000.

277. H. A. Haus and J. L. Pan. Photon spin and the paraxial wave equation. *American Journal of Physics*, **61**(9):818–821, September 1993.

278. M. Hawton. Photon position operator with commuting components. *Physical Review A*, **59**(2):954–959, February 1999.

279. M. Hawton. Photon wave functions in a localized coordinate space basis. *Physical Review A*, **59**(5):3223–3227, May 1999.

280. M. Hawton and W. E. Baylis. Photon position operators and localized bases. *Physical Review A*, **64**(1):012101, July 2001.

281. O. Heaviside. *Electrical Papers*, volumes 1 and 2. Macmillan, London, 1892.

282. O. Heaviside. On operators in physical mathematics: I. *Proceedings of the Royal Society of London*, **52**(320):504–529, February 2, 1893.

283. O. Heaviside. On operators in physical mathematics: II. *Proceedings of the Royal Society of London*, **54**(326):105–143, June 15, 1893.

284. O. Heaviside. *The Electrician*, **48**:861–863, March 21, 1902. Reprinted in [285] volume 3, pp 135–143. This particular citation can also be found in [461].

285. O. Heaviside. *Electromagnetic Theory*, volumes 1–3. Chelsea, New York, 1971.

286. G. C. Hegerfeldt. Remark on causality and particle localization. *Physical Review D*, **10**(10):3320–3321, November 15, 1974.

287. G. C. Hegerfeldt. Violation of causality in relativistic quantum theory? *Physical Review Letters*, **54**(22):2395–2398, June 3, 1985.

288. G. C. Hegerfeldt and S. N. M. Ruijsenaars. Remarks on causality, localization, and spreading of wave packets. *Physical Review D*, **22**(2):377–384, July 15, 1980.

289. R. W. Hellwarth and P. Nouchi. Focused one-cycle electromagnetic pulses. *Physical Review E*, **54**(1):889–895, July 1996.

290. J. Hendry. Monopoles before Dirac. *Studies in History and Philosophy of Science*, **14**(1):81–87, March 1983.

291. D. Hestenes. Vectors, spinors, and complex numbers in classical and quantum physics. *American Journal of Physics*, **39**:1013–1027, 1971.

292. S. Hilgenfeldt, S. Grossmann, and D. Lohse. A simple explanation of light emission in sonoluminescence. *Nature*, **398**:402–405, April 1, 1999.

293. E. A. Hinds. Cavity quantum electrodynamics. *Advances in Atomic Molecular and Optical Physics*, **28**:237–289, 1991.

294. K. C. Ho, P. T. Leung, A. Maassen van den Brink, and K. Young. Second quantization of open systems using quasinormal modes. *Physical Review E*, **58**(3):2965–2978, September 1998.

295. S. A. Hojman and L. C. Shepley. No Lagrangian? No quantization! *Journal of Mathematical Physics*, **32**(1):142–146, January 1991.

296. J. J. Hope. Theory of input and output of atoms from an atomic trap. *Physical Review A*, **55**(4):R2531–R2534, April 1997.

297. J. J. Hopfield. A quantum-mechanical theory of the contribution of excitons to the complex dielectric constant of crystals. Ph.D. dissertation. Cornell University, Ithaca, NY, June 1958.

298. J. J. Hopfield. Theory of the contribution of excitons to the complex dielectric constant of crystals. *Physical Review*, **112**(5):1555–1567, December 1, 1958.

299. P. Horak, K. M. Gheri, and H. Ritsch. Quantum dynamics of a single-atom cascade laser. *Physical Review A*, **51**(4):3257–3266, April 1995.

300. R. F. Hoskins. *Delta Functions: An Introduction to Generalised Functions.* Ellis Horwood, Chichester, West Sussex, England, 1999.

301. Y.-S. Huang and C.-L. Lin. A systematic method to determine the Lagrangian directly from the equations of motion. *American Journal of Physics*, **70**(7):741–743, July 2002.

302. R. J. Hughes. On Feynman's proof of the Maxwell equations. *American Journal of Physics*, **60**(4):301–306, April 1992.

303. R. G. Hulet, E. S. Hilfer, and D. Kleppner. Inhibited spontaneous emission by a Rydberg atom. *Physical Review Letters*, **55**(20):2137–2140, November 11, 1985.

304. J. C. Hull. *Options, Futures, and Other Derivatives.* Prentice Hall, Upper Saddle River, NY, 5th edition, 2002.

305. V. Hushwater. Repulsive Casimir force as a result of vacuum radiation pressure. *American Journal of Physics*, **65**(5):381–384, May 1997.

306. B. Huttner and S. M. Barnett. Dispersion and loss in a Hopfield dielectric. *Europhysics Letters*, **18**(6):487–492, March 15, 1992.

307. B. Huttner and S. M. Barnett. Quantization of the electromagnetic field in dielectrics. *Physical Review A*, **46**(7):4306–4322, October 1, 1992.

308. B. Huttner, J. J. Baumberg, and S. M. Barnett. Canonical quantization of light in a linear dielectric. *Europhysics Letters*, **16**(7):177–182, September 7, 1991.

309. D. Iannuzzi and F. Capasso. Comment on "repulsive Casimir forces". *Physical Review Letters*, **91**(2):029101, July 11, 2003.

310. T. Inagaki. Quantum-mechanical approach to a free photon. *Physical Review A*, **49**(4):2839–2843, April 1994. Reformulation of [119, 120] in terms of conventional quantum mechanics. See also the follow-up paper [311].

311. T. Inagaki. Physical meaning of the photon wave function. *Physical Review A*, **57**(3):2204–2207, March 1998. See also the Inagaki's previous paper on the same subject [311].

312. K. Itô. On stochastic differential equations. *Memoirs, American Mathematical Society*, **4**:1–51, 1951.

313. C. Itzykson and J.-B. Zuber. *Quantum Field Theory.* McGraw-Hill, New York, 1980.

314. J. D. Jackson. *Classical Electrodynamics.* Wiley, New York, 2nd edition, 1975.

315. J. D. Jackson and L. B. Okun. Historical roots of gauge invariance. *Reviews of Modern Physics*, **73**(3):663–680, July 2001.

316. M.-T. Jaekel, A. Lambrecht, and S. Reynaud. Quantum vacuum, inertia and gravitation. *New Astronomy Reviews*, **46**:727–739, 2002.

317. M. T. Jaekel and S. Reynaud. Casimir force between partially transmitting mirrors. *Journal de Physique I*, **1**(10):1395–1409, October 1991. Online at *http://www.edpsciences.org*.

318. R. C. Jaklevic, John Lambe, A. H. Silver, and J. E. Mercereau. Quantum interference effects in Josephson tunneling. *Physical Review Letters*, **12**(7):159–160, February 17, 1964.

319. J. Janszky and Y. Y. Yushin. Squeezing via frequency jump. *Optics Communications*, **59**(2):151–54, August 15, 1986.

320. J. M. Jauch. Gauge invariance as a consequence of Galilei-invariance for elementary particles. *Helvetica Physica Acta*, **37**:284–292, 1964.

321. J. M. Jauch. *Foundations of Quantum Mechanics*. Addison-Wesley, Reading, MA, 1968.

322. J. M. Jauch and C. Piron. Generalized localizability. *Helvetica Physica Acta*, **40**:559–570, 1967.

323. E. T. Jaynes and F. W. Cummings. Comparison of quantum and semiclassical radiation theory with application to the beam maser. *Proceedings of the IEEE*, **51**:89–109, 1963.

324. J. Jeffers and S. M. Barnett. Nonlocal effects in propagation through absorbing media. *Physical Review A*, **47**(4):3291–3298, April 1993.

325. J. L. Jiménez, L. de la Peña, and T. A. Brody. Zero-point term in cavity radiation. *American Journal of Physics*, **48**(10):840–846, October 1980.

326. S. John. Strong localization of photons in certain disordered dielectric superlattices. *Physical Review Letters*, **58**(23):2486–2489, June 8, 1987.

327. W. W. Johnson. Origin of the name "quaternions." *The Analyst*, **7**(2):52, March 1880.

328. B. Jones, S. Ghose, J. P. Clemens, P. R. Rice, and L. M. Pedrotti. Photon statistics of a single atom laser. *Physical Review A*, **60**(4):3267–3275, October 1999.

329. D. S. Jones. *Generalised Functions*, p. 71. McGraw-Hill, New York, 1966.

330. H. F. Jones. *Groups, Representations and Physics*. Adam Hilger, Bristol, UK, 1990.

331. T. F. Jordan. Why $-i\nabla$ is the momentum? *American Journal of Physics*, **43**(12):1089–1093, December 1975.

332. T. F. Jordan. Identification of the velocity operator for an irreducible unitary representation of the Poincaré group. *Journal of Mathematical Physics*, **18**(4):608–610, April 1977.

333. T. F. Jordan. Identification of the velocity operator in an irreducible unitary representation of the Poincaré group for imaginary mass or zero mass and variable helicity. *Journal of Mathematical Physics*, **19**(1):247–248, January 1978.

334. T. F. Jordan. Simple proof of no position operator for quanta with zero mass and nonzero helicity. *Journal of Mathematical Physics*, **19**(6):1382–1385, June 1978.

335. T. F. Jordan. Simple derivation of the Newton–Wigner position operator. *Journal of Mathematical Physics*, **21**(8), August 1980.

336. T. F. Jordan. What happens to the position, spin, and parity when the mass goes to zero. *Journal of Mathematical Physics*, **23**(12):2524–2528, December 1982.

337. T. F. Jordan. Simplified quantum oscillator solution. *American Journal of Physics*, **69**(10):1082–1083, October 2001.

338. H. Kaiser, E. A. George, and S. A. Werner. Neutron interferometric search for quaternions in quantum mechanics. *Physical Review A*, **29**(4):2276–2279, April 1984.

339. F. Kanai. On the quantization of the dissipative systems. *Progress in Theoretical Physics*, **3**:440–442, 1948.

340. I. Karatzas and S. E. Shreve. *Brownian Motion and Stochastic Calculus*, volume 113 of *Graduate Texts in Mathematics*. Springer-Verlag, Berlin, 2nd edition, 1997.

341. A. P. Kazantsev and G. I. Surdutovich. The quantum theory of the laser. *Progress in Quantum Electronics*, **3**(3):231–288, 1974.

342. O. Keller. Propagator picture of the spatial confinement of quantized light emitted from an atom. *Physical Review A*, **58**(5):3407–3425, November 1998.

343. O. Keller. Attached and radiated electromagnetic fields of an electric point dipole. *Journal of the Optical Society of America B*, **16**(5):835–847, May 1999.

344. O. Keller. Relation between spatial confinement of light and optical tunneling. *Physical Review A*, **60**(2):1652–1671, August 1999.

345. O. Keller. Near-field optics: The nightmare of the photon. *Journal of Chemical Physics*, **112**(18):7856–7863, May 8, 2000.

346. O. Keller. Space-time description of photon emission from an atom. *Physical Review A*, **62**(2):022111, August 2000. Keller takes the view that to fully understand the problem of photon localization we must consider the interaction with matter. See also [342].

347. O. Keller. Optical tunneling: A fingerprint of the lack of photon localizability. *Journal of the Optical Society of America B*, **18**(2):206–217, February 2001.

348. O. Keller. Nonlinear optics in the near-field zone of atoms. *Journal of Nonlinear Optical Physics and Materials*, **11**(3):275–301, 2002.

349. O. Keller. Polychromatic photons: Wave mechanics and spatial localization. In S. G. Pandalai, editor, *Recent Research Developments in Optics*, volume 2, part II, pp. 375–409. Research Signpost, Kerala, India, 2002.

350. O. Keller. Near-field optics and quantum optics: An assignation arranged by four kinds of photons. *Journal of Microscopy*, **209**(3):272–276, March 2003.

351. E. H. Kennard. Zur Quantenmechanik einfacher Bewegungstypen. *Zeitschrift für Physik*, **44**:326–352, 1927.

352. O. Kenneth, I. Klich, A. Mann, and M. Revzen. Repulsive Casimir forces. *Physical Review Letters*, **89**(3):33001, July 15, 2002.

353. O. Kenneth, I. Klich, A. Mann, and M. Revzen. Reply to comment on "Repulsive Casimir forces." *Physical Review Letters*, **91**(2):029102, July 11, 2003.

354. A. M. Khazanov, G. A. Koganov, and E. P. Gordov. Macroscopic squeezing in three-level laser. *Physical Review A*, **42**(5):3065–3069, September 1, 1990.

355. S. Ya. Kilin and T. B. Karlovich. Single-atom laser: Coherent and nonclassical effects in the regime of a strong atom–field correlation. *Journal of Experimental and Theoretical Physics*, **95**(5):805–819, November 2002.

356. H. J. Kimble. Comment on "Deterministic single-photon source for distributed quantum networking." *Physical Review Letters*, **90**(24):249801, June 20, 2003.

357. R. P. Kirshner. Throwing light on dark energy. *Science*, **300**:1914–1918, June 20, 2003.

358. P. E. Kloeden and E. Platen. *Numerical Solution of Stochastic Differential Equations*, volume 23 of *Applications of Mathematics: Stochastic Modelling and Applied Probability*. Springer-Verlag, Berlin, 3rd edition, 1999.

359. P. L. Knight and L. Allen. *Concepts of Quantum Optics*. Pergamon Press, Oxford, 1983.

360. P. L. Knight and P. W. Milonni. Long-time deviations from exponential decay in atomic spontaneous emission theory. *Physics Letters A*, **56**(4):275–278, April 5, 1976.

361. J. Knoester and S. Mukamel. Intermolecular forces, spontaneous emission, and superradiance in a dielectric medium: Polariton-mediated interactions. *Physical Review A*, **40**(12):7065–7080, December 15, 1989.

362. L. Knöll, W. Vogel, and D.-G. Welsch. Resonators in quantum optics: A first-principles approach. *Physical Review A*, **43**(1):543–553, January 1, 1991.

363. D. H. Kobe. Second quantization as a graded Hilbert space representation. *American Journal of Physics*, **34**:1150–1159, 1966.

364. D. H. Kobe. A relativistic Schrödinger-like equation for a photon and its second quantization. *Foundations of Physics*, **29**(8):1203–1231, 1999.

365. D. H. Kobe and G. Reali. Lagrangians for dissipative systems. *American Journal of Physics*, **54**(11):997–999, November 1986.

366. D. H. Kobe and A. L. Smirl. Gauge invariant formulation of the interaction of electromagnetic radiation and matter. *American Journal of Physics*, **46**(6):624–633, June 1978.

367. I. Yu. Kobzarev and L. B. Okun. On the photon mass. *Soviet Physics Uspekhi*, **11**(3):338–341, November–December 1968. English translation of the Russian original *Uspekhi Fizicheskikh Nauk*, 95, 131–137, May, 1968.

368. M. D. Kostin. On the Schrödinger–Langevin equation. *Journal of Chemical Physics*, **57**(9):3589–3591, November 1, 1972.

369. C. Kourkoumelis and S. Nettel. Operator functionals and path integrals. *American Journal of Physics*, **45**(1):26–30, January 1977.

370. A. M. Krall. *Linear Methods of Applied Analysis*. Addison-Wesley, Reading, MA, 1973, pp. 494–499.

371. B. K. Ksendal and B. Oksendal. *Stochastic Differential Equations: An Introduction with Applications*. Springer-Verlag, Berlin, 6th edition, 2003.

372. A. Kuhn, M. Hennrich, and G. Rempe. Deterministic single-photon source for distributed quantum networking. *Physical Review Letters*, **89**(6):067901, August 5, 2002.

373. A. Kuhn, M. Hennrich, and G. Rempe. Kuhn, Hennrich, and Rempe reply to comment on "Deterministic single-photon source for distributed quantum networking." *Physical Review Letters*, **90**(24):249802, June 20, 2003.

374. H. M. Lai, P. T. Leung, K. Young, P. W. Barber, and S. C. Hill. Time-independent perturbation for leaking electromagnetic modes in open systems with application to resonances in microdroplets. *Physical Review A*, **41**(9):5187–5198, May 1, 1990.

375. W. E. Lamb Jr., W. P. Schleich, M. O. Scully, and C. H. Townes. Laser physics: Quantum controversy in action. *Reviews of Modern Physics*, **71**(2):S263–S273, Centenary 1999.

376. W. E. Lamb Jr. and M. O. Scully. The photoelectric effect without photons. In *Polarisation, Matière, et Rayonnement: Volume Jubilaire en l'Honneur d'Alfred Kastler*, pp. 363–369. La Société Française de Physique, Presses Universitaires de France, Paris, 1969.

377. S. K. Lamoreaux. Resource letter cf-1: Casimir force. *American Journal of Physics*, **67**(10):850–861, October 1999.

378. C. Lamprecht, M. K. Olsen, M. Collett, and H. Ritsch. Excess-noise-enhanced parametric down conversion. *Physical Review A*, **64**(3):033811, September 2001.

379. C. Lamprecht, M. K. Olsen, P. D. Drummond, and H. Ritsch. Positive-P and Wigner representations for quantum-optical systems with nonorthogonal modes. *Physical Review A*, **65**(5):053813, May 2002.

380. C. Lamprecht and H. Ritsch. Quantized atom-field dynamics in unstable cavities. *Physical Review Letters*, **82**(19):3787–3790, May 10, 1999.

381. C. Lamprecht and H. Ritsch. Unexpected role of excess noise in spontaneous emission. *Physical Review A*, **65**(2):023803, February 2002.

382. C. Lamprecht and H. Ritsch. Theory of excess noise in unstable resonator lasers. *Physical Review A*, **66**(5):053808, November 2002.

383. C. Lanczos. *Die Funktionentheoretischen Beziehungen der Maxwellschen Aethergleichungen: Ein Beitrag zur Relativitäts- und Elektronentheorie*. Verlagsbuchhandlung Josef Németh, Budapest, 1919. This is Lanczos's Ph.D. dissertation (from Szeged University). It was reprinted in [145], volume VI, pp. A-1 to A-82.

384. C. Lanczos. Die Tensoranalytischen Beziehungen der Diracschen Gleichung. *Zeitschrift für Physik*, **57**:447–493, 1929. Reprinted and translated in [145], volume III, pp. 2–1133 to 2–1225.

385. C. Lanczos. The Poisson bracket. In A. Salam and E. P. Wigner, editors, *Aspects of Quantum Theory*, pp. 169–178. Cambridge University Press, Cambridge, 1972.

386. C. Lanczos. *The Variational Principles of Mechanics*. Dover, New York, 4th edition, 1986.

387. L. Landau and R. Peierls. Quantenelektrodynamik im Konfigurationsraum. *Zeitschrift für Physik*, **62**:188–200, 1930. An English translation of this paper appears in *Selected Scientific Papers of Sir Rudolf Peierls with Commentary*, edited by

R. H. Dalitz and Sir Rudolf Peierls, pp. 71–82, Imperial College Press, London, 1997.

388. L. Landau and R. Peierls. Erweiterung des Unbestimmtheitsprinzips für die relativistische Quantentheorie. *Zeitschrift für Physik*, **69**:56, 1931. English translation in *Collected Works of L. D. Landau*, pp. 83–94.

389. L. D. Landau and E. M. Lifshitz. *The Classical Theory of Fields*, volume 2 of *Course of Theoretical Physics*. Pergamon Press, Oxford, 2nd revised edition, 1962.

390. R. Lang and M. O. Scully. Fluctuations in mode locked "single-mode" laser oscillation. *Optics Communications*, **9**(4):331–337, December 1973.

391. R. Lang, M. O. Scully, and W. E. Lamb Jr. Why is the laser line so narrow? A theory of single-quasimode laser operation. *Physical Review A*, **7**(5):1788–1797, May 1973.

392. P. Langevin. Sur la théorie du mouvement brownien. *Comptes Rendus de l'Academie des Sciences (Paris)*, **146**:530–533, 1908. An English translation with comments appeared in 1997 in the *American Journal of Physics* [398].

393. I. R. Lapidus. Relativistic one-dimensional hydrogen atom. *American Journal of Physics*, **51**(11):1036–1038, November 1983.

394. C. K. Law. Effective Hamiltonian for the radiation in a cavity with a moving mirror and a time-varying dielectric medium. *Physical Review A*, **49**(1):433–437, January 1994.

395. C. K. Law. Interaction between a moving mirror and radiation pressure: A Hamiltonian formulation. *Physical Review A*, **51**(3):2537–2541, March 1995.

396. M. Lax. *Symmetry Principles in Solid State and Molecular Physics*. Dover, New York, 2001.

397. T.-Y. Lee. Quantum phases of electric and magnetic dipoles as special cases of the Aharonov–Bohm phase. *Physical Review A*, **64**(3):032107, September 2001.

398. D. S. Lemons and A. Gythiel. Paul Langevin's 1908 paper "On the theory of Brownian motion" [Sur la théorie du mouvement brownien, *Comptes Rendus de l'Academie des Sciences (Paris)* **146**, 530–533, 1908]. *American Journal of Physics*, **65**(11):1079–1081, November 1997.

399. U. Leonhardt. *Measuring the Quantum State of Light*, volume 17 of *Cambridge Series in Modern Optics*. Cambridge University Press, Cambridge, 1997.

400. P. T. Leung, S. Y. Liu, S. S. Tong, and K. Young. Time-independent perturbation theory for quasinormal modes in leaky optical cavities. *Physical Review A*, **49**(4):3068–3073, April 1994.

401. P. T. Leung, S. Y. Liu, and K. Young. Completeness and orthogonality of quasi-normal modes in leaky optical cavities. *Physical Review A*, **49**(4):3057–3067, April 1994.

402. P. T. Leung, S. Y. Liu, and K. Young. Completeness and time-independent perturbation of the quasinormal modes of an absorptive and leaky cavity. *Physical Review A*, **49**(5):3982–3989, May 1994.

403. P. T. Leung, W. M. Suen, C. P. Sun, and K. Young. Waves in open systems via a biorthogonal basis. *Physical Review E*, **57**(5):6101–6104, May 1998.

404. P. T. Leung and K. Young. Time-independent perturbation theory for quasinormal-mode solutions in quantum mechanics. *Physical Review A*, **44**(5):3152–3161, September 1, 1991.

405. B. G. Levi. Cavity lases when occupied, on average, by less than one atom. *Physics Today*, **48**(2):20–21, February 1995.

406. J.-M. Lévy-Leblond. Minimal electromagnetic coupling as a consequence of Lorentz invariance. *Annals of Physics*, **57**:481–495, 1970.

407. G. N. Lewis. The conservation of photons. *Nature*, **118**(2981):874–875, December 18, 1926. Online at *http://www.nobeliefs.com/photon.htm*.

408. M. Ley and R. Loudon. Quantum theory of high-resolution length measurement with a Fabry–Pérot interferometer. *Journal of Modern Optics*, **34**(2):227–255, February 1987.

409. E. H. Lieb. The stability of matter. *Reviews of Modern Physics*, **48**(4):553–569, October 1976.

410. M. Ligare and S. Becker. Simple soluble models of quantum damping applied to cavity quantum electrodynamics. *American Journal of Physics*, **63**(9):788–796, September 1995.

411. G. Lindblad. On the generators of quantum dynamical semigroups. *Communications in Mathematical Physics*, **48**:119–130, 1976.

412. M. Löffler, G. M. Meyer, and H. Walther. Spectral properties of the one-atom laser. *Physical Review A*, **55**(5):3923–3930, May 1997.

413. F. London. Die Theorie von Weyl und die Quantenmechanik. *Naturwissenschaften*, **15**:187, February 25, 1927.

414. F. London. Quantenmechanische Deutung der Theorie von Weyl. *Zeitschrift für Physik*, **42**:375–389, April 14, 1927.

415. H. A. Lorentz. *The Theory of Electrons*. Dover, New York, 1952. Originally published 1915.

416. R. Loudon and P. L. Knight. Squeezed light. *Journal of Modern Optics*, **34**:709–759, 1987.

417. L. A. Lugiato, M. O. Scully, and H. Walther. Connection between microscopic and macroscopic maser theory. *Physical Review A*, **36**(2):740–743, July 15, 1987.

418. J. Luo, L.-C. Tu, Z.-K. Hu, and E.-J. Luan. New experimental limit on the photon rest mass with a rotating torsion balance. *Physical Review Letters*, **90**(8):081801, February 28, 2003. This experiment sets an upper limit of 1.2×10^{-51} g to the photon's rest mass. The previous record was 2×10^{-50} g due to R. Lakes, *Physical Review Letters*, **80**(9):1826–1829, March 2, 1998.

419. J. Lützen. Heaviside's operational calculus and the attempts to rigorise it. *Archive for History of Exact Sciences*, **21**:161–200, 1979.

420. J. Lützen. *The Prehistory of the Theory of Distributions*, volume 7 of *Studies in the History of Mathematics and Physical Sciences*. Springer-Verlag, Berlin, 1982.

421. A. MacDonald and Q. Niu. New twist for magnetic monopoles. *Physics World*, **17**(1):18–19, January 2004.

422. E. Mach. *The Principles of Physical Optics: An Historical and Philosophical Treatment*. Dover, New York, 1961. Reprint of the 1926 edition of this book.

423. J. Maddox. The wonders of the microlaser. *Nature*, **373**(6510):101, January 12, 1995.

424. T. H. Maiman. Stimulated optical radiation in ruby. *Nature*, **187**(4736):493–494, August 6, 1960. Online at *http://www.coseti.org/maiman.htm*.

425. M. A. Maize and C. A. Burkholder. Electric polarizability and the solution of an inhomogeneous differential equation. *American Journal of Physics*, **63**(3):244–247, March 1995.

426. M. A. Maize, S. Paulson, and A. D'Avanti. Electric polarizability of a relativistic particle. *American Journal of Physics*, **65**(9):888–892, September 1997.

427. V. Majerník. Quaternionic formulation of the classical fields. *Advances in Applied Clifford Algebras*, **9**(1):119–130, 1999.

428. L. Mandel. Configuration-space photon number operators in quantum optics. *Physical Review*, **144**(4):1071–1077, April 29, 1966.

429. L. Mandel and E. Wolf. *Optical Coherence and Quantum Optics*. Cambridge University Press, Cambridge, 1995.

430. P. D. Mannheim. The physics behind path integrals in quantum mechanics. *American Journal of Physics*, **51**(4):328–334, April 1983.

431. M. A. M. Marte and S. Stenholm. Paraxial light and atom optics: The optical Schrödinger equation and beyond. *Physical Review A*, **56**(4):2940–2953, October 1997.

432. L. Martín-Moreno, F. J. García-Vidal, H. J. Lezec, K. M. Pellerin, T. Thio, J. B. Pendry, and T. W. Ebbesen. Theory of extraordinary optical transmission through subwavelength hole arrays. *Physical Review Letters*, **86**(6):1114–1117, February 5, 2001.

433. J. Mathews and R. L. Walker. *Mathematical Methods of Physics*. Addison-Wesley, Menlo Park, CA, 2nd edition, 1973.

434. J. C. Maxwell. On physical lines of force: I. The theory of molecular vortices applied to magnetic phenomena. *Philosophical Magazine, Fourth Series*, **21**(139):161–175, March 1861.

435. J. C. Maxwell. On physical lines of force: II. The theory of molecular vortices applied to electric currents. *Philosophical Magazine, Fourth Series*, **21**(140–141):281–291, 338–348, April–May 1861. See also the first part of this paper [434].

436. J. McKeever, A. Boca, A. D. Boozer, J. R. Buck, and H. J. Kimble. Experimental realization of a one-atom laser in the regime of strong coupling. *Nature*, **425**(6955):268–271, September 18, 2003.

437. J. McKeever, A. Boca, A. D. Boozer, J. R. Buck, and H. J. Kimble. Supplementary material to experimental realization of a one-atom laser in the regime of strong coupling. *Nature*, **425**(6955):268–271, September 18, 2003.

438. J. McKeever, A. Boca, A. D. Boozer, R. Miller, J. R. Buck, A. Kuzmich, and H. J. Kimble. Deterministic generation of single photons from one atom trapped in a cavity. *Science*, **303**(#5666):1992–1994, March 26, 2004.

439. B. H. J. McKellar and G. J. Stephenson Jr. Relativistic quarks in one-dimensional periodic structures. *Physical Review C*, **35**(6):2262–2271, June 1987.

440. D. Meschede, H. Walther, and G. Müller. One-atom maser. *Physical Review*, **54**(6):551–554, February 11, 1985.

441. A. Messiah. *Quantum Mechanics*. Dover, New York, 2000.

442. H. J. Metcalf and P. van der Straten. *Laser Cooling and Trapping*. Springer-Verlag, Berlin, 1999.

443. G. M. Meyer and H.-J. Briegel. Pump-operator treatment of the ion-trap laser. *Physical Review A*, **58**(4):3210–3220, October 1998.

444. G. M. Meyer, H.-J. Briegel, and H. Walther. Ion-trap laser. *Europhysics Letters*, **37**(5):317–322, February 10, 1997.

445. A. A. Michelson and E. W. Morley. On the relative motion of the Earth and the luminiferous ether. *American Journal of Science, Third Series*, **34**(203):333–345, 1887. Online at *http://www.aip.org/history/gap/PDF/michelson.pdf*.

446. R. Mignani, E. Recami, and M. Baldo. About a Dirac-like equation for the photon according to Ettore Majorana. *Lettere al Nuovo Cimento*, **11**:568–572, 1974. This paper describes Majorana's unpublished notes on a photon wavefunction.

447. A. I. Miller. *Early Quantum Electrodynamics: A Source Book*. Cambridge University Press, Cambridge, 1994.

448. P. W. Milonni. Casimir forces without the vacuum radiation field. *Physical Review A*, **25**(3):1315–1327, March 1982.

449. P. W. Milonni. Why spontaneous emission? *American Journal of Physics*, **52**(4):340–343, April 1984.

450. P. W. Milonni. Different ways of looking at the electromagnetic vacuum. *Physica Scripta*, **T21**:102–109, 1988.

451. P. W. Milonni. *The Quantum Vacuum*. Academic Press, Boston, 1994.

452. P. W. Milonni. Field quantization and radiative processes in dispersive dielectric media. *Journal of Modern Optics*, **42**(10):1991–2004, October 1995.

453. P. W. Milonni, J. R. Ackerhalt, and W. A. Smith. Interpretation of radiative corrections in spontaneous emission. *Physical Review Letters*, **31**(15):958–960, October 8, 1973.

454. P. W. Milonni and J. H. Eberly. *Lasers*. Wiley, New York, 1988.

455. P. W. Milonni and M.-L. Shih. Casimir forces. *Contemporary Physics*, **33**(5): 313–322, 1992.

456. P. W. Milonni and W. A. Smith. Radiation reaction and vacuum fluctuations in spontaneous emission. *Physical Review A*, **11**(3):814–824, March 1975.

457. G. T. Moore. Quantum theory of the electromagnetic field in a variable-length one-dimensional cavity. *Journal of Mathematical Physics*, **11**(9):2679–2691, September 1970.

458. P. M. Morse and H. Feshbach. *Methods of Theoretical Physics*, volumes 1 and 2. McGraw-Hill, New York, 1953.

459. Y. Mu and C. M. Savage. One-atom lasers. *Physical Review A*, **46**(9):5944–5954, November 1, 1992.

460. P. J. Nahin. *The Science of Radio*. Springer-Verlag, New York, 2001.

461. P. J. Nahin. *Oliver Heaviside*. Johns Hopkins University Press, Baltimore, 2002.

462. T. J. Nelson and R. H. Good Jr. Second-quantization process for particles with any spin and with internal symmetry. *Reviews of Modern Physics*, **40**(3):508–522, July 1968.

463. T. J. Nelson and R. H. Good Jr. Description of massless particles. *Physical Review*, **179**(5):1445–1449, March 25, 1969.

464. T. D. Newton and E. P. Wigner. Localized states for elementary systems. *Reviews of Modern Physics*, **21**(3):400–406, July 1949.

465. M. M. Nieto. The discovery of squeezed states in 1927. In D. Han, J. Janzky, Y. S. Kim, and V. I. Man'ko, editors, *Proceedings of the 5th International Conference on Squeezed States and Uncertainty Relations*. NASA, 1998. Available in the Los Alamos e-print archive under *quant-ph/9708012*.

466. A. Nisbet. Hertzian electromagnetic potentials and associated gauge transformations. *Proceedings of the Royal Society of London, Series A*, **231**:250–263, 1955.

467. G. Nogues, A. Rauschenbeutel, S. Osnaghi, M. Brune, J. M. Raimond, and S. Haroche. Seeing a single photon without destroying it. *Nature*, **400**(6741):239–242, July 15, 1999.

468. H. M. Nussenzveig. *Causality and Dispersion Relations*. Academic Press, New York, 1972.

469. M. Ohtsu. *Coherent Quantum Optics and Technology*. Advances in Optoelectronics. Kluwer Academic, Dordrecht, the Netherlands, 1992.

470. S. Okubo. Does the equation of motion determine commutation relations? *Physical Review D*, **22**(4):919–923, August 15, 1980.

471. R. Omnès. Localization of relativistic particles. *Journal of Mathematical Physics*, **38**(2):708–715, February 1997.

472. J. R. Oppenheimer. Note on light quanta and the electromagnetic field. *Physical Review*, **38**(4):725–746, August 15, 1931.

473. M. F. M. Osborne. *The Stock Market and Finance from a Physicist's Viewpoint*. Crossgar Press, Minneapolis, MN, 2001.

474. A. Pais. *Inward Bound: Of Matter and Forces in the Physical World*. Oxford University Press, Oxford, 1988.

475. L. Parker. Particle creation in expanding universes. *Physical Review Letters*, **21**(8):562–564, August 19, 1968.

476. W. Paul and J. Baschnagel. *Stochastic Processes: From Physics to Finance*. Springer-Verlag, Berlin, 1999.

477. W. Pauli. Einige die Quantenmechanik betreffenden Erkundigungsfragen. *Zeitschrift für Physik*, **80**:573–586, 1933.

478. W. Pauli. The connection between spin and statistics. *Physical Review*, **58**(8): 716–722, October 15, 1940.

479. W. Pauli. Matter. In H. Muschel, editor, *Man's Right to Knowledge, International Symposium Presented in Honor of the Two-Hundredth Anniversary of Columbia University, 1754–1954*, pp. 10–18, New York, 1954.

480. W. Pauli. *General Principles of Quantum Mechanics*. Springer-Verlag, Berlin, 1980.

481. W. Pauli. *Electrodynamics*, volume 1 of *Pauli Lectures on Physics*. Dover, Mineola, NY, 2000.

482. S. Pearlmutter et al. Measurements of ω and λ from 42 high-redshift supernovae. *Astrophysical Journal*, **517**:565–586, June 1999.

483. P. J. E. Peebles and B. Ratra. The cosmological constant and dark energy. *Reviews of Modern Physics*, **75**(3):559–606, April 2003.

484. D. T. Pegg and S. M. Barnett. Tutorial review: Quantum optical phase. *Journal of Modern Optics*, **44**(2):225–264, February 1, 1997.

485. R. Peierls. *Surprises in Theoretical Physics*. *Princeton Series In Physics*. Princeton University Press, Princeton, NJ, 1979.

486. R. E. Peierls, A. Salam, P. T. Matthews, and G. Feldman. A survey of field theory. *Reports on Progress in Physics*, **18**:423–477, 1955.

487. T. Pellizzari and H. Ritsch. Photon statistics of the three-level one-atom laser. *Journal of Modern Optics*, **41**(3):609–623, March 1994.

488. T. Pellizzari and H. Ritsch. Preparation of stationary Fock states in a one-atom Raman laser. *Physical Review Letters*, **72**(25):3973–3976, June 20, 1994.

489. A. Peres. Proposed test for complex versus quaternion quantum theory. *Physical Review Letters*, **42**(11):683–686, March 12, 1979.

490. V. Peřinová, A. Lukš, and J. Peřina. *Phase in Optics*, volume 15 of *World Scientific Series in Contemporary Chemical Physics*. World Scientific, Singapore, 1998.

491. K. Petermann. Calculated spontaneous emission factor for double-heterostructure injection lasers with gain-induced waveguiding. *IEEE Journal of Quantum Electronics*, **15**(7):566–570, July 1979.

492. E. R. Pike and S. Sarkar. *The Quantum Theory of Radiation*. *International Series of Monographs on Physics*. Clarendon Press, Oxford, 1995.

493. R. Podolny. *Something Called Nothing— Physical Vacuum: What Is It?* Mir, Moscow, 1986. English translation of the Russian original, 1983.

494. A. M. Polyakov. Particle spectrum in the quantum field theory. *JETP Letters*, **20**:194–195, 1974. English translation of the Russian original published in *Zhurnal Eksperimental'noi i Teoreticheskoi Fiziki Pis'ma*, **20**, 430–433, 1974. Reprinted in A. S. Goldhaber and W. P. Trower, *Magnetic Monopoles,* American Association of Physics Teachers, 1991, pp. 103–104; and in G. Soliani, *Solitons and Particles,* World Scientific, Singapore, 1985, pp. 522–523.

495. E. A. Power and T. Thirunamachandran. Quantum electrodynamics in a cavity. *Physical Review A*, **25**(5):2473–2484, May 1982.

496. R. E. Prange. Tunneling from a many-particle point of view. *Physical Review*, **131**(3):1083–1086, August 1, 1963.

497. A. Proca. Sur une explication possible de la différence de masse entre le photon et l'electron. *Journal de Physique Ser. VII*, **3**(2):83–101, February 1932.

498. A. Proca. Sur la théorie ondulatoire des électrons positifs et négatifs. *Journal de Physique*, **7**:347–353, 1936.

499. A. Proca. Sur les équations fondamentales des particules élémentaires. *Comptes Rendus*, **202**:1490–1492, 1936.

500. A. M. Prokhorov. Molecular amplifier and generator for submillimeter waves. *Soviet Physics JETP*, **7**:1140–1141, December 1958. English translation of the Russian original: *JETP (USSR)*, **34**:1658–1659, June 1958.

501. M. H. L. Pryce. The mass-centre in the restricted theory of relativity and its connexion with the quantum theory of elementary particles. *Proceedings of the Royal Society of London, Series A*, **195**:62–81, 1948.

502. E. M. Purcell. Spontaneous emission probabilities at radio frequencies. *Physical Review*, **69**:681, 1946.

503. J.-M. Raimond. Entanglement and decoherence studies in cavity QED experiments. Short course given in the NATO–Advanced Study Institute on *New Directions in Mesoscopic Physics (towards Nanoscience)* at the Ettore Majorana Centre in Erice (Sicily), Italy, July 20–August 1, 2002. The slides of this course are available on the conference Web site: *http://leopardi.cmp.sns.it/Erice/*.

504. T. C. Ralph and C. M. Savage. Squeezed light from conventionally pumped multilevel lasers. *Optics Letters*, **16**(14):1113–1115, July 1991.

505. T. C. Ralph and C. M. Savage. Squeezed light from a coherently pumped four-level laser. *Physical Review A*, **44**(11):7809–7814, December 1, 1991.

506. P. Rastall. Quaternions in relativity. *Reviews of Modern Physics*, **36**(3):820–832, July 1964. The definition of Hermitian conjugate of a quaternion used in this paper differs from ours by a multiplicative factor of −1.

507. J. R. Ray. Dissipation and quantum theory. *Lettere al Nuovo Cimento*, **25**(2):47–50, May 12, 1979. See also the erratum: J. R. Ray, *Lettere al Nuovo Cimento*, **26**(2):64, September 8, 1979.

508. J. R. Ray. Lagrangians and systems they describe: How not to treat dissipation in quantum mechanics. *American Journal of Physics*, **47**(7):626–629, July 1979.

509. J. W. S. Rayleigh. *The Theory of Sound*, volume 2. Dover, New York, 1945.

510. Lord Rayleigh. On maintained vibrations. *Philosophical Magazine*, **15**:229–235, 1883. Reprinted in *Scientific Papers by John William Strutt, Baron Rayleigh, D.Sc., F.R.S., Honorary Fellow of Trinity College, Cambridge, Professor of Natural Philosophy in the Royal Institution*, volume II, 1881–1887, Cambridge University Press, Cambridge, 1900, pp. 188–193. Online at the internet archive (under Strutt): *http://www.archive.org*.

511. S. Raynaud, A. Lambrecht, C. Genet, and M.-T. Jaekel. Quantum vacuum fluctuations. *Comptes Rendus de l'Academie des Sciences Série IV Physique Astrophysique*, **2**(9):1287–1298, November 2001.

512. M. Razavy. On the quantization of dissipative systems. *Zeitschrift für Physik B*, **26**:201–206, 1977.

513. M. Razavy. Hamilton's principal function for the Brownian motion of a particle and the Schrödinger–Langevin equation. *Canadian Journal of Physics*, **56**(3):311–320, March 1978.

514. E. Recami. Possible physical meaning of the photon wave-function, according to Ettore Majorana. In *Hadronic Mechanics and Non-potential Interactions*, pp. 231–238. Nova Science, New York, 1990.

515. F. Reif. *Fundamentals of Statistical and Thermal Physics. McGraw-Hill Series in Fundamentals of Physics: An Undergraduate Textbook Program*. McGraw-Hill, Auckland, New Zeeland, 1985.

516. G. Rempe, F. Schmidt-Kaler, and H. Walther. Observation of sub-Poissonian photon statistics in a micromaser. *Physical Review Letters*, **64**(23):2783–2786, June 4, 1990.

517. G. Rempe, H. Walther, and N. Klein. Observation of quantum collapse and revival in a one-atom maser. *Physical Review Letters*, **58**(4):353–356, January 26, 1987.

518. P. R. Rice and H. J. Carmichael. Photon statistics of a cavity-QED laser: A comment on the laser phase-transition analogy. *Physical Review A*, **50**(5):4318–4329, November 1994.

519. A. G. Riess, A. V. Filippenko, P. Challis, A. Clocchiatti, A. Diercks, P. M. Garnavich, R. L. Gilliland, C. J. Hogan, S. Jha, R. P. Kirshner, B. Leibundgut, M. M. Phillips, D. Reiss, B. P. Schmidt, R. A. Schommer, R. C. Smith, J. Spyromilio, C. Stubbs, N. B. Suntzeff, and J. Tonry. Observational evidence from supernovae for an accelerating universe and a cosmological constant. *Astronomical Journal*, **116**:1009–1038, September 1998.

520. F. Riewe. Mechanics with fractional derivatives. *Physical Review E*, **55**(3):3581–3592, March 1997.

521. H. Ritsch and P. Zoller. Dynamic quantum-noise reduction in multilevel-laser systems. *Physical Review A*, **45**(3):1881–1892, February 1, 1992.

522. H. Ritsch, P. Zoller, C. W. Gardiner, and D. F. Walls. Sub-Poissonian laser light by dynamic pump-noise suppression. *Physical Review A*, **44**(5):3361–3364, September 1, 1991.

523. M. D. Roberts. The orbit of Pluto and the cosmological constant. *Monthly Notices of the Royal Astronomical Society*, **228**(2):401–405, September 15, 1987. Online at *http://cosmology.mth.uct.ac.za/~roberts/*.

524. F. N. H. Robinson. The microscopic and macroscopic equations of the electromagnetic field. *Physica*, **54**(3):329–341, September 15, 1971.

525. F. N. H. Robinson. *Macroscopic Electromagnetism*. Pergamon Press, Oxford, 1973.

526. F. Rohrlich. *Classical Charged Particles. Addison-Wesley Series in Advanced Physics*. Addison-Wesley, Reading, MA, 1965.

527. C. A. Ronan. *The Cambridge Illustrated History of the World's Science*. Cambridge University Press, Cambridge, 1983.

528. D. M. Rosewarne. Ground-state quantum fluctuations in a scalar Hopfield dielectric. *Quantum Optics*, **3**:193–199, 1991.

529. T. D. Rossing. Chladni's law for vibrating plates. *American Journal of Physics*, **50**(3):271–274, March 1982.

530. C. L. Roy. Boundary conditions across a δ-function potential in the one-dimensional Dirac equation. *Physical Review A*, **47**(4):3417–3419, April 1993.

531. S. Saito and H. Hyuga. Dynamical Casimir effect without boundary conditions. *Physical Review A*, **65**(5):053804, May 2002.

532. D. J. Santos and R. Loudon. Electromagnetic-field quantization in inhomogeneous and dispersive one-dimensional systems. *Physical Review A*, **52**(2):1538–1549, August 1995.

533. M. Sargent III, M. O. Scully, and W. E. Lamb Jr. *Laser Physics*. Addison-Wesley, Redwood City, CA, 1987.

534. C. M. Savage. Quantum optics with one atom in an optical cavity. *Journal of Modern Optics*, **37**(11):1711–1725, November 1990.

535. C. M. Savage and H. J. Carmichael. Single atom optical bistability. *IEEE Journal of Quantum Electronics*, **24**(8):1495–1498, August 1988.

536. A. L. Schawlow and C. H. Townes. Infrared and optical masers. *Physical Review*, **112**(6):1940–1949, December 15, 1958.

537. W. Schleich. *Quantum Optics in Phase Space*. Wiley-VCH, Berlin, 2001.

538. B. P. Schmidt et al. The high-Z supernova search: Measuring cosmic deceleration and global curvature of the universe using type Ia supernovae. *Astrophysical Journal*, **507**:46–63, November 1998.

539. K. Schram. Quantum statistical derivation of the macroscopic Maxwell equations. *Physica*, **26**:1080–1090, 1960.

540. E. Schrödinger. Der stetig Übergang von der Mikro- zur Makromechanik. *Naturwissenschaften*, **14**:664–666, 1926.

541. E. Schrödinger. Quatisierung als Eigenwertproblem. *Annalen der Physik*, **79**:361–376, 489–527, 1926. The third and fourth parts were published in the same year, but in the next volume, **80**, of this journal, on pp. 437–490 and 109–139, respectively.

542. L. Schwartz. Sur l'impossibilité de la multiplication des distributions. *Comptes Rendus Hebdomadaires des Séances de l'Académie des Sciences*, **239**:847–848, October 11, 1954.

543. S. Schweber. *An Introduction to Relativistic Quantum Field Theory*. Row, Peterson, Evanston, IL, 1961.

544. J. Schwinger, editor. *Selected Papers on Quantum Electrodynamics*. Dover, New York, 1958.

545. J. Schwinger. A magnetic model of matter. *Science*, **165**:757–761, August 22, 1969.

546. J. Schwinger. Casimir light: A glimpse. *Proceedings of the National Academy of Sciences of the United States of America*, **90**:958–959, February 1993.

547. J. Schwinger. Casimir light: The source. *Proceedings of the National Academy of Sciences of the United States of America*, **90**:2105–2106, March 1993.

548. J. Schwinger. Casimir light: Photon pairs. *Proceedings of the National Academy of Sciences of the United States of America*, **90**:4505–4507, May 1993.

549. J. Schwinger. Casimir light: Pieces of the action. *Proceedings of the National Academy of Sciences of the United States of America*, **90**:7285–7287, August 1993.

550. J. Schwinger. Casimir light: Field pressure. *Proceedings of the National Academy of Sciences of the United States of America*, **91**:6473–6475, July 1994.

551. M. O. Scully. Enhancement of the index of refraction via quantum coherence. *Physical Review Letters*, **67**(14):1855–1858, September 30, 1991.

552. M. O. Scully and M. S. Zubairy. *Quantum Optics*. Cambridge University Press, Cambridge, 1997.

553. C. P. Search, S. Pötting, W. Zhang, and P. Meystre. Input–output theory for fermions in an atom cavity. *Physical Review A*, **66**(4):043616, October 2002.

554. H. P. Seidel. Quaternions in computer graphics and robotics. *Informationstechnik IT*, **32**:266–275, 1990.

555. J. Seke and W. N. Herfort. Long-time deviations from exponential decay in the Weisskopf–Wigner model of spontaneous emission. *Physics Letters A*, **126**(7): 422–426, January 18, 1988.

556. I. R. Senitzky. Semiclassical method and zero-point oscillations. *Physical Review Letters*, **20**(19):1062–1065, May 6, 1968.

557. I. R. Senitzky. Radiation-reaction and vacuum-field effects in Heisenberg-picture quantum electrodynamics. *Physical Review Letters*, **31**(15):955–958, October 8, 1973.

558. J. H. Shapiro, H. P. Yuen, and J. A. Machado Mata. Optical communication with two-photon coherent states: II. Photoemissive detection and structured receiver performance. *IEEE Transactions on Information Theory*, **IT-25**(2):179–192, March 1978.

559. B. W. Shore and P. L. Knight. The Jaynes–Cummings model. *Journal of Modern Optics*, **40**(7):1195–1238, July 1993.

560. A. E. Siegman. *Lasers*. University Science Books, Mill Valley, CA, 1986.

561. A. E. Siegman. Excess spontaneous emission in non-Hermitian optical systems: I. Laser amplifiers. *Physical Review A*, **39**(3):1253–1263, February 1, 1989.

562. A. E. Siegman. Excess spontaneous emission in non-Hermitian optical systems: II. Laser oscillators. *Physical Review A*, **39**(3):1264–1268, February 1, 1989.

563. A. E. Siegman. Laser beams and resonators: The 1960s. *IEEE Journal of Special Topics in Quantum Electronics*, **6**(6):1380–1388, November–December 2000. Online on Siegman's home page: *http://www-ee.stanford.edu/~siegman*.

564. L. Silberstein. Elektromagnetische Grundgleichungen in bivectorieller Behandlung. *Annalen der Physik*, **22**:579, 1907.

565. L. Silberstein. Nachtrag zur Abhandlung über "Elektromagnetische Grundgleichungen in bivectorieller Behandlung". *Annalen der Physik*, **24**:783, 1907.

566. L. Silberstein. Quaternionic form of relativity. *Philosophical Magazine*, **23**(137): 790–809, May 1912. Online at *http://home.pipeline.com/~hbaker1/quaternion/*.

567. C. C. Silva and R. de Andrade Martins. Polar and axial vectors versus quaternions. *American Journal of Physics*, **70**(9):958–963, September 2002.

568. J. E. Sipe. Photon wave functions. *Physical Review A*, **52**(3):1875–1883, September 1995.

569. R. E. Slusher, L. W. Hollberg, B. Yurke, J. C. Mertz, and J. F. Valley. Observation of squeezed states generated by four-wave mixing in an optical cavity. *Physical Review Letters*, **55**(22):2409–2412, November 25, 1985.

570. T. H. Solomon and S. Fallieros. Relativistic one-dimensional binding and two-dimensional motion. *Journal of the Franklin Institute*, **320**:323–344, 1985.

571. A. Sommerfeld. *Partial Differential Equations in Physics*. Academic Press, New York, 1949. Online at *http://mpec.sc.mahidol.ac.th/physmath/*.

572. G. Spavieri. Classical Lagrangian and quantum phase of the dipole. *Physics Letters A*, **310**(1):13–18, April 7, 2003.

573. M. B. Spencer and W. E. Lamb Jr. Laser with a transmitting window. *Physical Review A*, **5**(2):884–892, February 1972.

574. C. M. Sprenkle. Warrant prices as indicators of expectations and preferences. *Yale Economic Essays*, **1**(2):179–232, 1961. Reprinted in [122], pp. 412–474.

575. L. Spruch. Pedagogic notes on Thomas–Fermi theory (and on some improvements): Atoms, stars, and the stability of bulk matter. *Reviews of Modern Physics*, **63**(1):151–209, January 1991.

576. G. Squires. J. J. Thomson and the discovery of the electron. *Physics World*, **10**(4), April 1997.

577. P. Stehle. Atomic radiation in a cavity. *Physical Review A*, **2**(1):102–106, July 1970.

578. A. M. Steinberg, P. G. Kwiat, and R. Y. Chiao. Dispersion cancellation in a measurement of the single-photon propagation velocity in glass. *Physical Review Letters*, **68**(16):2421–2424, April 20, 1992.

579. A. M. Steinberg, P. G. Kwiat, and R. Y. Chiao. Dispersion cancellation and high-resolution time measurements in a fourth-order optical interferometer. *Physical Review A*, **45**(9):6659–6665, May 1, 1992.

580. R. L. Stratonovich. *Topics in the Theory of Random Noise*, volumes I and II. Gordon & Breach, New York, 1963.

581. E. C. G. Stückelberg. La mecanique du point matériel en théorie de relativité et en théorie des quanta. *Helvetica Physica Acta*, **15**:23–37, 1942.

582. C. Stuckens and D. H. Kobe. Quantization of a particle with a force quadractic in the velocity. *Physical Review A*, **34**(5):3565–3567, November 1986.

583. E. C. G. Sudarshan. Equivalence of semiclassical and quantum mechanical descriptions of statistical light beams. *Physical Review Letters*, **10**(7):277–279, April 1, 1963.

584. A. Sudbery. Quaternionic analysis. *Mathematical Proceedings of the Cambridge Philosophical Society*, **85**:199–225, 1979.

585. B. Sutherland and D. C. Mattis. Ambiguities with the relativistic δ-function potential. *Physical Review A*, **24**(3):1194–1197, September 1981.

586. O. Svelto. *Principles of Lasers*. Plenum Press, New York, 1989. Translated from the Italian original *Pricipi dei laser* by D. C. Hanna.

587. K. R. Symon. *Mechanics*. Addison-Wesley, Menlo Park, CA, 3rd edition, 1971.

588. P. G. Tait. Quaternions. In *Encyclopædia Britannica*, volume XX, pp. 160–164. 9th edition, 1886. Online at *http://home.pipeline.com/~hbaker1/quaternion/*.

589. S. M. Tan. A computational toolbox for quantum and atomic optics. *Journal of Optics B*, **1**(4):424–432, August 1999.

590. W. G. Teich and G. Mahler. Stochastic dynamics of individual quantum systems: Stationary rate equations. *Physical Review A*, **45**(5):3300–3318, March 1, 1992.

591. D. ter Haar, editor. *The Old Quantum Theory*. Pergamon Press, Oxford, 1967. *Selected readings in physics*.

592. R. J. Thompson, G. Rempe, and H. J. Kimble. Observation of normal-mode splitting for an atom in an optical cavity. *Physical Review Letters*, **68**(8):1132–1135, February, 24 1992.

593. S. P. Thompson. *Calculus Made Easy*. The Macmillan, London, 3rd edition, 1983. First published in 1910.

594. J. J. Thomson. On electrical oscillations and the effects produced by the motion of an electrified sphere. *Proceedings of the London Mathematical Society, Series 1*, **15**:197–218, April 3, 1884.

595. J. J. Thomson. *Notes on Recent Researches in Electricity and Magnetism (Intended as a Sequel to Prof. Clerk-Maxwell's Treatise on Electricity and Magnetism)*, Chap. IV, pp. 361–375. Clarendon Press, Oxford, 1893. This is a

textbook account of [594] by Thomson himself in the book that is often referred to as "the third volume of Maxwell." For another textbook account of [594], see app. II, pp. 214–224, of [571]. Thomson's book is available online at the Internet archive: *http://www.archive.org*.

596. J. J. Thomson. *Recollections and Reflections*. R. & R. Clark, Edinburgh, 1936. Online at the Internet archive: *http://www.archive.org*.

597. W. Thomson. On transient electric currents. *The London, Edinburgh, and Dublin Philosophical Magazine, Series 4*, **5**:393–405, June 1853.

598. G. 't Hooft. Magnetic monopoles in unified gauge theories. *Nuclear Physics B*, **79**:276–284, 1974.

599. E. C. Titchmarsh. *Introduction to Fourier Series*. Oxford University Press, Oxford, 1937.

600. J. S. Toll. Causality and the dispersion relation: Logical foundations. *Physical Review*, **104**(6):1760–1770, December 15, 1956.

601. S. Tomonaga. *The Story of Spin*. University of Chicago Press, Chicago, 1998.

602. C. H. Townes. *How the Laser Happened: Adventures of a Scientist*. Oxford University Press, Oxford, 1999.

603. K. Ujihara. Quantum theory of a one-dimensional optical cavity with output coupling: Field quantization. *Physical Review A*, **12**(1):148–158, July 1975.

604. K. Ujihara. Quantum theory of a one-dimensional optical cavity with output coupling: III. Cavity quasimodes as the field eigenvalue. *Physical Review A*, **20**(3):1096–1104, September 1979.

605. L. A. Vaĭnshteĭn. Open resonators for lasers. *Soviet Physics JETP*, **17**(3):709–719, September 1963.

606. L. A. Vaĭnshteĭn. Open resonators with spherical mirrors. *Soviet Physics JETP*, **18**(2):471–479, February 1964.

607. L. A. Vaĭnshteĭn. *Open Resonators and Open Waveguides*, volume 2 of *Golem Series in Electromagnetics*. Golem Press, Boulder, CO, 1969. English translation (by Petr Beckmann) of the 1966 Russian original.

608. R. W. F. van der Plank and L. G. Suttorp. Generalization of damping theory for cavities with mirrors of finite transmittivity. *Physical Review A*, **53**(3):1791–1800, March 1996.

609. B. van der Pol. The nonlinear theory of electric oscillations. *Proceedings of the IRE*, **22**(9):1051–1086, September 1934.

610. R. van der Pol and H. Bremmer. *Operational Calculus*. Cambridge University Press, Cambridge, 1955.

611. N. J. van Druten, Y. Lien, C. Serrat, S. S. R. Oemrawsingh, M. P. van Exter, and J. P. Woerdman. Laser with thresholdless intensity fluctuations. *Physical Review A*, **62**(5):053808, November 2000.

612. N. G. van Kampen. Stochastic differential equations. *Physics Reports*, **24**(3): 171–228, 1976.

613. N. G. van Kampen. Itô versus Stratonovich. *Journal of Statistical Physics*, **24**(1):175–187, 1981.

614. N. G. van Kampen. *Stochastic Processes in Physics and Chemistry*. Elsevier Science, Amsterdam, revised edition, 2001.

615. J. van Kranendonk and J. E. Sipe. Foundations of the macroscopic electromagnetic theory of dielectric media. *Progress in Optics*, **15**:245–350, 1977.

616. W. Vogel. Nonclassical states: An observable criterion. *Physical Review Letters*, **84**(9):1849–1852, February 28, 2000.

617. W. Vogel. Comment on "Nonclassical states: An observable criterion": Vogel replies. *Physical Review Letters*, **85**(13):2842, September 25, 2000.

618. W. Vogel and D.-G. Welsch. *Lectures on Quantum Optics*. Academie Verlag, Berlin, 1994.

619. A. R. von Hippel. *Dielectrics and Waves. Microwave Library Series*. Artech House, Norwood, MA, 1995.

620. J. A. von Neumann. Über die Grundlagen der Quantenmechanik. In *Von Neumann's Collected Works*, volume 1, pp. 104–133. Pergamon Press, Oxford, 1961.

621. F. S. G. von Zuben. Quantum time and spatial localization: An analysis of the Hegerfeldt paradox. *Journal of Mathematical Physics*, **41**(9):6093–6115, September 2000.

622. M. W. Walker, L. Shao, and R. A. Volz. Estimating 3-D location parameters using dual number quaternions. *CVGIP: Image Understanding*, **54**:358–367, 1991.

623. N. Wax, editor. *Selected Papers on Noise and Stochastic Processes*. Dover, New York, 1954.

624. D. L. Weaver, C. L. Hammer, and R. H. Good Jr. Description of a particle with arbitrary mass and spin. *Physical Review*, **135**(1B):B241–B248, July 13, 1964.

625. H. Weber. *Die partiellen Differential-Gleichungen der mathematischen Physik nach Riemann's Vorlesungen*, page 348. Friedrich Vieweg, Braunschweig, Germany, 1901.

626. S. Weinberg. Feynman rules for any spin. *Physical Review*, **133**(5B):B1318–B1332, March 9, 1964.

627. S. Weinberg. Feynman rules for any spin: II. Massless particles. *Physical Review*, **134**(4B):B882–B896, May 25, 1964.

628. S. Weinberg. Feynman rules for any spin: III. *Physical Review*, **181**(5):1893–1899, May 25, 1969.

629. S. Weinberg. The cosmological constant problem. *Reviews of Modern Physics*, **61**(1):1–23, January 1989.

630. V. Weisskopf. On the self-energy and the electromagnetic field of the electron. *Physical Review*, **56**:72–85, July 1, 1939.

631. V. Weisskopf and E. Wigner. Berechnung der natürlichen Linienbreite auf Grund der Diracschen Lichttheorie. *Zeitschrift für Physik*, **63**:54–73, 1930.

632. T. A. Welton. Some observable effects of the quantum-mechanical fluctuations of the electromagnetic field. *Physical Review*, **74**(9):1157–1167, November 1, 1948.

633. J. R. Wertz. *Spacecraft Altitude Determination and Control*. D. Reidel, Boston, 1980.

634. H. Weyl. Über die Randwertaufgabe der Strahlungstheorie und asymptotische Spektralgesetze. *Journal für die Reine und Angewandte Mathematik*, **143**:177–202, 1913. This and other papers by Weyl can be found in [101].

635. H. Weyl. Das asymptotische Verteilungsgesetz der Eigenschwingungen eines beliebig gestalteten elastischen Körpers. *Rendiconti del Circolo Matematico di Palermo*, **39**:1–50, 1915. This and other papers by Weyl can be found in [101].

636. H. Weyl. Eine neue Erweiterung der Relativitätstheorie. *Annalen der Physik*, **59**:101–133, 1919.

637. H. Weyl. *Gruppentheorie und Quantenmechanik*. S. Hirzel, Leipzig, 1928. English edition: [639].

638. H. Weyl. Elektron und Gravitation. *Zeitschrift für Physik*, **56**:330, 1929.

639. H. Weyl. *The Theory of Groups and Quantum Mechanics*. Dover, New York, 1950.

640. E. T. Whittaker. On an expression of the electromagnetic field due to electrons by means of two scalar potential functions. *Proceedings of the London Mathematical Society, Series 2*, **1**:367–372, 1904.

641. E. T. Whittaker. *Lives in Science*. Simon & Schuster, New York, 1957.

642. R. C. Whitten and P. T. McCormick. Elementary introduction to the Green's function. *American Journal of Physics*, **43**(6):541–543, June 1975.

643. A. S. Wightman. On the localizability of quantum mechanical systems. *Reviews of Modern Physics*, **34**(4):845–872, October 1962.

644. E. Wigner. On the quantum correction for thermodynamic equilibrium. *Physical Review*, **40**(5):749–759, June 1932.

645. E. Wigner. On unitary representations of the inhomogeneous Lorentz group. *Annals of Mathematics, Second Series*, **40**(1):149–204, January 1939.

646. E. P. Wigner. Do the equations of motion determine the quantum mechanical commutation relations? *Physical Review*, **77**(5):711–712, March 1, 1950.

647. E. P. Wigner. Relativistic invariance in quantum mechanics. *Il Nuovo Cimento*, **3**(3):517–532, March 1, 1956.

648. E. P. Wigner. Relativistic invariance and quantum phenomena. *Reviews of Modern Physics*, **29**(3):255–268, July 1957.

649. E. P. Wigner. On the time–energy uncertainty relation. In A. Salam and E. P. Wigner, editors, *Aspects of Quantum Theory*, Chap. 14, pp. 237–247. Cambridge University Press, Cambridge, 1972.

650. E. P. Wigner. *Group Theory and Applications to Quantum Mechanics*. Academic Press, Boston, 1997.

651. C. H. Wilcox. An expansion theorem for electromagnetic fields. *Communications on Pure and Applied Mathematics*, **IX**:115–134, 1956.

652. M. Wilkens and P. Meystre. Nonlinear atomic homodyne detection: A technique to detect macroscopic superpositions in a micromaser. *Physical Review A*, **43**(7):3832–3835, April 1, 1991.

653. P. Wilmott. *Paul Wilmott on Quantitative Finance*, volumes 1 and 2. Wiley, New York, 2000.

654. J. P. Woerdman, M. P. van Exter, and N. J. van Druten. Quantum noise of small lasers. *Advances in Atomic, Molecular, and Optical Physics*, **47**:205–248, 2001.

655. E. Yablonovitch. Inhibited spontaneous emission in solid-state physics and electronics. *Physical Review Letters*, **58**(20):2059–2062, May 18, 1987.

656. E. Yablonovitch. Accelerating reference frame for electromagnetic waves in a rapidly growing plasma: Unruh–Davies–Fulling–DeWitt radiation and the nonadiabatic Casimir effect. *Physical Review Letters*, **62**(15):1742–1745, April 10, 1989.

657. Y. Yamamoto and H. A. Haus. Preparation, measurement and information capacity of optical quantum states. *Reviews of Modern Physics*, **58**(4):1001–1020, October 1986.

658. C. N. Yang. *Selected Papers, 1945–1980*. Freeman and Company, New York, 1983.

659. G. J. Young and R. Chellappa. 3-D robot motion estimation using a sequence of noisy stereo images: Models, estimation, and uniqueness results. *IEEE Transactions on Pattern Analysis and Machine Intelligence*, **12**:735–759, 1990.

660. H. P. Yuen. Two-photon coherent states of the radiation field. *Physical Review A*, **13**(6):2226–2243, June 1976.

661. H. P. Yuen and J. H. Shapiro. Optical communication with two-photon coherent states: I. Quantum-state propagation and quantum-noise reduction. *IEEE Transactions on Information Theory*, **IT-24**(6):657–668, November 1978.

662. H. P. Yuen and J. H. Shapiro. Optical communication with two-photon coherent states: III. Quantum measurements realizable with photoemissive detectors. *IEEE Transactions on Information Theory*, **IT-26**(1):78–92, January 1980.

663. Ya. B. Zeldovich. Cosmological constant and elementary particles. *JETP Letters*, **6**(9):316, 1967. English translation of the Russian original *Zhurnal Eksperimental'noi i Teoreticheskoi Fiziki Pis'ma*, **6**:883, 1967.

664. D.-M. Zhang, Y.-J. Ding, and T. Ma. A note on the definition of delta functions. *American Journal of Physics*, **57**(3):281–282, March 1989.

665. A. S. Zibrov, M. D. Lukin, L. Hollberg, D. E. Nikonov, M. O. Scully, H. G Robinson, and V. L. Velichansky. Experimental demonstration of enhanced index of refraction via quantum coherence in *Rb*. *Physical Review Letters*, **76**(21):3935–3938, May 20, 1996.

666. R. W. Ziolkowski. Localized transmission of electromagnetic energy. *Physical Review A*, **39**(4):2005–2033, February 15, 1989.

Index

Absence of counterrotating terms, 269
Ad hoc coupling term, 256
Adding multiple reflections, 55, 258
Adiabatic elimination, 218
Aguirregabiria, 328
Aharonov–Bohm phase, 163
Aichelburg, 329
Akhiezer, 100
Album of the Mariner, 47
Allen, 17
Alternative photon position operators, 109
Ammonia-beam maser, 238
Ampère's law, 44
An, 252
Angular momentum commutation relations, 96
Annihilation operator
  free space, 45
Anti-Hermitian, 80, 115
Antiparticles, 107
  as a consequence of relativity, 107
  Stückelberg–Feynman–Wheeler idea, 102
Antiphoton, 99, 103, 108
Argand, 295
ASER idea, 217
Astronomical unit, **62**
Asymptotic condition, 270
Atom laser, 276
Atomic trap, 277
Average distance between the Sun and Earth, 62
Axial vector, 91

Bachelier, Louis, 252
Bag of tools, 313–320, 322
Baker–Hausdorff formula, 117, **311–312**
Bardeen, 276
Bardeen Hamiltonian, 276
Barnett, 262
Barton, 64
Bass, 109
Beam splitter, 140
Bennett, 257
Berestetsky, 100
Berry, 55, 67
Beta factor, 218
  typical order of magnitude, 219
Biagioni, 334
Bialynicki-Birula, 96, 99, 101
Biconjugation, 305
Biquaternions, **77, 304**
  gradient, **78**
  special relativity, 295, 304–310
    Lorentz boost, 310
Black and Scholes theory of option pricing, 252
Blackbody radiation, 63, 99, 238
Boca, 251
Boersma, 47
Bohm–Aharonov effect, 156
Bohr, 48
Boozer, 251
Born, 64
Bose–Einstein condensate, 276

Bose–Einstein statistics, 104
Boundary conditions, 272
Bouwkamp, 277
Breakdown of our current physical theories, 63
Bremmer, 277
Briegel, 250
Brill, 157
Brougham Bridge, 297, 299
Brownian motion, **220**
  Einstein's approach to, 222
  geometric, 252
Brown, Robert, 220
Buck, 251
Buée, 295
Canonical quantization, **18–23**
  properties of Poisson brackets, 19
Casimir effect, **47–63**, 65, 99
  Casimir's account of, 48
  cutoff function, 50
  design of nanomachinery, 48
  dynamic, 130
  expression for parallel plates, 51
  first experiment, 52
  maritime analog, 47
  Maxwell stress tensor approach, 54, 61
  order of magnitude, 51
  zero-point-potential-energy approach, 48, 53
Casimir, 277
  model for the electron, 65
  pressure, 60
Cathode ray tubes, 47
Cauchy, 324
  principal part, **325**
Cauchy–Riemann equations, 301
Causseé, 47
Caves, 136
Caves's diagrams, 139
Cavity and the outside, 256
Cavity modes, **256**
  as a superposition of external plane waves for
      which the cavity is transparent, 55, 67
  Fox–Li modes, 257
  in the Gardiner–Collett approach, 262
  in the Gardiner-Collett approach
      coupling with external modes, 268
  Rayleigh's remark, 255
  Thomson's approach, 271
      boundary conditions, 275
  Vaĭnshteĭn, 257
Cavity resonances
  shifted, 262
Cavity spectral mode density, 53
Cayley, 90, 306
Chan, 108
Chladni patterns, 28
Clausius–Mossotti relation, one-dimensional, 205

Clifford, 314
Closed cavities, 257
Coarse-grained photon wavefunction in
    configuration space, 107
Cohen, 276
Cohen-Tannoudji, 156–157
Coherent states, **114**
  as superpositions of number states, 117
  two-photon, 133
Colloidal stability, 48
Colombeau, 334
Comb function, 328
Commutator for position and momentum, 21
Completeness, 258
Completeness of Fox–Li modes, 257
Complex reflectivity, 262
Compton scattering, 47, 72
Compton wavelength, 93
Computer animation, 109
Conditional probability, 224
Configuration-space photon-number operator, 107
Continuous Markov process, 226–227
Continuum commutation relations, 57, 260
Contour integration, 278
Conway, 304
Cook, 99, 106
Coordinate-free system of mathematics, 294
Correlated emission laser, 252
Corson, 107
Cosmological constant, **63**
Costella, 107
Coulomb gauge, **157**
Coulomb self-energy, 155
Crandall, 109
Crandell, 94
Critical atom number, **233**
Curl integral theorem, **319**
Cylindrical cavity geometry, 272
Cylindrical coordinates, 49, 61
D'Alembertian, **78**
Damping
  direct description by Lagrangian or
      Hamiltonian, 276
Damping bases, 246
De Broglie wavelength of a two-photon wave
    packet, 108
De Broglie waves, 69
Delocalized, 256
Delocalization of absorption, 257
Delta function, 321–334
  normalization, 131
  product of two principal parts, 331
  representations, 322, 330
    comb function, 328
    Gaussian, 326
    general, 329

Laplacian of $1/r$, 327
Lorentzian, 324
  rectangular barrier, 323
  sinc function, 324
Density matrix, 114, **118–121**
  diagonal coherent-state representation, **121–124**
Deterministic "memoryless" process, 226
Deterministic source of single photons, 252
Deutsch, 107
Dielectric delta function mirror, 262
Diffusion equation, 222
Dirac, 18, 64, 107
  canonical quantization, **18–23**
  coordinate-free approach, 20
  delta function, 321–334
    normalization, 131
    product of two principal parts, 331
    representations, 322, 330
  development of canonical quantization, 63
  equation, 80, 100, 104–105
  hole theory, **100**
  magnetic monopoles, 108
  sea, **100**
Disagreement, the largest between theory and
  experiment, 48
Displacement operator, 116
Dissipation, 255
  in quantum mechanics, 255
Driven resonator, 257
Duplex form of Maxwell equations, 108
Dynamic Casimir effect, 130
Earth, 62
Eberly, 108, 251
Ehrenfest, 107
Einstein, 17, 62–63, 183, 217
  anno mirabilis, 221
  explanation of Brownian motion, 222
  Nobel prize, 47
Electric field
  positive-frequency part, 57, 60
  quantized in free space, expression for, 46
Electrical conductors, 291
Electrical oscillations on a spherical shell, 271
Electromagnetic field quantization
  canonical, **18–23**
    Dirac's account of, 63
    properties of Poisson brackets, 19
  classical limit of quantum radiation field, 113
  extreme quantum theory of light, **74–89**
  in Coulomb gauge not manifestly covariant, 157
  origin of the name *photon*, 42
  photon's wavefunction, 69
  the photon's wavefunction, 99
    problem with, 89–90
    recovering second quantization, 99, 106
  redundant degrees of freedom, 157

why only the radiation field is quantized, 24, 27
  with charges, 145
    arbitrary charge distribution, 154
    as a consequence of gauge invariance, 161
    dipole approximation, 146
    minimal coupling, **158**
  without charges, 17, 46
    in a cavity, 39
    in free space, 43
Electromagnetic potentials
  Bohm–Aharonov effect, 156
  gauge freedom, 161
  gauge-independent part, **17**
  quantization in free space without them, 43
  redundant degrees of freedom, 272
  Whittaker's two scalar potentials, 272
Electron–positron pair, 69
Energy conservation, 217, 260
Energy density inside the cavity, 260
Energy density of the vacuum, 62–63
Energy of a photon of visible light, 70
Ensemble averages versus macroscopic averages,
  195
Environment, 255
Euler's equation, 297
Euler–Maclaurin formula, 67
Excess quantum noise, 268
Existence of modes in open cavities, 257
Experimental confirmation of Maxwell's theory,
  215
Exponentials of operators, disentangling, 311
External vacuum fluctuations, 258
Extra phase gained on each trip inside the cavity,
  56
Extreme quantum theory of light, 69, **74–89**
Fabry–Pérot interferometer, 55, 257
Falicov, 276
Fang, 108
Fano diagonalization, **206**, 262
  dressed operator version, 265
  parallel with, 267
Faraday's law, 44, 66
Far-off resonance trap, 250
Feld, 252
Feld's single-atom laser, 252
Fermi–Dirac statistics, 104
Feynman, 64, 100, 102, 283, 297, 311
  account of spin-statistics theorem, 106
  formulation of quantum mechanics, 23, 64
  idea of antiparticles, 107
  proof of Maxwell equations, 276
  trick for vectorial expressions with ∇, 317
Fiat lux, 17
Field oscillator approach, 64
Field theory, 109
First quantization of light, 69, 99

problem with the photon's wavefunction, 89–90
recovering second quantization, 99, 106
First-order transitions, 234
FitzGerald, George Francis, 215
Fixing the gauge, 273
Fock, 107
  state, **107**
Fokker–Planck equation, 229
Four-momentum, 92
Four-vector, 79
Fox–Li modes, 257
Fraction of emission into the lasing mode, 218
Frahm, 327, 329
Franck–Condon transition, 130
Free-space spectral mode density, 53
Freely movable electrons, 291
Friction forces, 276
Gain saturation, 233
Gardiner–Collett approach, 257
Gardiner–Collett Hamiltonian, 236, **256**, 277
  breakdown, 256, 262
  first-principles derivation of, 258–269
  input–output theory, 276
Gardiner-Collett Hamiltonian
  as a first-order approximation, 262
  explicit expression for coupling strength, 268
Garrison, 107
Gauge invariance, 161
Gauss's divergence theorem, **318**
Gaussian, 326
Gaussian random variables, 227
General condition of radiation, **270**
General relativity, 62
General representations of the delta function, 329
Generalized functions, 321, 334
Generation of single-photon states on demand, 252
Geometric Brownian motion, 252
Gibbs, 294, 314
Glauber
  coherent states, **114**
    as superpositions of number states, 117
    two-photon, 133
  displacement operator, 116, 238
Glauber–Sudarshan diagonal representation, 114
Glauber-Sudarshan diagonal representation, **121**, **124**
Global operators, 263
Goldstein, 63
Good, 96, 100, 105–106
Goodman, 157
Gordon, 238, 257
Gottfried, 43
Goudsmit, 107
Gradient integral theorem, **319**
Grangier, 268
Grassmann, 314

Graves, 297
Gravitational force, 62
Gray, 17
Great quaternionic war, 294
Green's functions, 64
Griffiths, 332–333
Guitar modes, 28
Haken, 252
Half space, 57
Hamilton, 293, 314
  letter to Archibald, 297
  letter to Graves, 297
  The Tetractys, 293
Hammer, 100, 105
Han, 106
Harmonic functions, **301**
Haus, 64
Heaviside, 294, 313–314, 321
  magnetic monopoles, 108
  step function, **34**, **224**
Heisenberg's convention (or picture), 21
Helicity, **97**
  and the transversality condition, 99
  under parity transformation, 103
Helmholtz equation, 269, 274–275
  nonuniqueness of solutions in infinite domains, 270
HeNe laser, 219
Hermitian conjugation, 305
Hertz, 215, 270
Hibbs, 64
High-Q approximation, 258
Homodyne detection, 139
Huygens's principle, 326
Hypersurface, 303
Imprimitivity, 106
Impulsive, 321
Inagaki, 100, 106
Incoherent processes, 238
Independent degrees of freedom, 43
Independent reservoir assumption, 268
Index-guided edge-emitting semiconductor laser, 219
Infinitesimal Lorentz boost, 308
Instantaneous Coulomb scalar potential, 157
Invention of the laser, 257
Ion-trap laser, 250
Itô, 225
Itô's formula, 228
Jackson, 108
Janszky, 130
Jauch, 109
Jaynes–Cummings model, 47, **181**, 214, 241
  eigenstates, 246
Jaynes-Cummings model
  collapse, **181**

revival, **182**
Jordan, 64, 106
Keller, 108
Kim, 106
Kimble, 251
    deterministic source of single photons, 252
    one-and-the-same atom laser in the
        strong-coupling regime, 250
Kirchhoff, 326
Klein–Gordon equation, 70, 80
Kloeden, 252
Knight, 17
Kobe, 17, 107–108
Kronecker delta, 39, **286**
    determinants and products of Levy–Civita
        tensors, 316
Lamb shift, 47
Lanczos, 63, 304
Lanczos's equations, **79**
Lanczos's generalization of Dirac's method, 70
Landau, 106
Landau–Peierls wavefunction, 106
Langevin, Paul, 222
    delta-correlated force, 224
    modern version of his equation, 223
Laplacian of $1/r$, 327
Laser
    "back of the envelope" description, 218
    cavities, 218, 257
    effective temperature below threshold, 249
    idea of, 217
    ideal, 235
        quantum state, 238
    invention of, 257
    ion trap, 250
    is not a bomb, 233
    one-and-the-same atom, 250
    phase-transition analogy, 234
    problem with closed cavities, 257
    pumping operator, 251
    steady state, 232
    sub-Poissonian photon statistics, 251
    threshold, **234**, 249
        disappearance of, 249
Law, 108
Law of the norms, 300
LC oscillator as a radio-wave generator, 215
LC oscillators, 64
Leakage of electromagnetic radiation, 256
Leakage of radiation, 257
Left-handed coordinate system, 91
Levy–Civita tensor, 315
    products as a determinant of Kronecker deltas,
        316
Lewis, Gilbert N., 42
Leyden jar, 270

Limit of perfect reflectivity, 58, 60
Lindblad master equation, 241, 277
    for polarization decay, 245
Localization of relativistic particles, 71, 109
Localized polychromatic photon states, 107
London Mathematical Society, 271
Lord Kelvin (Sir William Thomson), 270
Lord Rayleigh (John William Strutt), 215, 255
Lorentz, Hendrik Antoon, 183
    physically infinitesimal volume elements, 194
Lorentz boost, 308
Lorentz force, 154, 276
Lorentz frames, 79
Lorentz gauge, 93
Lorentz transformation, 78
Lorentzian, 324
Lorentzian representation of the delta function, 324
Macroscopic averages
    compared with ensemble averages, 195
Macroscopic averaging, **184**
Macroscopic electrodynamics, **183**
Magnetic monopoles, 94, 108
Mandel, 107
Many-body theory of tunneling, 276
Marte, 107
Maser
    first, 238
    idea of, 217
Mass–energy equivalence principle, 62
Mathematical jewel, 297
MATLAB® quantum-optics toolbox, 251
Matrix algebra, 299
Maxwell, 270, 294, 313
    equations as limiting case of Proca's, 94, 99
    equations in reciprocal space, 43, 154
        generalization for cavity, 284, 289
    equations in symmetric form, 108
    equations without sources, 284
    equations, introduction of vector form, 313
    method of analogies, 40
    stress tensor, 54
McKeever, 251
McKellar, 107
Mechanism behind diffusion, 222
Melde's experiment, 215
Meschede, 238
Meyer, 250
Micromaser, 214
    average number of atoms crossing the cavity in
        a photon's lifetime, 239
    break with tradition, 244
    dimensionless pumping parameter, 242
    experimental parameters, 241
    first, 238
    optical, 245
    time-scale condition, 239

trapping states, **242**
  signature of, 242
  temperature sensitivity of, 243
  with atomic polarization damping, 245
Microwave ovens, 214
Milonni, 64, 107, 251
Minimal action in quantum mechanics, 23, 64
Minimal coupling, **158**
  as a consequence of gauge invariance, 161
Minimum-uncertainty states, 134
Minkowskian length, 304
Minkowsky's space-time, 78
Modes
  cavity, 27, 38, 256
    as a superposition of external plane waves for
      which the cavity is transparent, 55, 67
    calculation for perfect cavity, 64, **283–289**
    Fox–Li, 257
    Gardiner–Collett approach, 262,
    Gardiner-Collett approach, 268
    perfect reflector boundary conditions,
      291–292
    properties in perfect cavity, 39
    Rayleigh's remark, 255
    Thomson's approach, 271,
    Vaĭnshteĭn, 257
  delta-function normalized, 131
  guitar, 28
  modes-of-the-universe approach, 58, **256**, 262
    boundary problem divided into two, 276
    boundary problem, 275
  nonorthogonality, 276
  nonorthogonality and excess noise, 268
  outside, 264
  resonant, 283
  role of confinement, 34, 38
  violin (Chladni patterns), 28
Modulation factor, 260
Momentum space, 97
Monochromatic signal, 214
Monogenic forces, 276
Moon, 62
Moving mirrors, 130
Multiple scattered wave, 56
Multiple scattering, 55
Multiplication of distributions, 333
Müller, 238
Nahin, 294
Nanomachinery, 48
Natural modes, 258, **271**
  why they are not complete, 275
Negative-energy states, 100
Negative resistance voltaic arc, 215
Neptune, 63
Newton, Isaac, 276
Newton–Lorentz equations, 159

Newton–Wigner–Wightman group theory
  approach, 70
Newtonian formulation of classical mechanics, 276
Nisbet, 277
Nonabsorptive "black box" scatterer, 260
Nonclassical radiation field, 114
Noncommutability, 18
Noncommutative, 300
Non-Hermitian operators, 263
Normal modes, 256
Normal random variables, 227
Number of independent polarization modes, 49
Occupation-number wavefunction, 107
Old quantum theory, 64
One-dimensional cavity, 262
Open cavity mode, 256
Open systems, 255
Open-walled structures, 257
Oppenheimer, 69, 96, 99, 106
  extreme quantum theory of light with a twist,
    74, 89
Optical van der Pol oscillator, 219
Orbit of Pluto, 62
Osborne, 252
Overbeek, 47
Pais, 107
Parallel plates, 58
Parallelepiped box, 48
Parametric oscillator, 129, 215, 253
Parity operation, 103
Partial transmissions to the outside, 56
Path integrals, 23, 64
Pauli, 104, 145, 291
Pauli matrix, 245
Pauli's exclusion principle, 100, 277
Peierls, 106
Pendulum, 30
Perfect cavity boundary conditions, **291–292**
Perfect cavity modes, 64, **283–289**
Perfect intracavity modes, **265**
Perfect mirror, 260
Petermann, 268
Phenomenological Gardiner–Collett approach, 263
Phenomenological Hamiltonian, 258
Philips Research, 48
Phillips, 276
Photoelectric effect, 47
Photon
  antiparticle of, 99, 103
  deterministic source, 252
  in the paraxial approximation, 107
  negative-energy, 100, 102
  origin of the name, 42
  problem with configuration-space wavefunction,
    89–90
  quantum-nondemolition detection, 114

squeezed states, **124–125**
  movable cavity walls, 130
  squeezing operator, 125, 129
  sub-Poissonian statistics, 251
  transversality condition and nonexistence of
      photons at rest, 101
Pike, 106
Piron, 109
Planck, 47, 63
  constant divided by 2π, 20
  energy, 63
Plane wave, 43, 57
Platen, 252
Playground swing, 30, 215
Pluto, 62
Podolny, 64
Poincaré, Jules Henri, 252
Poisson, 324
Poisson brackets, **19**, 45, 63–64, 160
  formal analogy with commutator, 20
Poisson's sum formula, 261, 278
Poizat, 268
Polar vector, 91
Polarization decay
  effect on spontaneous emission, 246
Polarization vectors, 43
Polder, 48
Polyakov, 108
Polygenic forces, 276
Position operator in relativistic quantum
      mechanics, 71
Potential zero-point energy, 50
Poynting's theorem, 260
Prange, 276
Prediction that light is an electromagnetic wave,
      270
Problem of modes of open cavities, 257
  one-dimensional solution, 276
Proca, 106
Proca equations, 70, 93
  limit of vanishing rest mass, 94, 99
Prokhorov, 257
Pryce, 106
Pumping, 218
Pythagorean Tetractys, 294
Quantitative finance, 252
Quantum electronics, 257
Quantum gravity, 63
Quantum Poisson bracket, 20
Quasimodes, 277
Quasiprobability distribution, **124**
Quaternions, 109, 293, **295–300**, 314
  analog of Cauchy–Riemann equations, 304
  applications in robotics and animation, 294
  biconjugation or Hermitian conjugation, 305
  derivative, 300

directional derivative, 301
division algebra, 300
gradient, 78, 302
integral theorems, 303
left and right regularity, 303
Lorentz transformations, 304, 310
  Lorentz boost, 310
origin of the name, 294
three-dimensional rotation formula, 306
Rabi oscillations, **181**, 240
  vacuum, 245
Radio oscillators, 213, 251
Radmore, 262
Rastall, 304
Rate equations
  with noise, 232
  without noise, 218
Rawlinson, 107
Real cavity, 256
Reciprocal-space Maxwell equations, 43
  generalization for a cavity, 284, 289
Rectangular barrier, 323
Reduced density matrix, 241
Redundant degrees of freedom, 157
Reflection and transmission coefficients, 55
  fall-off with frequency, 51
  general conditions for, 260
Removing the sidewalls of the closed cavity, 257
Resonance, **29–34**
  parallel-plate cavity, 58
Revivals, 47
Robotics, 109
Rotating-wave approximation, 216, 269
Royal Netherlands Academy of Arts and Sciences,
      48
Royal Society, 271
Rydberg states, 241
Sampling property, **322**
Sarkar, 106
Saturation photon number, **233**
Scattering method, 58
Scattering of light by light, 73
Schawlow, 257
Schawlow–Townes linewidth, 268
Schmidt decomposition, 108
Schrödinger, 109
Schrödinger's convention (or picture), 22
Schrödinger's equation, 20
Schwartz, 333
Scully, 252
Second threshold, 251
Segerstad, 113
Selective increase of losses, 257
Semiclassical treatment, 257
Semitransparent mirror, 55, 260
Series summation result, 278

Sexl, 329
SI units versus CGS, 17
Siegman, 251, 268
Sifting property, **322**
Silberstein, 304
Simple choice of semitransparent mirror, 262
Sinc function, 324
Sinc-coupling strength, 268
Sipe, 99
Slowly varying envelope approximation, 215
Solar system, 62
Sommerfeld, 257, 271
  radiation condition, 275
  argument of, 269
  radiation condition, 257, **269**, 275
Source condition, 270
Sparnaay, 52
Special relativity, 295
Spherically symmetric radial oscillations, 269
Spin
  connection with light polarization, 108
  contribution to the total angular momentum, 95
  discovery, 107
  helicity, **97**
  matrices, 96
  spin-statistics theorem, **104**, 106
Spontaneous emission, 163, 217
  contribution, 218
  diffusion coefficient, 232
  irreversible, 246
Sprenkle, 252
Squeezed states, **124–125**
  detection, 139
  movable cavity walls, 130
  noise-ellipse representation, 135
  squeezing operator, 125, 129
Statistical mixtures, 124
Steady-state photon number, 232
Stenholm, 107
Step function, **34**
Stimulated emission, 214
  consequence of Bose–Einstein statistics, 217
Stochastic differential equation, 225
  Itô's, 228, 331
  Stratonovich's, 229, 331
Stochastic self-consistency condition, 226
Stochastic variable, 226
Stokes's theorem, 318
Stratonovich, 225
Strotonovich's chain rule, 230
Strutt, John William (Lord Rayleigh), 215, 255
Stückelberg, 102
Stückelberg–Wheeler–Feynman idea, 108
Sub-Poissonian light, 47
Sudden change, 131
Sum of Lorentzians, 262

Sun, 62
Superconducting microwave cavities, 241
Superoperator, **241**
  gain, 241
  Lindblad, 241
Supersymmetry, 62
Suspension of pollen grains, 220
Suspensions of quartz powder, 48
Svelto, 251
Swarming motion, 220
Tait, 294
Tan, 251
Tetractys, The, 293
Theory of multiplication of distributions, 334
Thomson
  Joseph John, **257**, 271
    natural modes, 271
  Sir William (Lord Kelvin), 270
't Hooft, 108
Three-momentum density, 93
Time–energy uncertainty, 64
Tomonaga, 107
Total energy, 260
Townes, 213, 238, 257
  maser idea, 217
Transversality condition, 99
  equivalent to no photons of zero energy, 101
Triode, 215
Tunneling of electrons between superconducting
    films, 276
Two scalar fields, 272
Two scalar potentials
  conditions on, 274
Two-level atom approximation, **175**
Two-photon coherent states, 133
Uhlenbeck, 107
Ultraviolet catastrophe, 63
Unification of algebra and geometry, 297
Unit velocity biquaternion, 308
Uranus, 63
Vacuum catastrophe, 48, 63, 65
Vacuum field strength, 57
Vacuum zero-point energy, 63
  parallel plates, 50
  physical consequence, 46
Vaĭnshteĭn, **257**
Van der Pol oscillator, 215, 251
  steady state built up from noise, 251
  steady state solution, 216
Van der Waals interaction, 48
Vector-calculus tools, 313–320
Velocity spread of the molecular beam, 238
Verwey, 48
Violin modes, 28
Voltaic arc, 215
Voyager spacecraft, 63

Walborn, 332–333
Wallis, 295
Walther, 238, 250
Wavefunction of a relativistic particle, 97
Weinberg, 106
Wheeler, 102
Whittaker, Sir Edmund, 277
    scalar potentials, 272
Wiener process, 227
Wightman, 106

Wigner, 106, 109
Wigner function, 124
Yuen, 133
Yukawa potential, 94
Yushin, 130
Zeiger, 238
Zeldovich, 63
Zero-point energy, **46**, 48
    physical consequence, 46
Zero-point fluctuations, 163
Zubairy, 252